# Stellar Astrophysics

## The Graduate Series in Astronomy

Series Editors: **R J Tayler**, University of Sussex
**R E White**, University of Arizona

The Graduate Series in Astronomy includes books on all aspects of theoretical and experimental astronomy and astrophysics. The books are written at a level suitable for senior undergraduate and graduate students, and will also be useful to practising astronomers who wish to refresh their knowledge of a particular field of research.

*Other books in the series*

**Dust in the Galactic Environment**
D C B Whittet

**Observational Astrophysics**
R E White (ed.)

*The front cover shows the 'Crab' Nebula in Taurus photographed in red light.*
*Photograph from the Hale Observatories.*

The Graduate Series in Astronomy

# Stellar Astrophysics

Edited by R J Tayler

*University of Sussex, UK*

Institute of Physics Publishing
Bristol and Philadelphia

*British Library Cataloguing-in-Publication Data*
A catalogue record for this book is available from the British Library

ISBN 0 7503 0200 3

*Library of Congress Cataloging-in-Publication Data are available*

**Citation**
When citing an item published in this book, authors are requested to cite the original journal of publication. The requisite information is given in the 'catchline' at the top of the title page of each article. If required, such a citation may be supplemented by the expansion 'Reprinted in:' followed by the book reference.



Published by IOP Publishing Ltd, a company wholly owned by
The Institute of Physics, London.

IOP Publishing Ltd
Techno House, Redcliffe Way, Bristol BS1 6NX, UK

US Editorial Office: IOP Publishing Inc., The Public Ledger Building,
Suite 1035, Independence Square, Philadelphia, PA 19106

Printed in Great Britain by Galliard (Printers) Ltd, Great Yarmouth

# Contents

# Introduction

This volume contains articles on Stellar Astrophysics which have appeared in *Reports on Progress in Physics* in the past few years. Where appropriate, authors have been given the opportunity to update their articles to take account of the most important developments in their subject since original publication.

Most of the discussion in the volume is concerned with late stages of stellar evolution. Stars can end their lives as black dwarfs, neutron stars or black holes or possibly as nothing, if an explosion can completely disrupt a star. A white dwarf is a dying star which will eventually become a black dwarf. In both white dwarfs and neutron stars the main force resisting gravity is provided by the pressure of a degenerate gas, which arises because the Pauli exclusion principle does not permit electrons and neutrons to be too closely packed in the six-dimensional phase space of position and momentum. This pressure only becomes important at extremely high densities, far greater than can be achieved in terrestrial experiments, and as a result white dwarfs and neutron stars can be regarded as laboratories for the study of matter at high densities. There is a maximum mass for each of white dwarfs and neutron stars; above the maximum mass they must become neutron stars and black holes respectively. The value of the white dwarf maximum mass is believed to be known accurately, but there is uncertainty in the neutron star maximum mass because there are problems in determining the equation of state of a neutron gas at densities higher than that of the atomic nucleus. White dwarf properties are in good qualitative agreement with those predicted using the ideal equation of state in which the pressure is a function of the density alone and is entirely due to the Pauli exclusion principle. In reality there are some deviations from this equation of state and they are discussed by Koester and Chanmugam in their article on the physics of white dwarfs.

The final mass of a star can be very much less than its initial mass. Although the maximum mass of a white dwarf is about $1.2M_{\odot}$ ($M_{\odot}$ is a solar mass), it is believed that stars with masses up to about $8M_{\odot}$ may become white dwarfs. Some stars which become neutron stars lose well over ninety per cent of their mass in doing so. Stars which are unable to lose sufficient mass to become white dwarfs or neutron stars must become black holes according to our present understanding of the laws of physics. Mass loss may be violent as in the explosion of a supernova or it may be more gentle and prolonged. Most, if not all, neutron stars are believed to be formed in the explosion of a supernova, while many white dwarfs result after mass loss from a star has formed what is known as a planetary nebula.

Mass loss from stars is very closely bound up with the origin of the chemical elements. Cosmological observations are at present in good agreement with the hot big bang cosmological theory in which the chemical composition of the Universe after the first few minutes was essentially a mixture of hydrogen and helium, with only a minimal quantity of other elements. Some additional helium and all of the heavier elements are then supposed to have been produced by nuclear reactions in stars. To understand the production of a particular isotope of an element it is necessary to identify both a nuclear reaction in which it is made and a type of star inside which conditions suitable for the reaction arise. If the current abundance of the isotope in the galaxy is to be explained, it is necessary to know how many of the relevant type of star have existed and how nuclearly processed material has reached the interstellar medium, and hence later generations of stars, as a result of mixing processes inside stars carrying it from the inner regions of a star to the outer regions and of mass loss from the star. Most elements and isotopes below iron in the periodic table are produced through nuclear fusion reactions between electrically charged nuclei. Above iron the main reactions involve the capture of neutrons either rapidly, probably during an explosion (r-process), or more slowly during a non-violent stage of stellar evolution (s-process). The s-process is the subject of the article by Käppeler, Beer and Wisshak.

In the later stages of stellar evolution, when the interior has a very high temperature or high density or both, there are many reactions which lead to the production of neutrinos and these neutrinos can escape freely into space. As the neutrino luminosity from the central regions of a star can greatly exceed the photon luminosity from the surface, the lifetime of a star is much reduced. This in turn means that there is a low probability of observing stars in such phases of evolution. There is a further production of neutrinos in the conversion of protons into neutrons immediately before the formation of a neutron star. At this stage the compact stellar interior may become opaque to neutrinos so that they have to diffuse out of it, but even so it is predicted that most of the energy released by neutron star formation is not converted into the visible and kinetic energy of a supernova but is radiated in the form of neutrinos. This prediction was strongly supported by observations of electron antineutrinos from the supernova in the Large Magellanic Cloud, SN 1987A. If allowance is made for the probability of antineutrino detection and for the fact that the supernova should have emitted electron neutrinos and muon and tauon neutrinos and their antiparticles, it does appear that essentially all of the energy released in forming a neutron star was released as neutrinos. This is important because it restricts the possible existence of further types of low mass neutrino which might also have been radiated by the star. It also suggests that there are as yet no other undiscovered types of elementary particle which could be produced inside an exploding supernova and then carry away a large fraction of the released energy. All aspects of the behaviour of SN 1987A both during and after its explosion are discussed in the article by Hillebrandt and Höflich.

One important question relating to neutrinos is whether they have a finite mass. The electron neutrino was first thought to be a massless particle but modern developments in particle physics suggest that all types of neutrino should have a small mass. Current experiments do not exclude this. If the neutrino does have a mass, this has important consequences in astrophysics and cosmology. The hot big bang cosmological theory requires an average $m^3$ of space to contain several hundred million neutrinos and, even if their masses were only of order $10 eV/c^2$, they would dominate the mass density of the Universe. If neutrinos have mass they do not travel at the speed of light. Instead, low-energy neutrinos travel more slowly than high-energy neutrinos. A spread of arrival times of the antineutrinos from SN 1987A may be at least partly due to neutrino mass and this observation has been used to place an upper limit to such masses.

Neutrinos are also present at early stages of stellar evolution because the first nuclear reaction of importance in stars, the conversion of hydrogen into helium, involves the conversion of two protons into neutrons with the emission of two neutrinos. These neutrinos have a minor effect on the course of stellar evolution but those emitted by the Sun have proved very interesting. An experiment designed to detect solar neutrinos has always detected fewer than was expected. Attempts to explain this result through errors in the input physics of the solar interior have not been successful. Instead, it is plausible that it might be explained through finite neutrino mass. One possible consequence of finite mass is that the different types of neutrino, electron, muon, tauon can interconvert so that some of the electron neutrinos leaving the Sun are no longer electron neutrinos when they reach the detector. As the experiment is not sensitive to muon and tauon neutrinos this might explain the results. It is sufficient to show that particle physics is important at many stages of stellar evolution.

There is now a second solar neutrino experiment in operation and its results may well clarify the extent to which neutrino properties are responsible for the discrepancy between theory and observation in the current experiment. Terrestrial experiments, either proton decay detectors or new neutrino detectors, will probably observe neutrinos from the next massive supernova in our Galaxy, whether or not it is a visible one. The number of neutrinos detected could be a hundred times more than those from SN 1987A and this would be bound to give very valuable information. It will be even more useful if the supernova is visible so that all of its properties can be correlated as in the case of SN 1987A. The last visible supernova in the Galaxy was in 1604 and estimates of the frequency of supernovae and of the fraction which should be visible encourage the hope that the next one might not be too long delayed, although it might not be a massive star like SN 1987A.

The visible energy from a supernova may have a light curve with an approximately exponential decay in time. More than 30 years ago it was first suggested that this was due to the radioactive decay of some unstable isotope produced by nuclear reactions during the supernova event. If the light output is to be dominated by radioactive decay, it is clear that the isotope in question must be a major product of the nuclear reaction chain. The original suggestion was that it was $^{254}$Cf, which had been observed to be produced in nuclear weapons tests. This is not, however, likely to arise in sufficient quantity to dominate the outburst. A much better candidate is $^{56}$Ni. Although this is unstable, it is a doubly magic number nucleus which can be thought of as being made up of $\alpha$ particles. At lower mass number, the $\alpha$ particle nuclei are the most stable and it is not surprising that nuclear fusion reactions build up $\alpha$ particle nuclei and lead to a significant production of $^{56}$Ni, which then decays successively to $^{56}$Co and $^{56}$Fe. The observations of SN 1987A have shown a light curve dominated by this decay chain and there is also good spectroscopic evidence of the presence of the relevant elements in the mass expelled by the explosion. This is another way in which the explosion of SN 1987A, the nearest supernova observed in outburst since the invention of the telescope, has served to confirm and also to refine views about the way in which stars explode as supernovae. Not only is the supernova light curve explained, but we also have confirmation that supernovae are an important source of iron and neighbouring elements in the periodic table.

Probably a majority of stars are members of binary or multiple stellar systems. In many cases the stars are sufficiently far apart that the only important consequence of their binary nature is that they move in orbits around one another. For some stars at some stages of their evolution, the interaction is very much more important. As a binary evolves it is possible for both mass and angular momentum to be lost from the system and for individual components to increase in size. Increase in component size and/or reduction in separation can lead to the outer layers of one star becoming more strongly attracted gravitationally to the other star. The resulting transfer of mass totally changes the subsequent evolution of the stars. Many of the most dramatic stellar events involve mass transfer in binaries. These include novae and X-ray binaries. The star receiving mass in a nova is a white dwarf, while X-ray binaries may contain a neutron star or a black hole. The observation of the properties of binaries enables an estimate to be made of the masses of the components and it is such an observation which has provided the best evidence for a stellar mass black hole, because a compact invisible star has a mass greater than the maximum neutron star mass. Some supernovae known as Type II, of which SN 1987A is an example, involve the explosion of a single massive star but there is another class, Type I, which is the subject of the article by Wheeler and Harkness. Type I supernovae were originally thought to be a single homogeneous class but recently they have been subdivided into Ia, Ib and Ic. It is possible that all Type I supernovae are in binary systems. In particular, the most common Type Ia are thought to involve the explosion of a carbon/oxygen white dwarf following mass transfer in a binary system. Although the general idea is accepted by most workers in the field there continues to be uncertainty about the precise nature of the explosion.

Most of the topics which I have mentioned above feature in one or more of the articles in this volume. In addition, Y Totsuka's article on neutrino astronomy is concerned with all aspects of the subject, some of which do not involve stars. One major topic of current research in stellar astrophysics, which is not covered in any of the articles, is that of mass loss from stars. When stellar structure and evolution was first studied, it was assumed that stars evolved at constant mass. There was at that time no observational evidence of significant mass loss, except in the explosion of a supernova at the end of a star's life and through the existence of planetary nebulae, apparently formed of mass expelled by a central star. There was, however, a suspicion that mass loss was more widespread in order, for example, to understand the number of stars which ended their lives as white dwarfs. There was, indeed, once a general unwillingness to accept the existence of either neutron stars or black holes, with the implication that all stars had to lose sufficient mass to become white dwarfs. Most of the evidence for stellar mass loss has followed the development of space astronomy through the observation of spectral lines in the ultraviolet, which cannot be observed from the Earth's surface. It is now clear that mass loss from stars is common at all stages of stellar evolution and the final masses of many stars are very much less than their initial masses. Stellar mass loss influences the nuclear evolution of a star and this in turn affects the contribution of the star to the origin of the chemical elements.

Another important factor in stellar evolution is the extent to which internal motions in a star mix material which is initially at different radial locations. It has been recognized for a long time that energy transport by convection is important in many stars and that convective motions are normally sufficiently rapid that a convection zone is kept chemically homogeneous. Convection is an instability and initially it was assumed that convective motions would occur only in a region in which an instability criterion was satisfied. It was then realized that material would be moving when it reached the edge of the unstable region and that as a result convection would penetrate into the stable region. There is still argument about the extent of this convective overshooting and of its influence on stellar evolution. Although any convective mixing occurs on a timescale very rapid compared to that of stellar evolution, there is also the possibility that other very much slower instabilities or mixing processes may occur in stellar interiors. One example arises because a rapidly rotating or strongly magnetic star cannot exist in hydrostatic equilibrium and this leads to a circulation which may be important in some stars. It is very difficult to give a theoretical discussion of all possible slow mixing processes in stars and it is therefore important to have some observational evidence. This is provided by the s-process isotopes discussed by Käppeler, Beer and Wisshak. The nuclear reactions producing these isotopes occur well inside stars but some of the reaction products appear at stellar surfaces. The discovery of the element technetium, which has no stable isotopes, in the atmosphere of a red giant star was one of the first triumphs of stellar nucleosynthesis. When s-process isotopes are observed in a stellar atmosphere, it is necessary to try to understand their abundances in terms of the physical conditions in which they are produced and the nuclear cross-sections influencing their production together with the mixing processes which carry them to stellar surfaces.

A major factor in studies of late stages of stellar evolution is that there may be important departures from spherical symmetry and also rapid time variations, by normal astronomical standards. In some cases, as in discussions of the explosion of a massive supernova, understanding of the basic physics involved may be stretched to the limit. Geometrical factors clearly complicate a complete discussion of the transfer of mass between components in a close binary system. Progress has been dependent on the development of powerful computers and of efficient numerical methods to take advantage of them. Discussion of the explosion of a Type II supernova, at least in the spherically symmetric approximation, has been simplified by the discovery that the behaviour of the core of the star is to a good approximation independent of the total mass of the star. For quite a large mass range it appears that the ultimate end product is a neutron star of about $1.4M_{\odot}$, although above some mass stars may not form a neutron star and may collapse into a black hole. Having said that, it has proved difficult to follow the evolution of a Type II supernova right through to the explosion. It is not too difficult to understand this. As has already been explained, most of the energy released in forming a neutron star is carried away by neutrinos. The energy of the explosion is a small fraction of one per cent of the total. This means that whether or not a star does explode involves a very fine energy balance. It is not at all surprising that many calculations failed to produce an explosion. It is also clear that an apparently small amount of energy tied up in rotation or a magnetic field, which would destroy the spherical symmetry of the system, might influence the result.

Although the articles in this volume do not cover the whole of stellar astrophysics, they should be sufficient to show the considerable progress which has been made in studying the structure and evolution of stars. At the same time, we are still a little way from being able 'to understand so simple a thing as a star' to quote the final words of Eddington's 1926 book *The Internal Constitution of the Stars*.

R J Tayler

Astronomy Centre, Division of Physics & Astronomy

University of Sussex, Falmer, Brighton BN1 9QH

March 1992

Rep. Prog. Phys. **52** (1989) 945–1013. Printed in the UK

# s-process nucleosynthesis—nuclear physics and the classical model

F Käppeler, H Beer and K Wisshak
Kernforschungszentrum Karlsruhe, Institut für Kernphysik, D-7500 Karlsruhe, Federal Republic of Germany

**Abstract**

Among the various processes responsible for the formation of the heavy elements in stars, the slow neutron capture process (s-process) is distinguished by the fact that it involves mostly stable isotopes. Therefore, the relevant nuclear physics data can be determined by experiments. With this rather reliable data basis, s-process nucleosynthesis offers an important testground of models for the late stages of stellar evolution, which are supposed to be the s-process site. The empirical counterpart for such models is the so-called classical s-process, a purely phenomenological picture, that is successfully used to derive the resulting abundances as well as information on the physical conditions during the s-process. The status of this classical approach is reviewed with emphasis on the implications for various stellar models of the s-process and in the light of results obtained by stellar spectroscopy. A brief account of the potential s-process chronometers is also presented.

This review was received in January 1989.

**Contents**

## 1. Introduction

This review deals with stellar nucleosynthesis by slow neutron captures (the s-process). It concentrates on the classical model and tries to delineate the essential nuclear physics which is required for input as well as the main results which represent completely unbiased constraints for stellar models of the s-process.

The origin of the chemical elements in nature constitutes a fascinating problem of nuclear astrophysics. Among the various theories of element formation, stellar nucleosynthesis has become most successful and is now commonly accepted. It assumes that only hydrogen, helium and the rare light isotopes with $A < 12$ are produced in the big bang. These isotopes are stepwise converted to heavier nuclei by the energy-generating fusion reactions in stars. In successive stages of stellar evolution a sequence of burning processes converts H to He, He to C, and so on until the isotopes around iron are reached in Si burning. These are the most stable nuclei in nature, exhibiting the highest binding energies per nucleon, and their stability results in a large abundance maximum around $A = 56$. Beyond that point, fusion of charged particles quickly ceases to contribute to the observed abundances because of the increasing coulomb barriers and the decreasing binding energies per nucleon. Heavier nuclei can only be created by successive neutron capture reactions and beta decays.

Stellar nucleosynthesis was greatly stimulated by the comprehensive survey of Burbidge *et al* (1957) who formulated a consistent picture of element formation in stars. Meanwhile, three decades of intense research confirmed many of these classical ideas though on a considerably refined level. While the static burning situations seem to be fairly well understood, explosive burning still represents a great challenge for nuclear astrophysics. The s-process takes an 'intermediate' position: the classical approach allows for a successful description of the observed abundances and many other aspects with the assumption of a *steady* neutron flux. On the other hand, stellar s-process models indicate that the s-process may occur predominantly in a *dynamic* environment. This duality will be repeatedly dealt with in this article because it is the interplay between classical approach and observations on one side and stellar models on the other side, which will provide the clues for identifying the true mechanisms of the s-process.

The classical s-process as it was first outlined by Burbidge *et al* (1957) represents a purely phenomenological approach to neutron capture nucleosynthesis. The only assumption on the astrophysical scenario was that it should be associated with helium burning in the red-giant stage of stellar evolution in order to define the s-process temperature. Evidence for such a site came from the observation of Tc lines in the spectra of certain S stars (Merrill 1952), and from the fact that under these burning conditions $(\alpha, n)$-reactions may provide an efficient neutron source. This empirical ansatz has two attractive features:

(i) Practically all important features of the s-process, such as abundances, neutron economy and physical conditions, are described surprisingly well.

(ii) By its empirical nature it is completely unbiased by any stellar model, thus yielding independent constraints for the more comprehensive treatment in terms of stellar evolution.

The recipe for the classical s-process is simply to expose a pre-existing distribution of seed abundances to a suited neutron irradiation. The associated neutron flux is assumed to be steady and low enough that the beta decay rates dominate over the neutron capture rates. Consequently, the neutron capture chain of the s-process follows the valley of beta stability. This is best illustrated at the section of the chart of nuclides sketched in figure 1. The inset of figure 1 shows the observed abundance distribution (see § 3.1) with a strong maximum around iron. Consequently, neutron capture nucleosynthesis of the heavy chemical elements starts from the iron group isotopes. Neutron captures on a slow timescale of the order 1 to 10 yr follow the thick solid line in the valley of beta stability; if an unstable nucleus is encountered, usually beta decay to the next higher element is much faster than neutron capture. The resulting abundances are, therefore, determined by the respective neutron capture cross sections. Isotopes with small cross sections act as bottle necks in the neutron capture chain and build up large abundances, whereas isotopes with large cross sections are more easily burnt and hence develop small abundances only. This explains naturally the sharp maxima in the inset of figure 1 as being due to the s-process. At these points, the s-process path encounters nuclei with magic neutron numbers $N = 50$, 82, 126, which all exhibit particularly small cross sections.

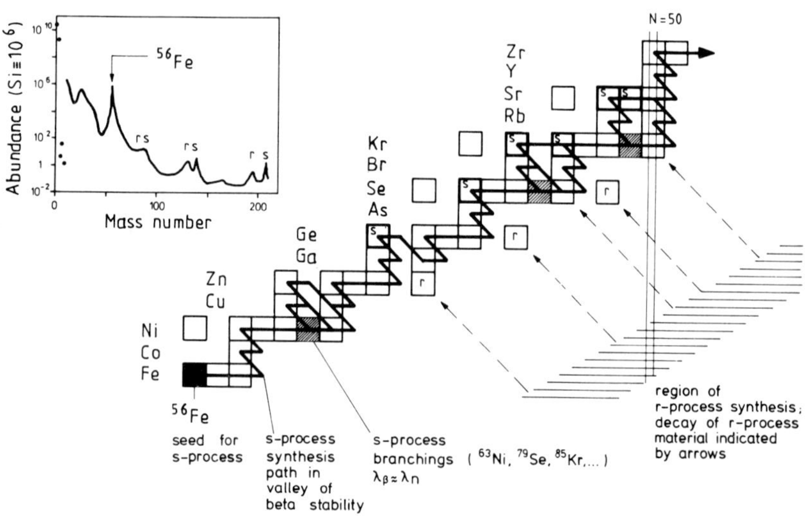


**Figure 1.** Section of the chart of nuclides for a discussion of the main features of neutron capture nucleosynthesis (see text).

Another important feature of the s-process is the branchings of the neutron capture chain—indicated by hatched boxes in figure 1—where the rates for beta decay and neutron capture are comparable. At these isotopes, e.g. at $^{63}$Ni, $^{79}$Se and $^{85}$Kr, competition between neutron capture and beta decay leads to a branching of the synthesis path which is determined by the s-process neutron density and the respective beta half-life. Analysis of the abundance patterns in these branchings (§ 5) can provide for estimates of the neutron density, $n_n$, but also for the temperature, $T_s$, and even for the electron density during the s-process. The latter quantities manifest themselves through their influence on the beta decay rates of several branching isotopes. This field was recently investigated in detail (§ 3.3). Branchings may also serve for investigating the conditions of a pulsed s-process.

In addition to the s-process at least two other mechanisms contributed to the observed abundances in the mass range $A > 70$. The neutron-poor isotopes are attributed to the so-called p-process, where p stands for proton capture and/or photodisintegration. The related abundances are usually smaller by more than one magnitude compared to those built by the other two processes, and this may be one of the reasons why the p-process is still not well understood. However, p-process corrections to the s-only isotopes (§ 4.1) become increasingly important in view of the improved s-process treatment. Hence, a quantitative model for the p-process is badly needed. At present, p-process corrections are either neglected or somehow interpolated from the p-only nuclei. Already both approximations lead to uncertainties similar to those of the classical model, and may severely hamper further improvements.

The existence of the neutron-rich isotopes which cannot be reached by the s-process because of their short-lived neighbours calls for a rapid neutron capture process. This r-process is supposed to occur on a timescale of some hundred milliseconds, presumably during explosive stellar burning. The associated neutron densities of $\sim 10^{19}$ cm$^{-3}$ or more lead to a synthesis path close to the neutron drip line. This is indicated in figure 1 by the hatched area. The isotopic abundances produced in the r-process are determined by the beta decay times on the r-process path. As the half-lives are longest for nuclei with magic neutron numbers, these build up large abundances similar to the s-process, but at smaller mass number $A$ (see figure 1). Therefore, the shift between the s- and r-process peaks in the abundance distribution marks the atomic numbers, $Z$, where the r-process path reached the magic neutron numbers. If the respective beta decay times are known from experiments or calculations, this information can be interpreted in terms of the prevailing r-process neutron densities and temperatures (Kratz *et al* 1988).

The isotopes created in the r-process are all short-lived and decay back to the valley of beta stability, so that most observed isotopes represent mixtures of abundance contributions from the s- and r-processes with contaminations from the p-process which are usually small. Fortunately, there are some 40 isotopes that can be attributed entirely to either process: in all cases where the beta decay chain ends at an r-only isotope that is not reached by the s-process because of its short-lived neighbour, e.g. at $^{80}$Se and $^{82}$Se, the respective isobars, $^{80}$Kr and $^{82}$Kr, are in turn shielded from the r-process. The abundances of this ensemble of s-only nuclei can therefore be used as a test for any s-process model.

The present work concentrates on the classical s-process; further information on related topics of nucleosynthesis may be found in the following references. Neutron capture nucleosynthesis: Mathews and Ward (1985); p-process: Rayet *et al* (1988); stellar neutron capture cross sections: Bao and Käppeler (1987); stellar weak-interaction rates: Takahashi and Yokoi (1987); solar abundances: Anders and Grevesse (1989); isotopic anomalies: Begemann (1980) abundances in stellar atmospheres: Lambert (1988); stellar models: Iben and Renzini (1983); nucleocosmochronology: Clayton (1984), Cowan *et al* (1987).

## 2. s-process models

### *2.1. The classical s-process*

Under neutron irradiation the change in abundance of a particular isotope $A$ can be formally written as

$$\mathrm{d}N_s(A)/\mathrm{d}t = \lambda_n(A-1)N_s(A-1) - (\lambda_n(A) + \lambda_\beta(A))N_s(A). \tag{2.1}$$

The neutron capture rate, $\lambda_n = \Phi\sigma$, is proportional to the neutron capture cross section averaged over the stellar spectum, $\sigma$, and to the neutron flux $\Phi = n_n v_T$, where $n_n$ denotes the neutron density and $v_T = (2kT/m)^{1/2}$ the thermal neutron velocity, with $m$ being the reduced mass. The beta decay rate $\lambda_\beta = \ln 2/t_{1/2}$ must be considered if isotope $A$ is radioactive. Equation (2.1) defines a system of coupled differential equations comprising all isotopes on the s-process path. In the most general case the coefficients $\lambda$ are not constant but depend on the (possibly dynamic) stellar environment. In the *classical* model the situation is simplified by two assumptions:

(i) The s-process temperature, $T_s$, is constant, allowing for well defined cross sections. This is a good approximation because the neutron capture rates are almost independent of temperature: the neutron cross sections fall with $\sim 1/\sqrt{E_n}$ and this is compensated in the expression for $\lambda_n$ by the thermal neutron velocity $v_T$.

(ii) Radioactive nuclei on the synthesis path are either treated as stable ($\lambda_\beta \ll \lambda_n$) or as sufficiently short-lived that the capture chain continues with the daughter nucleus ($\lambda_\beta \gg \lambda_n$). In this way, $\lambda_\beta$ is eliminated from equation (2.1). This assumption is justified for most unstable nuclei with the few exceptions of the s-process branchings; these have to be treated individually as shown below.

With these assumptions, equation (2.1) can be rewritten with time being replaced by the time-integrated neutron flux $\tau = \int \Phi \, \mathrm{d}t$,

$$\mathrm{d}N_s(A)/\mathrm{d}\tau = \sigma(A-1)N_s(A-1) - \sigma(A)N_s(A). \tag{2.2}$$

Sufficiently long irradiation results in equilibrium between the production and destruction terms in equation (2.2), leading to a constant product, $\sigma N_s$, of cross section and s-process abundance. Though this condition is not completely reached, the product $\sigma N_s$—which is the characteristic quantity of the classical s-process—exhibits at least a smooth dependence on mass number.

Clayton *et al* (1961) showed that the observed abundances of the s-only isotopes cannot be reproduced by a single irradiation of an iron seed. A satisfactory solution was found by Seeger *et al* (1965) assuming an exponential distribution of neutron exposures, $\tau$, of the form

$$\rho(\tau) = \frac{fN_{56}}{\tau_0} \exp(-\tau/\tau_0). \tag{2.3}$$

This ansatz even allows for a simple analytical solution of the system of equations (2.2):

$$\sigma(A)N_s(A) = \frac{fN_{56}}{\tau_0} \prod_{i=56}^{A} [1 + (\sigma(i)\tau_0)^{-1}]^{-1} \tag{2.4}$$

The product, $\sigma N_s(A)$, is completely determined by the cross sections, $\sigma(i)$, after the parameters $f$ and $\tau_0$ are fixed by a fit to the empirical $\sigma N_s$ values of the s-only isotopes. In this way, equation (2.4) yields the abundances produced by the s-process and, via the parameters $f$ and $\tau_0$, information on the seed abundance and the neutron economy. A useful quantity in that respect is the number of neutrons captured per $^{56}$Fe seed nucleus (Clayton *et al* 1961)

$$n_c(A_{\text{seed}}) = \left( \sum_{A=A_{\text{seed}}}^{209} (A - A_{\text{seed}}) N_s(A) \right) \Big/ f N_\odot(^{56}\text{Fe}). \tag{2.5}$$

The total number of neutrons captured during the s-process is $\Sigma n_c(i)$, where $i$ runs over all seeds (for a discussion see § 4.2).

When sufficiently detailed input data became available, it turned out that three different exponential distributions of neutron exposures are required for a complete description of the observed s-process abundances (Clayton and Rassbach 1967, Clayton and Ward 1974, Beer and Macklin 1985). The bulk of s-process material in the mass range $90 < A < 204$ is produced by what is called the *main* s-process component in the following. Below $A \sim 90$ this component fails to describe the steep increase of the $\sigma N_s$ curve towards the iron seed. Therefore, a *weak* component is added which is characterised by a smaller mean neutron exposure. Finally, a *strong* component is to be postulated in order to account for the abundance maximum at lead. These three components will be distinguished by adding indices 1, 2 and 3, respectively, to the parameters $f$ and $\tau_0$. An exponential distribution for $\rho(\tau)$ appears to provide for an excellent fit of the main s-process component, but the weak component may better be described by single irradiations (Beer 1986, Beer and Macklin 1989). The strong component is only needed to explain a significant part of the $^{208}$Pb abundance (Beer and Macklin 1985, Ratzel *et al* 1989).

Within the classical s-process model the description of the mass flow through the above-mentioned branchings requires the additional assumption that (iii) the s-process neutron flux is constant in time, and was formulated by Ward *et al* (1976). If we consider the branching at $^{79}$Se in figure 2, the corresponding split of the $\sigma N_s$ curve to the neutron-rich side follows from

$$\sigma N(^{79}\text{Se}) = \zeta'(^{79}\text{Se}) \sigma N(^{78}\text{Se}). \tag{2.6a}$$

The beta decay side of the branching is expressed by

$$\sigma N(^{79}\text{Br}) = \zeta(^{79}\text{Br})[f(^{79}\text{Se})/(1 - f(^{79}\text{Se}))] \sigma N(^{79}\text{Se}) \tag{2.6b}$$

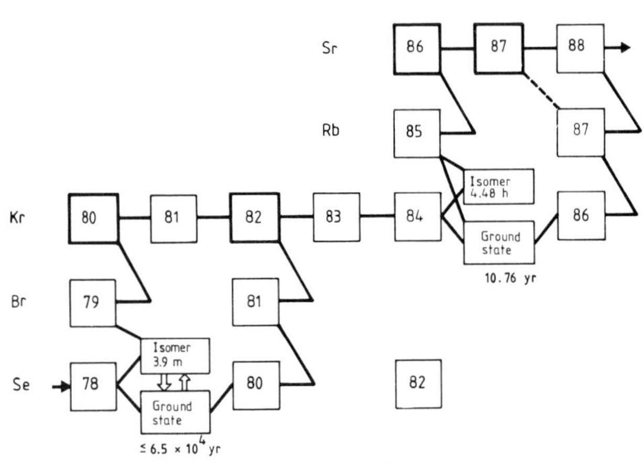


**Figure 2.** The s-process branchings at $^{79}$Se and $^{85}$Kr. The $\sigma N_s$-values of the s-only isotopes imply almost equal branching ratios in both cases, suggesting a strong temperature dependence of the $^{79}$Se half-life.

and the propagators in equations (2.6) are defined as

$$\zeta(^{A}Z)=[1+(\tau_0\sigma(^{A}Z))^{-1}]^{-1} \tag{2.7a}$$

$$\zeta'(^{A}Z)=[(f_n(^{A}Z))^{-1}+(\tau_0\sigma(^{A}Z))^{-1}]^{-1} \tag{2.7b}$$

with the branching factor

$$f_\beta(^{79}\mathrm{Se})=\lambda_\beta(^{79}\mathrm{Se})(\lambda_\beta(^{79}\mathrm{Se})+\lambda_n(^{79}\mathrm{Se}))^{-1}. \tag{2.8}$$

The s-process flow follows the two branches in the traditional way unless there is another branching point, as for example $^{80}$Br, which decays to both $^{80}$Kr and $^{80}$Se. Finally, the two branches merge into the main s-process path at $^{82}$Kr. Among the isotopes involved the s-only nuclei $^{80,82}$Kr hold a key position as normalisation points for the $^{79}$Se branching. Obviously, a fit of equations (2.6) to the empirical $\sigma N_s$ values of $^{80,82}$Kr would yield $\lambda_n$ via equation (2.8) and hence the neutron density, if the beta decay rate of $^{79}$Se were not affected by temperature. As this is true in case of the $^{85}$Kr branching, the neutron density thus obtained can be used to derive the effective stellar beta decay rate, $\lambda_\beta$, from the analysis of those branchings which exhibit a significant temperature dependence of the beta half-life, as for example at $^{79}$Se.

Another aspect of the s-process branchings is the problem of a possible time-dependence of neutron density and temperature. A pulsed rather than a steady s-process was proposed by the conditions during helium shell burning which also provides for a natural explanation of the exponential form of the exposure distribution $\rho(\tau)$ (§ 2.2). An analytic solution of this problem within the classical model has been worked out by Ward *et al* (1978) for rectangular pulses, but a more realistic treatment requires numerical integration of equation (2.1) (Howard *et al* 1986, Beer and Macklin 1989).

It is evident from the above that each of the simplifying assumptions for the classical s-process bears immediate consequences for the s-process branchings. Systematic branching analyses are therefore capable of revealing many details of the s-process environment (§ 5).

### 2.2. *Stellar models*

Instead of a complete review of stellar s-process models (for recent references see Mathews and Ward (1985), Arnould (1986), Lambert (1988)) this section emphasises those models which are detailed enough to allow for quantitative comparisons with the results of the classical model as well as stellar observations (§ 6). From considerations concerning the classical model it was inferred that helium burning zones in red-giant stars are the most likely sites for the s-process.

*2.2.1. Intermediate and low mass stars.* The first assignment of a true stellar environment was achieved by the work of Weigert (1966) and of Schwarzschild and Härm (1967). These authors found that stars with masses $M<9M_\odot$ ($M_\odot$ = mass of sun) develop a convective helium burning shell during their AGB stage of evolution. This convection might reach outward far enough to mix hydrogen-rich material from the envelope into the He shell. There, neutrons could be produced by the reaction sequence $^{12}\mathrm{C}(\mathrm{p},\gamma)^{13}\mathrm{N}(\mathrm{e}^+\nu)^{13}\mathrm{C}(\alpha,\mathrm{n})^{16}\mathrm{O}$. Helium burning in the convective shell occurs in rather short pulses which last for some years, followed by an interpulse period of several hundred years with hydrogen burning quiescently at the bottom of the envelope. If enough helium is replenished in that way, helium shell burning reignites and the next pulse episode starts. The fact that the mass zones comprised of successive pulses overlap was shown by Ulrich (1973) to necessarily result in the exponential distribution of neutron exposures, $\rho(\tau)$, which was already found empirically by the classical model (Seeger *et al* 1965).

The pulsating helium shell model was further developed by Iben (1975a, b, 1976) who showed that mixing of protons into the helium shell is prohibited by an entropy barrier, but that the $^{22}Ne(\alpha, n)^{25}Mg$ reaction may be an alternative neutron source. This is illustrated in figure 3 which shows a very schematic sketch of the s-process operating in the helium burning shell of an AGB star. Note that the mass of the helium shell is greatly exaggerated; the true relations for a $7M_\odot$ star are such that the carbon-oxygen core has $M_c = 0.95\ M_\odot$ followed by a very thin helium shell of $M_{sh} = 10^{-7}\ M_\odot$. All of the remaining mass is contained in the stellar envelope. With each pulse the helium shell grows outward leaving behind the ashes of the $3\alpha$ reaction and comprising fresh material from the hydrogen burning region. Figure 4 illustrates this evolution for the 15th and 16th pulse of the $7M_\odot$ model star (Iben 1976). The positions of the convective envelope and the helium shell are shown in mass coordinates as a function of time. The shaded parts are the convective regions, i.e. the envelope and the helium shell during the pulses. As can be seen, the pulses last for about 12 yr with interpulse periods of ~3000 yr.

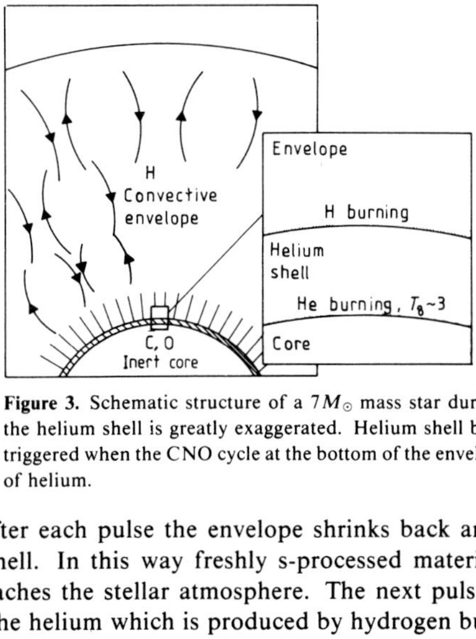


**Figure 3.** Schematic structure of a $7M_\odot$ mass star during shell burning. For illustration, the helium shell is greatly exaggerated. Helium shell burning occurs in pulses, which are triggered when the CNO cycle at the bottom of the envelope has produced a critical amount of helium.

Shortly after each pulse the envelope shrinks back and covers the upper part of the helium shell. In this way freshly s-processed material enters the envelope and ultimately reaches the stellar atmosphere. The next pulse already starts further out, powered by the helium which is produced by hydrogen burning via the CNO cycle in the interpulse period. The second most abundant product of the CNO cycle is $^{14}N$ to which almost all of the initial carbon, nitrogen and oxygen is converted. At the higher temperature of the helium shell the reaction sequence $^{14}N(\alpha, \gamma)^{18}F(e^+\nu)^{18}O(\alpha, \gamma)^{22}Ne$ produces $^{22}Ne$ for the s-process to operate during the next helium shell flash via $^{22}Ne(\alpha, n)^{25}Mg$ reactions.

The fractional overlap of successive convective shells results in an exponential distribution of neutron exposures (Ulrich 1973): if $\Delta\tau$ is the neutron exposure per pulse, and if $r$ denotes the overlap of the $N$th and the $(N-1)$th convective shell, then

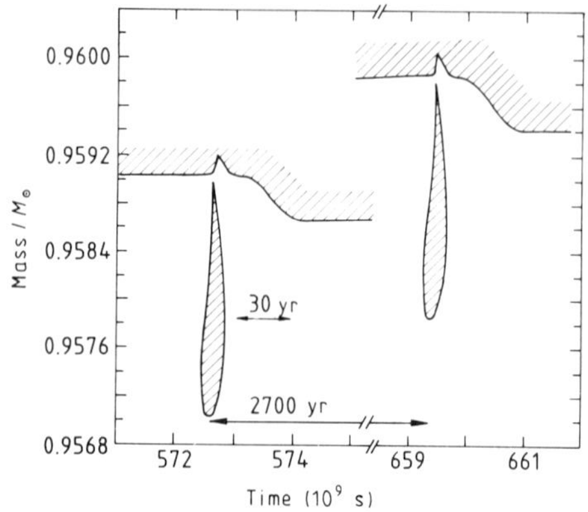


**Figure 4.** Position and size of the convective helium shell and of the convective stellar envelope for the 15th and 16th shell flash of a $7M_{\odot}$ mass star. Pulse duration is ~12 yr followed by interpulse periods of about 3000 yr. After each pulse, freshly processed material from the helium shell is taken up by the envelope and eventually transported to the stellar atmosphere. During the interpulse period the helium shell grows outward with new material from the CNO cycle being added. This fractional substitution of the helium shell results in an exponential distribution of neutron exposures in the s-process (see text).

the fraction of matter, $\rho(\tau)$, which experienced a total exposure $\tau = N\Delta\tau$ will be

$$\rho(\tau) = r^N = r^{(\tau/\Delta\tau)} \qquad \text{or} \qquad \rho(\tau) = \exp(\tau \ln(r)/\Delta\tau). \tag{2.9}$$

The parameter $\tau_0$ from the classical model, equation (2.3), is hence related to the exposure per pulse, $\Delta\tau$, and the fractional overlap between pulses, $r$. According to Iben (1982), $r$ depends on the core mass and decreases with increasing pulse number until a limiting value $r = 0.6 - 0.555(M_H - 0.6)$ is reached, $M_H$ being the stellar mass up to the hydrogen burning shell. For the $7M_{\odot}$ mass star with $M_H = \sim 0.96$ one thus obtains $r = \sim 0.4$. During their AGB evolution, stars experience enough helium shell flashes to establish an exponential distribution $\rho(\tau)$.

Another feature of the s-process associated with helium shell burning is the correlation between the number of seed nuclei and the number of available neutrons. The abundance distribution of the light elements in population I stars being practically solar yields a fixed ratio $^{56}Fe/\Sigma CNO = 1/39$ (Anders and Ebihara 1982). Nearly all of the C, N and O is transformed to $^{14}N$ during the CNO cycle, so that $^{56}Fe$ and $^{14}N$ are mixed into the helium shell in the same ratio. This restricts the number of available neutrons per $^{56}Fe$ seed to $<39$, even if all of $^{14}N$ is used up for neutron production.

For stellar masses $<3M_{\odot}$ the $^{22}Ne(\alpha, n)^{25}Mg$ reaction is not efficient enough for significant s-processing because the temperatures during the shell flash are too low. However, as stars in this mass range do show s-process enhancements, other mechanisms activating the $^{13}C(\alpha, n)^{16}O$ neutron source may become important, such as semiconvective mixing or diffusion (Iben and Renzini 1983). Therefore, low mass stars have received increased attention (Gallino *et al* 1988, Boothroyd and Sackmann 1988, Hollowell and Iben 1988, Gallino 1989), and it was found that low mass stars are indeed efficient sites for the main s-process component.

The advantage of the helium shell models is that they have been worked out in great detail, describing the relevant s-process conditions for a wide mass-range of AGB stars, e.g. temperature and neutron density as a function of time. A critical discussion of these results will be given later in comparison with the classical model (§ 6).

*2.2.2. Massive stars.* Core helium burning in massive stars has been identified as another site for the s-process (Peters 1968). Towards the end of core helium burning in stars with $M > 10 M_\odot$ temperatures become sufficiently high that a substantial neutron density can be established via the $^{22}Ne(\alpha, n)$ reaction. Recent calculations (Arnett and Thielemann 1985, Busso and Gallino 1985, Langer *et al* 1989) based on an improved stellar model and updated nuclear data confirm the earlier results of Lamb *et al* (1977) that significant s-process overproduction is obtained for $A < 90$. Arnett and Thielemann (1985) also found significant s-processing in that same mass region during hydrostatic carbon burning. For very massive stars Prantzos *et al* (1987) showed that the s-process yields are almost independent of stellar mass for $50 < M < 100 M_\odot$. The abundances produced in all these scenarios appear to be compatible with the weak s-process component.

## 3. The data base

### *3.1. Abundances*

The success of any theory of nucleosynthesis has to be measured by comparison with the abundance patterns observed in nature. The enormous efforts invested in abundance determinations cannot be adequately summarised in this review. Instead we will try to sketch the present status and to illustrate recent trends with a few examples of importance to the s-process. More detailed information may be found in the literature cited.

Our knowledge of abundances has been established essentially during the past five decades, based on the hypothesis that primitive meteorites might have preserved the original composition of the early solar system (Goldschmidt 1937). This idea has been confirmed by numerous measurements, and meteorite analyses now constitute the main source of information on solar abundances. Ten years earlier, the first quantitative abundance determinations were obtained from stellar spectra, in particular from that of the sun (Russell and Adams (1928); for a historic account see Unsöld (1979)). Analyses of stellar spectra yield—on average—surprisingly good agreement with solar values. This led people to think of generalising solar abundances to what were called 'cosmic abundances', assuming a homogeneous abundance distribution throughout our galaxy with the solar system being a representative sample. In the discussion about the site of nucleosynthesis, this homogeneity was considered to support the hypothesis of a cosmic event which created all matter in the universe. It was supposed that these element abundances were only marginally changed afterwards by stellar nuclear burning. However, significant improvements in observation techniques led to the discovery of numerous deviations from the composition of solar material, and the abundance patterns in objects like red giants, supernova remnants, interstellar matter etc are now considered as direct clues to stellar nucleosynthesis.

With respect to the s-process, it is clear from § 2 that the *isotopic* pattern derived from meteoritic samples provides for the most sensitive test of the classical model. On the other hand, the pronounced s-process enhancements in the surface abundances of red giants represent a direct link to stellar models and deserve therefore equal attention.

The status of the abundance information from various sources is briefly summarised below.

*3.1.1. Meteoritic abundances.* In the first compilation of Suess and Urey (1956) typical uncertainties were estimated to be not better than a factor of about 1.5. Subsequent compilations (Cameron 1973, Suess and Zeh 1973, Palme *et al* 1981, Cameron 1982, Anders and Ebihara 1982, Anders and Grevesse 1989) were based on continuously improved measurements and showed increasingly good agreement. In particular, the comprehensive work of Anders and collaborators confirmed earlier assumptions that the original composition of the solar system is best preserved in the relatively rare C1 carbonaceous chondrites. All other types of meteorites differ more or less from the C1 composition, especially with respect to volatile elements. These differences are readily ascribed to processing in the early solar nebula. From the variance of the available data obtained by different authors from different meteorites and with different techniques, it is found that the precision of solar abundances has reached the 5–10% level in the meantime.

More recent measurements, e.g. on rare earth elements in C1 chondrites (Patchett 1983, Beer *et al* 1984), have succeeded in improving the meteoritic abundance determinations for these elements to an accuracy of about 2%, and a similar precision has been reported for Fe and Cr. Figure 5 shows a set of Fe/Cr abundance ratios in carbonaceous chondrites (Palme 1985). Type 2 and type 3 chondrites define an Fe/Cr ratio which is about 3% lower than that observed in C1 chondrites, and it is not clear yet which of the two values represents the Fe/Cr ratio of the solar photosphere. The higher ratio in the C2 meteorite Nogoya reflects inhomogeneities in the Nogoya parent body due to extensive weathering, while the larger spread in the C3V chondrites is due to the larger grain size in these meteorites. The samples used for analysis were not big enough to obtain a reliable average. This problem is also evident from the variations in the Allende results. Nevertheless, the example of figure 5 demonstrates impressively the recent progress in solar abundance data, and this development provides great motivation to make an effort to improve the accuracy of neutron capture cross sections (§ 3.2).

In summary, meteoritic abundances are of twofold importance for s-process studies: they represent a complete abundance pattern for direct comparison with model calculations, and their (potential) precision allows or will allow sensitive tests of s-process models which afterwards can be applied to analyses of stellar abundances.

*3.1.2. Solar spectrum.* Compilations of solar abundances from the spectrum of the sun have been reported by Ross and Aller (1976), Holweger (1979) and Grevesse (1984). In general, somewhat larger uncertainties of 15 to 25% are reported for these data compared to meteoritic results. The difficulties in spectral analyses arise from uncertainties in the model atmosphere of the sun (deviations from local thermodynamic equilibrium, microturbulence, temperature inhomogeneities) as well as from uncertain atomic physics data (atomic transition probabilities, collision cross sections, line broadening parameters). For 38 elements of very different character the close correspondence of meteoritic and photospheric abundances is shown in figure 6 (Holweger 1984). Discrepancies between the two data sets are mainly due to depletion effects. In meteorites this holds for the volatile elements, especially for C, N and O. In all critical cases, the range of depletion from carbonaceous chondrites of type 1 towards types 2 and 3 is indicated by horizontal bars. In the sun, lithium is efficiently depleted, because this light element can undergo reactions with thermal protons in the convective zone. A similar comparison of meteoritic and solar abundances is given by Wasson (1985). A more puzzling discrepancy was noted for iron, where solar values exceed

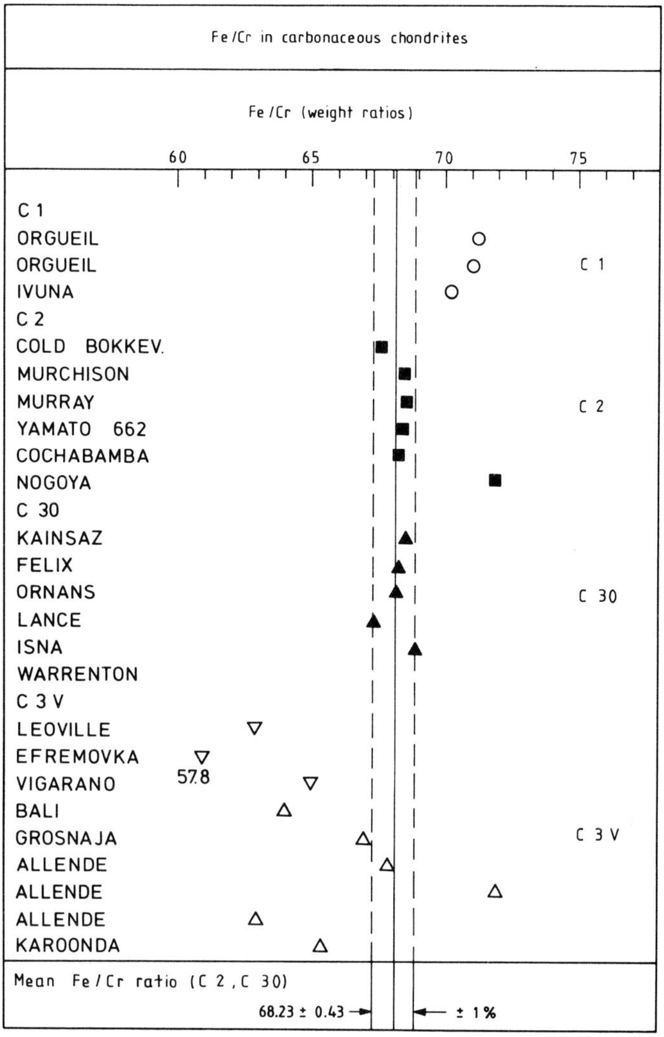


**Figure 5.** Recent precise measurements (black squares) of Fe/Cr ratios for a variety of meteorites showing remarkable consistency on the 1% level (Palme 1985).

the meteoritic abundances by ~40% (Grevesse 1987); possible reasons for this difference are currently being investigated (Anders and Grevesse 1989).

An example of a refined solar abundance determination is the work of Kwiatkowski *et al* (1984) on osmium. By using a greater variety of Os lines together with new and more accurate experimental transition probabilities, the osmium abundance in the solar photosphere ($N_{\mathrm{Os}}^{\odot} = 0.77 \pm 20\%$) could be improved significantly, and is now in excellent agreement with the carbonaceous chondrite result ($N_{\mathrm{Os}}^{\mathrm{met}} = 0.717 \pm 9.8\%$,

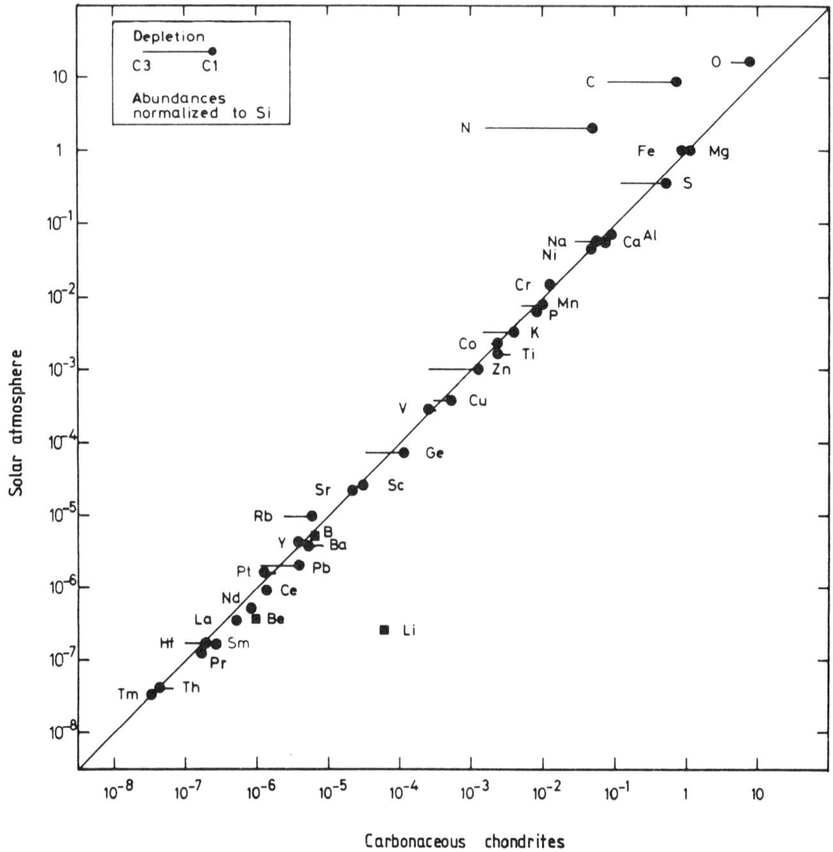


**Figure 6.** Comparison of elemental abundances in the sun with those in carbonaceous chondrites.

Anders and Ebihara (1982)). Note that all abundances quoted in this review correspond to the normalisation $N_{Si} \equiv 10^6$. Solar and, in general, also stellar abundances represent *elemental* data as they are derived from atomic transitions. This feature has consequences for the interpretation of s-process patterns in red giant stars (§ 6).

*3.1.3. Stellar spectra.* The accurate new results for the solar photosphere appear most promising for the prospects of stellar abundances. The study of stellar spectra has, on average, confirmed the close relationship of abundance patterns in the sun, in main sequence stars similar to the sun and in interstellar matter. But, in addition, many deviating peculiarities were discovered reflecting galactic chemical evolution or ongoing nucleosynthesis. (For a comprehensive review see Trimble (1975).) Most important with respect to neutron capture reactions are the s-process enhancements in red giant stars.

A breakthrough for the idea of stellar nucleosynthesis was the discovery of technetium lines in the S star, R Andromedae, by Merrill in 1952 and in several other red giants in the following years. As the longest-lived isotope of Tc has a half-life of four million years, it was immediately obvious that the observed technetium was produced in the stars themselves. By 1955, Cameron had explained quantitatively the production of Tc and other overabundant heavy elements in these stars in terms of a neutron capture process with neutrons released from the $^{13}C(\alpha, n)^{16}O$ reaction.

In his detailed review on abundances in evolved stars Baschek (1979) estimated a typical uncertainty for stellar abundance data of about three tenths of a decade. More recent observations claim uncertainties as small as 20% for favourable cases, and this does not appear to be a limit yet. Among the many publications the work of Tomkin and Lambert (1983) deserves a special mention as a nice example of a relatively complete set of abundances in the red giant HR774. Figure 7 shows how well these data can be reproduced by the classical s-process model (Käppeler (1986a); see also § 6).

It has to be noted at this point that analyses of stellar spectra do not yield data for a number of elements (As, Se, Kr, Te, Xe, I, Ta) either because these are rare or because their lines are masked by more abundant species.

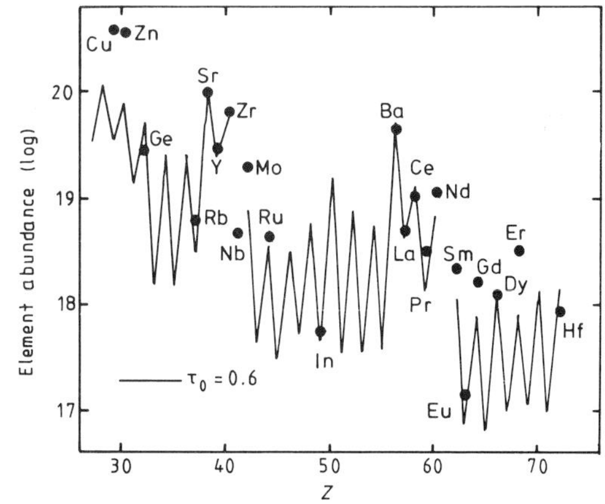


**Figure 7.** Observed s-process enhancements in HR774 fitted with the classical model (full curve). From Tomkin and Lambert (1983).

*3.1.4. Cosmic rays.* Galactic cosmic rays are an important independent channel of abundance information as they constitute the only accessible sample of matter from outside the solar system. This sample is rather young since the measured lifetime from source to observer is about $10^7$ yr (Wiedenbeck and Greiner 1980), and it is also complete, containing all elements from hydrogen to the actinides. Cosmic rays have been extensively analysed in the past 15–20 yr with an impressive progress in detection techniques. Isotopes up to $Z \sim 28$ can be completely resolved, and at higher masses this is limited by decreasing yields rather than by detector resolution. Spacecraft experiments may significantly improve the exposure factors of present balloon flights so that isotopic abundances may also be expected in the mass region $Z > 28$.

Although cosmic-ray abundances can be measured with a precision of less than 10%, it is not easy to obtain a simple representation because the yields differ for different energy ranges. Modulations are due to interactions with interstellar matter or within the solar system, involving spallation reactions, energy loss by ionisation, radioactive decay, etc. A cautious interpretation (Simpson 1983, Meyer 1985) suggests a mixture of aged r-process material and interstellar matter as the main component of cosmic-ray abundances. For a quantitative discussion in terms of neutron capture nucleosynthesis, the presently available data are not detailed enough. This holds in particular for the isotopic abundances in the range $Z > 28$.

### *3.2. Neutron capture rates*

*3.2.1. Laboratory cross-section measurements.* As the s-process capture path follows the valley of beta stability, the neutron capture cross sections of most of the involved nuclei can be measured experimentally. Figure 8 illustrates the signature of a neutron capture event. The highly excited compound system decays promptly to the ground state of the product nucleus by emission of a gamma-ray cascade of total energy $E_\gamma^{\text{tot}} = (B_n + E_n)$ with $B_n$ being the neutron separation energy and $E_n$ the kinetic energy of the captured neutron. If the product nucleus is unstable the capture event can also be determined via its decay. The experimental techniques for neutron capture measurements will be briefly sketched in this section to outline their current status and possible improvements. In view of the role of reliable neutron capture rates as the main input data for the s-process this is an important point.

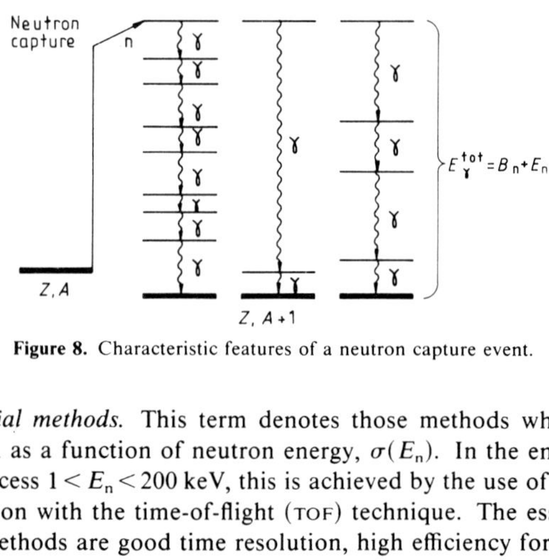


**Figure 8.** Characteristic features of a neutron capture event.

*Differential methods.* This term denotes those methods which yield the capture cross section as a function of neutron energy, $\sigma(E_n)$. In the energy range of interest for the s-process $1 < E_n < 200$ keV, this is achieved by the use of a pulsed neutron flux in combination with the time-of-flight (TOF) technique. The essential features of the respective methods are good time resolution, high efficiency for detecting the prompt gamma-ray cascade independent of its multiplicity and low sensitivity to scattered neutrons. The second point is important because the gamma-ray cascade is completely determined by statistics for practically all isotopes on the s-process path. (There are very few exceptions, where valence or direct capture dominates over the compound nucleus mechanism.) The many possible transitions in the decay of the compound nucleus lead to a smooth capture gamma-ray spectrum without prominent, isolated peaks which might be used for identification of capture events by high resolution spectroscopy, e.g. with Ge detectors. As neutron scattering is, in general, about ten times more likely than capture, detector systems for neutron capture studies should be insensitive to scattered neutrons to avoid backgrounds, which are very difficult to distinguish from true events.

In principle, the best technique would be to sum up all gamma rays in the cascade to obtain a signal proportional to $E_\gamma^{tot}$. An ideal detector covering the entire solid angle of $4\pi$ with 100% efficiency and good energy resolution would then yield a spectrum of capture events as indicated in figure 9(*a*). This was first approximated by means of large liquid scintillator tanks of 300–3000 l volume, but these detectors are far from being ideal: the efficiency is limited to ~70%, the large volume results in a significant cosmic ray background and the energy resolution is poor. Moreover, these detectors are sensitive to neutrons scattered in the sample, which are thermalised in the scintillator and subsequently captured in hydrogen. The resulting 2.2 MeV gamma-ray line is overwhelming the spectra, as sketched in figure 9(*b*), and has to be suppressed electronically. The corresponding correction for events falling below that threshold must be determined by extrapolation of the spectrum to zero pulse height. This procedure is affected by large systematic uncertainties of 5–8% thus limiting the overall accuracy to ~10% (Chrien 1975).

In order to circumvent the background problems with scintillator tanks, an alternative detector was developed by Moxon and Rae (1963). This very simple set-up is shown in figure 10(*a*). It consists of a graphite disc acting as a converter of gamma rays into electrons, followed by a thin plastic scintillator and a photomultiplier. The idea was to obtain a detector efficiency proportional to gamma-ray energy: electrons from low-energy gamma rays can reach the scintillator only from a thin layer of the converter, whereas high-energy gamma rays produce electrons energetic enough to penetrate the entire converter. As the efficiency of this Moxon–Rae detector is small

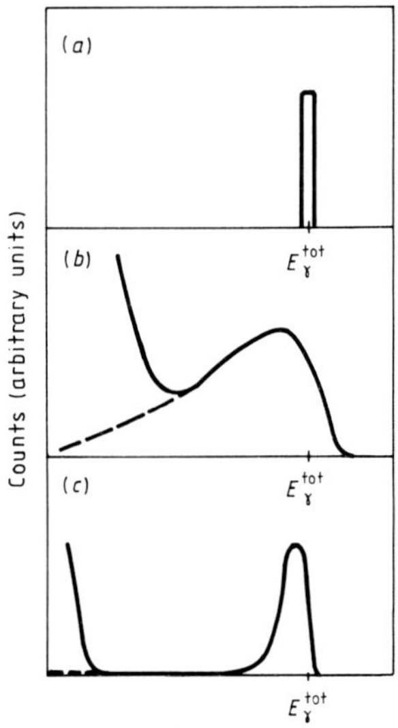


**Figure 9.** (*a*) Spectrum of capture events recorded with an ideal $4\pi$ detector. (*b*) Typical spectrum taken with a liquid scintillator tank. The peak around $E_\gamma^{tot}$ is broadened and obscured by an intense background. (*c*) Expected spectrum for a $4\pi$ detector of $BaF_2$.

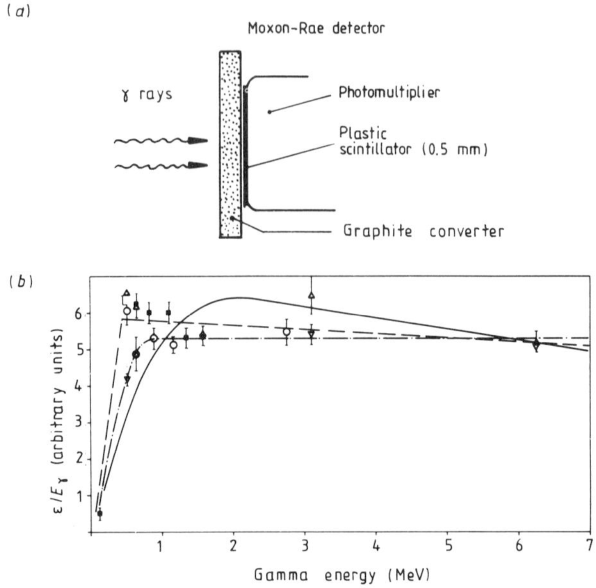


**Figure 10.** (*a*) Schematics of a Moxon–Rae detector. (*b*) Efficiency of a Moxon–Rae detector versus gamma-ray energy (Wisshak and Käppeler 1981). Graphite converter: ——, calculated; – – –, experimental. Bi-C converter, — · —, experimental.

($\leqslant$1%), at most one gamma ray is detected per event, and the cascade efficiency

$$\varepsilon_{\text{casc}} = \sum_{i=1}^{m} \varepsilon_i(E_{\gamma i}) = \sum k \, E_{\gamma i} = kE_{\gamma}^{\text{tot}} \tag{3.1}$$

is in good approximation independent of the multiplicity and only determined by the total energy, $E_{\gamma}^{\text{tot}}$ (Chrien 1975). With the set-up of figure 10(*a*) the proportionality between gamma-ray energy and efficiency is almost fulfilled as shown in figure 10(*b*) except for very low energies, where the efficiency drops steeply. This can be a significant advantage for the investigation of radioactive targets because low-energy gamma rays from the natural decay are automatically suppressed. This has been demonstrated with several examples by Wisshak and Käppeler (1983 and references therein). These authors also studied the systematic uncertainties of the Moxon–Rae measurements in great detail and found that accuracies of typically 5% can be achieved (Reffo *et al* 1982, 1983).

A generalisation of the Moxon–Rae detector was developed by Macklin and Gibbons (1967). In this approach the converter in figure 10(*a*) is eliminated and the scintillator thickness is increased to ~10 cm. Even then, the volume of ~1 l is small enough that the ordinary organic scintillator can be replaced by the corresponding deuterated compound to avoid the problem of hydrogen capture. This so-called total-energy detector offers a much better efficiency per cascade of ~20%, but no longer has an efficiency proportional to the gamma-ray energy. As was shown experimentally, the gamma-ray energy can be approximated by the observed pulse height, and a weighting function $W(E_{\gamma})$ can be applied to each event in order to restore the relation $\varepsilon \sim E_\gamma$ artificially (Macklin and Gibbons 1967). This type of detector was extensively used in capture measurements, mostly in Oak Ridge by Macklin and collaborators (see Winters and Macklin (1982) for an example) but also in many other laboratories (e.g. in Karlsruhe, Livermore, Geel and Tokyo). Similar to the Moxon–Rae detector one finds that inherent systematic uncertainties restrict the accuracy of this technique to ~5%. This value seems to be a general limit for the differential techniques available at present, which have been optimised over more than a decade. Therefore, further improvements of neutron capture rates for the s-process can only be achieved by new techniques.

*A new 4π detector.* For a significant improvement of the differential techniques described above, one has to return to the idea of detecting in the total gamma-ray energy of the capture cascade (Käppeler *et al* 1983). Figure 9(*a*) shows the gamma-ray spectrum for the ideal detector with 100% efficiency. While an ideal detector is only roughly approximated by the liquid scintillator tank (figure 9(*b*)), the use of scintillation crystals appears more promising because of their much better energy resolution. The gamma-ray spectrum expected for a 4π arrangement of scintillation crystals is sketched in figure 9(*c*). Most of the events fall in a line corresponding to the sum energy of the entire gamma-ray cascade, and only a minor fraction appears at smaller energies, because part of the gamma rays may escape from the detector or are absorbed in the detector cannings or in the sample.

At present, $BaF_2$ exhibits the best scintillation properties for a detector of neutron capture rates. The scintillation light of $BaF_2$ contains an extremely fast component (decay time = 600 ps) which is emitted in the UV region at 220 nm (Laval *et al* 1983). Even with large crystals an excellent time resolution of ~400 ps can be obtained, while the energy resolution of $BaF_2$ is close to that of NaI (Wisshak and Käppeler 1984).

Because the size of the available crystals is limited, a 4π detector must be composed of many elements. So far, two concepts for a 4π geometry have been pursued. A honeycomb structure of equally shaped hexagonal crystals arranged in axial symmetry offers the advantage of flexibility (Twin 1983). The arrangement can be changed by inclusion of more and more elements, all of the same shape. The second approach, a crystal-ball-type detector (Habs *et al* 1979) is based on a spherical arrangement of detector elements all covering the same solid angle with respect to the sample, thus facilitating multiplicity measurements. Such a detector for the accurate determination of neutron capture rates has now been completed at Karlsruhe (Wisshak *et al* 1989). It is composed of 42 $BaF_2$ crystals, 12 pentagons and 30 hexagons forming a spherical shell of 15 cm thickness. A sketch of the set-up at the accelerator is given in figure 11. The openings of the detector for the neutron beam and sample changer together with the spacings between the detector elements reduce the 4π solid angle by less than 5%.

The response of such a 4π detector for monoenergetic gamma rays was investigated by Wisshak *et al* (1984). With this information the efficiency for neutron capture events was calculated for three extreme examples ($^{56}$Fe, $^{197}$Au and $^{241}$Am) characterised by very different gamma-ray cascades with respect to total energy, $E_\gamma^{\rm tot}$, and to multiplicity. Figure 12 presents the resulting efficiency per gamma-ray energy interval. One finds a total efficiency close to 95% for all cases if a realistic threshold of 2.5 MeV is used for background suppression. This is remarkable as it means that the difference in efficiency even for very different capture cascades is only ~1%, leading to very small systematic uncertainties in the measurements. The high efficiency and the expected low backgrounds of this detector will allow for much shorter measuring times compared to present techniques. This, in turn, enables the careful experimental investigation of

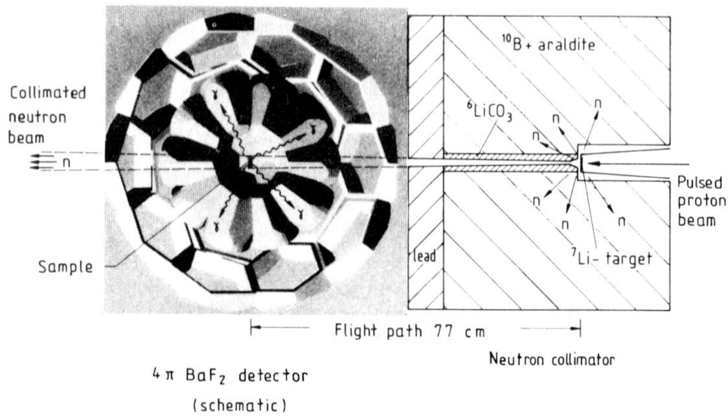


**Figure 11.** Schematic set-up of the Karlsruhe $4\pi$ detector at the accelerator. The detector consists of 12 pentagonal and 30 hexagonal $BaF_2$ crystals forming a spherical shell with 20 cm inner diameter and 15 cm thickness.

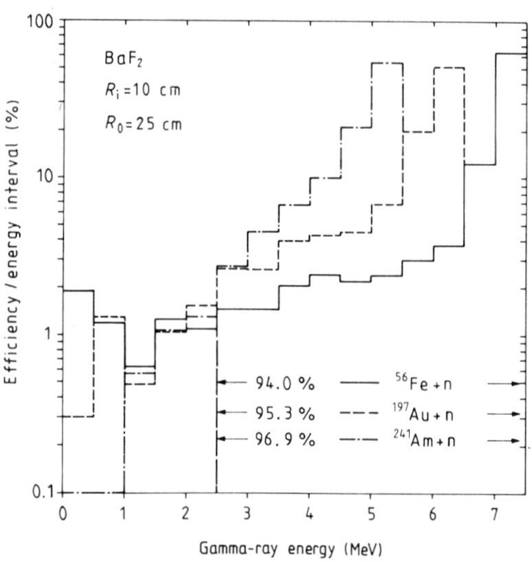


**Figure 12.** Calculated efficiency of the $4\pi$ detector for neutron capture cascades from $^{56}$Fe, $^{197}$Au and $^{241}$Am. Note that for a threshold of 2.5 MeV one obtains the same efficiency of 95% for all three isotopes although these exhibit extremely different gamma-ray cascades.

remaining uncertainties which are related to the capture sample itself (neutron multiple scattering and self-shielding as well as gamma-ray self-attenuation). From the properties of the new detector and from the experience with present techniques an overall accuracy close to 1% can be expected for this new experimental method.

*Activation technique.* The potential of the activation technique in conjection with the $^{7}$Li(p, n)$^{7}$Be reaction for the determination of stellar capture rates was outlined by Beer and Käppeler (1980). An important point is the possibility of producing a neutron energy spectrum very similar to the Maxwell–Boltzmann distribution during the s-process. This advantage adds to the well known features of the activation technique (i.e. selectivity, sensitivity and accuracy) resulting in a very successful method for neutron capture cross-section measurements. However, this method can only be applied if neutron capture leads to a radioactive isotope with a convenient half-life between a few seconds and about 400 days. On the other hand, it is the only means for measuring the partial capture cross sections for populating long-lived isomeric states which can decisively influence s-process branchings (§ 5).

The set-up during neutron irradiation is shown in figure 13(*a*). The energy of the incoming protons is chosen as 31 keV above the threshold of the $^7$Li(p, n)$^7$Be reaction.

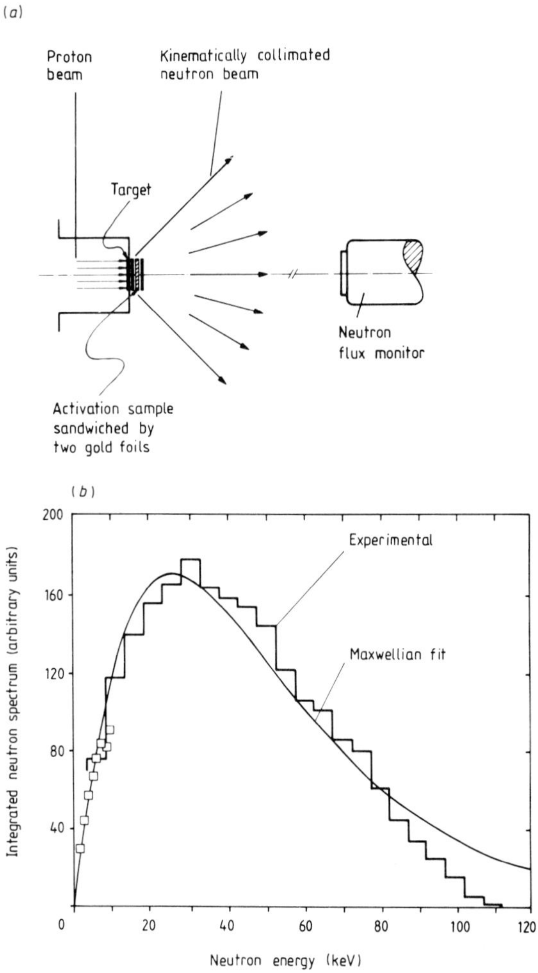


**Figure 13.** (*a*) Principle of the activation technique and (*b*) simulation of a Maxwellian energy spectrum ($kT = 25.0$ keV) via the $^7$Li(p, n) reaction. $E_p = 30$ keV above reaction threshold.

Then all neutrons are collimated by the reaction kinematics in a forward cone of 120° opening angle. The angle-integrated neutron spectrum imitates almost perfectly a Maxwell–Boltzmann distribution for a thermal energy of $kT = 25$ keV, as is illustrated in figure 13(*b*) (Ratynski and Käppeler 1988). With the same arrangement, a spectrum for $kT = 52$ keV can be produced via the $^3H(p, n)^3He$ reaction (Käppeler *et al* 1987). This may be a useful complement, if the $kT = 25$ keV cross section has to be extrapolated to higher energies.

During the irradiation the samples are usually located very close to the neutron target and are sandwiched between two gold foils, which are used for determination of the neutron flux via the $^{197}Au$ as a cross section standard. A monitor at a distance of ~1 m from the neutron target records the time dependence of the neutron yield as the target degrades during the activation. From this time dependence the fraction of activated nuclei that decay during irradiation is calculated. The resulting correction can be significant if the half-lives of the investigated isotope and the gold reference are very different. After irradiation the activated samples are counted with high-resolution gamma or beta detectors as a function of time. This allows us to check whether the background in the activity measurement was properly subtracted, because only then are the correct half-lives reproduced.

The features of the activation technique are illustrated by the example of the very small cross section of the neutron magic isotope $^{48}Ca$. Figure 14 shows the gamma-ray spectrum from a $^{48}Ca$ sample after irradiation with neutrons from the $^7Li(p, n)^7Be$ reaction (Käppeler *et al* 1985). The sample consisted of a 43 mg $CaCO_3$ tablet (6 mm diameter, 82% $^{48}Ca$) between two gold foils (15 mg each). The irradiation and counting

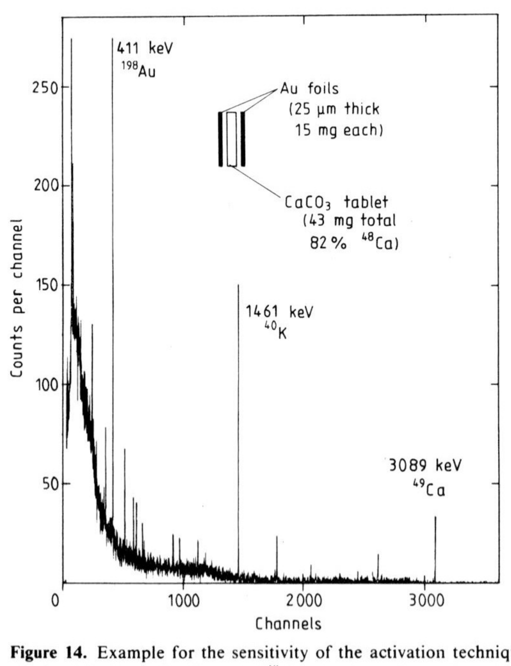


**Figure 14.** Example for the sensitivity of the activation technique: gamma-ray spectrum obtained for the difficult case of $^{48}Ca$. The inset shows the activation sample.

times of 1000 s correspond roughly to two half-lives of $^{49}$Ca. The spectrum exhibits the strong 411 keV line from $^{198}$Au, the background line from $^{40}$K and the 3089 keV line from $^{49}$Ca. Although statistics are rather poor after one irradiation, it is obvious that backgrounds are almost negligible. This means that an accurate cross section ratio to $^{197}$Au can be obtained by repeated irradiations. The good energy resolution in the gamma-ray spectrum allows identification of the gamma-ray transition in the particular isotope of interest, which makes the method selective: no highly enriched samples are required. The method is also very sensitive because of the low background: rather small sample amounts are sufficient and/or very small cross sections can be determined. For example, the $^{48}$Ca cross section was found to be only $0.95 \pm 0.09$ mb.

Another important aspect of the activation technique is its suitability for very precise measurements. This has been demonstrated in conjunction with a recent improvement of the capture cross section of $^{197}$Au (Ratynski and Käppeler 1988). As gold is often used as a cross-section standard, the sizeable spread in existing data called for an additional measurement with an independent method in order to avoid previous systematic uncertainties. The experiment was carried out with the activation technique by irradiating a gold sample in the kinematically collimated neutron cone (figure 13) such that all neutrons pass through the sample. Then the time-integrated neutron flux can be evaluated directly from the $^7$Be activity of the neutron target. By systematic variation of the relevant experiment parameters the remaining uncertainties were carefully studied. In this way, the precision of the gold cross section could be improved to 1.5% (figure 15).

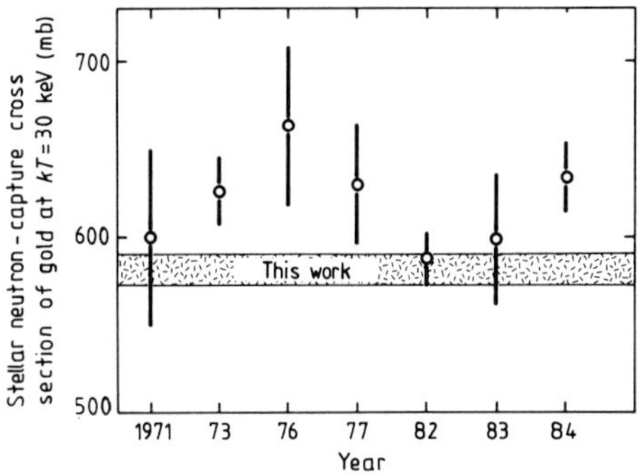


**Figure 15.** The capture cross section of $^{197}$Au measured via the activation technique to $\pm 1.5\%$ (Ratynski and Käppeler 1988) compared to previous results.

The characteristic uncertainties of the activation technique are listed in table 1. While practically all effects can be defined with good accuracy, the spectroscopic information on the decay of the activated isotope (to be taken from literature) is often a limiting factor. In those cases it is advantageous to determine the induced sample activity via the beta-decay electrons, e.g. by the use of a $4\pi$ Si(Li) detector system (Ratzel *et al* 1989). In summary, the activation technique is indispensable for s-process investigations particularly because of its combination of high sensitivity and high accuracy.

*Radioactive isotopes.* Although the vast majority of nuclei involved in the s-process are stable, there are a number of radioactive isotopes which play a key role in the

**Table 1.** Systematic uncertainties of the activation technique.

| Effect | Uncertainty (%) |
|---|---|
| Efficiency of activity measurement | 1–2 |
| Self absorption correction | <0.5–5 |
| Sample mass | <1 |
| Correction for decay during activation | <0.5 |
| Relative intensity of detected radiation | 0.5–10 |
| Gold reference cross section | 1.5 |

analysis of s-process branchings (§ 5). Cross-section measurements on these radioactive species in the laboratory will be extremely difficult in most cases because of the short half-lives of the order of 10 yr or less. The availability of sample material, sample handling and gamma-ray backgrounds from sample decay represent the main problems in this respect. Therefore, all existing measurements on radioactive isotopes of relevance to the s-process (Macklin 1982, 1985a,b) were carried out for the long-lived isotopes $^{99}$Tc ($2.1\times10^5$ yr), $^{107}$Pd ($6.5\times10^6$ yr) and $^{93}$Zr ($1.5\times10^6$ yr), respectively. These nuclei could be measured with a conventional set-up because of their low specific activity and the very soft gamma-radiation associated with their decay. Measurements on shorter-lived nuclei have been reported for a series of actinide isotopes by Wisshak and Käppeler (1983). With a dedicated set-up these authors obtained a good signal-to-noise ratio even for the comparatively strong gamma-ray emitter $^{243}$Am. By extrapolation from this example, some cases of interest to the s-process might be within the reach of conventional techniques.

An exciting new approach has become feasible, however, with the completion of the proton storage ring at the LAMPF accelerator in Los Alamos, which allows for such a high neutron flux that backgrounds from radioactive targets can be overridden about 1000 times better than at existing accelerators. In addition, this laboratory would also be capable of producing the required radioactive targets directly at the site (Bowman and Talbert 1985). A programme to systematically investigate the radioactive branching point isotopes at this facility would provide the information necessary to reveal the physical conditions during the s-process with good accuracy.

*3.2.2. Model calculations.* The calculation of cross sections represents an important complement to experimental studies for two reasons:

(i) Many of the radioactive species on the s-process path which constitute the important branching points (§ 5) are unaccessible to direct experiments at present. Hence, branching analyses depend heavily on reliable model calculations.

(ii) The typical s-process temperatures of 300–400 million degrees (corresponding to thermal energies of $kT\sim30$ keV) are sufficient to establish thermal equilibrium between the low-lying nuclear states. This affects mostly the beta decay rates (§ 3.3) but also the stellar reaction rates. In general, the calculated laboratory and stellar neutron capture rates (Holmes *et al* 1976, Harris 1981) agree within ~10% because the population probabilities of excited states are usually small (Walter and Beer 1982). Nevertheless, such differences are significant when experimental rates with 1–2% precision become available. A famous example, where capture from excited states accounts for a large effect is the chronometric pair $^{187}$Re/$^{187}$Os (§ 7).

Neutron capture rates for the s-process are usually calculated with the statistical model of nuclear reactions (Hauser and Feshbach 1952, Vogt 1969, Hodgson 1971).

This theory can be applied successfully only if two conditions are met: the level density in the compound nucleus at the interaction energy must be high enough so that many (>10) resonances contribute to the reaction rate and the energy of the incident particle must be sufficiently low that other than compound processes (direct or non-statistical reactions) are negligible. The second condition is practically always satisfied for s-process neutron energies, because non-statistical effects start to be important only above a few MeV. The first condition, however, is in question for all nuclei at or close to magic nucleon numbers, and must be considered individually.

By far the most calculations of stellar neutron capture rates are based on the Hauser–Feshbach formula. For the (n, $\gamma$) reaction the cross section is given by

$$\sigma = \frac{\pi \hbar^2}{(2J_I+1)(2J_j+1)} \sum_{j,\pi} (2J+1) \frac{T_{\mathrm{f}}(J^\pi)\,T_{\mathrm{d}}(J^\pi)}{T_{\mathrm{tot}}(J^\pi)}\, W \tag{3.2}$$

where $\lambda$ is the wavenumber of the incoming neutron, and $J_I$, $J_j$ are the spins of the capturing state in the target nucleus and of the neutron, respectively. The transmission functions $T_{\mathrm{f}}$ and $T_{\mathrm{d}}$ describe the formation and decay of the compound nucleus. The summation runs over all states in the compound nucleus which can be reached by a combination of target and projectile spins $I$ and $j$. The factor $W$ accounts for width fluctuations.

The probability for formation of the compound nucleus $T_{\mathrm{f}}$ is calculated by means of the optical model. The corresponding parameters which characterise the potential well of the nucleus as felt by the incoming neutron, are the well depth, the radius and some information on the absorptive properties of the potential. Information on these parameters can be derived from a fit of total neutron cross sections, strength functions and scattering radii. Variation of the optical model parameters within reasonable limits does not affect the resulting cross section by more than about 10%; hence, this part of the cross section calculation is not as sensitive as the treatment of the exit channel.

The decay of the compound nucleus through the photon channel, i.e. the photon transmission function is calculated using the Brink–Axel hypothesis: If one assumes that dipole transitions dominate the photon channel, the photon transmission function can be expressed as

$$T_\gamma(J^\pi) = \sum_{\nu=0}^{\omega} (T_\gamma^\nu(\mathrm{E1}) + T_\gamma^\nu(\mathrm{M1})) + \int_{E_\omega}^{E_{\mathrm{max}}} \int_{J^\nu \pi^\nu} (T_\gamma^\nu(\mathrm{E1}) + T_\gamma^\nu(\mathrm{M1})) \times \rho(E^\nu, J^\nu, \pi^\nu)\,\mathrm{d}E^\nu\,\mathrm{d}\pi^\nu\,\mathrm{d}J^\nu \tag{3.3}$$

where the superscript $\nu$ indicates all excited states in the compound nucleus and E1, M1 are the electric and magnetic dipole transmission functions. The two terms in equation (3.3) account for transitions between discrete levels and transitions in and from the 'continuum' above the last known level at $E_{\mathrm{max}}$. This continuum of levels is described by a level density $\rho(E)$. All transitions are subject to the appropriate dipole selection rules.

The dominant electric-dipole transitions can be described through the cross section for the inverse reaction. This means that the available information about the giant dipole resonance can be used to determine $T_\gamma$ by extrapolation to the energy of the respective gamma transition in the decay of the compound nucleus. A global parameterisation for the calculation of $T_\gamma$ is given e.g. by Holmes *et al* (1976) and Woosley *et al* (1978). More refined treatments have been reported by Reffo (1980) or Kopecky (1982).

Most important in the calculation of $\sigma_{n\gamma}$ are the probabilities for formation and decay of the compound nucleus, which depend sensitively on the number of accessible levels. Thus the level density $\rho$ in the particular compound nucleus is the determining quantity in any cross-section calculation. Experimental information on the level density is available at low excitation energy from the known discrete levels (ground state up to about 1 MeV, e.g. Lederer and Shirley (1978)) and at the neutron separation energy from high-resolution neutron-resonance studies. In practical applications, a theoretical approach as for instance the back-shifted Fermi gas formula is parameterised and adjusted to the respective experimental values for $\rho$ (Holmes *et al* 1976). Until this point, an accurate description of the level density throughout the periodic table could not be achieved, and cross-section calculations which relied on global parameter sets for $\rho$ suffered from systematic uncertainties of typically 50%. Therefore, *local* parameter studies were carried out to account for shell and deformation effects on the level density (Reffo 1980, Reffo *et al* 1982, 1983), which used as much experimental information on discrete levels and neutron resonance schemes as were available. This method can be extended to all parameters in order to achieve a consistent set for a group of isotopes around the investigated sample, for which the relevant parameters can then be *inter*polated from the systematics.

The example of figure 16 shows the systematics in the mass region $145 < A < 155$, which was established to calculate the cross sections of radioactive $^{147,148}$Pm (Winters *et al* 1986a). These nuclei belong to the important branching points on the s-process path (see § 5). One finds from figure 16 that the level density parameter $a$ depends critically on the neutron number in the compound nucleus; beside a pronounced odd–even effect there is a clear discontinuity at $N = 89$, which would not be so evident in a global parameterisation. By sensitivity studies, starting from the scatter of the experimental values around the smooth systematics (e.g. all points in figure 16 except the interpolated stars for the Pm isotopes), an uncertainty of 20–25% was estimated for this calculational technique.

Whether the uncertainties of such calculations can be further reduced is difficult to judge at present. This problem deserves further investigation as better input data and more accurate experimental capture cross sections become available.

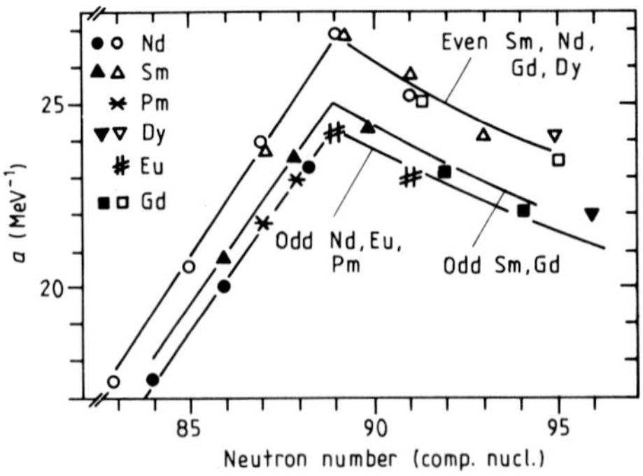


**Figure 16.** Systematics of the level density parameter $a$ in the mass region $145 < A < 155$. Open and full symbols represent the level densities for even and odd target isotopes as derived from experimental information. Interpolated values for $^{147,148}$Pm are denoted by stars.

*3.2.3. Stellar neutron capture rates.* As helium shell burning in red giant stars is commonly considered as the astrophysical site for the s-process, one deals with typical matter densities of $\sim 1000\ \mathrm{g\ cm^{-3}}$, neutron densities of $\sim 10^8\ \mathrm{cm^{-3}}$, and temperatures between 300 and 400 million degrees. In such an environment neutrons are in complete thermal equilibrium, exhibiting a Maxwell–Boltzmann spectrum

$$\Phi \sim \sqrt{E_n}\, \exp(-E_n/kT) \tag{3.4}$$

with thermal energies around $kT = 30$ keV. The effective stellar capture cross section, $\langle\sigma\rangle$, is an average of the product of the differential cross section, $\sigma(E_n)$, and the relative velocity of neutron and target, $v$, over the thermal flux spectrum of equation (3.4). For astrophysical applications, $\langle\sigma\rangle$ is defined as

$$\langle\sigma\rangle = \frac{\langle\sigma v\rangle}{v_T} = \frac{2}{\sqrt{\pi}} \frac{\int_0^\infty \sigma(E_n) E_n \exp(-E_n/kT) \mathrm{d}E_n}{\int_0^\infty E_n \exp(-E_n/kT) \mathrm{d}E_n} \tag{3.5}$$

where $E_n$ is the total kinetic energy in the centre-of-mass system (Clayton 1968). The factor $2/\sqrt{\pi}$ comes from the fact that the mean thermal velocity $v_T = \sqrt{(2kT/m)}$ with $m$ being the reduced mass was used for normalisation instead of the average velocity. The definition of $\langle\sigma\rangle$ implies that the stellar neutron capture rate, $\lambda_n$, is rather insensitive to temperature because most differential cross sections $\sigma(E_n)$ scale with $1/v$. Therefore, s-process calculations are usually performed for a *standard* thermal energy of 30 keV.

It was noted recently that the stellar neutron spectrum may not be completely thermalised but may exhibit a small epithermal component, depending on the $Q$-value of the neutron-producing reaction (Petrov and Shlyakhter 1988). This component is certainly too small to affect the stellar neutron capture cross sections, but it may well cause an enhancement of (n, p) or (n, $\alpha$) rates due to the rather sharp thresholds in these cross sections. Whether this effect can be verified by a typical signature in the observed abundances is still open at present.

*Neutron capture rates.* Previous compilations of experimental neutron capture cross sections for the s-process (Allen *et al* 1971, Conrad 1976, Newman 1978) have recently been extended and updated (Bao and Käppeler 1987, Beer *et al* 1989a). The present status is briefly summarised in table 2, and numerical values are included in table 4 (below). Though the majority of all required cross sections is determined experimentally, the respective uncertainties are often much larger than the 5 to 10% range of present techniques. Further measurements will have to concentrate on improving the accuracy, especially for the s-only isotopes, and on tackling the problem of radioactive nuclei.

In making the step from the laboratory to stellar capture rates it is not sufficient to determine the proper average over the stellar neutron spectrum. In addition, neutron capture in excited states and competition of superelastic scattering (by which the neutron de-excites a nucleus and gains in energy) must be considered as well. Model

**Table 2.** Maxwellian-averaged neutron capture cross sections at $kT = 30$ keV.

| Classification | Remarks |
|---|---|
| Stable isotopes on or near s-process path | ~240 cross sections; only 22 not yet measured, but only 40% with uncertainties better than 8% |
| Stable s-only isotopes or isotopes with minor r-process contributions | ~40 cross sections; all measured except for $^{128,130}$Xe and $^{192}$Pt; in 9 cases still uncertainties between 12 and 20%: $^{87}$Rb, $^{89}$Y, $^{96}$Mo, $^{110}$Cd, $^{116}$Sn, $^{122,124}$Te, $^{134,136}$Ba |
| Radioactive isotopes | 50 cross sections, only 6 measured so far |

calculations indicate that the stellar cross sections differ by less than 5% from the laboratory values in most cases (Harris 1981, Holmes *et al* 1976); but there are important s-only isotopes, where this difference is estimated to reach the 10–20% level, e.g. $^{154}$Gd, $^{160}$Dy, $^{170}$Yb and $^{176}$Hf, as well as the osmium isotopes 186 and 187 which are decisive for the chronometric pair $^{187}$Re/$^{187}$Os. Such corrections are already significant for the present $\sigma N(A)$ curve (Käppeler 1989), which is considered to be accurate to about 10%. As the model calculations available differ from each other by that same percentage, further efforts to refine the classical s-process model have therefore to improve on that aspect. As far as dynamic s-process models are concerned, superelastic scattering has been shown to be an important effect (Petrov and Shlyakhter 1984). For neutron densities $>10^{10}$ cm$^{-3}$ it is expected to influence the observed abundances significantly. As Ward and Fowler (1980) have pointed out, the stellar capture rates can become complex functions of the physical conditions during the s-process, when the population of excited states does not reach thermal equilibrium (see also § 3.3).

*Other neutron-induced reactions.* For a number of isotopes in the mass range $A<90$, the (n, $\alpha$) and (n, p) cross sections become significant and may well exceed that for neutron capture, leading to another type of s-process branchings. Important examples are the radioactive nuclei $^{35}$Cl, $^{40}$K and $^{41}$Ca which all affect the abundances of the potential chronometric pair $^{40}$K/$^{40}$Ar (Beer and Penzhorn 1987). Another example for an important (n, p) reaction is that on $^{26}$Al, which is not related to the s-process, but determines the $^{26}$Al abundance in explosive nuclear burning (Trautvetter *et al* 1986).

### 3.3. Beta decay rates

Branchings in the s-process path are determined by the competition between neutron capture and $\beta$ decay. The branching ratio can be expressed by equation (2.8) in terms of the neutron capture rate, $\lambda_n = n_n\sigma v_T$, and the beta decay rate, $\lambda_\beta = \ln 2/t_{1/2}$, of the respective branch-point isotope. The beta decay rate stands for all possible decay processes: $\beta^-$decay, electron capture or $\beta^+$decay. For each s-process branching it has to be checked whether $\lambda_n$ or $\lambda_\beta$ could be affected by temperature. As was noted in § 3.2.3, neutron capture rates depend only slightly on temperature. In contrast, beta decay rates may undergo dramatic changes at typical s-process temperatures, either through beta decays from excited states or through atomic effects in the highly ionised, dense stellar plasma. As nuclear matrix elements remain unchanged, the stellar beta decay rates can be calculated from pure phase space arguments (Bahcall 1961), at least for the important allowed and first non-unique forbidden transitions.

How aggravatingly s-process temperatures may affect beta decay rates can be illustrated by the example of the branchings at $^{79}$Se and $^{85}$Kr. Figure 2 shows the s-process flow through the mass region $75<A<90$. The branchings manifest themselves by different $\sigma N$-values of the respective s-only isotopes, i.e. $^{80,82}$Kr and $^{86,88}$Sr. (Formally, $^{88}$Sr is not a pure s-process isotope but the abundance contribution from the r-process is only about 20%; Walter *et al* (1986a).) While the $\sigma N$-ratios of $^{80,82}$Kr yield a good measure for the branching factor at $^{79}$Se, the $^{85}$Kr branching is more complicated due to the short-lived isomeric state in $^{85}$Kr and due to the neutron shell closure at $N=50$. About 57% of all neutron captures populate the isomer $^{85m}$Kr, which then decays promptly (compared to the neutron capture timescale) with 79% probability to $^{85}$Rb. In other words, only 54% of the s-process flow proceeds through the long-lived ground state of $^{85}$Kr. Furthermore, the $\sigma N$-values of $^{86,88}$Sr have to be corrected for the effect of the small cross sections at $N=50$, which cause a significant step in the $\sigma N$ curve. The observation that the resulting branching ratios, both for $^{79}$Se and $^{85}$Kr, are almost equal seems to be in conflict with the fact that the terrestrial beta decay rates of $^{79}$Se and $^{85}$Kr differ by roughly four orders of magnitude.

This puzzle is resolved if beta decay from thermally populated excited states is taken into account. The ground-state decay of $^{79}$Se is first unique forbidden, but allowed decays are possible from the isomeric state at 95.7 keV as well as from the states at 128 and 137.1 keV. These states are populated to 1.0, 0.4 and 1.2% at $kT = 30$ keV, and dominate the stellar beta decay rate of $^{79}$Se completely:

$$\lambda_\beta^* = \sum_i \left( p_i \sum_j \lambda_{\beta ij} \right). \tag{3.6}$$

The index $i$ runs over all excited states with population probabilities, $p_i$, and $\lambda_{ij}$ is the decay rate from state $i$ in the unstable nucleus to level $j$ in the daughter nucleus. In thermal equilibrium the $p_i$ are given by a Boltzmann distribution

$$p_i = (2J_i + 1) \exp(-E_i/kT) \Big/ \sum_m (2J_m + 1) \exp(-E_m/kT), \tag{3.7}$$

the sum in the denominator being the nuclear partition function. Population probabilities of excited states in radioactive nuclei of relevance for s-process studies have been published in tabular form by Walter and Beer (1982).

Apart from the beta decay of excited states, stellar decay rates are also influenced by atomic effects via the high degree of ionisation in the hot stellar plasma. Under such conditions electrons can also be emitted into unoccupied atomic orbits, thus adding the respective binding energy to the decay energy. This effect is always important when the decay energy is comparable to or smaller than the K-shell binding energy. The relevant nuclear parameter in this case is the decay energy itself, which is not always sufficiently well known as in the important example of the chronometric pair $^{205}$Tl/$^{205}$Pb (Yokoi *et al* 1985). Even in the case of $^{79}$Se, where the beta decay energy is about 20 times larger than the K-shell ionisation energy, bound-state decay contributes ~25% to the stellar rate (Conrad 1976, Takahashi and Yokoi 1987).

One of the major uncertainties in the stellar beta decay rates is due to the log *ft* values for transitions from excited states, which have been estimated by comparison with known analogous transitions in neighbouring isotopes. Experimental log *ft* values can hardly be expected for these states, because gamma decay by far dominates over beta transitions. $^{79}$Se represents an important exception, as the stellar beta decay rate is determined by the isomeric state at 96 keV which has a half-life of 3.9 m (figure 17). At $kT = 30$ keV this isomer is populated to ~1%, and its decay represents the only significant term in equation (3.6). The 10% uncertainty of the estimated log $ft = 5.0$, however, results in an uncertainty of the stellar decay rate of a factor of three.

The rather long half-life of 3.9 m for the gamma decay to the ground state enabled a successful measurement of the beta decay branching ratio (Klay and Käppeler 1988). The 96 keV isomer was produced by activation in the high flux reactor of the Institut Laue–Langevin in Grenoble, using a sample of highly enriched $^{78}$Se implanted into a thin, ultrapure graphite disc. The subsequent decay was investigated by using a Mini-Orange beta spectrometer with a low-energy cut-off near 100 keV for suppression of conversion electrons. With the experimental log *ft* value the temperature dependence of the $^{79}$Se half-life was recalculated with the formalism of Takahashi and Yokoi (1983). Figure 17 compares this result with previous work using different estimates for the log *ft* value (Takahashi and Yokoi 1987, Conrad 1976, Cosner and Truran

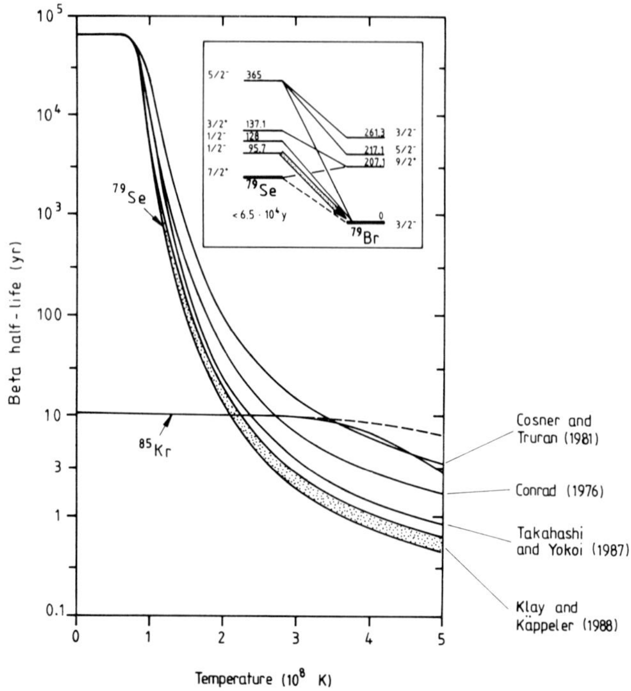


**Figure 17.** The beta decay half-lives of $^{79}$Se and $^{85}$Kr as a function of temperature in units of $T_8 = 10^8$ K. The decay of $^{79}$Se is dramatically enhanced for $T_8 > 1$ due to decay from excited states.

1981). One finds that the decay rate is constant for $T < 10^8$ K but drops quickly to the order of a few years when the first excited states start to be thermally populated. Note that the calculated curves were uncertain by a factor of three due to the estimated 10% uncertainty in the log *ft* value.

For comparison, figure 17 also contains the beta decay rate of $^{85}$Kr, which appears to be insensitive to temperature, at least under helium burning conditions, because the first excited states in this nucleus occur at 305 and 1107 keV (Yokoi and Takahashi 1985). This curve should be taken as a lower limit as it was calculated using the beta decay rate ($t_{1/2} = 4.5$ h) as the effective destruction rate of the isomer instead of the smaller equilibration rate by which the isomer is fed from the ground state (Ward and Fowler 1980; broken curve).

A summary of thermal effects on stellar beta decay rates is given in table 3 (Conrad 1976). Most important is beta decay from excited states including the possible back decay from excited states in the daughter nucleus as in the example of $^{79}$Se. Bound-state beta decay is important for small decay energies below $Q_\beta < 300$ keV, where the electron binding energy has a significant impact. Quantitative treatment of bound-state beta decay requires knowledge of the degree of ionisation, i.e. a model of the stellar plasma. Such a model is also needed to calculate the stellar electron capture rates as these depend critically on the electron density at the nuclear surface. At typical s-process

**Table 3.** Thermal effects on beta decay rates under typical s-process conditions ($kT = 30\,\text{keV}$, $n_c = 10^{27}\,\text{cm}^{-3}$).

| Effect | Comment |
|---|---|
| 1. $\beta^-$ decay through excited states | Involves excited states below 200 keV; experimental decay rates very scarce |
| 2. Bound state $\beta^-$ decay | Significant for $E_\beta < 300$ keV |
| 3. Hindrance of electron capture due to high degree of ionisation | Typical factor 10? |
| 4. Electron capture from continuum | Significant for $E_\beta < 30$ keV |
| 5. Weakening of screening corrections due to high degree of ionisation | Affects low-energy $\beta^+$ decays |
| 6. Change of competing internal conversion rates due to high degree of ionisation | Applies for small transition energies and high multiplicity (isomers) |
| 7. Other effects | Negligible for the s-process; see text |

densities, electron concentrations are of the order of $10^{27}\,\text{cm}^{-3}$, which means that electron capture from the continuum is no longer negligible, especially at decay energies below 30 keV. Another consequence of the high degree of ionisation is that the Fermi function is no longer modified by screening from the atomic electron cloud; this is important for low-energy $\beta^+$ decays.

Finally, minor effects are mentioned in entry 7 of table 3 which are not significant for the s-process but may become important at higher temperatures (photon-induced beta decay; Shaw *et al* (1965)), or at higher densities (Fermi statistics, reduction of electron phase space in a degenerate environment, neutronization of matter; Conrad (1976), Fuller *et al* (1982)).

Global calculations of beta decay rates for the s-process have been published by Newman (1973), and by Cosner and Truran (1981) using schematic assignments of unknown log *ft* values for transitions from excited states and neglecting bound-state decays. The work of Conrad (1976) included a simplified model for bound-state decay and log *ft* values derived from similar transitions in neighbouring nuclei. A thorough and comprehensive investigation was then carried out by Takahashi and Yokoi (1983) with due consideration of ionic states in the stellar plasma, transition probabilities and decay energetics in a wide domain of possible temperatures ($10^6$–$10^9$ K) and densities ($10^{-4}$–$10^4\,\text{g cm}^{-3}$). An important outcome of this work was that bound-state decay—which was neglected in some previous calculations—is of great importance to s-process analyses. Remarkable examples are $^{163}$Dy, a terrestrially stable isotope which becomes unstable at high temperature (§ 5.2.2), and $^{187}$Re, which was considered as a cosmochronometer until it was found that its decay is affected by temperature (§ 7).

These improved beta decay rates for the s-process are tabulated in numerical form for all relevant isotopes along the s-process path (Yokoi and Takahashi 1985, Takahashi and Yokoi 1987). Figure 18 is adopted from this work and illustrates impressively the influence of bound-state $\beta^-$ decays. It shows the ratio between the rates of bound-state decay and continuum decay for allowed and non-unique first forbidden transitions versus $Q_n$, the neutral atomic mass difference between the initial and final nucleus. For small values of $Q_n$ the rates are completely dominated by bound-state decay. As expected, this effect exhibits a strong $Z$ dependence. Consequently, Yokoi and Takahashi noted severe discrepancies to earlier calculations, where bound-state decay was not considered (Newman 1973, Cosner and Truran 1981).

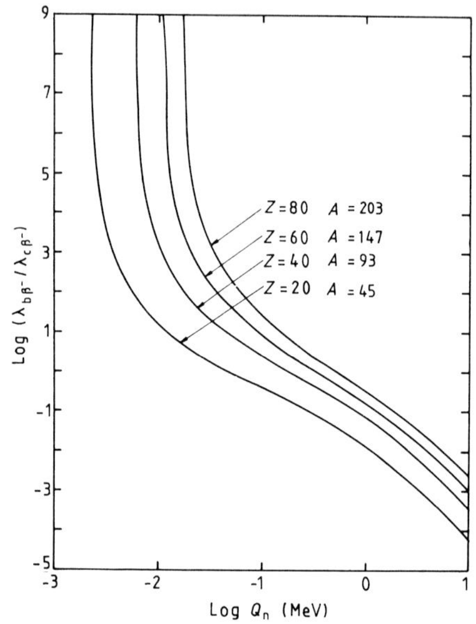


**Figure 18.** Ratio of bound-state $\beta^-$ decay rate and continuum $\beta^-$ decay rate versus $Q_n$, the neutral atomic mass difference between the initial and final nucleus (Yokoi and Takahashi 1985) for allowed and non-unique first forbidden transition.

For the study of s-process branchings it is sometimes also important to consider the timescale for reaching stationary conditions. This is of special importance in connection with long-lived isomeric states. These states are characterised by quantum numbers which are very different from those of the ground state, so that direct gamma-ray transitions are hindered by selection rules. During the s-process isomers can be populated by neutron capture or beta decay.

In principle, nuclei in the isomeric state must be treated as a separate species in parallel to the nuclei in the ground state. This concerns the effective capture cross sections (§ 3.2) as well as the stellar decay rates. However, separate treatment of the isomer and the ground state is only valid if the initial population of the isomeric state remains unchanged under s-process conditions. This needs to be verified in each case, as isomeric states may well be thermalised at high temperatures through transitions via mediating states at higher excitation energy (Ward 1977, Ward and Fowler 1980).

There are examples where the level spacing is too large for a mediating state to occur within the available thermal energy, e.g. in $^{121}$Sn. Such isotopes fall under the category of *un*thermalised cases for which the equilibration time is much longer than the time required to capture a neutron. Consequently, the isomer and the ground state have to be treated separately (Ward 1977). In most cases, thermal equilibrium is rapidly achieved, as the equilibration time for a particular state is

$$\tau_{eq} = (\lambda\uparrow + \lambda\downarrow)^{-1} \tag{3.8}$$

where $\lambda\uparrow$ and $\lambda\downarrow$ denote the upward and downward transition rates. The downward rate is the sum of the natural decay rate plus the rate for stimulated emission, while the upward rate equals the downward rate times the Boltzmann factor. In any case, the equilibration time is shorter than the natural half-life of a state. Therefore, the majority of all isotopes involved in the s-process are in *complete* thermal equilibrium, a situation which is convenient to calculate (Takahashi and Yokoi 1987).

Between these extremes one finds those isotopes which are only *partially* thermalised because the equilibration timescale is comparable to the timescale for neutron capture and/or changes in the astrophysical conditions. These cases are difficult to handle as the quantitative verification requires knowledge of the mediating transitions that provide the link between the ground state and the isomer. In view of the complicated structure of these nuclei, experimental transition rates are usually not available, and even level schemes are often incomplete. Under stellar conditions, terrestrial decay rates can additionally be modified by suppression of internal conversions, and inelastic collisions in the hot stellar plasma (Shaw and Clayton 1967, Ward 1981). An example for such a partially thermalised isomer is possibly $^{176}$Lu, which is discussed in § 7.4.

The influence of temperature on alpha decay rates was also investigated (Perrone and Clayton 1971) with the result that these rates are affected relatively little under s-process conditions. Weak interaction rates for more extreme density and temperature regimes, e.g. conditions during stellar collapse, have been tabulated by Fuller *et al* (1982).

In summary, it is to be emphasised that the understanding of many details in the s-process depends on the reliability of relevant stellar beta decay rates. Further improvement of these rates mainly represents a theoretical challenge, but information from the experimental side remains an essential complement. Only then can the signatures in the observed abundance patterns be deciphered in terms of the s-process neutron density and temperature. This aspect will be discussed in detail in § 5.

## 4. Status of the classical s-process

### *4.1. The $\sigma N_s(A)$ curve*

The credibility of the classical model is based on the successful reproduction of the empirical products of cross section times s-process abundance, $\sigma N_s$, for the s-only isotopes. This property was the better confirmed the more reliable the relevant cross sections, and abundances were determined with time. A historic account of this observation is given by Mathews and Ward (1985). Figure 19 shows a comparison of the calculated $\sigma N_s$ curve obtained by Käppeler *et al* (1982) and a present fit together with the respective empirical $\sigma N_s$ products for those isotopes which are of pure s-process origin or which have only minor r-process contributions. The main component is plotted as a heavy full curve showing the contribution of the weak component explicitly. The present curve was determined with the code SPEED.CLAS (Beer 1985). The major improvement compared with the previous calculation is that now all significant s-process branchings are considered as indicated by the respective splits of the $\sigma N_s$ curve. Within error bars—which include the uncertainties from the abundances and from the cross sections as well—the empirical $\sigma N_s$-values of all s-only nuclei are perfectly reproduced: a striking success of the classical model!

Differences between the two curves are due to improved abundances and cross sections as well:

(i) The abundance compilations of Anders and Ebihara (1982) and of Anders and Grevesse (1989) contain several changes with respect to that of Cameron (1982), which were important for the $\sigma N_s$ curve. For example, the Te abundance was lowered by

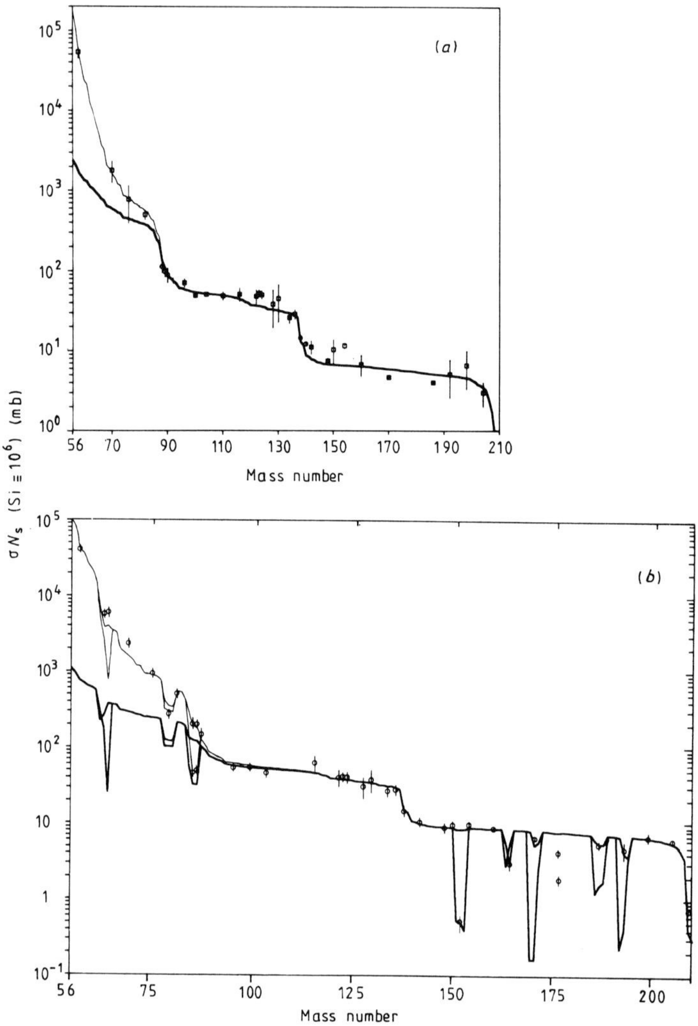


**Figure 19.** The characteristic product of cross section times s-process abundance versus mass number. We compare the status of this curve in 1982 (*a*) (with $N_s$ from Cameron 1981) with the present situation (*b*) ($N_s$ from Anders and Ebihara 1982). Symbols with error bars denote the empirical products of pure or almost pure s-isotopes.

25%, leading to good agreement of the $^{122,123,124}$Te values with the model curve. Even more important was the revision of the rare-earth abundances. While the Gd abundance was lowered by 20%, most of the others were raised by ~10%. These changes yielded an empirical $\sigma N_s$-value of Gd, which is now consistent with the model curve, and reduced also the step in the $\sigma N$ curve at the magic neutron number 82. As a result, the ratio between seed abundance and integral s-process yield for the main component has decreased, resulting in consequences for the neutron economy (§ 4.2). The abundances of volatile elements that had previously been determined by interpolation between neighbouring isotopes are now determined from the $\sigma N_s$ systematics, i.e. for Ar (Beer and Penzhorn 1987), Kr (Walter *et al* 1986a), Xe (Beer *et al* 1983) and Hg (Beer and Macklin 1985).

(ii) The set of neutron capture cross sections was also considerably improved during the last years (Bao and Käppeler 1987). Prominent examples are the s-nuclei $^{70}$Ge, $^{128,130}$Xe, $^{142}$Nd, $^{148,150}$Sm, $^{160}$Dy, $^{198}$Hg and $^{204}$Pb, where the cross sections have been either measured for the first time or reinvestigated with considerably reduced uncertainties.

With all this new information, the reliability of the empirical $\sigma N_s$ products could be greatly enhanced, the combined uncertainties in cross section and abundance now reaching ~10% for many of these isotopes. Consequently, it is fair to say that the present $\sigma N_s$ curve is also accurate to ±10% because it does reproduce all these normalisation points within their uncertainties. Hence, the classical model allows for a consistent description of the observed s-process abundances including the various branchings (§ 5). This result is somewhat surprising at first sight, because it means that the classical model cannot be as wrong as could be expected from the very schematic assumptions of that approach. On the other hand, this result implies that the true nature of the s-process may still be hidden behind the present uncertainties. As these are mostly due to the nuclear physics involved, experimentalists are challenged to further improve the relevant input data.

However, at the present level two corrections to the empirical $\sigma N_s$ values can already no longer be neglected: the stellar enhancement of neutron capture rates as well as the p-process contributions to the observed s-abundances. It has been mentioned in § 3.2.3 that thermally populated excited states may exhibit very different capture cross sections, and this effect has the more impact on the total rate the higher the population probabilities are, i.e. the higher the temperature is. Being negligible for thermal energies below ~20 keV, these corrections can be of the order of 10% at $kT = 30$ keV. Among the s-only isotopes there are a few examples which are sensitive to this enhancement, but the present uncertainties are just at the limit for it to become manifest: the empirical $\sigma N_s$-values of $^{160}$Dy or $^{170}$Yb need to be increased by 3 to 5% at the currently estimated s-process temperature of $2.5 \times 10^8$ K. (The only exception showing a much stronger temperature dependence is $^{187}$Os, but this case must be considered separately; see § 7.) Principally, the stellar enhancement of the neutron capture rates of a variety of isotopes constitutes a potential thermometer for the s-process in addition to those represented by s-process branchings (§ 5). Though a quantitative estimate cannot be derived at present, it is at least possible to consider $kT = 30$ keV as an upper limit for the thermal energy characteristic of the s-process; any higher temperature would lead to stellar enhancement factors larger than the 10% uncertainty of the present $\sigma N_s$ curve.

It is certainly plausible that the p-process is not restricted to the relatively few isotopes that are not accounted for by neutron capture reactions, but that it also contributes to the s-abundances. At present, the corresponding corrections are either neglected or somehow interpolated from the p-only nuclei. Both approximations lead to uncertainties similar to those discussed above; hence, the future development of a quantitative model for the p-process is much needed.

The results of figure 19 are presented in numerical form in table 4. The cross sections quoted in column 3 are based on the compilation of Bao and Käppeler (1987),

**Table 4.** $\sigma N_s$ values from the classical model compared to empirical $\sigma N_\odot$ products of s-only nuclei corrected for p-process contributions; r-process residuals and r-only abundances are listed in the last two columns. The numerical values of this table correspond to figures 19 and 20. Note that the r-residuals in the mass range $A<90$ are slightly different if the single flux solution for the weak s-process component is adopted ($\sigma N_s$-values in brackets).

| Mass number | Element | $\sigma \pm \Delta\sigma$ (mb) | p-correction to s-only abundance (%) | $\sigma N_\odot$ s-only nuclei (mb, Si = $10^6$) | $\sigma N_s$ main component (mb, Si = $10^6$) | $\sigma N_s$ all components (mb, Si = $10^6$) | $N_{\odot,r}$ r-only nuclei (Si = $10^6$) | $N_r = N_\odot - N_s$ (Si = $10^6$) |
|---|---|---|---|---|---|---|---|---|
| 56 | Fe | 14.0 ± 0.7 | | | 1097 | 104 859 (3561) | | |
| 57 | Fe | 44.7 ± 4.5 | | | 946 | 80 840 (3968) | | |
| 58 | Fe | 15.6 ± 1.6 | | 40 720 ± 4220 | 765 | 46 323 (7720) | | |
| 59 | Co | 43.3 ± 4.3 | | | 705 | 36 463 (8662) | | |
| 60 | Ni | 34.9 ± 1.7 | | | 638 | 25 549 (8792) | | |
| 61 | Ni | 92.9 ± 9.0 | | | 614 | 21 860 (9063) | | |
| 62 | Ni | 43.4 ± 4.9 | | | 566 | 16 332 (9311) | | |
| 63 | Ni | 30 | | | 224 | 6 864 (5693) | | |
| 63 | Cu | 70.9 ± 10.6 | | | 298 | 4 647 (3410) | | 60 ± 44 |
| 64 | Ni | 11.8 ± 1.3 | | | 261 | 3 798 (4533) | | 150 ± 39 |
| 64 | Zn | 70.2 ± 5.9 | | | 176 | 2 580 (1944) | | |
| 65 | Cu | 60.5 ± 5.7 | | | 375 | 3 942 (5363) | | 49 ± 21 |
| 65 | Zn | 148.5 | | | 25 | 763 (444) | | |
| 66 | Zn | 39.5 ± 3.4 | | | 366 | 3 461 (4675) | | 265 ± 16 |
| 67 | Zn | 182 ± 18 | | | 358 | 3 207 (4426) | | 34 ± 2.7 |
| 68 | Zn | 21.9 ± 2.7 | | | 307 | 1 951 (2692) | | 145 ± 14 |
| 69 | Ga | 161 ± 7 | | | 300 | 1 793 (2466) | | 11.6 ± 1.6 |
| 70 | Ge | 105.4 ± 6.2 | 9.7 | 2 305 ± 256 | 290 | 1 581 (2121) | 7.81 ± 0.79 ($^{70}$Zn) | |
| 71 | Ga | 138 ± 9 | | | 283 | 1 437 (1894) | | 4.7 ± 1.2 |
| 72 | Ge | 65 | | | 268 | 1 191 (1489) | | 14.0 ± 3.0 |
| 73 | Ge | 324 | | | 265 | 1 144 (1410) | | 5.67 ± 0.86 |
| 74 | Ge | 60 ± 8 | | | 249 | 944 (1072) | | 27.3 ± 4.4 |
| 75 | As | 636 ± 39 | | | 248 | 926 (1042) | | 5.33 ± 0.75 |
| 76 | Se | 186 ± 23 | 9.7 | 940 ± 130 | 243 | 866 (945) | 9.20 ± 0.95 ($^{76}$Ge) | |
| 77 | Se | 504 | | | 241 | 846 (910) | | 3.04 ± 0.30 |
| 78 | Se | 102 | | | 233 | 755 (765) | | 7.21 ± 0.94 |
| 79 | Se | 249 ± 57 | | | 104 | 412 (537) | | 3.87 ± 1.13 |
| 79 | Br | 702 ± 46 | | | 127 | 316 (242) | | |

| | | | | | | | | |
|---|---|---|---|---|---|---|---|---|
| 80 | Se | 46.6 ± 2.6 | | | 102 | 347 (383) | | 24.3 ± 1.99 |
| 80 | Kr | 287 ± 14 | 15.0 | 270 ± 45 | 120 | 293 (225) | | |
| 81 | Br | 702 ± 46 | | | 101 | 336 (361) | | 4.64 ± 1.11 |
| 81 | Kr | 779 | | | 119 | 289 (222) | | |
| 82 | Kr | 89.1 ± 6.8 | 3.0 | 495 ± 88 | 212 | 553 (449) | 5.71 ± 0.47 ($^{82}$Se) | |
| 83 | Kr | 267 ± 16.0 | | | 209 | 531 (422) | | 3.75 ± 0.47 |
| 84 | Kr | 40.0 ± 4.6 | | | 191 | 423 (245) | | 18.00 ± 2.45 |
| 85 | Kr | 76 ± 19 | | | 55.3 | 126 (69.1) | | |
| 85 | Rb | 397 ± 20 | | | 132 | 274 (168) | | 2.79 ± 0.47 |
| 86 | Kr | 5.2 ± 0.4 | | 45 ± 8 | 32.4 | 49.3 (40.6) | | <0.93 |
| 86 | Sr | 90 ± 4.0 | 5.6 | 199 ± 19 | 123 | 190 (140) | | |
| 87 | Rb | 20.6 ± 2.0 | | | 31.8 | 45.6 (31.7) | | 0.10 ± 0.22 |
| 87 | Sr | 123.4 ± 8.0 | | 201 ± 21 | 119 | 178 (126) | | |
| 88 | Sr | 7.3 ± 0.6 | | 142 ± 17 | 100 | 124 (101) | | 2.55 ± 2.84 |
| 89 | Y | 32.0 ± 3.4 | | 148 ± 20 | 90.0 | 107 | | 1.31 ± 0.50 |
| 90 | Zr | 18.7 ± 1.1 | | 103 ± 15 | 75.2 | 84.7 | | 0.99 ± 0.77 |
| 91 | Zr | 68.5 ± 9.1 | | | 71.4 | 79.3 | | 0.04 ± 0.21 |
| 92 | Zr | 57.1 ± 6.9 | | | 67.1 | 73.6 | | 0.54 ± 0.28 |
| 93 | Nb | 309 ± 17 | | | 65.0 | 70.8 | | 0.11 ± 0.12 |
| 94 | Zr | 37.7 ± 5.7 | | | 59.2 | 63.6 | | 0.17 ± 0.34 |
| 95 | Mo | 333 ± 14 | | | 58.6 | 62.9 | | 0.213 ± 0.024 |
| 96 | Mo | 128 ± 9 | | 53.9 ± 4.8 | 57.0 | 60.8 | 0.300 ± 0.039 ($^{96}$Zr) | |
| 97 | Mo | 387 ± 16 | | | 56.4 | 60.2 | | 0.087 ± 0.015 |
| 98 | Mo | 113.1 ± 8.0 | | | 54.6 | 58.1 | | 0.093 ± 0.051 |
| 99 | Tc | 893 ± 57 | | | 5.27 | 5.55 | | 0.172 ± 0.014 |
| 99 | Ru | 1063 | | | 49.2 | 52.3 | | |
| 100 | Ru | 235 ± 15 | | 55.1 ± 4.6 | 53.6 | 56.8 | 0.242 ± 0.013 ($^{100}$Mo) | |
| 101 | Ru | 1138 ± 46 | | | 53.4 | 56.5 | | 0.266 ± 0.017 |
| 102 | Ru | 212 ± 13 | | | 52.5 | 55.4 | | 0.327 ± 0.036 |
| 103 | Rh | 996 ± 80 | | | 52.3 | 55.2 | | 0.289 ± 0.028 |
| 104 | Pd | 330 ± 33 | 9.0 | 46.0 ± 5.8 | 51.7 | 54.5 | 0.348 ± 0.019 ($^{104}$Ru) | |
| 105 | Pd | 1369 ± 69 | | | 51.6 | 54.3 | | 0.269 ± 0.024 |
| 106 | Pd | 288 ± 29 | | | 50.9 | 53.5 | | 0.193 ± 0.035 |
| 107 | Ag | 913 ± 27 | | | 50.8 | 53.3 | | 0.239 ± 0.026 |
| 108 | Pd | 232 ± 23 | | | 50.0 | 52.4 | | 0.145 ± 0.037 |
| 109 | Ag | 890 ± 26 | | | 49.8 | 52.1 | | 0.196 ± 0.024 |

**Table 4.** (continued)

| Mass number | Element | $\sigma \pm \Delta\sigma$ (mb) | p-correction to s-only abundance (%) | $\sigma N_\odot$ s-only nuclei (mb, Si = $10^6$) | $\sigma N_s$ main component (mb, Si = $10^6$) | $\sigma N_s$ all components (mb, Si = $10^6$) | $N_{\odot,r}$ r-only nuclei (Si = $10^6$) | $N_r = N_\odot - N_s$ (Si = $10^6$) |
|---|---|---|---|---|---|---|---|---|
| 110 | Cd | 289 ± 34 | 7.0 | 53.5 ± 7.2 | 49.2 | 51.4 | 0.164 ± 0.013 ($^{110}$Pd) | |
| 111 | Cd | 1214 ± 143 | | | 49.0 | 51.2 | | 0.162 ± 0.014 |
| 112 | Cd | 254 ± 34 | | | 48.3 | 50.4 | | 0.184 ± 0.037 |
| 113 | Cd | 831 ± 91 | | | 48.1 | 50.1 | | 0.134 ± 0.014 |
| 114 | Cd | 171 ± 29 | | | 47.1 | 49.0 | | 0.170 ± 0.056 |
| 115 | In | 849 ± 85.0 | | | 46.9 | 48.7 | | 0.119 ± 0.013 |
| 116 | Sn | 114 ± 22 | 2.6 | 62.9 ± 13.3 | 45.5 | 47.0 | 0.119 ± 0.008 ($^{116}$Cd) | |
| 117 | Sn | 459 ± 88 | | | 45.1 | 46.6 | | 0.194 ± 0.034 |
| 118 | Sn | 72.0 ± 13.7 | | | 42.9 | 44.2 | | 0.315 ± 0.145 |
| 119 | Sn | 282 ± 54 | | | 42.4 | 43.6 | | 0.175 ± 0.043 |
| 120 | Sn | 44.5 ± 8.0 | | | 39.1 | 40.0 | | 0.341 ± 0.199 |
| 121 | Sb | 845 ± 114 | | | 39.0 | 39.8 | | 0.155 ± 0.037 |
| 122 | Te | 337 ± 69 | 3.6 | 40.0 ± 9.4 | 38.5 | 39.4 | 0.174 ± 0.016 ($^{122}$Sn) | |
| 123 | Te | 939 ± 57 | | 41.3 ± 5.6 | 38.4 | 39.2 | 0.150 ± 0.027 ($^{123}$Sb) | |
| 124 | Te | 185 ± 24 | 2.0 | 41.0 ± 7.2 | 37.6 | 38.4 | 0.215 ± 0.020 ($^{124}$Sn) | |
| 125 | Te | 507 ± 50 | | | 37.4 | 38.1 | | 0.269 ± 0.042 |
| 126 | Te | 91.4 ± 13.7 | | | 35.9 | 36.6 | | 0.518 ± 0.126 |
| 127 | I | 725 ± 34 | | | 35.7 | 36.4 | | 0.850 ± 0.189 |
| 128 | Xe | 284 ± 67 | | 30.7 ± 9.5 | 35.3 | 35.9 | 1.560 ± 0.187 ($^{128}$Te) | |
| 129 | Xe | 537 ± 127 | | | 35.0 | 35.6 | | 1.314 ± 0.277 |
| 130 | Xe | 175 ± 48 | | 37.9 ± 12.9 | 34.3 | 34.9 | 1.690 ± 0.203 ($^{130}$Te) | |
| 131 | Xe | 397 ± 99 | | | 34.0 | 34.5 | | 0.995 ± 0.218 |
| 132 | Xe | 65.7 ± 4.3 | | | 32.2 | 32.7 | | 0.829 ± 0.268 |
| 133 | Cs | 581 ± 24 | | | 31.9 | 32.3 | | 0.316 ± 0.023 |
| 134 | Ba | 252 ± 40 | 4.2 | 27.0 ± 4.5 | 31.0 | 31.4 | 0.482 ± 0.097 ($^{134}$Xe) | |
| 135 | Ba | 522 ± 90 | | | 30.7 | 31.1 | | 0.242 ± 0.017 |
| 136 | Ba | 78.8 ± 11.4 | 1.3 | 28.2 ± 4.3 | 29.9 | 30.3 | 0.391 ± 0.078 ($^{136}$Xe) | |
| 137 | Ba | 65.1 ± 11.4 | | | 28.3 | 28.6 | | 0.074 ± 0.082 |

| | | | | | | | | |
|---|---|---|---|---|---|---|---|---|
| 138 | Ba | 4.4 ± 0.3 | | 15.5 ± 1.6 | 15.4 | 15.6 | | <0.29 |
| 139 | La | 42.5 ± 3.0 | | | 14.2 | 14.3 | | 0.111 ± 0.035 |
| 140 | Ce | 12.3 ± 0.6 | | | 10.9 | 11.0 | | 0.131 ± 0.075 |
| 141 | Pr | 136 ± 17 | | | 10.6 | 10.8 | | 0.095 ± 0.016 |
| 142 | Nd | 51.7 ± 4.3 | 9.3 | 10.7 ± 1.2 | 9.92 | 10.0 | 0.129 ± 0.007 ($^{142}$Ce) | |
| 143 | Nd | 273 ± 10 | | | 9.79 | 9.91 | | 0.0647 ± 0.0073 |
| 144 | Nd | 123 ± 6.3 | | | 9.51 | 9.63 | | 0.121 ± 0.015 |
| 145 | Nd | 554 ± 114 | | | 9.45 | 9.56 | | 0.0521 ± 0.0060 |
| 146 | Nd | 179 ± 46 | | | 9.26 | 9.37 | | 0.092 ± 0.017 |
| 147 | Sm | 1148 ± 114 | | | 2.65 | 2.77 | | 0.0331 ± 0.0030 |
| 148 | Sm | 300 ± 13 | 1.1 | 9.0 ± 0.8 | 9.12 | 9.23 | 0.0477 ± 0.0033 ($^{148}$Nd) | |
| 149 | Sm | 1638 ± 66 | | | 9.10 | 9.21 | | 0.0316 ± 0.0028 |
| 150 | Sm | 504 ± 25 | 1.7 | 9.8 ± 0.9 | 9.03 | 9.15 | 0.0468 ± 0.0033 ($^{150}$Nd) | |
| 151 | Sm | 2207 | | | 8.49 | 8.59 | | |
| 151 | Eu | 4988 ± 200 | | | 0.53 | 0.54 | | 0.0426 ± 0.0038 |
| 152 | Sm | 451 ± 25 | | | 8.44 | 8.55 | | 0.0523 ± 0.0054 |
| 152 | Gd | 1125 ± 70 | 33.0 | 0.50 ± 0.04 | 0.50 | 0.51 | | |
| 153 | Eu | 3620 ± 362 | | | 8.54 | 8.65 | | 0.0481 ± 0.0042 |
| 153 | Gd | 3152 | | | 0.40 | 0.41 | | |
| 154 | Gd | 1503 ± 116 | 5.0 | 9.9 ± 0.9 | 8.92 | 9.03 | 0.0605 ± 0.0045 ($^{154}$Sm) | |
| 155 | Gd | 3198 ± 320 | | | 8.91 | 9.02 | | 0.0462 ± 0.0023 |
| 156 | Gd | 730 ± 73 | | | 8.86 | 8.98 | | 0.0559 ± 0.0035 |
| 157 | Gd | 1757 ± 176 | | | 8.84 | 8.96 | | 0.0469 ± 0.0025 |
| 158 | Gd | 244 ± 22 | | | 8.71 | 8.83 | | 0.0460 ± 0.0052 |
| 159 | Tb | 2056 ± 114 | | | 8.70 | 8.81 | | 0.0546 ± 0.0071 |
| 160 | Dy | 941 ± 41 | 2.4 | 8.7 ± 0.5 | 8.66 | 8.78 | 0.0722 ± 0.0034 ($^{160}$Gd) | |
| 161 | Dy | 2461 ± 74 | | | 8.65 | 8.77 | | 0.0746 ± 0.0035 |
| 162 | Dy | 540 ± 57 | | | 8.59 | 8.71 | | 0.0888 ± 0.0050 |
| 163 | Dy | 1294 ± 44 | | | 2.85 | 2.97 | | 0.0982 ± 0.0005 |
| 163 | Ho | 2937 | | | 5.72 | 5.83 | | |
| 164 | Dy | 306 ± 24 | | | 5.44 | 5.56 | | 0.0977 ± 0.0056 |
| 164 | Er | 846 ± 69 | 9.0 | 3.34 ± 0.31 | 3.05 | 3.16 | | |
| 165 | Ho | 1505 ± 75 | | | 8.47 | 8.58 | | 0.0818 ± 0.0047 |
| 166 | Er | 694 ± 70 | | | 8.42 | 8.54 | | 0.0722 ± 0.0041 |
| 167 | Er | 1757 ± 176 | | | 8.40 | 8.52 | | 0.0530 ± 0.0027 |
| 168 | Er | 417 ± 50 | | | 8.33 | 8.45 | | 0.0483 ± 0.0040 |

**Table 4.** (continued)

| Mass number | Element | $\sigma \pm \Delta\sigma$ (mb) | p-correction to s-only abundance (%) | $\sigma N_\odot$ s-only nuclei (mb, Si = $10^6$) | $\sigma N_s$ main component (mb, Si = $10^6$) | $\sigma N_s$ all components (mb, Si = $10^6$) | $N_{\odot,r}$ r-only nuclei (Si = $10^6$) | $N_r = N_\odot - N_s$ (Si = $10^6$) |
|---|---|---|---|---|---|---|---|---|
| 169 | Er | 696 | | | 0.17 | 0.28 | | |
| 169 | Tm | 1289 ± 64 | | | 8.14 | 8.26 | | 0.0320 ± 0.0025 |
| 170 | Er | 255 ± 38 | | | 0.17 | 0.28 | 0.0376 ± 0.0017 ($^{170}$Er) | |
| 170 | Tm | 2581 | | | 2.80 | 2.81 | | |
| 170 | Yb | 876 ± 27 | 4.3 | 6.61 ± 0.31 | 5.28 | 5.39 | | |
| 171 | Tm | 1125 | | | 2.37 | 2.48 | | |
| 171 | Yb | 1618 ± 59 | | | 5.82 | 5.94 | | 0.0292 ± 0.0013 |
| 172 | Yb | 468 ± 39 | | | 8.12 | 8.24 | | 0.0356 ± 0.0025 |
| 173 | Yb | 988 ± 82 | | | 8.09 | 8.21 | | 0.0311 ± 0.0016 |
| 174 | Yb | 201 ± 18 | | | 7.95 | 8.07 | | 0.0367 ± 0.0048 |
| 175 | Lu | 1364 ± 47 | | | 7.92 | 8.04 | | 0.0309 ± 0.0037 |
| 176 | Lu | 1755 ± 68 | | | 3.50 | 3.55 | 0.0300 ± 0.0011 ($^{176}$Yb) | |
| 176 | Hf | 545 ± 24 | | | 4.39 | 4.46 | | |
| 177 | Hf | 1560 ± 70 | | | 7.87 | 7.99 | | 0.0241 ± 0.0020 |
| 178 | Hf | 371 ± 12 | | | 7.80 | 7.91 | | 0.0215 ± 0.0030 |
| 179 | Hf | 1166 ± 35 | | | 7.77 | 7.89 | | 0.0146 ± 0.0015 |
| 180 | Hf | 219 ± 6.2 | | | 7.64 | 7.76 | | 0.0197 ± 0.0040 |
| 181 | Ta | 828 ± 46 | | | 7.61 | 7.72 | | 0.0133 ± 0.0023 |
| 182 | W | 313 ± 9.0 | | | 7.52 | 7.64 | | 0.0116 ± 0.0029 |
| 183 | W | 588 ± 17 | | | 7.47 | 7.59 | | 0.0067 ± 0.0016 |
| 184 | W | 256 ± 8.0 | | | 7.37 | 7.48 | | 0.0128 ± 0.0034 |
| 185 | W | 907 | | | 1.24 | 1.26 | | |
| 185 | Re | 1749 ± 228 | | | 6.11 | 6.21 | | 0.0141 ± 0.0024 |
| 186 | W | 201 ± 6.0 | | | 1.57 | 1.57 | | 0.0314 ± 0.0029 |
| 186 | Os | 482 ± 18 | 1.1 | 5.48 ± 0.58 | 5.57 | 5.66 | | |
| 187 | Re | 1793 ± 114 | | | 1.75 | 1.77 | 0.0333 ± 0.0041 ($^{187}$Re) | |
| 187 | Os | 1013 ± 32 | | | 5.51 | 5.61 | | |
| 188 | Os | 456 ± 17 | | | 7.20 | 7.32 | | 0.0793 ± 0.0094 |
| 189 | Os | 1754 ± 53 | | | 7.19 | 7.30 | | 0.1108 ± 0.0113 |

| | | | | | | | |
|---|---|---|---|---|---|---|---|
| 190 | Os | 337 ± 51 | | | 7.11 | 7.22 | | 0.1676 ± 0.0188 |
| 191 | Os | 1245 | | | 0.23 | 0.23 | | |
| 191 | Ir | 1536 ± 77 | | | 6.86 | 6.98 | | 0.2413 ± 0.0155 |
| 192 | Os | 355 ± 51 | | | 0.40 | 0.40 | 0.2929 ± 0.0288 ($^{192}$Os) | |
| 192 | Ir | 3837 | | | 2.53 | 2.57 | | |
| 192 | Pt | 443 ± 150 | 1.8 | 4.66 ± 1.64 | 4.12 | 4.19 | | |
| 193 | Ir | 988 ± 49 | | | 3.27 | 3.29 | 0.4084 ± 0.0261 ($^{193}$Ir) | |
| 193 | Pt | 1738 | | | 3.74 | 3.81 | | |
| 194 | Pt | 358 | | | 6.96 | 7.08 | | 0.4312 ± 0.0451 |
| 195 | Pt | 1212 | | | 6.94 | 7.06 | | 0.4572 ± 0.0463 |
| 196 | Pt | 200 ± 19 | | | 6.82 | 6.93 | | 0.3123 ± 0.0349 |
| 197 | Au | 665 ± 10 | | | 6.78 | 6.89 | | 0.1756 ± 0.0279 |
| 198 | Hg | 193 ± 17 | | 6.8 ± 1.0 | 6.65 | 6.77 | 0.0986 ± 0.0099 ($^{198}$Pt) | |
| 199 | Hg | 431 ± 27 | | | 6.59 | 6.71 | | 0.0434 ± 0.0056 |
| 200 | Hg | 127 ± 13 | | | 6.41 | 6.53 | | 0.0298 ± 0.0094 |
| 201 | Hg | 325 ± 17 | | | 6.34 | 6.45 | | 0.0265 ± 0.0045 |
| 202 | Hg | 78.1 ± 6.3 | | | 6.05 | 6.17 | | 0.0256 ± 0.0120 |
| 203 | Tl | 142 ± 9.1 | | | 5.80 | 5.91 | | 0.0115 ± 0.0059 |
| 204 | Pb | 99.1 ± 6.1 | | 6.1 ± 0.6 | 5.50 | 5.61 | 0.0204 ± 0.0023 ($^{204}$Hg) | |
| 205 | Tl | 61.7 ± 4.6 | | | 3.66 | 3.71 | | 0.0412 ± 0.0146 |
| 205 | Pb | 61.7 ± 13.7 | | | 1.71 | 1.76 | | |
| 206 | Pb | 15.4 ± 0.8 | | | 4.58 | 5.02 | | 0.2233 ± 0.0434 |
| 207 | Pb | 12.7 ± 0.8 | | | 3.55 | 3.98 | | 0.2800 ± 0.0471 |
| 208 | Pb | 0.46 ± 0.04 | | | 0.40 | 0.73 | | 0.1184 ± 0.1672 |
| 209 | Bi | 12.4 ± 1.23 | | | 0.31 | 0.63 | | 0.0930 ± 0.0127 |

but are extrapolated to $kT = 23$ keV according to the s-process temperature derived from branching analyses; in a few significant cases, the cross sections were also corrected for stellar enhancement factors. Branchings are considered only if they are strong enough to stand out on figure 19; the weaker branchings do not affect the abundances significantly, but can be important for estimating neutron density and temperature (see for example the branchings at $A = 147,148$ and at $^{134}$Cs, $^{154}$Eu, (§ 5.2.1). The branchings connected with the s-only isotopes $^{176}$Lu, $^{176}$Hf and $^{187}$Os are complicated by long-lived radioactive decays. These decays are potential chronometers for the age of the s-process elements and will be discussed in some detail in § 7.

The weak and strong s-process components can alternatively be assumed to result from single neutron exposures. Such an assumption allows for a better reproduction of the s-only isotopes $^{70}$Ge and $^{76}$Se (Beer 1986, 1988, Beer and Macklin 1989); this is illustrated in table 4, where numerical $\sigma N_s$-values for a single flux solution of the weak component are given in brackets in column 7. For possible stellar scenarios see § 6.

### 4.2. Neutron economy

The main difference between the $\sigma N_s$ curves in figure 19 comes from the revision of the rare-earth abundances by Anders and Ebihara (1982), giving rise to a much less pronounced step in the new $\sigma N_s$ curve at the magic neutron number 82. This corresponds to an increase of the mean neutron exposure, $\tau_0$, and to a decrease of the required seed abundance, $f$. A comparison with the results for the 1982 curve (Käppeler *et al* 1982; Almeida and Käppeler 1983) is given in table 5.

Given the perfect agreement between the empirical $\sigma N_s$-values and the calculated curve, one can be rather confident about the present s-process abundances. This statement is additionally supported by the smoothness of the related r-process abun-

**Table 5.** Comparison of mean neutron exposure, $\tau_0$, and fractional seed abundance, $f$, with values based on abundance compilations prior to Anders and Ebihara (1982). For a discussion of the neutron balance condition for the $^{22}$Ne($\alpha$, n) source see the text.

| | Käppeler *et al* (1982) | Present |
|---|---|---|
| Mean neutron exposure $\tau_0$ (mb$^{-1}$) | | |
| Main component | $0.24 \pm 0.01$ | $(0.30 \pm 0.01)\left(\frac{kT(\text{keV})}{30}\right)^{1/2}$ |
| Weak component | $0.056 \pm 0.005$ | $(0.068 \pm 0.007)\left(\frac{kT(\text{keV})}{30}\right)^{1/2}$ |
| Strong component | — | 7.0 |
| Seed abundance, $f$ (% of $N_\odot(^{56}\text{Fe})$) | | |
| Main component | $0.092 \pm 0.015$ | $0.043 \pm 0.002$ |
| Weak component | $2.7 \pm 0.2$ | 1.6 |
| Strong component | — | $1.2 \times 10^{-4}$ |
| Total number of neutrons captured per $^{56}$Fe seed nucleus, $\Sigma\, n_c$ (equation (4.1)) | | |
| Main component | 13.0[a] | $15.1 \pm 0.8$ |
| Weak component | — | 2.8 |
| Strong component | — | 141.0 |

[a] Almeida and Käppeler (1983).

dance distribution (§ 4.3). As the number of neutrons, $n_c$, consumed during the s-process by the various seed nuclei is directly linked to the abundances (equation (2.5)), the values of $n_c$ are equally reliable. This is necessary but not sufficient for a discussion of the neutron balance. Knowledge of the neutron-producing reaction(s) and of the stellar environment is also needed. These conditions are uncertain for the weak and strong components, but are rather well defined for the main component by the model for helium shell burning in intermediate mass stars.

As discussed in § 2.2, there is a detailed model for the main component to be attributed to helium shell burning with the $^{22}Ne(\alpha, n)^{25}Mg$ reaction as the neutron source. For that case, no more than 39 neutrons are available per $^{56}Fe$ nucleus if C, N, O and Fe occur in the standard abundance ratios typical for population I stars. This condition was checked by Almeida and Käppeler (1983) via the neutron balance: number of neutrons produced = number of neutrons captured, i.e.

$$\begin{aligned} q(fN_\odot(\text{CNO})) &= \textstyle\sum n_c(A_{\text{seed}}) + (1-q)n_c(^{22}\text{Ne}) + qn_c(^{25}\text{Mg}) \\ 39.0q &= 13.0 \qquad\quad + (1-q)12.2 \qquad\quad + 24.7q \end{aligned} \tag{4.1}$$

where $q$ is the fraction of $^{22}Ne$ that undergoes an $(\alpha, n)$ reaction. On the right-hand side of equation (4.1) one has to consider all possible seeds. As the metallicity is assumed to be that for population I stars, all elements are expected to occur in their solar ratios. This means that the fraction $f = 0.043\%$ (table 5) that was obtained for the seed abundance of $^{56}Fe$ to reproduce the empirical values can also be used for the lighter elements. Though these do not contribute noticeably in the mass range above $^{56}Fe$, they act to some extent as neutron poisons. Therefore, the sum over $n_c(A_{\text{seed}})$ in equation (4.1) runs from $A = 20$ to 56. Extra terms are required for the abundant species $^{22}Ne$ and $^{25}Mg$ which take part in the $(\alpha, n)$ reaction and which therefore depend on $q$. The numbers in equation (4.1) follow from the s-process abundances derived from the $\sigma N_s$ curve of Käppeler *et al* (1982), yielding $q = 0.95$. The remaining 5% of $^{22}Ne$ were attributed to the competing $(\alpha, \gamma)$ reaction. As the reaction rate ratio $\lambda(\alpha, n)/\lambda(\alpha, \gamma)$ depends strongly on temperature, the result of the neutron balance condition could be used to estimate a lower limit of $3.3 \times 10^8$ K for the s-process temperature.

However, this nicely consistent picture no longer holds (Käppeler 1986b). With the larger $\tau_0$ values of the revised $\sigma N_s$ curve, the number of neutrons captured per $^{56}Fe$ seed nucleus increased by $\sim 30\%$, which changes the neutron balance equation into

$$39.0q = 18.0 + (1-q)16.7 + 29.7q$$

leading to the physically unreasonable solution $q > 1$. The above neutron balance is based on the abundance compilation of Anders and Ebihara (1982); it also depends sensitively on the neutron capture rates of $^{22}Ne$ and $^{25}Mg$. Even if these data were modified within the quoted uncertainties, however, one would still be left with $q > 1$. This result implies that the s-process during helium shell burning requires the C, N and O abundances to be enriched by about 40% relative to $^{56}Fe$ in order to re-establish $q = 0.95$, consistent with the prevailing temperature. In fact, an enrichment in C and O may result naturally from the helium burning zone itself: according to § 2.2, a fraction $r \sim 0.4$ of the material exposed to helium burning is mixed to the envelope after each pulse episode, thus adding freshly produced carbon and oxygen to the CNO cycle, which burns hydrogen during the quiescent interpulse phases. This feeding of the C and O abundances might explain the extra supply of s-process neutrons. A quantitative treatment may ultimately tell something about the mixing mechanism between the helium shell and envelope, provided that the $^{22}Ne(\alpha, n)^{25}Mg$ reaction is the only neutron source for the main s-process component.

The neutron balance argument does not apply for scenarios with $^{13}C(\alpha, n)^{16}O$ as the main neutron-producing reaction, e.g. in low mass stars (Gallino *et al* 1988). In that case, the number of neutrons is determined by the number of protons mixed into the helium zone to yield $^{13}C$, and therefore no longer correlated to the seed abundances.

### 4.3. r-process residuals

The improved s-process abundances from the updated $\sigma N_s$ curve can be used for a revision of the r-process residuals $N_r = N_\odot - N_s$. The resulting differences between solar and s-process abundances are plotted in figure 20 together with the ensemble of r-only isotopes. The refined s-process treatment and the more accurate solar abundances allow calculation of the uncertainties for the r-process residuals thus derived. These are reasonably small except near magic neutron numbers where large s-process abundances occur as a consequence of the small capture cross sections. These r-abundances are included in table 4. Some of the quoted r-only nuclei are corrected for minor s-process contributions if these did not exceed ~5%.

First of all, it is most remarkable that the agreement between the calculated distribution of r-process yields and the r-only isotopes became considerably better with the improved s-process analysis, if compared with a previous study (Käppeler *et al* 1982). This convincingly supports the s-process abundances obtained with the classical

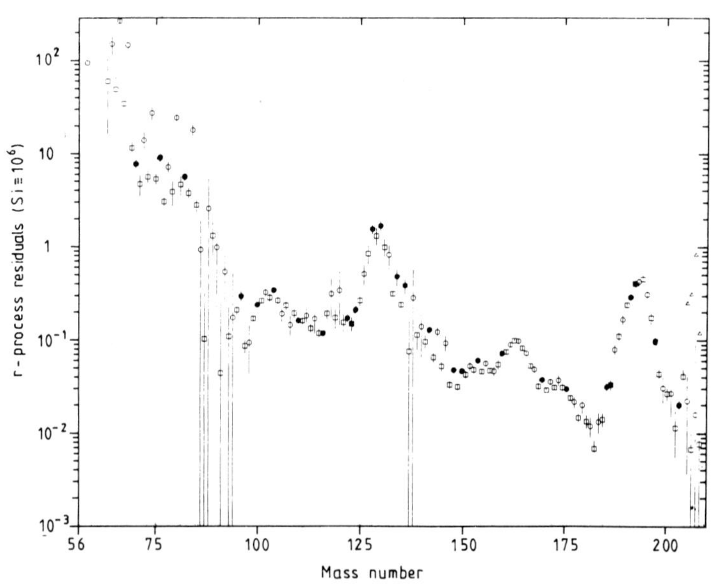


**Figure 20.** r-Process yields: the distribution of r-only isotopes (full points) and the abundances obtained by subtraction of s-process abundances from solar values (open symbols).

model, because it means that two independent pieces of information are perfectly matching. It also ensures that figure 20 represents the best information on the actual r-process abundances, and hence can be recommended as a reliable basis for model calculations of the r-process.

In particular, one finds that the abundance maxima around $A = 130$ and 195 appear well defined. They indicate those mass numbers where the r-process path reached the magic neutron shells $N = 82$ and 126. Their position and shape are important to identify the 'waiting points' during the r-process. With this information and with the recently measured beta decay rates of some key isotopes, Kratz *et al* (1988) were able to show that $(n,\gamma)$-$(\gamma, n)$ equilibrium was achieved during the r-process. The r-process abundance peak related to the magic neutron number 50 could be studied for the first time by the explicit treatment of the s-process branchings at $A = 79$ and 85 (Walter *et al* 1986a, b; § 5). It is not as outstanding as the other two and appears as a sharp step in the abundance distribution rather than as a peak. This could indicate that abundance contributions from other processes might have to be considered up to mass 80. Currently, the related uncertainties are still too large for any definite conclusions. In this context, it is interesting to note the outstanding abundances at $A = 74$, 80 and 84. A possible explanation via strong feeding of the respective r-process decay chains by beta-delayed neutron emission (Kratz *et al* 1988) appears doubtful, as the neighbouring abundances do not seem to be very depleted.

At the end of the distribution, the need for a strong s-process component becomes evident. The triangles at $A = 206$, 207, 208 and 209 would be obtained, if the main component only were subtracted from the solar values; but by inclusion of the strong component one obtains the smooth r-process yields compatible with the general phenomenology and with the abundances of the long-lived actinides (Beer and Macklin 1985). The reduction of the lead abundance suggested by solar spectroscopy (Grevesse and Meyer 1985) would eliminate the need for a strong component (Ratzel *et al* 1989); however, a recent spectroscopic study (Youssef and Khalil 1989) confirmed the meteoritic lead abundance (Anders and Grevesse 1989), leaving the $^{208}$Pb problem unsolved.

The distribution of figure 20 exhibits the smoothness expected for the r-process yields. Any initial abundance fluctuations due to odd–even effects in the beta decay rates on the r-process path are effectively smeared out due to beta-delayed neutron emission in the decay chains to the stability valley. Within uncertainties, there are no deviations from a smooth distribution, though some indications for an odd–even effect exist in the mass range $145 < A < 180$. The only significant deviation from the smooth distribution occurs at the tin isotopes 117, 118, 119 and 120. In this case, one might suspect that the tin abundance of Anders and Ebihara (1982) is too high by 23%. This suspicion is based on the fact that the r-process abundances become smoothly distributed if the tin abundance is normalised by the $\sigma N$-value of s-only $^{116}$Sn (Beer *et al* 1989b).

## 5. *s*-process branchings

### *5.1. General discussion*

As shown in §§ 2.1 and 3.3, s-process branchings can be analysed to yield estimates for neutron density and temperature during the s-process. The branching factor $f_\beta$

(equation (2.8)) can be expressed by the rates for neutron capture and beta decay

$$\lambda_n = n_n \sigma v_T \qquad \text{and} \qquad \lambda_\beta = \ln 2/t_{1/2}(T) \tag{5.1}$$

of the branching point isotope. On the other hand, $f_\beta$ can be derived from the classical model as outlined in § 2.1 by comparison of the empirical $\sigma N$-value of the involved s-only isotope with the smooth $\sigma N(A)$ curve.

If the beta decay rate of the branching point isotope is known, equation (2.8) yields the stellar neutron capture rate and ultimately the neutron density during the s-process via equation (5.1). However, the beta decay rates of some of the branching point isotopes are affected by temperature as discussed in § 3.3. This means that all possible branchings need to be investigated systematically, starting with those cases which allow for an estimate of the neutron density. In a second step, this information can be used to derive the effective beta decay rates from the temperature-dependent branchings. With the tabulated temperature dependence of the relevant s-process isotopes of Takahashi and Yokoi (1987), these branchings can be interpreted as thermometers to estimate the s-process temperature.

The branching analyses represent an important test for the validity of the classical model. If the solutions for neutron density and temperature derived from as many branchings as possible are found to be consistent with each other, then the s-process was indeed smooth enough to justify the simple assumption of constant neutron density; otherwise, the abundance patterns of the various branchings should be determined by freeze-out effects, leading to inconsistent results in the classical approach. This would be important information in itself, and would allow the study of s-process dynamics.

Two classes of branchings can be distinguished: those which are characterised by a single s-only isotope partially bypassed by the s-process flow, and others where an additional s-only isotope of the same element occurs that experiences the total s-process flow. The advantage of the second class is that the branching factor can be derived via the well known cursive ratio of the s-only isotopes, while in the first case the uncertainty of the respective *elemental* abundance must be considered, which is usually much larger. Therefore, branchings with two s-only isotopes can be analysed more precisely, e.g. those at $A = 79$, $A = 147{,}148$ and $A = 151$.

As the classical model deals with three components, these have to be investigated separately. There are six to eight branchings that can be used to characterise the main component, but only two for the weak, and possibly two others for the strong component. Consequently, a systematic study has to start with the main component. With the results thus obtained, the contribution of the main component can be corrected in the mass regions $A < 90$ and $A > 205$ to derive the isotopic patterns representative for the weak and strong components. Branchings are not only sensitive to neutron density and temperature (§ 5.2.1), but to electron density and to a possible time dependence of the neutron density as well; these latter aspects will be discussed in §§ 5.2.2 and 5.2.3.

## *5.2. The physical conditions during the s-process*

*5.2.1. Neutron density and temperature.* The first step in a systematic branching analysis is to investigate the main s-process component in the mass range $100 < A < 200$. By combination of the branchings at $^{151}$Sm, $^{170}$Yb and $^{185}$W, Beer *et al* (1984) determined limits for neutron density and temperature characteristics of the main s-process component, as indicated by the full outline in figure 21. (In the following text, thermal energies, $kT$, are always given in keV, and temperatures, $T_8$, in units of $10^8$ K; conversion

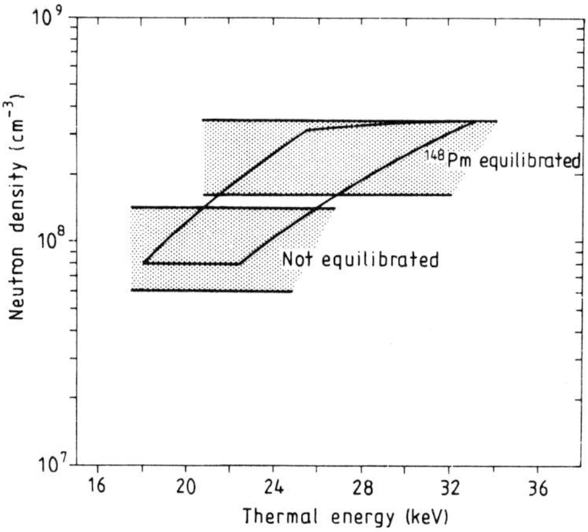


**Figure 21.** Allowed regions of neutron density and thermal energy for the main s-process component as obtained from analyses of various branchings. Note the ambiguity due to the problem of thermal equilibration in $^{148}$Pm. ——, Beer *et al* (1984); shaded areas, Winters *et al* (1986a).

between these quantities is provided by the relation $kT = 8.62\ T_8$.) The allowed region in the neutron density–temperature plot of figure 21 could be reduced by inclusion of the branchings at $A = 147$ and 148 (Winters *et al* 1986a). However, the influence of temperature is ambiguous in this case, depending on whether the ground state and the 137 keV isomer in $^{148}$Pm—which are about equally populated by neutron capture in $^{147}$Pm—are thermally equilibrated on the beta decay timescale of a few days. This was a longstanding question, because the level scheme of $^{148}$Pm was almost unknown: beside the ground state (5.4 d) and the isomer (41 d) only one additional level is reported at 76 keV. As direct transitions between the isomer and the other two states are too slow, equilibrium could only be achieved via mediating states at higher excitation. Such states are now reported by Lesko *et al* (1989), so that the high-temperature–high-neutron-density solution of figure 21 is confirmed.

In principle, the branchings at $^{134}$Cs and $^{154}$Eu could be used to derive additional lower limits for the s-process temperature. The beta-decay rates of these branch point nuclei and hence the branching factors depend strongly on temperature. This is shown in figure 22 (Käppeler 1988), where the shaded bands are calculated using two values for the neutron density ($1.0 \times 10^8$ and $2.5 \times 10^8$ cm$^{-3}$). Given the present cross section uncertainties, there is no evidence for an effective branching, neither for $^{134}$Cs nor for $^{154}$Eu; instead, the related uncertainties can be used for estimating a lower limit for the respective branching ratios. In the case of $^{134}$Cs only the 15% uncertainty of the cross-section ratio $\sigma(^{134}\mathrm{Ba})/\sigma(^{136}\mathrm{Ba})$ must be considered, but for $^{154}$Eu the uncertainties of the cross section, the abundance and the $\sigma N$ curve (8, 5 and 10%) contribute. The upper limits for the branching factor thus obtained (indicated in figure 22) are not yet stringent enough to allow for a conclusion on the $^{148}$Pm ambiguity, but seem to support the possibility of complete equilibration. If the allowed region in figure 21 can be further restricted by refined branching analyses, this would have consequences for the s-process neutron source: a low temperature would favour the $^{13}\mathrm{C}(\alpha, \mathrm{n})^{16}\mathrm{O}$ reaction, whereas the $^{22}\mathrm{Ne}(\alpha, \mathrm{n})^{25}\mathrm{Mg}$ reaction requires higher temperatures to operate.

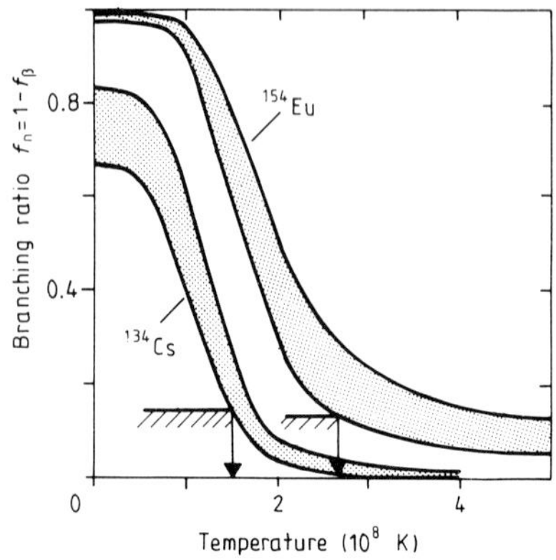


**Figure 22.** The branching factors of $^{134}Cs$ and $^{154}Eu$ plotted against temperature. Though no effective branchings are evident from s-process analyses, the respective uncertainties (shaded bars) are used to estimate lower limits for the s-process temperature.

Further efforts to identify the neutron source of the s-process should therefore concentrate on improving the stellar neutron capture and beta decay rates for the relevant isotopes in the various branchings, including

(i) a confirmation of the level scheme of $^{148}Pm$;

(ii) measurements of the capture cross sections of the s-only nuclei to 1–2% accuracy to reliably define the branching factors;

(iii) measurements of the capture cross sections of the unstable branching point isotopes as proposed by Bowman and Talbert (1985);

(iv) more precise estimates for the log *ft* values of excited states.

*Weak component.* The two branchings to characterise the weak s-process component are those at $^{79}Se$ and $^{85}Kr$. Figure 2 shows the s-process path in the mass region $75 < A < 90$ with the terrestrial half-lives indicated at the branch points. The fact that both branching ratios are about equal (Walter *et al* 1986a, b) implies that the beta decay rate of $^{79}Se$ must be greatly enhanced at s-process temperatures. As was shown in § 3.3, the temperature dependence of the $^{79}Se$ half-life is now well described by the experimental log *ft* value (Klay and Käppeler 1988), hence allowing for a more reliable analysis of the $^{79}Se$ thermometer. Figure 23 (Käppeler 1988) illustrates how the possible region for neutron density and temperature typical for the weak s-process component was reduced with the experimental log *ft* value compared to a previous estimate, the remaining uncertainty being dominated by cross-section uncertainties; an updated version of figure 23 yields slightly higher neutron densities for the weak component (Klay and Käppeler 1988). Note that this analysis is still based on the assumption that the decay of $^{148}Pm$ is not affected by temperature; but if the ground state and the isomer in $^{148}Pm$ are thermally equilibrated (Lesko *et al* 1989), then the region for the main component centres around $n_n = 2.5 \times 10^8\ \mathrm{cm}^{-3}$ and $kT = 27$ keV, while the region for the weak component will move to somewhat lower neutron densities; these results have recently been confirmed and updated (table 6; Toukan and Käppeler 1989, Käppeler *et al* 1989a, b, Beer *et al* 1989c).

Principally, the weak s-process component is difficult to analyse in terms of the classical model because of two problems:

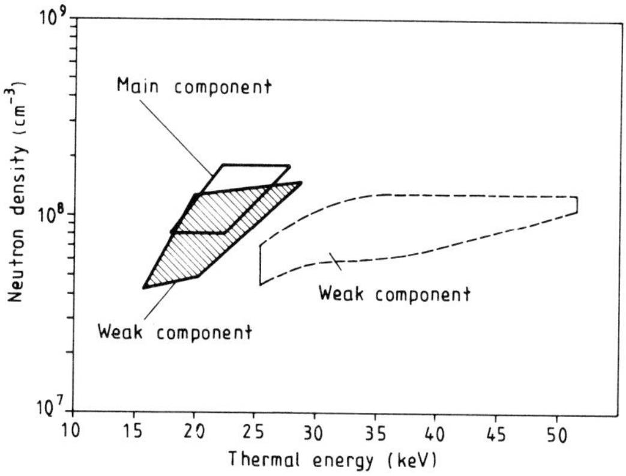


**Figure 23.** Allowed regions of neutron density and thermal energy for the two s-process components (Käppeler 1988); this analysis was still based on the assumption that the decay of $^{148}$Pm was not affected by temperature. The new results of Lesko *et al* (1988) suggest that the region for the main component centres at $2.5 \times 10^8$ cm$^{-3}$ and 27 keV.

(i) The use of an exponential distribution of neutron exposures is not as successful in reproducing the s-only isotopes as it is for the main component. Besides the tendency for underproducing $^{70}$Ge and $^{76}$Se, there is also a tendency for overproduction of the light Zr isotopes. Therefore, a single exposure solution was suggested (Beer 1986, Beer and Macklin 1989), which accounts for the observed abundances in the mass range $A > 65$ (see table 4).

(ii) The contribution of the weak component to the branchings at $^{79}$Se and $^{85}$Kr depends significantly on the type of the adopted neutron exposure. Analysis of these branchings is more complicated than outlined above, because the contributions from the main component depend critically on whether and how the main component is characterised by a pulsed s-process (see § 5.2.3). At present, the only conclusion can be the need to improve the input data for these analyses to better identify the possible deficiencies of the classical model.

*Strong component.* At the end of the s-process chain, the $\alpha$-unstable isotopes $^{210,211}$Po and $^{211}$Bi cause a recycling of the neutron capture flow to $^{206,207}$Pb as illustrated in figure 24. The analytic treatment (Clayton and Rassbach 1967, Ward and Clayton 1982, Beer and Macklin 1985) shows that this recycling is possibly influenced by two branchings, which depend strongly on the physical conditions. For the temperature and neutron-density regimes typical for the main and weak components, one finds that the $\alpha$-decay of $^{210}$Po dominates, and that the branchings at $^{210}$Bi and $^{210}$Po are almost negligible: the small capture rate of neutron-magic $^{210}$Po prevents a significant production of $^{211}$Po at neutron densities below $3 \times 10^9$ cm$^{-3}$. Similarly, the production of $^{211}$Bi is hampered because the long-lived isomer $^{210m}$Bi is quickly destroyed in the stellar photon bath at temperatures higher than about $1.5 \times 10^8$ K.

Whether the strong component took place under such conditions as to make these branchings efficient may possibly be deduced from the observed abundance ratio $^{206}$Pb/$^{207}$Pb. If the branchings were not significant, the only decay is that of $^{210}$Po, leading to $^{206}$Pb; otherwise the decays from $^{211}$Po and/or $^{211}$Bi are feeding the $^{207}$Pb abundance. A determination of the s-process abundances of $^{206,207}$Pb is complicated,

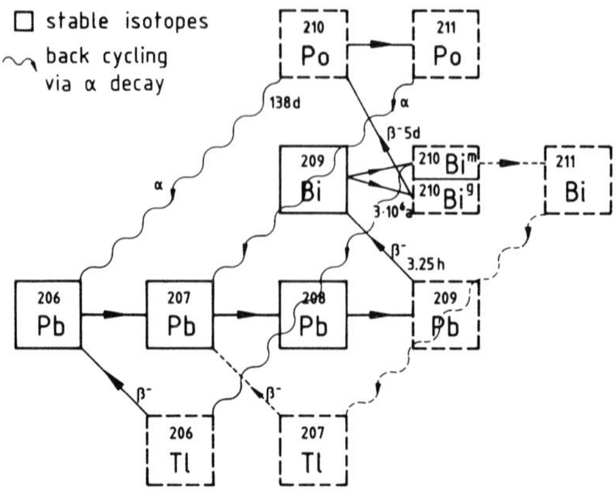


**Figure 24.** The s-process flow at the end of the neutron capture chain. Various possibilities for $\alpha$ recycling to $^{206,207}$Pb are indicated, but are difficult to identify because of the different r-process contributions involved. For details see text.

however, because their separation from the respective r-process contributions has to consider not only the normal r-process yields, but also the radiogenic component from the decay of various trans-bismuth progenitors and of $^{235,238}$U. At present, this separation is still too uncertain (see error bars in figure 20 for $A>205$) for any conclusion on the possible branchings characteristic for the strong component.

In a recent study of the s-process in the lead–bismuth region, Ratzel *et al* (1989) noted that the lower lead abundance from the solar photosphere (Grevesse and Meyer 1985) would question the need for a strong component; however, this possibility is superseded by the measurement of Youssef and Khalil (1989). In this context it would be important to reconsider the radiogenic $^{207}$Pb lead abundance as a chronometer for the r-process as well as the question of whether the decay of $^{204}$Tl was thermally enhanced as assumed previously, or whether spins and parities of the low-lying states were such as to make $^{204}$Tl an effective branching point. This would be a necessary condition in order to reconcile a lower lead abundance with the $\sigma N(A)$ systematics of s-only $^{204}$Pb.

*5.2.2. Mass density.* For the identification of the stellar s-process site the mass density is an important parameter. As described in § 3.2.3, the electron capture rates in the highly ionised stellar plasma are dominated by capture of continuum electrons, and hence depend on the electron density. The electron-to-baryon ratio being almost exactly 0.5 allows determination of the mass density, if a branching can be identified that is characterised by such a density-dependent electron capture rate. Yokoi and Takahashi (1981) have pointed out that terrestrially stable $^{163}$Dy becomes unstable at s-process temperatures due to bound-state decays to $^{163}$Ho. This isotope may either capture an electron (which means a decay back to $^{163}$Dy) or it may capture a neutron. The latter possibility leads via $^{164}$Ho to $^{164}$Er, thus explaining the relatively large abundance of this isotope, which does not seem at first glance to belong to the s-process chain. The competition between electron and neutron capture on $^{163}$Ho was studied in terms of an s-process branching by Beer *et al* (1985). In order to reproduce the observed $^{164}$Er abundance (corrected for a possible p-process contribution, see table 4) these authors derived an estimate for the electron density via the strong dependence of the half-lives of $^{163}$Dy, $^{163}$Ho and $^{164}$Ho on that quantity (Takahashi and Yokoi 1987); the neutron density and temperature in this analysis were as inferred from the other branchings (§ 5.2.1). The result for the electron density, $0.8\times10^{27}<n_e<4\times10^{27}$ cm$^{-3}$ is equivalent to mass densities between 2600 and 13 000 g cm$^{-3}$ for the s-process site, in reasonable agreement with stellar models for helium shell burning in intermediate mass stars (Iben and Renzini 1983).

*5.2.3. Limits for pulsed s-process scenarios.* Stellar s-process scenarios based on helium shell burning in intermediate mass stars (§ 2.2) imply a pulsed s-process with pulses of a few years followed by quiescent intervals of several hundred years. According to Ward and Newman (1978) such a sequence should have affected some of the observed branchings. The unstable branch point isotope can decay during the interpulse period and has to build up again in the subsequent pulse before the neutron-deficient branch becomes operative. The criterion for whether a branching is sensitive to a pulsed s-process is that the total lifetime, $\tau=(\lambda_n+\lambda_\beta)^{-1}$, of the branch point isotope should be shorter than or comparable to the pulse duration, $\Delta t$. For $\Delta t \ll \tau$ the branching does not exist at all, whereas $\Delta t \gg \tau$ means that the pulsed s-process approaches the classical steady-state solution.

Of all observed branchings, the case of $^{151}$Sm is best suited for deriving a lower limit of the pulse duration (Beer 1986, Beer and Macklin 1988). Under the physical conditions derived in the previous section, the lifetime of this branch point is determined by the neutron capture rate (via its large capture cross section). This means that the $^{151}$Sm abundance is quickly rebuilt to its equilibrium value at the beginning of each pulse. In addition, this branching is well defined by the two s-only nuclei $^{152,154}$Gd. Figure 25 shows how the $\sigma N_s$-values of the isotopes involved in that branching depend on the pulse duration $\Delta t$. The calculation assumes the interpulse periods to be long enough so that all the unstable $^{151}$Sm and $^{153}$Gd decays completely. Then one finds

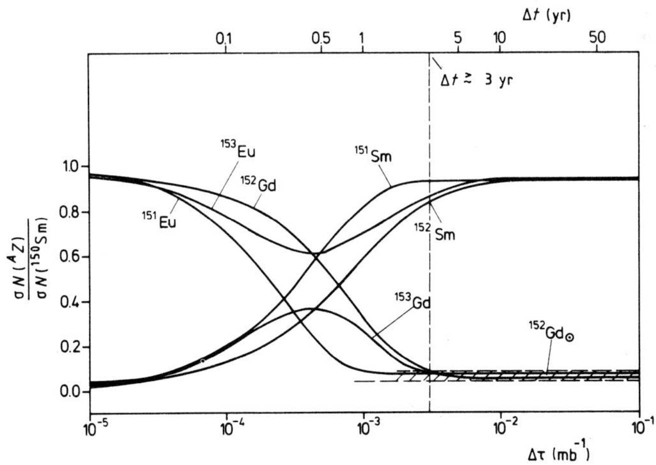


**Figure 25.** The $\sigma N_s$-values of the isotopes in the $^{151}$Sm branching as a function of the pulse duration during the s-process. The empirical value for the s-only isotope $^{152}$Gd (hatched area) is reproduced for $\Delta t > 3$ yr.

that the empirical $\sigma N_s$-values of $^{152,154}$Gd are correctly reproduced only if the pulse duration is longer than about 3 yr.

A similar study of the $^{85}$Kr branching points to an upper limit for $\Delta t$ of ~10–15 yr, because longer pulses would tend to cause an overproduction of $^{86}$Kr (Beer and Macklin 1989). The difficulty of this analysis comes from the fact that the contribution of the weak component to the abundance pattern of the $^{85}$Kr branching is strongly model-dependent, and that there is no reliable way to deduce the respective parameters, except via the branchings at $^{79}$Se and $^{85}$Kr as well. The situation in the mass region $A < 90$ therefore appears ambiguous. Another obvious branching for a test of the maximum pulse length is not in sight. The delicate balance between the influence of neutron density, temperature and pulse duration requires, however, significant improvements on the relevant input data for these branching analyses; otherwise, it will not be possible to separate true effects from ambiguities in the model or in the input values.

### 5.3. *Branchings to rare nuclei: the case of* $^{180}$*Ta*

Isomeric states often play an important role in s-process branchings. This was demonstrated in §§ 3.3 and 5.2 for the examples $^{79m}$Se, $^{147m}$Pm and $^{210m}$Bi, and in more detail by Ward (1977), Ward and Fowler (1980) and Beer and Käppeler (1982). A particularly interesting case is the problem of the $^{180}$Ta abundance, which has received considerable attention during the past decade. The abundance of nature's rarest stable isotope $^{180}$Ta does not fit into the general s-, r- and p-process systematics. There seem to be several competing possibilities to produce $^{180}$Ta, all of which pose exciting questions for nuclear astrophysics. The various production mechanisms for $^{180}$Ta are indicated in figure 26 by the heavy arrows. Following the normal s-process capture chain in the hafnium isotopes, Beer and Ward (1981) proposed a branching in the decay of $^{180}$Hf to be responsible for the observed $^{180}$Ta abundance. Beer and Macklin (1982) showed that under s-process conditions 1.24% of all neutron captures in $^{179}$Hf populate the high-spin isomer in $^{180}$Hf, which can beta decay by an allowed transition to the long-lived isomer in $^{180}$Ta. With plausible estimates for the log *ft* value of that transition, it appeared as if all the $^{180}$Ta could have been made in this way. Meanwhile, there exist two measurements of this log *ft* value (Barden 1985, Rauch 1987), which set a limit of 1.6% for this beta decay branch, thus restricting that particular s-process contribution to 22% of the observed $^{180}$Ta abundance.

A second possibility for populating the high-spin isomer in $^{180}$Hf was r-process beta decays in the $A = 180$ mass chain. If the beta decay of $^{180}$Lu feeds into this isomer, then the $^{180}$Ta production could have occurred in this way as well. Eschner *et al* (1984) reported a branching factor of 0.46% for this decay, in contradiction to another experiment (Kellogg and Norman 1986), which yielded $0.005 \pm 0.018\%$, compatible with zero. This discrepancy seemed to indicate the existence of an additional isomer in $^{180}$Lu with a half-life of less than 300 s, which would have been detected only in the experiment of Eschner *et al* (1984). However, the existence of such an isomer can now be excluded, since neither Lesko *et al* (1986) nor Runte *et al* (1987) succeeded in finding gamma transitions related to this hypothetical isomer. Therefore, any r-process contribution to the $^{180}$Ta abundance seems to be at least questionable.

An alternative route to $^{180}$Ta has been proposed by Yokoi and Takahashi (1983), who found that $^{179}$Hf becomes unstable under s-process conditions against beta decay from excited states. Accordingly, a finite $^{179}$Hf/$^{179}$Ta ratio will develop during the s-process, which means that $^{180}$Ta is now produced directly by neutron captures in

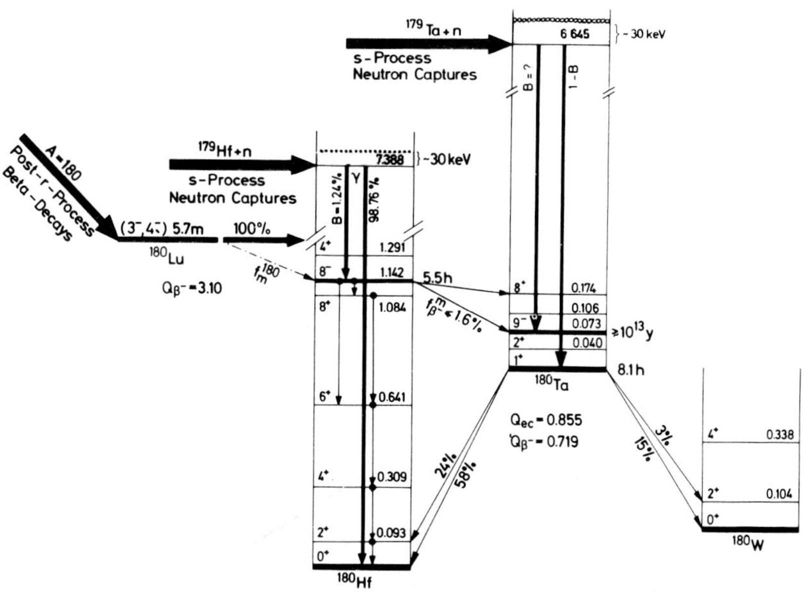


**Figure 26.** The nuclear level schemes along the $A = 180$ isobaric chain of significance to the stellar nucleosynthesis of $^{180m}$Ta. The main production paths to $^{180m}$Ta are indicated by the heavy arrows.

$^{179}$Ta. As part of these captures (probably the majority) will populate the short-lived ground state of $^{180}$Ta, this provides an explanation for the $^{180}$W abundance, which was previously attributed to the p-process. In fact, comparison with the observed $^{180}$W abundance could help to further constrain the neutron density and temperature during the s-process as emphasised by Yokoi and Takahashi (1983). For this analysis, however, one needs to know which fraction B of the captures in $^{179}$Ta populates the long-lived isomer, i.e. one needs the partial cross section to that isomer. As such a measurement involves radioactive $^{179}$Ta ($t_{1/2} = 665$ d) as a target; this problem remains a challenge to date. It is interesting to note that the s-process branching at $A = 179$ could explain all of the observed $^{180}$Ta, which means that this isotope might well be of pure s-process origin.

A similar example for an s-process production of rare, neutron-deficient isotopes is the branching at $^{113}$Cd and $^{114}$In, which accounts for most of $^{113}$In and $^{114,115}$Sn via beta decays from isomeric states (Ward and Beer 1981).

## 6. Comparison to stellar models and to astronomical observations

The question for the neutron source led to the assumption that the s-process is associated with helium burning zones during the late stages of stellar evolution when the stars expand to become red giants (Burbidge *et al* 1957). This was supported by astronomical observations, which showed that s-elements were enhanced in the atmospheres of these stars. A particularly convincing argument for this hypothesis was the discovery of technetium lines in the spectra of certain S stars (Merrill 1952). As there are no stable Tc isotopes, and as the longest Tc half-life is much shorter than the age of red giants, this was direct evidence for ongoing s-processing in the interior of these stars. Various stellar models have been worked out in sufficient detail to allow for quantitative studies (§ 2.2). Comparison with results from the classical model shows, however, that there are still many open problems, and that both approaches have to be further improved to solve the quest for the stellar s-process.

### 6.1. Stellar models

The three s-process components outlined in § 2.1 may be attributed to various stellar sites. According to Truran and Iben (1977) these are

(i) massive stars ($>15M_\odot$), which were shown to produce most of the abundances in the range $A<90$ during core helium burning;

(ii) intermediate mass stars (2–$8M_\odot$), which may synthesise the isotopes between $A=90$ and 204 in their thermally unstable helium burning shell;

(iii) stars of comparably low mass ($<1M_\odot$), which could produce the lead and bismuth abundances in core helium flashes.

This concept is still useful, but requires some modifications outlined below.

*6.1.1. Weak component.* Helium burning in the cores of massive stars has been identified as a possible s-process site by Peters (1968). Towards the end of core helium burning in stars of 10–$25M_\odot$, temperatures become sufficiently high that a substantial neutron density can be established via the $^{22}$Ne($\alpha$, n)$^{25}$Mg reaction. Recent calculations (Arnett and Thielemann 1985, Busso and Gallino 1985, Prantzos *et al* 1987, Langer *et al* 1989) with an improved stellar model and with updated nuclear information confirm earlier results (Lamb *et al* 1977) reporting significant s-process production in the mass range $A<90$. In all these models, $^{58}$Fe is produced in similar abundance ratios compared to other s-process nuclei. The respective yields depend on the helium core masses such that in lighter cores ($<4M_\odot$) only isotopes near $^{56}$Fe are significantly overproduced, while the overproduction is fairly constant in the mass range $60<A<80$ for larger helium cores (Arnett and Thielemann 1985). The $^{58}$Fe abundance calculated by these models is obviously in contradiction to the results obtained if the weak component is described by a single neutron exposure in the classical s-process (§ 5).

Very massive stars ($M>\sim50M_\odot$) are known to develop strong stellar winds, which efficiently remix the s-processed material into the interstellar medium. In this scenario, all s-only isotopes in the mass region $56<A<90$ are well reproduced if the respective contributions from the main component are considered (Prantzos *et al* 1987). The integral neutron dose derived from the stellar model also agrees reasonably well with the mean neutron exposure $\tau_0$ obtained from the classical model. However, the results for neutron density and temperature are somewhat discrepant. The stellar models predict neutron densities around $\sim10^7$ cm$^{-3}$ and temperatures above $2.5\times10^8$ K, while the classical model yields $n_n$ $\sim10^8$ cm$^{-3}$ and somewhat lower temperatures. These problems do not originate from the simplifying assumptions of the classical model, but were identified by Prantzos *et al* (1987) to be the consequence of the abundances in the $^{79}$Se branching. On the other hand, it was noted in § 5 that previous estimates of neutron density and temperature for the weak component have to be modified according to the changes for the main component. In addition, due consideration of a pulsed main component also affects the resulting neutron density and temperature for the weak component. These changes may ultimately lead to better agreement between the classical s-process and the stellar models.

*6.1.2. Main component.* The first model that was able to predict s-process abundances was that of helium shell burning in intermediate mass stars (Weigert 1966, Schwarzschild and Härm 1967, Iben 1976, Iben and Renzini 1983). The operation of the $^{22}Ne(\alpha, n)^{25}Mg$ source in thermally pulsing stars was shown to produce the s-process abundances in the mass range $90 < A < 204$ in solar proportions (Truran and Iben 1977) as well as to be efficient enough to explain the abundance level of s-process nuclei in the galaxy (Iben and Truran 1978). However, it was argued that the high neutron densities associated with the pulsed model may lead to non-solar abundance ratios at s-process branchings (Despain 1980). This problem was quantified by Howard *et al* (1986) in a parametric study of the pulsed model for a range of core masses between 1–1.4$M_\odot$. In all these cases, poor fits of the empirical $\langle\sigma\rangle N_s$ values were obtained, because of the extremely rapid decay of the neutron density during the pulses. The resulting average neutron density is much too high to reproduce the branchings properly, even though the effects of this high neutron density are partly smoothed out during captures at the end of the pulse (Cosner *et al* 1980). As an alternative approach, Howard *et al* (1986) investigated a schematic model with pulses of constant neutron density and temperature. With this assumption, they arrived at conclusions very similar to the results of the classical model, e.g. neutron densities around $10^8$ cm$^{-3}$, temperatures of $2.7 \times 10^8$ K and a mean neutron exposure of $\tau_0 =$ 0.26 mb$^{-1}$ (§§ 4 and 5).

The conditions typical of the pulsed model are compared to the results of the classical model in table 6, showing that the neutron density is significantly smaller in the latter case. The pulse duration predicted by the stellar model appears to be close to the lower limit derived from the classical approach, but is much shorter than $\Delta t$ ~450 yr estimated by Howard *et al* (1986) from their schematic model pulses via the relation $\tau_0 = n_n v_T \Delta t$. For the stellar scenario, these 450 yr represent an upper limit, because they correspond to the accumulated exposure from several pulses: as the helium burning shell moves outward, only a fraction $(1-r) = \sim 70\%$ of processed matter is removed from the shell after each pulse (§ 2.2). Reasonably good agreement is obtained for the mean neutron exposure, at least for the core mass of 1.16$M_\odot$, for which the predicted $\tau_0 = 0.32$ mb$^{-1}$ is very close to the result of the classical model ($\tau_0 = 0.30\ (kT/30)^{1/2}$ mb$^{-1}$).

The above problems with the $^{22}Ne(\alpha, n)^{25}Mg$ source in intermediate mass stars and the fact that significant s-process enhancements are only observed in low mass, metal-poor AGB stars have increased the interest in these stars, particularly since Iben and

**Table 6.** The physical conditions during the s-process—results from the classical model compared to predictions of stellar models

| Parameter | Main component | Helium shell burning (2–8$M_\odot$; $M_{core} = 1.16M_\odot$) | Weak component | Core helium burning (>10$M_\odot$) |
|---|---|---|---|---|
| Temperature ($10^8$ K) | 2.8–3.9 | 1.6–3.7 | 1.8–3.0 | >2.5 |
| Neutron density ($10^8$ cm$^{-3}$) | 2.3–4.5 | 1–5000 | <(0.8–1.9) | ~0.1 |
| Matter density ($10^3$ g cm$^{-3}$) | 2.6–13 | 2–10 | | |
| Duration of pulse (yr) | >3 | 3.2 | | |

Renzini (1982, 1983) had shown that semiconvective mixing of protons into the helium burning shell could produce enough $^{13}$C to make the $^{13}C(\alpha, n)^{16}O$ reaction an efficient neutron source. This possible site for the s-process was investigated in a number of recent studies (Boothroyd and Sackmann 1988, Hollowell and Iben 1988, Gallino *et al* 1988). Again, first estimates yielded high neutron densities of $10^{11}$ to $10^{12}$ cm$^{-3}$ (Iben and Renzini 1982, Malaney 1987, Picchio *et al* 1988) in contradiction to the classical model or to observations, which show little or no $^{96}$Zr (Zook 1978). In reanalysing the pulse model of Iben (1982) with a large reaction network, Gallino *et al* (1988) obtained a strong reduction in neutron density to values below $3\times10^9$ cm$^{-3}$, and hence a remarkably constant s-process overproduction in the entire mass range from $A = 90$ to 200.

In this model, there are two neutron bursts in each thermal pulse. The main burst is caused by the $^{13}C(\alpha, n)\ ^{16}O$ reaction at $\sim1.5\times10^8$ K, when a $^{13}$C pocket left from the previous pulse is engulfed in the convective helium shell. A smaller burst follows at the end of the pulse via the $^{22}Ne(\alpha, n)^{25}Mg$, when the temperature at the bottom of the shell rises sharply to $\sim3\times10^8$ K. In a way, this second burst adjusts the s-process thermometers (§ 5), which otherwise would reflect the lower temperature during the main burst.

*6.1.3. Strong component.* The idea that the core helium flash in stars of less than $1M_\odot$ could be responsible for the strong component (Truran and Iben 1977), has not been investigated quantitatively yet. To date it is not clear whether this core flash is a single event or whether it may occur repeatedly (Wood 1985). However, it is to be expected that neutrons are produced via the $^{13}C(\alpha, n)^{16}O$ reaction, and that these flashes may be characterised by a sufficiently large neutron exposure to account for the observed lead and bismuth abundances. On the other hand, the reservations noted by Ratzel *et al* (1988) need first to be clarified before the strong component can be considered as established.

In summary, one finds that the present results from the classical s-process only allow for a rough comparison with stellar models, indicating a number of inconsistencies. Consequently, both the classical approach as well as comprehensive studies of s-process enhancements in AGB stars need to be improved in order to establish reliable constraints for stellar nucleosynthesis. Stellar models yield an increasingly quantitative picture of the s-process. The weak component seems to be well described by helium burning in massive stars, while the pulsed s-process associated with helium shell burning remains an attractive model for the main component. At present, low mass AGB stars appear as the most promising candidates for this s-process component.

### *6.2. Astronomical observations*

Enhancements of s-process elements in stellar atmospheres have been found in two classes of red giants, in relatively cool stars of type MS and S with surface temperatures between 2000 and 3800 K, and in so-called barium stars with rather normal surface temperatures of 4000 to 5500 K. For many years, the s-enhancements in barium stars were not understood, because these stars have not yet reached the asymptotic giant branch (AGB stage), i.e. they were not expected to exhibit convective mixing of material from the deep interior to the surface. Detailed studies (McClure 1984, Böhm-Vitense *et al* 1984) revealed that barium stars are members of binary systems that have received considerable mass overflow from the companion star during or after its AGB phase. This means that the s-abundances in barium stars are the products of the more evolved companion. Therefore, the MS and S giants can be considered as the only true, living examples of AGB stars with a pulsating helium shell, from where freshly produced s-process material is mixed at the surface.

Consequently, the MS and S giants are of greatest interest as their composition provides for the most direct access to the actual s-process. Though their low temperatures make them difficult to observe, recent investigations succeeded in deriving remarkably complete and accurate abundance patterns (e.g. Tomkin and Lambert 1983, Smith and Lambert 1986, 1988). As only elemental abundances can be derived spectroscopically, most aspects of the s-process cannot be resolved with as much detail as for solar material. Nevertheless, the abundances of typical s-elements, such as Sr, Y, Zr, Ba and Nd, can be used to determine the mean neutron exposures characteristic for these stellar abundance patterns. One finds the corresponding values for $\tau_0$ to range between 0.15 and 0.5 $mb^{-1}$, compatible with the result obtained for the main component. Estimates for the neutron density can be derived from the Sr/Rb or Y/Rb ratios (Tomkin and Lambert 1983, Käppeler 1986a, Malaney and Lambert 1988, Beer and Macklin 1989) or from the isotopic abundances of Zr, which can be determined via molecular bands for sufficiently cool stars (Zook 1978, Malaney 1986, Smith 1988). These studies yield neutron densities around $10^8\ cm^{-3}$ similar to the result obtained with the classical model.

A discrepancy with respect to the s-process in the pulsed helium shell concerns the neutron source. So far, all observed Mg isotope ratios are close to solar and do not show any $^{25}$Mg enhancements due to the $^{22}Ne(\alpha, n)^{25}Mg$ reaction. Furthermore, estimates of the luminosities indicate that the available observations refer to low mass stars ($<2M_\odot$), whereas in more massive AGB stars s-process enhancements were not yet observed. Recent studies even consider the possibility that stars with $M>3M_\odot$ may not reach the AGB stage because of excessive mass loss in stellar winds (Mould and Reid 1987); this result would totally question intermediate mass stars as a possible s-process site!

Though stellar observations of s-process enhancements are not yet sufficiently accurate or representative to allow for detailed conclusions, it is clear that this type of information is invaluable as it represents a direct link to actual s-processing. Further studies with improved accuracy and more complete abundance patterns promise to yield important constraints on stellar dredge-up mechanisms (§ 7), and on the systematics of the s-process as a function of stellar evolution.

## 7. Chronometers

The chronological aspect has fascinated many researchers since the beginning of quantitative studies of nucleosynthesis. A large number of publications have been devoted to interpreting the naturally occurring radioactive nuclei in terms of a cosmic timescale, either by studying the abundances of nuclei that are unstable but still alive, e.g. $^{176}$Lu, $^{187}$Re and the long-lived Th and U isotopes, or the abundances of the respective decay products. Instead of a thorough discussion, which would require a review of its own, this section is restricted to a selection of s-process chronometers ordered according to their timescales. Principally, any unstable isotope on the s-process path may be considered as a potential chronometer, provided that it is not obscured by abundance contributions from the r-process, and that its decay had significant consequences for the isotopic abundance pattern observed today. The available chronometers relate to different phenomena, depending on their half-lives.

$t_{1/2} < 10^6$ *years.* This timescale is typical for the dynamics in helium shell burning, which is the likely scenario for the main component of the s-process; it concerns for example the duration of shell flashes and the time required for dredge-up of processed material to the stellar atmosphere.

$t_{1/2} \sim 10^6\text{–}10^8$ *years.* This timescale relates to the question of whether the formation of the solar system was triggered by a nearby supernova explosion. There is evidence from the discovery of isotopic anomalies in meteorites that the protosolar cloud was peppered by freshly synthesised material originating from such an event.

$t_{1/2} > 10^9$ *years.* This timescale is close to that for the age of the universe as deduced from the age of the oldest stars or from the red shift of distant galaxies, and it would be most important to complement these estimates with an age for the chemical elements.

With this classification, one example will be given for each epoch in the following subsections. More information can be found in the references cited.

### *7.1. Short timescale*

An example of this type was presented in § 5.2.3, where it was shown that $^{151}$Sm may be used as a clock for the duration of the helium shell flashes, which provide the natural explanation for the exponential distribution of neutron exposures characteristic of the main s-process component.

Another outstanding example is $^{99}$Tc, which is produced in the s-process, and which was discovered in the atmospheres of S stars, a subclass of red giants (Merrill 1952). This observation provided an important argument in favour of stellar nucleosynthesis. As all Tc isotopes are unstable with half-lives much shorter than the age of these stars, they must have been produced within these stars. By improved observational techniques and atmospheric models, technetium abundances can be determined even quantitatively (Smith and Wallerstein 1983, Dominy and Wallerstein 1986, Smith and Lambert 1986, 1988). These direct observations were all the more exciting since it was found that the decay rate of $^{99}$Tc is drastically enhanced at high temperature (Cosner and Truran 1981, Schatz 1983, Takahashi and Yokoi 1987): the half-life decreases from $2.1 \times 10^5$ yr at $T < 10^8$ K to a few years at $T \sim 3 \times 10^8$ K, the estimated temperature for the s-process (§ 5.2.1).

That $^{99}$Tc survives and is observed at the surface therefore implies that it is quickly removed from the s-process site to cooler zones by efficient convective motions. Hence, the short lifetime of $^{99}$Tc at s-process temperatures represents a clock for the dynamics of helium shell burning in these stars. As soon as $^{99}$Tc has reached a zone with $T < 10^8$K, it practically decays with its terrestrial half-life. This change in the timescale of the $^{99}$Tc clock can be accounted for by means of $^{93}$Zr ($t_{1/2} = 1.5 \times 10^6$ yr), a neighbouring unstable s-process nucleus with a half-life almost insensitive to temperature. While $^{93}$Zr is not easily detected in astronomical observations due to the presence of the stable Zr isotopes, the decay can be followed via the daughter nucleus $^{93}$Nb. As this is the only stable isotope of niobium, and as it is *not* produced in the s-process, its abundance is a sensitive indicator for the decay of $^{93}$Zr. Note that with time $^{93}$Nb increases in abundance and $^{99}$Tc decreases.

This change of abundances during the transport of $^{99}$Tc and $^{93}$Zr/$^{93}$Nb from the helium burning zone to the surface and in the stellar atmosphere itself, was studied on the basis of existing stellar models (Mathews *et al* 1986, Takahashi *et al* 1986). It was found that the observed Tc and Nb abundances relative to each other as well as to their neighbouring elements Zr and Mo may indeed be considered as a clock for the lifetime of stars in the thermally pulsing third dredge-up phase. Further investigations of this possibility require more accurate and systematic field studies of AGB stars and improved stellar models that are able to reproduce the observed s-process abundances (§ 6).

### *7.2. Intermediate timescale*

Isotopes with half-lives between $\sim 10^6$ and $10^8$ yr are particularly interesting in connection with isotopic anomalies in meteorites. Whereas isotopic ratios are fairly constant in most materials or are at most modified by physical effects such as diffusion, very pronounced anomalous isotope patterns were discovered in small inclusions of certain meteorites. Some of these anomalies were shown to be the result of *in situ* decay of now-extinct radionuclides, e.g. $^{26}$Al, $^{107}$Pd, $^{129}$I and $^{244}$Pu. This means that these radionuclides were alive during the formation of solid bodies in the early solar system, and hence were synthesised shortly before. One possible explanation for their origin was to assume a local supernova, which triggered the collapse of the protosolar cloud and also produced these radionuclides. The resulting isotopic anomalies could therefore be analysed in terms of the different half-lives to establish a chronometry of the early solar system.

However, this interpretation is questioned by Clayton (1979), who pointed out that the anomalies need not be the result of a local event. Instead, presolar grains could have formed much earlier and could have been ingested by the protosolar cloud. Such a model, of course, would deny any chronological information with respect to the solar system, but would rather yield information on the timescale for condensation of supernova ejecta. For a detailed presentation of the field see the reviews by Clayton (1979), Begemann (1980), Wasserburg and Papanastassiou (1982), for example.

The s-process aspect of this problem may be interesting insofar as it depends on the stellar mass, and whether and how much of s-processed matter is released in a supernova; alternatively, s-process material could have been ejected in stellar winds during the red-giant phase prior to the supernova. Significant isotopic anomalies have been reported for xenon and krypton (Srinivasan and Anders 1978, Alaerts *et al* 1980, Ott *et al* 1988) and for neodymium (Lugmair *et al* 1983a, b) and were interpreted as being characteristic for s-process nucleosynthesis (Clayton 1978, 1979, 1983, Beer *et al* 1983, Mathews and Käppeler 1984, Walter *et al* 1986b). In this context, it would be important to identify an anomaly associated with an unstable s-process isotope, which could serve as an appropriate clock in order to test these ideas.

Yokoi *et al* (1985) have shown that $^{205}$Pb would be such a case. This clock was long overlooked, because it was assumed that the decay of $^{205}$Pb was strongly enhanced at stellar temperatures so that it could not be produced in significant amounts during the s-process. However, the small neutral atomic mass difference between $^{205}$Pb and $^{205}$Tl causes the latter isotope to become unstable in the ionised stellar plasma against bound-state decay (§ 3.3). Under certain conditions, the decay rate of $^{205}$Tl can even exceed that of $^{205}$Pb as shown in figure 27. Though $^{205}$Tl is unstable only at temperatures above $\sim 1.5 \times 10^8$ K and therefore does not counterbalance the decay of $^{205}$Pb in the

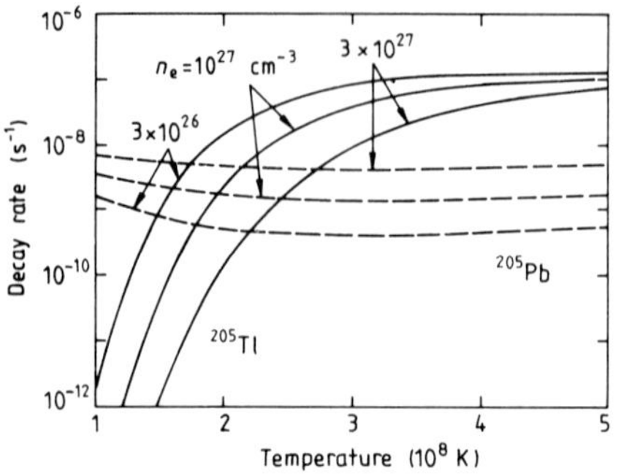


**Figure 27.** The stellar decay rates of $^{205}$Tl (full curves) and $^{205}$Pb (broken curves) versus temperature (Takahashi and Yokoi 1987). Note the strong dependence on electron density, $n_e$.

range $2\times10^6 < T < 1.5\times10^8$ K, Yokoi *et al* (1985) estimated that in intermediate mass stars enough $^{205}$Pb may survive the convective transport from the helium burning zone to sufficiently cool the outer layers. The resulting $^{205}$Pb/$^{204}$Pb ratios were estimated to range between $10^{-5}$ and unity, so that one could expect isotopic anomalies in excess of those reported by Huey and Kohman (1972) who found an upper limit for the $^{205}$Pb/$^{204}$Pb ratio of $9\times10^{-5}$ at the time of solidification of the meteoritic material. A renewed search for extinct $^{205}$Pb seems therefore to be justified and possibly rewarding.

### *7.3. Long timescale*

There are numerous attempts to derive an estimate of the age of our galaxy by interpretation of the observed long-lived radionuclides as chronometers for galactic nucleosynthesis. The main difficulty in these attempts originates from the fact that they have to be based on solar system material. This material is known to represent the result of element production and modification up to 4.6 Gyr ago. Since that time there has been free decay of the respective radionuclides; but what was before? Obviously, this question has two aspects:

(i) What are the isotopic yields of the various nucleosynthetic processes, e.g. in the s- or r-process?

(ii) What is the time dependence of galactic nucleosynthesis for each of these processes?

Both aspects are still affected by sizeable uncertainties with the consequence that the galactic timescale set by astronomical methods cannot be improved significantly by present nucleocosmochronology.

In first attempts, the rate of element production was simply assumed to have decreased exponentially with time due to the decreasing density in the universe (Fowler 1973). Later, galactic evolution models were found to yield additional constraints, and were considered as well (Tinsley 1980, Clayton 1984). In all these studies, the history of element production had to be described by suited parameters, and it has been emphasised that the combined analysis of several long-lived radionuclides may yield enough information for a reasonably good fit of these parameters, provided the respective half-lives were sufficiently different.

Studies of this type concentrated on the r-process chronometers $^{232}$Th and $^{235,238}$U. Depending on the particular assumptions for the rate of element production and on the calculated r-process yields for the long-lived radionuclides, galactic ages between 10 and 20 Gyr were reported (Fowler 1978, Fowler and Meisl 1986, Thielemann and Truran 1986, Cowan *et al* 1987). This range is in reasonable agreement with, but no more accurate than, current estimates derived from the Hubble time (Sandage and Tammann 1984, Huchra 1987, Branch 1987) or from the age of globular clusters (Sandage 1982, VandenBerg 1983, 1985).

Beside these r-process chronometers, $^{176}$Lu is the only long-lived isotope of pure s-process origin that seemed to be suited as a cosmic clock (Audouze *et al* 1972, Arnould 1973). However, quantitative studies revealed that this clock may be in question due to temperature effects (see § 7.4). The remaining long-lived species on the chart of nuclides, e.g. $^{40}$K (Beer and Penzhorn 1987), $^{87}$Rb (Beer and Walter 1984) and $^{187}$Re (Yokoi *et al* 1983, Arnould *et al* 1984, Winters *et al* 1986b), are produced in at least two nucleosynthetic processes. For this reason they are difficult to interpret, apart from problems related to a likely enhancement of the stellar decay rates for the latter two isotopes (Takahashi and Yokoi 1987).

Recent progress in observational techniques meanwhile offers the exciting possibility of escaping the restriction to solar system material by direct determination of long-lived radionuclides in old stars. In this way, the timescale for nucleosynthesis can be directly related to that for stellar and galactic evolution. In a systematic study of $^{232}$Th abundances in G-dwarf stars of different ages relative to the stable elements Ba, Nd and Eu (Butcher 1987, 1988), no evidence was found for an evolution of the thorium abundance with stellar age. These results are plotted in figure 28 together with results obtained by galactic evolution models (Clayton 1987, Mathews and Schramm 1988), which consider a gradually increasing abundance contribution from the s-process to an initial r-process yield. While both models are able to fit the observed Th/Nd ratios quite well, they fail to explain the more sensitive Eu/Ba ratios.

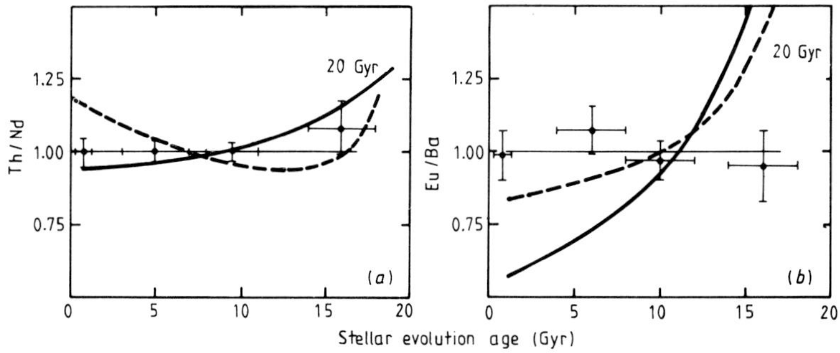


**Figure 28.** The abundance ratios (*a*) Th/Nd and (*b*) Eu/Ba as obtained for G-dwarfs of different ages (Butcher 1987, 1988) compared to results obtained with galactic evolution models (Clayton 1987, full curve; Mathews and Schramm 1988, broken curve) which postulate a gradually increasing contribution from the s-process to an initial r-process abundance. The good fits to the Th/Nd ratio for galactic ages of 20 and 15 Gyr are questioned by the fact that the Eu/Ba ratio cannot be reproduced.

It seems as if current ideas about galactic evolution are in conflict with Butcher's observations and as if this G-dwarf problem indicates that nucleosynthesis peaked at early epochs. This scenario fits to models concerning a quark–hadron phase transition in the big bang, which may have caused strong inhomogeneities in the proton–neutron distribution during primordial nucleosynthesis. Under such conditions even a big-bang r-process seems to be possible (Applegate and Hogan 1985, Alcock *et al* 1987, Malaney and Fowler 1989). On the other hand, this picture would have difficulty in explaining the galactic metallicity gradient. In any case, this fascinating chronometry is just starting and may produce new insights in the history of our galaxy as observational uncertainties are further reduced.

### 7.4. $^{176}$Lu—cosmic clock or stellar thermometer?

The suited half-life of 36 Gy and the fact that it is shielded against the r-process (figure 29) makes $^{176}$Lu a good candidate for studying the age of the s-process elements (Audouze *et al* 1972, Arnould 1973). The only difficulty seemed to be the proper treatment of the branching at $^{176}$Lu as it is indicated in figure 29. This branching is due to the isomeric state at $125\pm2$ keV, which is significantly populated by neutron capture in $^{175}$Lu, and which decays with a half-life of 3.68 h to $^{176}$Hf. Because of the large difference in spin and $K$ quantum number, electromagnetic transitions are highly forbidden by selection rules and the ground state and the isomer could be considered as independent of each other. Therefore, the s-process branching could be followed according to the relative population of ground state and isomer in the (n, $\gamma$) reaction.

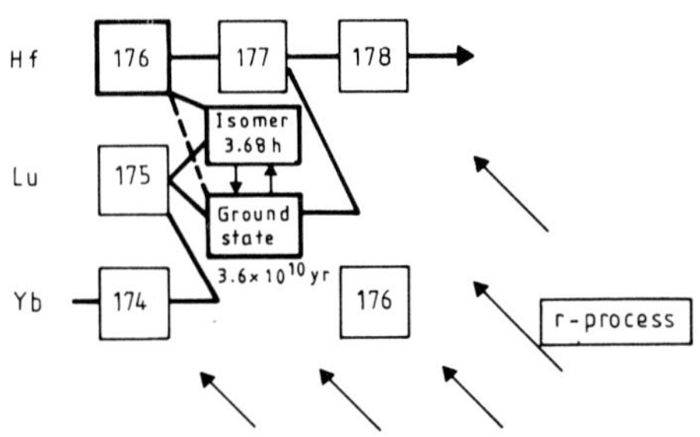


**Figure 29.** The s-process capture chain around $^{176}$Lu.

Soon after the first quantitative study of $^{176}$Lu as a cosmic clock (Beer and Käppeler 1980), it was realised that the decay of $^{176}$Lu migh be accelerated at high temperatures (Ward 1980, Beer *et al* 1981). Though direct transitions are forbidden, ground state and isomer could well be connected via transitions through higher lying states. This means that any quantitative treatment requires a detailed knowledge of the level scheme and the relevant transition rates. In 1980, Ward was left with two completely different families of excited states: those with high $K$ quantum numbers related to the ground state and the others with small $K$ connected with the isomer. The only possible links between these groups were $K$-forbidden transitions. By combination of estimated intra- and interband transition rates with the population probabilities of the excited states it was found that the levels at 239, 376 and 537 keV with $J^{\pi}=3^{-}$, $4^{+}$ provided for the most efficient links between ground state and isomer. As a result, thermal equilibrium in the population of these states was expected to appear already above $T \sim 2\times10^8$ K, so that $^{176g}$Lu is efficiently destroyed via isomer decay in the temperature range of the s-process. However, this result was considered to be rather uncertain because of the schematic estimates for transition rates and hindrance factors for $K$-forbidden transitions.

That the decay of $^{176}$Lu was indeed thermally enhanced during the s-process was also inferred from an extensive analysis of the s-process path in the mass range $160 < A < 180$ (Beer *et al* 1984). If one just follows the neutron capture chain, the branching ratio at $^{176}$Lu can be obtained from the ratio of the partial cross section to the isomer and the total (n, $\gamma$) cross section of $^{175}$Lu, neglecting any thermally induced transitions. This simple solution yields branching ratios of $0.32\pm0.05$, $0.17\pm0.05$ or almost zero, depending on the partial cross section adopted from literature (Beer and Käppeler 1980, Allen *et al* 1982, or Allen *et al* 1981, Stecher-Rasmussen *et al* 1988). In contrast, the complete analysis including s-only $^{176}$Hf gives

$$0.39 < B_n^{eff} < 0.67$$

indicating that in fact more of the s-process flow bypassed $^{176}$Hf. This means that at least some of the lutetium produced in the isomeric state was transformed to the long-lived ground state via thermally induced transitions. If the mechanism for these thermal effects could be described quantitatively, $^{176}$Lu could possibly be used as a thermometer for the s-process. So far, only the extreme case of complete thermal equilibration between ground state and isomer can be treated (Beer *et al* 1984).

This problem of whether, to what extent and how the isomer and the ground state of $^{176}$Lu are thermally connected, initiated a number of investigations. First, experimental evidence for an electromagnetic link between ground state and isomer was obtained by photoactivation of Lu samples in intense gamma-ray fields (Veres and Pavlicsek 1970, Norman and Kellogg 1985). In these measurements, a positive effect was found in irradiations with $^{60}$Co ($E_\gamma = 1332$, 1173 keV), but not with $^{137}$Cs ($E_\gamma =$ 662 keV), a clear hint that excitation energies well above 500 keV were required. The quantitative interpretation of such measurements in terms of a stellar reaction rate (Norman and Kellogg 1985) is somewhat speculative, as neither the reaction mechanism nor the photon spectrum are well enough known to reliably infer the stellar photoactivation cross section. For example, figure 30 illustrates how critically the half-life of $^{176}$Lu depends on the unknown energy of the mediating level (Norman and Kellogg 1985).

An improved theoretical approach was presented by Gardner *et al* (1988), who complemented the experimentally determined levels in $^{176}$Lu by postulated states inferred from systematic trends in neighbouring nuclei. With this 'complete' level scheme, which extended up to 1.5 MeV, they obtained a strong temperature dependence of the $^{176}$Lu half-life, but now using only *unhindered* transitions. Comparison with the estimates from photoactivation shows an even shorter half-life! If thermalisation is indeed provided by allowed transitions, there is a chance for the experimental assessment of level energies and branching ratios of the relevant interband transitions, e.g. from the $5^-$ band head found by Coulomb excitation (Elze 1984). Presently, a comprehensive set of experiments for the completion of the level scheme of $^{176}$Lu is in progress (Klay *et al* 1988), using the high-resolution spectrometers GAMS and BILL at the high flux reactor of the Institut Laue-Langevin in Grenoble as well as additional information from (d, p)-reaction studies (Mayerhofer *et al* 1987).

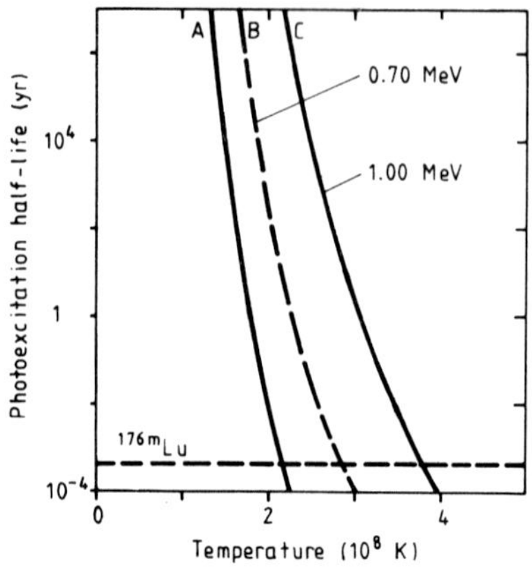


**Figure 30.** Estimates for the half-life of $^{176}$Lu against photoactivation as a function of temperature based on experimental cross sections (curves B and C) (Norman and Kellogg 1985) and on a postulated level scheme (curve A) (Gardner *et al* 1988). The broken curve illustrates how sensitively the half-life depends on the energy of the activation level.

Annihilation of a positron with a K-shell electron has been proposed as another excitation mechanism for $^{176}$Lu, that may populate mediating levels above 1 MeV. This effect was experimentally verified by Watanabe *et al* (1981), but the interpretation of Norman and Kellogg (1985) probably overestimated its influence in a stellar environment; a more conservative treatment indicates that the corresponding decay rate is lower than that due to photoexcitation (Käppeler 1988).

Though $^{176}$Lu is likely not to be a useful chronometer for the s-process, the stellar fate of $^{176}$Lu remains one of the fascinating problems of nuclear astrophysics—and there is a good chance that it may represent a sensitive stellar thermometer instead.

## 8. Summary

The quest for the s-process origin of the heavy elements stands for a lively field of research with numerous fascinating problems and many different aspects. The present status of this field is promising and rewarding in the sense that the overall picture seems fairly consistent, but that the many parts of the puzzle do not yet fit together. It appears hopeful that the aim of understanding the s-process within the next decade is in reach, and the rewarding aspect is that the present concept still allows or even demands substantial improvements and modifications. This mostly concerns the data base, i.e. the observed abundances and the relevant nuclear physics, which are the important input for the empirical approach via the classical s-process as well as for the stellar models. The results obtained from these models need definitely to be complemented with and checked against the s-process abundance patterns, which are directly obtained by spectroscopy of individual red giants. Our ultimate understanding of s-process nucleosynthesis therefore requires tightening of the circle indicated opposite:

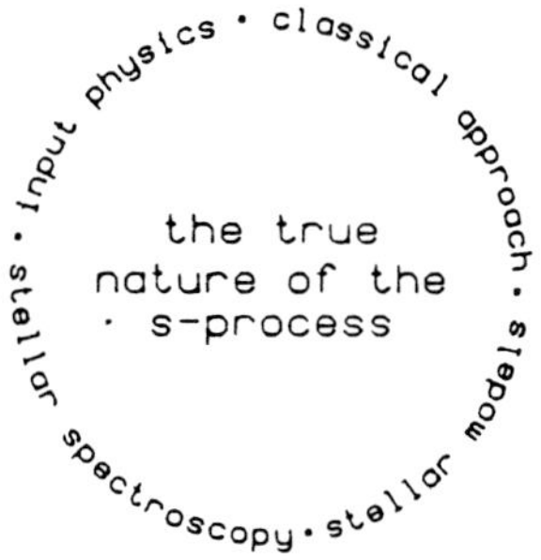


**Acknowledgments**

We gratefully acknowledge the help of all colleagues who contributed to this work in many discussions, by critical comments on and amendments of the manuscript, and by allowing us to use their results prior to publication. In particular we thank J-P Arcoragi, R Gallino, H Holweger, H Palme, J T Wasson, K Yokoi, S Yorka, A Mengoni, G Reffo and N Grevesse. We also thank the editors for their patience during the delays in the preparation of this manuscript.

## ADDENDUM
### Recent measurements and improved analyses

The following outline briefly summarizes the improvements of s-process nucleosynthesis that have been achieved since the preceding article was completed in December 1988. This concerns the input data—and here mainly the stellar $(n,\gamma)$ cross sections—as well as the progress in stellar s-process models.

*Input data*

To start with the input data, there is not only a great number of new measurements, but also considerable progress in the quality and applicability of the experimental techniques. The use of $4\pi$ gamma-ray detection with $BaF_2$-crystals was shown to reduce the uncertainties of stellar capture rates to the 1 to 2% level. Measurements on radioactive branch point isotopes became feasible, both with the activation method and by time-of-flight (TOF) techniques. In addition, these methods also allowed for the determination of very small $(n,\gamma)$ cross sections at magic neutron numbers and for light nuclei of importance for primordial nucleosynthesis. The stellar beta decay rates of relevance to the s-process have been improved experimentally, and were extended to higher densities and temperatures.

*$4\pi$ $BaF_2$ Detectors.* The set-up at KfK Karlsruhe (p 19) became operational in 1989 (Wisshak *et al* 1990a) and was demonstrated to reach an accuracy of 1% for cross section ratios and of 2% for absolute cross sections (Wisshak *et al* 1990b). First measurements for the s-process were carried out on the stable tellurium isotopes (Wisshak *et al* 1991). Confirmation of the 'local approximation' for the three s-only

isotopes, i.e. the equality of the $[\sigma N]$-values of $^{122}$Te, $^{123}$Te and $^{124}$Te within the 1.2% uncertainty of the cross section ratio allowed to derive an *upper* limit for the s-process neutron density. Studies with this detector are continued for the important s-only nuclei, with $^{128}$Xe, $^{130}$Xe, $^{134}$Ba, $^{136}$Ba, $^{148}$Sm, and $^{150}$Sm being the next examples. A similar, but somewhat less sophisticated $BaF_2$ detector array is being set up at Los Alamos (Koehler *et al* 1987). In combination with the very high neutron flux at the LANSCE facility, the set-up in Los Alamos will be particularly useful for measurements on radioactive samples.

*Other* TOF *techniques.* The set-up of Moxon–Rae detectors for measurements at very short flight paths (p 18) has been considerably improved with respect to time resolution, and was adapted to the geometry of the quasi-stellar neutron spectrum (p 21) (Jaag 1990). It was demonstrated at the examples of $^{87}$Rb and $^{209}$Bi that this set-up is suited for measurements of small cross sections in the range of a few mbarn (Jaag and Käppeler 1990). On the other hand, this set-up could also be used to study the $(n,\gamma)$ cross section of $^{147}$Pm, a radioactive branch point nucleus with a half-life of only 2.6 yr (Gerstenhöfer *et al* 1991). In this case, the good sensitivity was important to compensate for the fact that the sample mass was restricted to $\sim 5$ mg in order to keep the radiation hazard at an acceptable level.

A similar set-up with a flight path of a few cm was used by Nagai *et al* (1990, 1991) for measurements in the $\mu$barn range. These authors were able to study the $^{7}\text{Li}(n,\gamma)^{8}\text{Li}$ and the $^{12}\text{C}(n,\gamma)^{13}\text{C}$ reactions by recording the prompt capture gamma-rays by means of a well-shielded NaI(Tl) detector that was located 80 cm from the sample at an angle of 125° to the beam axis. The sensitivity achieved in these measurements is impressive despite the fact that rather thick samples had to be used.

The sensitivity of the above techniques is based on the use of high neutron fluxes at short flight paths, and is achieved by renouncing on energy resolution. While this approach works well for the quasi-stellar spectrum yielding the steller cross section at $kT = 25$ keV, extrapolation to lower or higher temperatures becomes more uncertain the smaller the respective cross sections are. In these cases, widely spaced resolved resonances may dominate the stellar rate, a prominent example being the temperature dependence of $^{208}$Pb: the stellar cross section *rises* with increasing neutron energy until the thermal energy exceeds the position of the strong resonance at 71.2 keV (Macklin *et al* 1977). For the small cross sections of neutron magic nuclei, it is, therefore, indispensable to complement average cross sections for $kT = 25$ keV with high-resolution data in order to obtain reliable extrapolation to other stellar temperatures. A similar aspect holds for the determination of stellar $(n,\gamma)$ rates at the comparably low temperature phases in low-mass stars (see p 57 and comments below) that are characterized by $kT \sim 12$ keV. Since the energy range of resolved resonances may extend to a few keV, it accounts for a sizable part of the stellar rate at these temperatures. Therefore, the individual resonances should be considered instead of extrapolating the smooth cross section in calculating the Maxwellian average (Beer *et al* 1991a).

*Activation method.* Compared to the above TOF techniques, activation measurements in the quasi-stellar neutron spectrum for $kT = 25$ keV (p 21) offer the advantage of superior sensitivity. This aspect has been exploited recently for a number of rather different investigations. Examples for measurements of particularly small cross sections are the neutron magic nuclei with $N = 50$, where significantly improved data were obtained for $^{86}Kr$ (Beer 1991a), $^{87}Rb$ (Neuberger *et al* 1991), as well as for $^{88}Sr$ and $^{99}Y$ (Käppeler *et al* 1990a). Even smaller cross sections in the $\mu$barn range could be determined for $^{22}Ne$ (Beer *et al* 1991b), which represents an important neutron poison in the s-process, and the light isotopes $^{7}Li$ (Wiescher *et al* 1989) and $^{14}C$ (Beer *et al* 1991c) that are of relevance with respect to a possible primordial r-process. In all these cases, a cyclic activation technique has to be used due to the short half-lives of the respective product nuclei (Beer 1991a).

Apart from the determination of small cross sections, the high sensitivity of the activation technique has also been used for measurements on small samples and in cases where the induced activity was low. One such example was the very first measurement of the stellar rate for the s-only isotope $^{192}Pt$ (Tepe and Käppeler 1991), where the induced activity is difficult to detect because of the long half-life of the product nucleus $^{193}Pt$ ($t_{1/2}=50$ yr) and because only $L$ X-rays are emitted in the electron capture decay of that nucleus. Another example is the radioactive isotope $^{147}Pm$ ($t_{1/2}=2.6$ yr), where the sample mass had to be restricted for keeping the radiation hazard manageable (Gerstenhöfer *et al* 1991).

Extrapolation of the experience with the activation technique led to a proposal for measuring a number of the important radioactive branch point nuclei via this method (Käppeler *et al* 1991a). The high sensitivity of this method allows the radiation hazard to be reduced, since it requires sample masses of only 10–100 ng, which can be made by means of ISOL techniques (Naumann *et al* 1987).

*Model calculations.* Many of the above mentioned studies have been complemented by theoretical calculations based on the statistical model (Reffo and Fabbri 1991). This method will remain an indispensable tool in future, but can be refined considerably by the availability of more experimental information. In particular, this technique is important for determining the stellar enhancement factors that may cause a considerable modification of the experimental rates (p 35).

*Weak interaction rates.* The comprehensive table of stellar decay rates by Takahashi and Yokoi (1987) was prepared under the assumption that all excited nuclear states are populated in thermal equilibrium with the ground state. While this assumption holds for the majority of the relevant cases, it is in question for nuclei with long-lived isomeric states. The influence of the isomer on the temperature dependence of the stellar decay rate could be settled experimentally for the three important nuclei $^{148}Pm$ (Lesko *et al* 1989), $^{176}Lu$ (Klay *et al* 1991a,b; Lesko *et al* 1991), and $^{180}Ta$ (Németh and Käppeler 1991), all with exciting consequences for the s-process site (see below).

In addition to these experimental findings, the decay rates of Takahashi and Yokoi (1987) have been extended to higher electron densities and temperatures in order to fit the needs of stellar s-process models (Neuberger 1991).

### *s-process studies*

Considerable progress was achieved in s-process nucleosynthesis, with the classical approach as well as with the stellar models sketched in sections 2.2 and 6.1. In particular, certain stellar s-process sites appear now to be well characterized, since these models are capable of reproducing the s-process abundances within 5 to 10%. In combination with the constraints derived from branching analyses via the classical approach, the stellar models may be further improved. Another important constraint for the stellar models arises from their direct link to the observation of s-process enhancements in AGB stars. The consideration of these aspects in a consistent description of the s-process will eventually allow for an improved picture of the burning phases during the AGB stage.

*Main component—phenomenological approaches and low mass stars.* The mass region $90 < A < 209$ that is dominated by the main component contains a fascinating variety of s-process branchings. The isotopic abundance patterns in these branchings can be analysed in terms of the physical conditions of the stellar plasma, yielding information on neutron density, temperature, electron density, and—presumably—on the s-process dynamics as well.

With the improved level scheme of $^{148}$Pm (Lesko *et al* 1989) the ambiguity in the neutron density derived from the branchings at $A = 147,148$ could be resolved. The new scheme clearly showed that thermal equilibrium between ground state and isomer can, indeed, be established within the 5.4 d half-life of the ground state. Accordingly, the mean neutron density had to be increased by a factor of 3 compared to the preceding article, where ground state and isomer were treated independently. This higher neutron density was confirmed by the first analysis of the $^{95}$Zr branching (Toukan and Käppeler 1990), by an improved study of the branching at $A = 185,186$ (Käppeler *et al* 1991b), and by the recent measurement of the $^{192}$Pt cross section (Tepe and Käppeler 1991) that allowed for analysing the branching at $A = 191,192$. A weighted average of these individual results yields a mean neutron density of $n_n = (3.4 \pm 1.1) \times 10^8 \mathrm{cm}^{-3}$.

In analysing the temperature-sensitive branchings, the higher neutron density then implied higher temperatures as well. In terms of thermal energies, analysis of the branchings at $A = 151,152$ and $A = 154,155$ yields $kT = 29 \pm 5$ keV (Käppeler *et al* 1990b), in agreement with the analysis of the $^{176}$Lu/$^{176}$Hf pair, where $kT = 27 \pm 6$ keV was obtained (Klay *et al* 1991b).

By simultaneously fitting all branchings by a $\chi^2$-minimization, similar results were obtained [$n_n = (2.0 \pm^{1.4}_{0.8}) \times 10^8$ cm$^{-3}$, $kT = 25 \pm 2$ keV, and an electron density, $n_e = (2.1 \pm 0.2) \times 10^{27}$ cm$^{-3}$] which can also be interpreted as effective values of the respective time-dependent quantities (Beer 1991a). Given the present uncertainties, all branchings can be described with one single set of parameters ($n_n$, $kT$, $n_e$). Though this corresponds to the assumption of the phenomenological approaches, it is clear that the accuracy of the input data needs to be further improved in order to test for the fine structure in the abundance patterns as it is suggested by stellar models.

The predictive power of stellar models has been considerably improved. Detailed calculations of Gallino *et al* (1988) and Gallino (1989), who followed the s-process in low-mass, low-metallicity stars (LMS) using the model by Hollowell and Iben (1989 and references therein) have shown for the first time, that a stellar model is capable of reproducing the s-abundances to better than 10%, and to yield a satisfactory description of the branchings as well. The comparison with the phenomenological approaches (Käppeler *et al* 1990; Beer 1991a) revealed a number of problems that turned out to relate to important phenomena. Aspects related to the neutron pulse structure of the stellar scenario are explained by considering the analytic solution of this double pulse mechanism (Beer 1991b). Another persisting problem refers to the observation that the s-only nuclei involved in branchings seem to be systematically underproduced by the stellar model. If it can be quantified by improved cross sections, this indication of too high a neutron density may have consequences for the treatment of semiconvection at the He–H interface. At this point, it is important to emphasize again the need for experimental $(n,\gamma)$ rates of the radioactive branch points, which may refine the present analyses sufficiently in order to study the convective processes during He shell burning.

So far, the LMS model for the s-process has been demonstrated for a limited range of masses and metallicities. If it can be shown to apply in general, it will represent an appealing tool for a variety of phenomena, e.g. for the interpretation of the s-abundances that are directly observed in the atmospheres of red giants (Busso *et al* 1991) as well as for the isotopic anomalies discovered in SiC inclusions of meteorites (Gallino *et al* 1991), yielding a huge amount of information about the AGB phase of stellar evolution.

*Weak component.* The stellar models for the weak component have been complemented recently by Langer *et al* (1989), Prantzos *et al* (1990), El Eid and Baraffe (1990), and Raiteri *et al* (1991a, 1991b). The latter authors followed the s-process in a $25M_\odot$ mass star up to carbon shell burning. Confirming the previous scenario of core helium burning in massive stars with neutron densities below $10^6$ cm$^{-3}$ and a total exposure of about 0.21 mb$^{-1}$ (section 6.1), it was found that the very high neutron densities during carbon shell burning do affect the final abundances of, for example, $^{70}$Ge, $^{76}$Se, $^{80}$Kr, and $^{82}$Kr. This result was not obvious from the beginning, since the total exposure in carbon shell burning is comparably small (0.06 mb$^{-1}$). In addition, the products of carbon shell burning are mixed back to the interstellar medium only for a narrow range of stellar masses.

For the empirical approach, Beer and Macklin (1988a) have shown that the weak component can be fitted more satisfactorily by adopting a single exposure rather than an exponential distribution of neutron exposures as for the main component. With this solution—that is also equivalent to the situation in massive stars—the abundances of the s-only isotopes $^{70}$Ge and $^{76}$Se could be perfectly fitted. However, this treatment affects the branching at $^{85}$Kr in a somewhat different way compared to the stellar model (Beer *et al* 1991d, Neuberger 1991).

*Isotopic anomalies.* Primitive meteorites are known to contain inclusions of atypical isotopic composition, which may reflect the signature of specific nucleosynthetic events, such as the s-process (section 7.2). Particularly large effects were discovered in SiC grains with respect to C, N, and Si (Zinner *et al* 1989), and these same grains were shown to carry also noble gases with isotopic patterns typical of the s-process (Zinner *et al* 1987; Ott *et al* 1988; Lewis *et al* 1990). Other new anomalies related to the s-process were found for Sr and Ba (Ott 1990). The interpretation of these anomalies via the s-process model for low-mass stars may offer a new source of information on stellar evolution during the AGB phase (Gallino *et al* 1990).

*Primordial nucleosynthesis.* This aspect of neutron capture nucleosynthesis deals with the possibility of a primordial r-process that may take place in the low-density neutron-rich regions predicted as a consequence of the quark hadron phase transition in the early universe (Applegate *et al* 1987; Fuller *et al* 1987; Kawano *et al* 1991). Though it is not directly connected with the s-process, this mechanism should be mentioned here because of the similarity in the temperature range. The measurements of reaction rates for this process concentrate on the mass region below $A=28$. Recent results were reported for $^7$Li$(n,\gamma)^8$Li (Wiescher *et al* 1989; Nagai *et al* 1990) and for $^{14}$C (Beer *et al* 1991).

## References


Alaerts L, Lewis R S, Matsuda J-I and Anders E 1980 *Geochim. Cosmochim. Acta* **44** 189

Alcock C R, Mathews G J and Fuller G M 1987 *Astrophys. J.* **320** 439

Allen B J, Lowenthal G C, Boldeman J W and de Laeter J R 1982 *Neutron Capture Gamma Ray Spectroscopy and Related Topics 1981* (Inst. Phys. Conf. Ser. 62) p 573

Allen B J, Lowenthal G C and de Laeter J R 1981 *J. Phys. G: Nucl. Phys.* **7** 1271

Allen B J, Macklin R L and Gibbons J H 1971 *Adv. Nucl. Phys.* **4** 205

Almeida J and Käppeler F 1983 *Astrophys. J.* **265** 417

Anders E and Ebihara M 1982 *Geochim. Cosmochim. Acta* **46** 2363

Anders E and Grevesse N 1989 *Geochim. Cosmochim. Acta* **53** 197

Applegate J H and Hogan C 1985 *Phys. Rev.* D **30** 3037

Applegate J H, Hogan C J and Scherrer R J 1987 *Phys. Rev.* D **35** 3037

Arnett W D and Thielemann F-K 1985 *Astrophys. J.* **295** 589

Arnould M 1973 *Astron. Astrophys.* **22** 311

—— 1986 *Advances in Nuclear Astrophysics* ed. E. Vangioni-Flam *et al* (Gif sur Yvette: Editions Frontières) p 113

Arnould M, Takahashi K and Yokoi K 1984 *Astron. Astrophys.* **137** 51

Audouze J, Fowler W A and Schramm D N 1972 *Nature Phys. Sci.* **238** 8

Bahcall J N 1961 *Phys. Rev.* **124** 495

Bao Z Y and Käppeler F 1987 *At. Data Nucl. Data Tables* **36** 411

Barden R 1985 *Diploma Thesis* Mainz University, FRG

Baschek B 1979 *Les Élements et leurs Isotopes dans l'Univers* (Liège: University of Liège) p 327

Beer H 1985 *Kernforschungszentrum Karlsruhe internal report*

—— 1986 *Advances in Nuclear Astrophysics* ed. E Vangioni-Flam *et al* (Gif sur Yvette: Editions Frontières) p 375

—— 1988 *Origin and Distribution of the Elements* ed. G J Mathews (Singapore: World Scientific) p 505

—— 1991a *Astrophys. J.* at press

—— 1991b *Astrophys. J.* at press

Beer H and Käppeler F 1980 *Phys. Rev.* C **21** 534

—— 1982 *Neutron Capture Gamma Ray Spectroscopy and Related Topics 1981* (Inst. Phys. Conf. Ser. 62) p 558

Beer H, Käppeler F and Arcorag J-P 1989a *Proc. 5th Workshop on Nuclear Astrophysics* ed. W Hillebrandt and E Müller *Report MPA/PI* MPZ für Physik und Astrophysik, Munich

Beer H, Käppeler F, Reffo G and Venturini G 1983 *Astrophys. Space Sci.* **97** 95

Beer H, Käppeler F, Wisshak K and Ward R A 1981 *Astrophys. J. Suppl.* **46** 295

Beer H and Macklin R L 1982 *Phys. Rev.* C **26** 1404

—— 1985 *Phys. Rev.* C **32** 738

—— 1988 *Astrophys. J.* **331** 1047

—— 1988a *Astrophys. J.* **339** 962

—— 1989 *Astrophys. J.* **339** 962

Beer H and Penzhorn R-D 1987 *Astron. Astrophys.* **174** 323

Beer H, Rupp G and Käppeler F 1991b *Astrophys. J.* at press

Beer H, Voss F and Winters R R 1991a in preparation

Beer H and Walter G 1984 *Astrophys. Space Sci.* **100** 243

Beer H, Walter G and Käppeler F 1989b *Astron. Astrophys.* **211** 245

—— 1989c in preparation

—— 1991d in preparation

Beer H, Walter G and Macklin R L 1985 *Capture Gamma-Ray Spectroscopy and Related Topics 1984* ed. S Raman (New York: AIP) p 778

Beer H, Walter G, Macklin R L and Patchett P J 1984 *Phys. Rev.* C **30** 464

Beer H and Ward R A 1981 *Astron. Astrophys.* **103** 189

Beer H, Wiescher M, Käppeler F, Koehler P E and Görres J 1991c in preparation

Begemann F 1980 *Rep. Prog. Phys.* **43** 1309

Böhm-Vitense E, Nemec J and Proffitt C 1984 *Astrophys. J.* **278** 726

Boothroyd A I and Sackmann I-J 1988 *Astrophys. J.* **328** 632, 641, 653, 671

Bowman C D and Talbert W C 1985 *Neutron Capture Measurements on Radioactive Targets at the Proton Storage Ring, Los Alamos National Laboratory internal report*

Branch D 1987 *Astrophys. J. Lett.* **320** L23

Burbidge E M, Burbidge G R, Fowler W A and Hoyle F 1957 *Rev. Mod. Phys.* **29** 547

Busso M and Gallino R 1985 *Astron. Astrophys.* **151** 205

Busso M, Gallino R, Picchio G and Raiteri C M 1990 *Proc. Confrontation Between Stellar Pulsation and Evolution* (Bologna) May 28–31

Butcher H R 1987 *Nature* **328** 127

—— 1988 *The ESO Messenger* **51** 1

Cameron A G W 1955 *Astrophys. J.* **121** 144

—— 1973 *Space Sci. Rev.* **15** 129

—— 1982 *Essays in Nuclear Astrophysics* ed. C A Barnes *et al* (Cambridge: Cambridge University Press) p 23

Chrien R E 1975 *Nuclear Cross Sections and Technology* Vol. 1 ed. R A Schrack and C D Bowman (Washington DC: NBS) (Special Publication 425) p 139

Clayton D D 1968 *Principles of Stellar Evolution and Nucleosynthesis* (New York: McGraw Hill)

—— 1978 *Astrophys. J.* **224** 1007

—— 1979 *Space Sci. Rev.* **24** 147

—— 1983 *Astrophys. J. Lett.* **271** L101

—— 1984 *Astrophys. J.* **285** 411

—— 1987 *Nature* **329** 397

Clayton D D, Fowler W A, Hull T E and Zimmerman B A 1961 *Ann. Phys.* **12** 331

Clayton D D and Rassbach M E 1967 *Astrophys. J.* **148** 69

Clayton D D and Ward R A 1974 *Astrophys. J.* **193** 397

Conrad J H 1976 *PhD Thesis* University of Heidelberg

Cosner K, Iben I Jr and Truran J W 1980 *Astrophys. J. Lett.* **238** L91

Cosner K and Truran J W 1981 *Astrophys. Space Sci.* **78** 85

Cowan J J, Thielemann F-K and Truran J W 1987 *Astrophys. J.* **323** 543

Despain K H 1980 *Astrophys. J. Lett.* **236** L165

Dominy J F and Wallerstein G 1986 *Astrophys. J.* **310** 371

El Eid M and Baraffe I 1990 *Nuclei in the Cosmos* ed. H Oberhummer and W Hillebrandt (Garching: MPA) p 238

Elze Th W 1984 private communication

Eschner W, Schmidt-Ott W-D, Gippert K L, Runte E, Beer H, Walter G, Kirchner R, Klepper O, Roeckl E and Schardt D 1984 *Z. Phys.* A **317** 281

Fowler W A 1973 *Explosive Nucleosynthesis* ed. D N Schramm and W D Arnett (Austin: University of Texas)

—— 1978 *Proc. 21st Welch Foundation Conf. on Chemical Research* ed. W D Milligan (Houston: Robert A Welch Foundation) p 61

Fowler W A and Meisl C C 1986 *Cosmogonical Processes* ed. W D Arnett *et al* (Singapore: VNU) p 83

Fuller G M, Fowler W A and Newman M J 1982 *Astrophys. J. Suppl.* **48** 279

Fuller G M, Mathews G J and Alcock C R 1987 in *The Origin and the Distribution of the Elements* ed. G J Mathews (Singapore: World Scientific) p 36

Gallino R 1989 in *The Evolution of Peculiar Red Giants* IAU Symposium 106 ed. H L Johnson and B Zuckerman (Cambridge: Cambridge University Press) p 176

Gallino R, Busso M, Picchio G and Raiteri C M 1990 *Nature* **348** 298

Gallino R, Busso M, Picchio G, Raiteri C M and Renzini A 1988 *Astrophys. J. Lett.* **334** L45

Gardner D G, Gardner M A and Hoff R W 1988 *Capture Gamma-Ray Spectroscopy 1987* (Inst. Phys. Conf. Ser. 88) p 315

Gerstenhöfer Th W, Käppeler F, Wisshak K and Reffo G 1991 *Int. Conf. Nuclear Data for Science and Technology* (Jülich, Germany) May 13–15

Goldschmidt V M 1937 *Skr. Nor. Vidensk.-Akad. Oslo I* no 4

Grevesse N 1984 *Phys. Scr.* **T8** 49

—— 1987 private communication

Grevesse N and Meyer J-P 1985 *19th Int. Cosmic Ray Conf., La Jolla*

Habs D, Stephens F S and Diamond R M 1979 *Lawrence Berkeley Laboratory* LBL-8945

Harris M J 1981 *Astrophys. Space Sci.* **77** 357

Hauser W and Feshbach H 1952 *Phys. Rev.* **78** 366

Hodgson P E 1971 *Nuclear Reactions and Nuclear Structure* (Oxford: Clarendon Press)

Hollowell D E and Iben I Jr 1988 *Astrophys. J. Lett.* **333** L25

—— 1989 *Astrophys. J.* **340** 966

Holmes S E, Woosley S E, Fowler W A and Zimmerman B A 1976 *At. Data Nucl. Data Tables* **18** 306
Holweger H 1979 *Les Élements et leurs Isotopes dans l'Univers* (Liège: University of Liège) p 117
—— 1984 update of figure 6 in Holweger (1979) private communication
Howard W M, Mathews G J, Takahashi K and Ward R A 1986 *Astrophys. J.* **309** 633
Huchra J P 1987 *13th Texas Symp. on Relativistic Astrophysics* ed. M P Ulmer (Singapore: World Scientific) p 1
Huey J M and Kohman T P 1972 *Earth Planet. Sci. Lett.* **16** 401
Iben I Jr *Astrophys. J.* 1975a **196** 525
—— *Astrophys. J.* 1975b **196** 549
—— *Astrophys. J.* 1976 **208** 165
—— *Astrophys. J.* 1982 **260** 821
Iben I Jr and Renzini A 1982 *Astrophys. J. Lett.* **259** L79
Iben I Jr and Truran J W 1978 *Astrophys. J.* **220** 980
—— 1983 *Ann. Rev. Astron. Astrophys.* **21** 271
Jaag S 1990 *Diploma Thesis* University of Karlsruhe
Jaag S and Käppeler F 1990 *Nuclei in the Cosmos* ed. H Oberhummer and W Hillebrandt (Garching: MPA) p 250
Käppeler F 1986a *Nucleosynthesis and its Implications on Nuclear and Particle Physics* ed. J Audouze and N Mathieu (Dordrecht: Reidel) p 253
—— 1986b *Advances in Nuclear Astrophysics* ed. E Vangioni-Flam *et al* (Gif sur Yvette: Editions Frontières) p 355
—— 1988 *Capture Gamma-Ray Spectroscopy 1987* (Inst. Phys. Conf. Ser. 88) p 297
—— 1989 *Nuclear Science and Technology* ed. S Igarasi (Tokyo: Saikon) p 1107
Käppeler F, Bao Z Y, Reffo G and Wang S N 1989b *Astrophys. J.* submitted
Käppeler F, Beer H, Wisshak K, Clayton D D, Macklin R L and Ward R A 1982 *Astrophys. J.* **257** 821
Käppeler F, Gallino R, Busso M, Picchio G and Raiteri C 1989a *Astrophys. J.* submitted
—— 1990b *Astrophys. J.* **354** 630
Käppeler F, Jaag S, Bao Z Y and Reffo G 1991b *Astrophys. J.* **366** 605
Käppeler F, Naqvi A A and Al-Ohali M 1987 *Phys. Rev.* C **35** 936
Käppeler F, Schatz G and Wisshak K 1983 *Kernforschungszentrum Karlsruhe* KfK-3472
Käppeler F, Walter G and Matthews G J 1985 *Astrophys. J.* **291** 319
Käppeler F, Wiescher M and Koehler P E 1991a *KfK Internal Report*
Käppeler F, Zhao W R, Beer H and Ratzel U 1990a *Astrophys. J.* **355** 348
Kawano L H, Fowler W A, Kavanagh R and Malaney R A 1991 *Astrophys. J.* **372** 1
Kellogg S E and Norman E B 1986 *Phys. Rev.* C **34** 2248
Klay N, Beer H, Börner H, Hoyler F, Käppeler F, Krusche B, Robinson S, Schatz G and Schreckenbach K 1988 *Kernforschungszentrum Karlsruhe* KfK-4405 p 7
Klay N and Käppeler F 1988 *Phys. Rev.* C **38** 295
Klay N, Käppeler F, Beer H and Schatz G 1991b *Phys. Rev.* C at press
Klay N *et al* 1991a *Phys. Rev.* C at press
Koehler P E, O'Brian H A and Bowman C D 1987 *Los Alamos National Laboratory Report* LA-UR 87-4104
Kopecky J 1982 *Neutron Capture Gamma-Ray Spectroscopy and Related Topics 1981* (Inst. Phys. Conf. Ser. 62) pp 423, 426
Kratz K-L, Thielemann F-K, Hillebrandt W, Möller P, Harms V, Wöhr A and Truran J W 1988 *Capture Gamma-Ray Spectroscopy 1987* (Inst. Phys. Conf. Ser. 88) p 331
Kwiatkowski M, Zimmermann P, Biémont E and Grevesse N 1984 *Astron. Astrophys.* **135** 59
Lamb S A, Howard W M, Truran J W and Iben I Jr 1977 *Astrophys. J.* **217** 213
Lambert D L 1989 *Evolution of Peculiar Red Giant Stars* ed. H R Johnson (Cambridge: Cambridge University Press)
Langer N, Arcoragi J-P and Arnould M 1989 *Astron. Astrophys.* **210** 187
Laval M, Moszynski M, Allemand R, Cormoreche E, Guinet P, Odru R and Vacher J 1983 *Nucl. Instrum. Methods* **206** 169
Lederer C M and Shirley V S 1978 *Table of Isotopes* (New York: Wiley)
Lesko K T, Norman E B, Larimer R M, Bacelar J C and Beck E M 1989 *Phys. Rev.* C **39** 619
Lesko K T, Norman E B, Larimer R M and Sur B 1991 *Phys. Rev.* C at press
Lesko K T, Norman E B, Moltz D M, Larimer R M, Crane S G and Kellogg S E 1986 *Phys. Rev.* C **34** 2256
Lewis R, Amari S and Anders E 1990 *Nature* **348** 293
Lugmair G W, Shimamura T, Lewis R S and Anders E 1983a *Science* **222** 1015

—— 1983b *Lunar Planet. Sci.* **14** 448
Macklin R L 1982 *Nucl. Sci. Eng.* **81** 520
—— 1985a *Nucl. Sci. Eng.* **89** 79
—— 1985b *Astrophys. Space Sci.* **115** 71
Macklin R L and Gibbons J H 1967 *Phys. Rev.* **159** 1007
Macklin R L, Halperin J and Winters R R 1977 *Astrophys. J.* **217** 222
Malaney R A 1986 *Advances in Nuclear Astrophysics* ed. E Vangioni-Flam *et al* (Gif sur Yvette: Editions Frontières) p 407
—— 1987 *Astrophys. J.* **321** 832
Malaney R A and Fowler W A 1989 *Mon. Not. R. Astron. Soc.* **237** 67
Malaney R A and Lambert D L 1988 *Mon. Not. R. Astron. Soc.* **235** 695
Mathews G J and Käppeler F 1984 *Astrophys. J.* **286** 810
Mathews G J and Schramm D N 1988 *Astrophys. J. Lett.* **324** L67
Mathews G J, Takahashi K, Ward R A and Howard W M 1986 *Astrophys. J.* **302** 410
Mathews G J and Ward R A 1985 *Rep. Prog. Phys.* **48** 1371
Mayerhofer U, Hlawatsch G and von Egidy T 1987 private communication
McClure R D 1984 *Astrophys. J.* **280** 231
Merrill P W 1952 *Science* **115** 484
Meyer J-P 1985 *Astrophys. J. Suppl.* **57** 173
Mould J and Reid N 1987 *Astrophys. J.* **321** 156
Moxon M C and Rae E R 1963 *Nucl. Instrum. Methods* **24** 445
Nagai Y, Igashira M, Kitazawa H, Mukai N, Uesawa F, Takeda K, Ando T, Osaki T, Hirasawa J, Shibata S and Kubono S 1990 *Int. Conf. Capture Gamma-Ray Spectroscopy and Related Topics* (Asilomar, Pacific Grove, California) October 14–19
Nagai Y, Igashira M, Takeda K, Mukai N, Motoyama S, Uesawa F, Kitazawa H and Fuketa T 1991 *Astrophys. J.* **327** 683
Naumann R A, Burke D, Ravn H L, Hlawatsch G, von Egidy T and de Boer J 1987 *Nucl. Instr. Meth.* B **26** 59
Nemeth Zs and Käppeler F 1991 *Nuclear Astrophysics* ed. E Müller and W Hillebrandt (Garching: MPA)
Neuberger D 1991 *Diploma Thesis* University of Karlsruhe
Neuberger D, Tepe M and Käppeler F 1991 *Int. Conf. on Nuclear Data for Science and Technology* (Jülich, Germany) May 13–15
Newman M J 1973 *MSc Thesis* Rice University, Houston
—— 1978 *Astrophys. J.* **219** 976
Norman E B and Kellogg S 1985 *Capture Gamma-Ray Spectroscopy and Related Topics 1984* ed. S Raman (New York: AIP) p 753
Ott U 1990 *Nuclei in the Cosmos* ed. H Oberhummer and W Hillebrandt (Garching: MPA) p 151
Ott U, Begemann F, Yang J and Epstein S 1988 *Nature* **332** 700
Palme H 1985 data from MPI für Chemie, Abteilung Kosmochemie, Mainz, FRG, private communication
Palme H, Suess H E and Zeh H D 1981 *Landolt Börnstein New Series VI* vol. 2 ch. 3, 4
Patchett P J 1983 *46th Annual Meeting of the Meteoritical Soc.* (Mainz, FRG) September 5–9
Perrone F A and Clayton D D 1971 *Astrophys. Space Sci.* **11** 451
Peters J G 1968 *Astrophys. J.* **154** 224
Petrov Yu V and Slyakhter A I 1984 *Astrophys. J.* **278** 385
—— 1988 *Astrophys. J.* **327** 294
Picchio G, Busso M, Gallino R and Raiteri C M 1988 *Mass Outflow from Stars and Galactic Nuclei* ed. L Bianchi and R Gilmozzi (Dordrecht: Kluwer Academic.) p 279
Prantzos N, Arnould M and Arcoragi J-P 1987 *Astrophys. J.* **315** 209
Prantzos N, Hashimoto M and Nomoto K 1990 *Astron. Astrophys.* **234** 211
Raiteri C M, Busso M, Gallino R and Picchio G 1991b *Astrophys. J.* **371** 665
Raiteri C M, Busso M, Gallino R, Picchio G and Pulone L 1991a *Astrophys. J.* **367** 228
Ratynski W and Käppeler F 1988 *Phys. Rev.* C **37** 595
Ratzel U, Beer H and Käppeler F 1989 in preparation
Rauch M 1987 *Diploma Thesis* University of Mainz, FRG
Rayet M, Prantzos N and Arnould M 1988 *Origin and Distribution of the Elements* ed. G J Mathews (Singapore: World Scientific) p 625
Reffo G 1980 *Parameter Systematics for Statistical Theory Calculations of Neutron Reaction Cross Sections IAEA-SMR-43* (Vienna: International Atomic Energy Agency) p 205
Reffo G and Fabbri F 1991 private communication

Reffo G, Fabbri F, Wisshak K and Käppeler F 1982 *Nucl. Sci. Eng.* **80** 630

—— 1983 *Nucl. Sci. Eng.* **83** 401

Ross J E and Aller L H 1976 *Science* **191** 1223

Runte E, Schmidt-Ott W-D, Eschner W, Rosner I, Kirchner R, Klepper O and Rykaczewski K 1987 *Z. Phys.* A **328** 119

Russell H N and Adams W S 1928 *Astrophys. J.* **68** 9

Sandage A 1982 *Astrophys. J.* **252** 553

Sandage A and Tammann G A 1984 *Nature* **307** 326

Schatz G 1983 *Astron. Astrophys.* **122** 327

Schwarzschild M and Härm R 1967 *Astrophys. J.* **150** 961

Seeger P A, Fowler W A and Clayton D D 1965 *Astrophys. J. Suppl.* **11** 121

Shaw P B and Clayton D D 1967 *Phys. Rev.* **160** 1193

Shaw P B, Clayton D D and Michel F C 1965 *Phys. Rev.* **140** B1433

Simpson J A 1983 *Ann. Rev. Nucl. Part. Sci.* **33** 323

Smith V V 1988 *Origin and Distribution of the Elements* ed. G J Mathews (Singapore: World Scientific) p 535

Smith V V and Lambert D L 1986 *Astrophys. J.* **311** 843

—— 1988 *Astrophys. J.* at press

Smith V V and Wallerstein G 1983 *Astrophys. J.* **273** 742

Srinivasan B and Anders E 1978 *Science* **201** 51

Stecher-Rasmussen F, Abrahams K, Kopecky J, Lindner J and Polak P 1988 *Capture Gamma-Ray Spectroscopy 1987* (Inst. Phys. Conf. Ser. 88) p 754

Suess H E and Urey H C 1956 *Rev. Mod. Phys.* **28** 53

Suess H E and Zeh H D 1973 *Astrophys. Space Sci.* **23** 173

Takahashi K, Mathews G J, Ward R A and Becker S A 1986 *Nucleosynthesis and its Implications on Nuclear and Particle Physics* ed. J Audouze and N Mathieu (Dordrecht: Reidel) p 285

Takahashi K and Yokoi K 1983 *Nucl. Phys.* A **404** 578

—— 1987 *At. Data Nucl. Data Tables* **36** 375

Tepe M and Käppeler F 1991 *KfK Report* KFK-4875 p 1

Thielemann F-K and Truran J W 1986 *Galaxy Distances and Deviations from Universal Expansion* ed. B F Madore and R B Tully (Dordrecht: Reidel) p 185

Tinsley B A 1980 *Fund. Cosmic Phys.* **5** 411

Tomkin J and Lambert D L 1983 *Astrophys. J.* **273** 722

Toukan K A and Käppeler F 1989 *Astrophys. J.* at press

—— 1990 *Astrophys. J.* **348** 357

Trautvetter H P *et al* 1986 *Z. Phys.* A **323** 1

Trimble V 1975 *Rev. Mod. Phys.* **47** 877

Truran J W and Iben I Jr 1977 *Astrophys. J.* **216** 797

Twin P J 1983 *Proc. Int. Conf. on Nuclear Physics, Florence, Italy* (Bologna: Tipografia Compositori) p 527

Ulrich R K 1973 *Explosive Nucleosynthesis* ed. D N Schramm and W D Arnett (Austin: University of Texas) p 139

Unsöld A 1979 *Les Élements et leurs Isotopes dans l'Univers* (Liège: University of Liège) p 7

VandenBerg D A 1983 *Astrophys. J. Suppl.* **51** 29

—— 1985 *Astrophys. J. Suppl.* **58** 711

Veres A and Pavlicsek I 1970 *Acta Phys. Hung.* **28** 419

Vogt E W 1969 *Advances in Nuclear Physics* vol. 1 ed. M Baranger and E Vogt (New York: Plenum) p 261

Walter G and Beer H 1982 *Kernforschungszentrum Karlsruhe* KfK-3327

Walter G, Beer H, Käppeler F and Penzhorn R-D 1986b *Astron. Astrophys.* **155** 247

Walter G, Beer H, Käppeler F, Reffo G and Fabbri F 1986a *Astron. Astrophys.* **167** 186

Ward R A 1977 *Astrophys. J.* **216** 540

—— 1980 private communication

—— 1981 *Astron. Astrophys.* **97** 154

Ward R A and Beer H 1981 *Astron. Astrophys.* **103** 189

Ward R A and Clayton D D 1982 unpublished

Ward R A and Fowler W A 1980 *Astrophys. J.* **238** 266

Ward R A and Newman M J 1978 *Astrophys. J.* **219** 195

Ward R A, Newman M J and Clayton D D 1976 *Astrophys. J. Suppl.* **31** 33

Wasserburg G J and Papanastassiou D A 1982 *Essays in Nuclear Astrophysics* ed. C A Barnes *et al* (Cambridge: Cambridge University Press) p 77
Wasson J T 1985 *Meteorites—Their Record of Early Solar System History* (New York: Freeman)
Watanabe Y, Mukoyama T and Katano R 1981 *Phys. Rev.* C **23** 695
Weigert A 1966 *Z. Astrophys.* **64** 395
Wiedenbeck M E and Greiner D E 1980 *Astrophys. J. Lett.* **239** L139
Wiescher M, Steininger R and Käppeler F 1989 *Astrophys. J.* **344** 464
Winters R R, Carlton R F, Harvey J A and Hill N W 1986b *Phys. Rev.* C **34** 840
Winters R R, Käppeler F, Wisshak K, Mengoni A and Reffo G 1986a *Astrophys. J.* **300** 41
Winters R R and Macklin R L 1982 *Phys. Rev.* C **25** 208
Wisshak K, Guber K, Käppeler F, Krisch J, Müller H, Rupp G and Voss F 1990a *Nucl. Instr. Meth.* A **292** 595
Wisshak K and Käppeler F 1983 *Nucl. Sci. Eng.* **85** 251
—— 1984 *Nucl. Instrum. Methods* **227** 91
Wisshak K, Käppeler F and Schatz G 1984 *Nucl. Instrum. Methods* **221** 385
Wisshak K, Käppeler F, Voss F and Guber K 1989 *IEEE Trans. Nucl. Sci.*, **NS-36** 101
Wisshak K, Voss F, and Käppeler F 1991 *Nuclear Astrophysics* ed. E Müller and W Hillebrandt (Garching: MPA)
Wisshak K, Voss F, Käppeler F and Reffo G 1990b *Phys. Rev.* C **42** 1732
Wood P R 1985 *Cool Stars with Excesses of Heavy Elements* ed. M Jaschek and P C Keenan (Dordrecht: Reidel) p 357
Woosley S E, Fowler W A, Holmes J A and Zimmerman B A 1978 *At. Data Nucl. Data Tables* **22** 371
Yokoi K and Takahashi K 1981 *Proc. 4th Int. Conf. on Nuclei Far from Stability (Helsingør) CERN rep 81-09* p 351
—— 1983 *Nature* **305** 198
—— 1985 *Kernforschungszentrum Karlsruhe* KfK-3849
Yokoi K, Takahashi K and Arnould M 1983 *Astron. Astrophys.* **117** 65
—— 1985 *Astron. Astrophys.* **145** 339
Youssef N H and Khalil N M 1989 *Astron. Astrophys.* **208** 271
Zinner E, Tang M and Anders E 1987 *Nature* **330** 730
—— 1989 *Geochim. Cosmochim. Acta* **53** 1235
Zook A C 1978 *Astrophys. J. Lett.* **221** L113

Rep. Prog. Phys. **53** (1990) 837–915. Printed in the UK

# Physics of white dwarf stars

Detlev Koester and Ganesar Chanmugam

Department of Physics and Astronomy, Louisiana State University, Baton Rouge, LA 70803, USA

**Abstract**

White dwarf stars, compact objects with extremely high interior densities, are the most common end product in the evolution of stars. In this paper we review the history of their discovery, and of the realisation that their structure is determined by the physics of the degenerate electron gas. Spectral types and surface chemical composition show a complicated pattern dominated by diffusion processes and their interaction with accretion, convection and mass loss. While this interaction is not completely understood in all its detail at present, the study may ultimately lead to important constraints on the theory of stellar evolution in general. Variability, caused by non-radial oscillations of the star, is a common phenomenon and is shown to be a powerful probe of the structure of deeper layers that are not directly accessible to observation. Very strong magnetic fields detected in a small fraction of white dwarfs offer a unique opportunity to study the behaviour of atoms under conditions that cannot be simulated in terrestrial laboratories.

This review was received in its present form in December 1989.

0034-4885/90/070837+79$14.00 

**Contents**

## 1. Introduction

The theory of stellar structure knows three final states for a star: black holes, neutron stars and white dwarfs (for a recent review see e.g. Shapiro and Teukolsky 1983). Candidates for black holes have been proposed in certain binary star systems, at the centres of galaxies (possibly even our own Milky Way), and as the energy sources in quasi-stellar objects, but to date their actual existence has not been proven. The existence of neutron stars, extremely compact objects with masses similar to the sun but radii of about 10 km, was proven with the discovery of pulsars and their interpretation, and we may at the present time witness the birth of a new object of this class as the remnant of the supernova explosion in the Large Magellanic Cloud in 1987. Neutron stars and black holes are exotic objects with spectacular properties and certainly deserve the great attention they get in the current astrophysical research.

The vast majority (of the order of 90%, see Koester and Weidemann 1980) of all stars, however, including our sun, will finally evolve towards the third final state, that of white dwarfs. White dwarfs have about the same masses as neutron stars, but sizes of the order of the earth; typical interior densities are therefore $\sim 10^6\ \mathrm{g\,cm^{-3}}$. The evolution towards white dwarfs is the dominant channel in galaxies, which determines the evolution and final fate of most of their mass. For the same reasons the population of white dwarfs contains a wealth of information on the evolution of individual stars from birth to death, on the previous history of the galaxy and on the rate of star formation (Koester and Weidemann 1980).

According to our current understanding, the primary parameter that determines the final fate of a star is its mass at birth. All stars below a critical value, estimated to be in the range of 6 to 8 solar masses ($M_\odot$), will finally become white dwarfs, while the more massive stars will become neutron stars or black holes or even possibly be totally disrupted. Because the maximum mass of a white dwarf ($\simeq 1.4\ M_\odot$) is much smaller than this value a large fraction of the original stellar mass must be lost to the surrounding interstellar medium. This mass loss is indeed confirmed by observations of stars in the progenitor stages, although a theoretical derivation is still missing and an estimate of total mass loss during the life of a star is difficult.

Observations of white dwarfs can at least partly fill this gap. By determining the masses of white dwarfs and tracing their evolution back to their progenitors, it is possible at least in a statistical sense to estimate the total mass loss. This provides valuable constraints for theoretical attempts to understand this phenomenon as well as for calculations of stellar evolution.

On the other hand, the difference between initial and final mass must have been given back, almost unchanged, to the interstellar medium and is available for the formation of new stars. The study of white dwarfs therefore provides important insights into the mass budget of the galaxy: how much is locked up forever in the interior of white dwarfs and how much is given back to be used again in a new cycle?

Although the white dwarf state is a final state in the sense that no nuclear energy generation occurs, the observable white dwarfs still evolve. They are 'born' with high luminosities ($L$ = total radiative energy loss per second), and gradually, over 5 to 10 billion years fade away into invisibility. The faintest observed white dwarfs are therefore very old and contain information about the early phases of our galaxy. Recently, it was even proposed that these observations indicate that star formation in the galactic disk began about 9 billion years ago (Winget *et al* 1987a).

Besides these implications for the galaxy, however, white dwarfs are also extremely interesting objects in their own right. At their extremely high densities the electrons become degenerate and their quantum mechanical nature, through the Pauli principle, determines the equation of state, the structure of white dwarfs and the existence of a limiting mass.

These extreme densities cannot be simulated in terrestrial laboratories. The properties of matter under these conditions must therefore be calculated theoretically, and to the degree that the predictions can be confirmed by observations white dwarfs can be considered to be a 'physics laboratory' for matter at extreme densities and pressures.

The case for white dwarfs as physical laboratories can be made even more convincing. At the present time 26 white dwarfs are known to have magnetic fields in the range $10^6$ to $10^9$ G (Schmidt 1989). By comparison, the largest steady field that can be produced in the laboratory is only $10^5$ G because for higher fields the magnetic pressure exceeds that of steel. Transient fields up to 10 MG may be produced in implosions. Thus white dwarfs provide a unique opportunity to study the effect of strong magnetic fields in great detail from the ultraviolet to the infrared. These observations have provided the main motivation in recent years to extend theoretical calculations of energy eigenvalues and oscillator strengths in hydrogen atoms into this very difficult intermediate range, where the Coulomb forces are of the same order of magnitude as the magnetic interaction (Henry and O'Connell 1985, Wunner *et al* 1987).

Pulsating stars in general have provided extremely important tests for theoretical calculations of stellar structure and evolution. This is also true for white dwarfs: several classes of pulsating white dwarfs are now known (Winget 1986, Cox 1986). In white dwarfs, the oscillations are non-radial, complicating the analysis considerably as compared to radial oscillations, but nevertheless providing important information about the structure of the deeper layers below the visible surface. These observations and their subsequent analysis with theoretical models have, for example, confirmed that the outer, non-degenerate envelopes of white dwarfs show a layered structure: in most cases a very thin almost pure hydrogen layer floating on top of a helium layer. It has even been possible to estimate the thickness of these layers, although at present there exists some controversy regarding the interpretation of the data.

The locations of the 'instability strips' in the Hertzsprung–Russell diagram (HRD) have also been used to constrain the free parameter (mixing-length/pressure scale height) in the mixing-length approximation that is used in stellar evolution calculations to model the convective energy transport (Bradley *et al* 1989).

Oscillation periods depend on the structure of the pulsating star. In principle, periods and their rates of change can be measured with very high accuracy, although in practice the observations are often complicated for white dwarfs because many periods are present simultaneously and the necessary frequency resolution may require long observing runs. Nevertheless, it may become possible in the near future to compare the structural changes in a pulsating white dwarf, derived from observed period changes, with theoretical calculations for the evolution and thus for the first time directly observe stellar evolution in white dwarfs (Kawaler and Hansen 1989).

The spectra emitted by white dwarf atmospheres show a quite varied pattern, unlike any other stellar group. The majority have lines of hydrogen only, extremely broadened by the high gravitational fields and correspondingly high pressures in the atmospheres; others show only the spectral lines of neutral helium, or only the Swan bands of the $C_2$ molecule. Yet, others show only indications of calcium, magnesium and iron, especially in the ultraviolet spectra. Detailed analysis, however, reveals an underlying simplicity: with very few exceptions the main constituent of the atmosphere is either hydrogen or helium, the two most abundant elements in the universe. Only traces of other elements are present (Koester 1987a).

The basic mechanism responsible for the unusual composition was proposed by Schatzman (1958). Under the combined influence of gravitational and electric fields in the outer layers, elements with different atomic weights separate by diffusion: heavy elements move downwards, while the lightest one present (either hydrogen or helium) floats to the top of the atmosphere. The time scales for this diffusion process are always short compared to the evolutionary time scales of white dwarfs (Paquette *et al* 1986a), we can therefore in most cases assume that the equilibrium state has been reached in observed objects. This simple picture is, however, complicated by the action of competing processes such as convection, stellar winds and accretion. The study of the very complicated interaction between these processes and its implication for the observed white dwarf population is currently at the forefront of active research. A basic understanding has already emerged, but many problems remain to be solved (Koester 1987a).

Progress in this and other fields of white dwarf research is of course intimately connected to progress in observational methods and instrumentation. The last decade has seen tremendous technical advances due to the construction of large telescopes (white dwarfs are very faint objects), the availability of new electronic detectors (CCD), and especially satellite projects that made accessible the ultraviolet and soft X-ray range of the electromagnetic spectrum (International Ultraviolet Explorer (IUE), Einstein, Exosat). In this paper, however, we will concentrate on the physics that govern the interior structure and outer layers of white dwarfs. The coverage of the wealth of observational data will be very incomplete; the reader interested in this aspect is referred to the excellent reviews by, for example, Liebert (1980), Sion (1986), the recent conference proceedings of IAU Colloquium No 114 (Wegner 1989a), Weidemann (1990) and D'Antona and Mazzittelli (1990).

A good introduction to the basic physics of white dwarfs can be found in chapters 2 to 4 of the textbook by Shapiro and Teukolsky (1983), which places the white dwarfs in the more general context of 'compact objects'. A very useful source will certainly be the monograph on white dwarfs currently being prepared by Van Horn and Liebert (1989).

We have also restricted the scope of this paper to the discussion of single white dwarfs. The first two white dwarfs to be discovered (see below) were members of binary systems, and today many more are known. But in these cases the binaries are wide systems, meaning that the structure and evolution of the white dwarf is not significantly influenced by the presence of a companion. We wish to exclude, however, cataclysmic variables which include novae, where the interaction between the two stars, normally by the transfer of matter and the formation of an accretion disk surrounding the white dwarf, dominates the evolution and the observational features. Recent reviews of this large and interesting field can be found in Patterson (1984), Bath (1985) and Mauche (1990).

## 2. History: Early observations and theoretical interpretation

### 2.1. Discovery

The first realisation that a new class of stars quite different from 'ordinary' stars exists can be traced back to 1910. Schatzman (1958) gives a lively description of that event in the words of Henry Norris Russell. What Russell had detected is shown in figure 1, one of the first examples of the HRD, which was to become one of the most important tools in astrophysics. This particular version of the diagram is of the 'observational' form, it uses the observed astronomical parameters of spectral type and absolute magnitude. This diagram is topologically equivalent to a more physical diagram, although the transformations are sophisticated and non-linear, where the vertical axis corresponds to the total luminosity of the star (increasing upwards) and the horizontal axis to a surface temperature (increasing to the left).

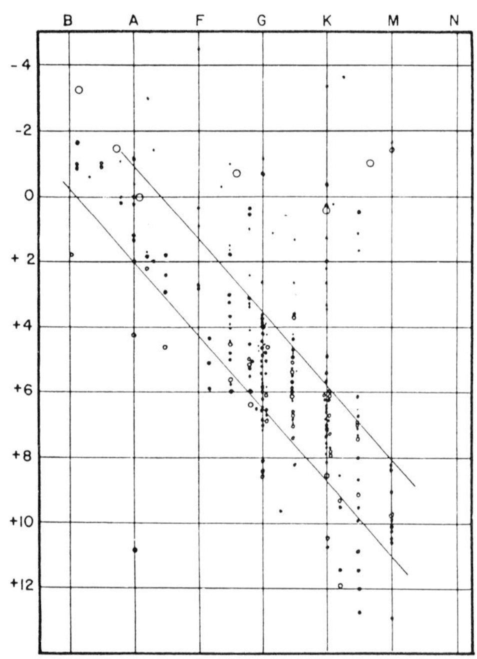


**Figure 1.** Absolute magnitude versus spectral type for stars of known distance. After Russell (1914), reproduced from Lang and Gingerich (1979).

This surface temperature is defined in astronomy as the effective temperature $T_{\mathrm{eff}}$ by means of Stefan's law, so that

$$L = 4\pi R^2 \sigma T_{\mathrm{eff}}^4 \tag{2.1}$$

where $L$ is the luminosity, $R$ the radius of the star and $\sigma$ the radiation constant from Stefan's law. Thus $T_{\mathrm{eff}}$ is a measure of the energy flux at the surface and not a real temperature, but it nevertheless constitutes a useful measure of the atmospheric temperature of the star. From the definition it is obvious that stars in the lower left part of this diagram are hot (white) and have small radii, while stars in the upper right corner are large and red ('red giants'). The position of our sun in this diagram would be near (G,4.8), close to the centre of the figure. By studying this diagram, Russell noticed, that all white stars of classes B and A are bright, far exceeding the luminosity of the sun; all very faint stars are red — the single exception being the star now known under the name 40 Eridani B, a member of the triple stellar system 40 Eridani. The position implies, as noted above, a very small radius, hence the name 'white dwarf' for the new class.

A similar puzzle was provided by the star Van Maanen 2, for which Van Maanen (1913) had determined a spectral type that made the star appear too hot for its faint luminosity. Though it was clear that these new objects had the size of a planet, their masses were not known at that time. Further physical insight into the problem was obtained from the third of the so-called 'classical white dwarfs' to be discovered, Sirius B, the binary companion of the bright star Sirius.

Between 1834 and 1844 the mathematician and astronomer Bessel had observed oscillations in the apparent path of Sirius across the sky, which had finally enabled him to show that Sirius was a double star. The companion, Sirius B, was detected visually by the American telescope maker Alvan Clark in 1861, and the first orbit determination published shortly thereafter. In 1910 the mass of the companion was determined as 0.94 $M_\odot$ (Boss 1910; a more recent value is 1.053 $M_\odot$ (Gatewood and Gatewood 1978). From the spectral type (Adams 1915) an effective temperature of 8000 K was assigned, which together with the known luminosity provided the first estimate of the radius according to equation (2.1) and a density of $5 \times 10^4$ g cm$^{-3}$, to which it was considered proper, as Eddington (1926) puts it in his famous book, to append the statement 'this is absurd'. The derived effective temperature was a severe underestimate, due to the observational problems posed by the close proximity of the much brighter Sirius A. A more modern value is 27 000 K (Thejll and Shipman 1986), implying a mean density of $3 \times 10^6$ g cm$^{-3}$.

An independent test of the high density became feasible around 1920, when Sirius A and B reached the greatest apparent separation in their orbit. Einstein (1907) had predicted the redshift of light originating in a strong gravitational field, depending on the stellar parameters as $\Delta\lambda/\lambda = GM/Rc^2$. With the values of mass and radius for Sirius B assumed at that time, the expected redshift of H$\alpha$ was $\sim$ 0.4 Å, corresponding to a velocity of recession via the Doppler effect of $\sim$ 25 km s$^{-1}$. When Adams (1925) reported the detection of a shift equivalent to $\sim$ 19 km s$^{-1}$ this was considered as a triumphant confirmation of general relativity as well as of the high densities in white dwarfs.

Somewhat embarrassingly, the modern values for $M$ and $R$ require a much larger shift, equivalent to $\simeq$ 85 km s$^{-1}$, which was confirmed by later observations (Greenstein *et al* 1971). Greenstein *et al* (1985) have recently given an excellent historical account of the circumstances surrounding Adam's measurement and their own subsequent attempts to determine the shift.

Thus in 1920 only three white dwarfs were known, but it was already evident to Eddington (1926) that they must be among the most common objects in our galaxy. All three were very close to the sun that is within 5 pc (1 pc = 3.26 light years), and only their intrinsic faintness limited further observations. By 1939, a total of 18 was known, and by 1950 the number had increased to 111 (Schatzman 1958).

Systematic searches for white dwarfs have capitalised on two properties of the class. Because white dwarfs are intrinsically faint, they are discovered only relatively close to the sun and should therefore show large tangential motions with respect to the sun ('proper motion' in astronomical nomenclature). Searches for this motion were initiated in 1929 by Luyten (Bruce Proper Motion Survey) and the results published in a series of papers (e.g. Luyten 1970, 1977 and references therein). A similar search was conducted at the Lowell Observatory (e.g. Giclas *et al* 1980, and references to earlier papers). White dwarfs are also 'too blue' for their luminosities, when compared with ordinary main sequence stars. This property was first exploited by Humason and Zwicky (1947) and more recently with the Palomar–Green Survey of faint blue objects (Green *et al* 1986). Major advances in the understanding of white dwarfs have been made by the extensive spectroscopic and photometric observational work of Greenstein and Eggen (cf Greenstein 1984a, Eggen 1985 and references therein).

An up-to-date list of 1279 spectroscopically identified white dwarfs is the catalogue by McCook and Sion (1987). The study of white dwarfs has thus come a long way from three exotic objects to an important part of stellar astrophysics, as documented very recently at the IAU Colloquium No 114 on 'White Dwarfs' (Wegner 1989a).

### *2.2. The degenerate electron gas*

Taking the current best estimates for the mass and radius of Sirius B, its average density is $\sim 3 \times 10^6$ g cm$^{-3}$. Assuming for simplicity an interior composition of pure carbon we can estimate a typical distance $r_c$ between carbon nuclei from $(4\pi/3)n_c r_c^3 = 1$, where $n_c$ is the number density of heavy particles. The value is $r_c = 1.2 \times 10^{-10}$ cm, much less than the Bohr radius even for a carbon ion with only one remaining electron. There is thus no room for bound orbits, and the matter in white dwarfs must be completely 'pressure ionised'. Similarly a typical distance between the electrons is $r_e \simeq 6.4 \times 10^{-11}$ cm, which is smaller than the thermal de Broglie wavelength of the electrons for all but extremely high temperatures ($> 10^9$ K), so that the correct thermodynamic description of matter under these conditions requires quantum mechanics.

In 1926, it was shown that the electrons obey what is now called Fermi–Dirac statistics (Fermi 1926, Dirac 1926). This rule was immediately applied to solve the white dwarf puzzle by Fowler (1926), who realised that the pressure supporting these stars against gravity is supplied by the (almost) completely degenerate electrons. This pressure remains even at zero temperature, due to zero-point motion, thus establishing that white dwarfs are really stable final configurations and opening up the field of study of zero-temperature configurations.

Anderson (1929) and Stoner (1930) noticed that at the extreme densities in the most massive white dwarfs the velocities must become relativistic and calculated relativistic corrections to Fowler's equation of state. Chandrasekhar (1931) discovered that the relativistic 'softening' of the equation of state leads to the existence of a limiting mass, now called the 'Chandrasekhar limit', where gravitational forces overwhelm the pressure support, and no stable white dwarf can exist, a conclusion independently arrived at by Landau (1932). This provided the first clue to a fundamental difference in the evolution and final stages of low and high mass stars.

During the next eight years Chandrasekhar worked out the complete theory of the equation of state (EOS) at all densities, including the effects of finite temperatures and special relativity, and the resulting structure of zero-temperature models and their mass–radius relation (Chandrasekhar 1939).

## 3. Equation of state, interior structure and mass–radius relation

### 3.1. Equation of state for the ideal electron gas

Derivations of the EOS for ideal, non-interacting electrons are given in many sources. In this section we loosely follow the approach of Cox and Giuli (1968) where, in a thorough discussion, many tables of useful functions and further references can be found.

In statistical physics a complete description of a system in equilibrium is provided by the dimensionless distribution function in phase space, $f(\boldsymbol{x}, \boldsymbol{p})$, defined by

$$\frac{\mathrm{d}n(\boldsymbol{x}, \boldsymbol{p})}{g\mathrm{d}^3x\mathrm{d}^3p/h^3} = f(\boldsymbol{x}, \boldsymbol{p}). \tag{3.1}$$

Here $\mathrm{d}n$ denotes the number of electrons in the volume element of phase space $\mathrm{d}^3x\mathrm{d}^3p$, $g$ the number of internal degrees of freedom (two for electrons), and the denominator on the left is the number of 'cells' corresponding to the phase space volume. The function $f$ thus gives the average occupation number of a cell in phase space. From this distribution function all interesting quantities (number density $n$, pressure $p$, energy density $u$) can be derived using the relativistic relation between momentum and kinetic energy

$$\varepsilon(p) = mc^2\left[\sqrt{1 + (p/mc)^2} - 1\right] \tag{3.2}$$

and the distribution function for an ideal Fermi gas in equilibrium

$$f(\varepsilon) = \frac{1}{\exp\left[(\varepsilon - \mu)/kT\right] + 1}. \tag{3.3}$$

Here $k$ is Boltzmann's constant and $\mu$ the chemical potential. For sufficiently low particle densities and high temperatures, $f(\varepsilon)$ reduces to the Maxwell–Boltzmann distribution. For completely degenerate fermions ($T \to 0$), $\mu$ is called the Fermi energy $\varepsilon_\mathrm{F}$, and

$$f(\varepsilon) = \begin{cases} 1 & \varepsilon \le \varepsilon_\mathrm{F} \\ 0 & \varepsilon > \varepsilon_\mathrm{F}\ . \end{cases} \tag{3.4}$$

The momentum corresponding to $\varepsilon_\mathrm{F}$ according to (3.2) is called the Fermi momentum $p_\mathrm{F}$. The EOS can be derived from the distribution function in parametrised form through functions of two dimensionless variables, the 'degeneracy parameter' $\eta = \mu/kT$ and the 'relativity parameter' $\beta = kT/mc^2$. In terms of these variables

$$P = \frac{16\pi\sqrt{2}}{3}\frac{m^4c^5}{h^3}\beta^{5/2}\left[F_{3/2}(\eta, \beta) + (\beta/2)F_{5/2}(\eta, \beta)\right] \tag{3.5}$$

$$n = 8\pi\sqrt{2}\frac{m^3c^3}{h^3}\beta^{3/2}\left[F_{1/2}(\eta, \beta) + \beta F_{3/2}(\eta, \beta)\right] \tag{3.6}$$

$$u = 8\pi\sqrt{2}\frac{m^4c^5}{h^3}\beta^{5/2}\left[F_{3/2}(\eta, \beta) + \beta F_{5/2}(\eta, \beta)\right] \tag{3.7}$$

where $P$, $n$ and $u$ are the pressure, number density and energy density. The functions $F_k$ appearing here are defined by the integrals

$$F_k(\eta, \beta) = \int_0^\infty \frac{x^k\left(1 + (\beta/2)x\right)^{1/2}\mathrm{d}x}{\exp(\eta + x) + 1} \tag{3.8}$$

and can in general only be evaluated numerically. Cox and Giuli (1968) give tables for some of the functions as well as very useful expansions for the different physical regimes (high versus low degeneracy, relativistic versus non-relativistic, etc).

Fortunately, it is almost always possible to work with these much simpler expansions, because in real white dwarfs low degeneracy always implies the non-relativistic regime, whereas relativistic effects are only important at very high degeneracy. The most significant simplification results from the assumption of complete degeneracy ($T \to 0$), the realm of 'zero-temperature stars'. In this case $\beta$ is no longer useful as a measure of the effects of relativity, and could be replaced by $\eta\beta = \mu/mc^2$ with the limit $\varepsilon_F/mc^2$ for complete degeneracy. It is customary, however, to use $x = p_F/mc$ instead, related to $\eta\beta$ through (3.2)

$$1 + x^2 = (1 + \eta\beta)^2. \tag{3.9}$$

Asymptotic expansions for $\eta \to \infty$ can be obtained for the $F_k$ (Cox and Giuli 1968), it is, however, simpler to use directly the simple form of $f(\varepsilon)$ for complete degeneracy (3.4). The number density is

$$n = \frac{2}{h^3}\int_0^{p_F} 4\pi p^2\,\mathrm{d}p = \frac{8\pi}{3h^3}p_F^3 = \frac{8\pi m^3c^3}{3h^3}x^3 = n_0x^3 \qquad \text{with } x = p_F/mc \tag{3.10}$$

and the pressure

$$P = \frac{8\pi}{3h^3}\int_0^{p_F}\frac{(p^4/m)\,\mathrm{d}p}{\sqrt{1+(p/mc)^2}} = \frac{\pi m^4c^5}{3h^3}f(x) = P_0f(x) \tag{3.11}$$

where $f(x)$ stands for

$$f(x) = x(x^2+1)^{1/2}(2x^2-3) + 3\ln(x+\sqrt{1+x^2}) \tag{3.12}$$

and should not be confused with $f$ in (3.4). It is interesting to note that the constants $P_0$ and $n_0$ in the above expressions involve only fundamental physical constants, and especially $h$, emphasising the role of quantum mechanics in the structure of white dwarfs.

For the construction of stellar models we need the matter density $\rho$ instead of the electron number density $n$. These two are related through the requirement of electrical neutrality. In completely ionised matter of atomic mass number $A$ and charge $Z$ the number of heavy ions is $n/Z$ and the mass density $\rho = nAH/Z$, where $H$ is the atomic mass unit. Equation (3.10) can therefore be replaced by

$$\rho = \rho_0\mu_e x^3 \qquad \text{with } \rho_0 = n_0H \qquad \text{and} \qquad \mu_e = A/Z. \tag{3.13}$$

Here $\mu_e$ is called the 'molecular weight per electron' and should not be confused with the symbol $\mu$ for the chemical potential. It has the value 2 for realistic compositions. For hydrogen it would be 1, for completely ionised iron 2.15. If the composition is a mixture of several elements, appropriate averages must be taken.

Finally we consider the limiting cases $x \to 0$ (non-relativistic) and $x \to \infty$ (extremely relativistic). In both cases expansions of $f(x)$ allow elimination of the variable $x$, resulting in very simple equations:

$$P = \begin{cases} \dfrac{8P_0}{5\rho_0^{5/3}}\left(\dfrac{\rho}{\mu_e}\right)^{5/3} = \dfrac{3^{2/3}h^2}{8^{2/3}5\pi^{2/3}mH^{5/3}}\left(\dfrac{\rho}{\mu_e}\right)^{5/3} & x \ll 1 \\[2ex] \dfrac{2P_0}{\rho_0^{4/3}}\left(\dfrac{\rho}{\mu_e}\right)^{4/3} = \dfrac{2\,3^{1/3}hc}{8^{4/3}\pi^{1/3}H^{4/3}}\left(\dfrac{\rho}{\mu_e}\right)^{4/3} & x \gg 1\,. \end{cases} \tag{3.14}$$

### 3.2. Simple models, the mass–radius relation and the Chandrasekhar limit

The structure of a spherically symmetric star is completely determined by four basic equations, derived from the conservation laws of energy, mass and momentum, and the law describing the transport of energy in the presence of a temperature gradient (e.g. Clayton 1968, Cox and Giuli 1968). For zero-temperature stars only two of these are needed, the equation of hydrostatic equilibrium

$$\frac{\mathrm{d}P}{\mathrm{d}r} = -\frac{Gm\rho}{r^2} \tag{3.15}$$

and of mass conservation

$$\frac{\mathrm{d}m}{\mathrm{d}r} = 4\pi r^2 \rho. \tag{3.16}$$

Here $r$ is the radial variable ( $= 0$ at the centre), and $m$ the mass inside a sphere of radius $r$. The system of equations is completed with the EOS $P = P(\rho, \mu_e)$, and the boundary conditions:

$$m(0) = 0 \quad \text{at the centre} \qquad \text{and} \qquad P(R) = 0 \quad \text{at the surface.} \tag{3.17}$$

Chandrasekhar (1939) has shown how the system can be reformulated into a second-order differential equation for a dimensionless function, with two parameters: $\mu_e$ and $\rho_c$, the central density. This function must be determined numerically and is tabulated in Chandrasekhar (1939) and Cox and Giuli (1968). It is, however, easy to see how the solution can be obtained in principle from equations (3.15) to (3.17). For a fixed value of $\mu_e$ (chemical composition) a value of $\rho_c$ is chosen. The integration of (3.15) and (3.16) then proceeds from the centre until $P(r) = 0$ is reached. This determines the radius $R$ and the mass $M = M(R)$ of the model. For a given value of $\mu_e$ a one-parameter family of models is thus obtained, defining implicitly a relation between mass and radius, the famous mass–radius relation (Chandrasekhar 1935). In the limit $\rho_c \rightarrow \infty$, $R$ becomes zero, while the mass approaches a limiting value $M_{\mathrm{Ch}} = 5.826/\mu_e^2\, M_\odot$, the 'Chandrasekhar limiting mass'.

It is possible to understand these basic features of the mass–radius relation without actually solving the above equations from a simple dimensional analysis. From (3.15) and (3.16) we can derive the general relations

$$P_c \propto \frac{GM^2}{R^4} \qquad \text{and} \qquad \rho_c \propto \frac{M}{R^3} \tag{3.18}$$

for the physical quantities at the centre. For low density models ($x \ll 1, \rho \ll \rho_0\mu_e$ the non-relativistic form of the EOS (3.14) together with (3.18) leads to

$$P_c \propto \frac{P_0}{(\rho_0\mu_e)^{5/3}} \frac{M^{5/3}}{R^5}. \tag{3.19}$$

In equilibrium the 'gravitational pressure' (3.18) must be balanced by the electron pressure (3.19). This can always be achieved by adjusting the radius, leading to a mass–radius relation

$$R \propto \frac{P_0}{G\,(\rho_0\mu_e)^{5/3}} M^{-1/3}. \tag{3.20}$$

The neglected constants in the above expressions are dimensionless numbers of order unity. Equation (3.20) shows that for low-density white dwarfs the radius decreases with increasing mass as $M^{-1/3}$.

For the highly relativistic case (high central densities) on the other hand, one gets

$$P_c \propto \frac{P_0}{(\rho_0 \mu_e)^{4/3}} \frac{M^{4/3}}{R^4}. \tag{3.21}$$

In this case both pressures depend on the radius in the same way and hydrostatic equilibrium is possible only for a uniquely determined mass

$$M_{\mathrm{Ch}} \propto \frac{(P_0/G)^{3/2}}{(\rho_0 \mu_e)^2} \propto \frac{(hc)^{3/2}}{G^{3/2} H^2} \mu_e^{-2}. \tag{3.22}$$

The radius has dropped from the result; equilibrium is therefore possible at arbitrary radius. This, however, is not realistic, because the assumption of the extremely relativistic EOS requires infinite density and $R = 0$. The physical interpretation is as follows:

Only for $M \equiv M_{\mathrm{Ch}}$ can gravitational and electron pressure be in balance. For a slightly larger mass, the gravitational pressure wins and the star collapses dynamically to a singularity. For a slightly smaller mass, the electron pressure dominates. The star starts to expand, lowering the density and approaching the non-relativistic EOS in parts of the interior, until equilibrium can be obtained at a finite radius. It is interesting to note that the electron mass $m$ has also dropped from (3.22). Similar arguments for the degenerate neutrons in neutron stars therefore lead to the prediction of a limiting mass for neutron stars of the same order of magnitude. In figure 2 the Chandrasekhar mass–radius relation for white dwarfs with $\mu_e = 2$ and $\mu_e = 2.15$ is displayed, together with results for more realistic EOSs discussed below.

### *3.3. Refinements to the equation of state — Hamada and Salpeter*

Although the ideal, non-interacting electron gas accounts for the dominant contribution to the EOS at high densities, a real stellar plasma consists of electrons together with heavy, positively charged ions of one or more species. Kothari (1931, 1936, 1938) studied the effects of Coulomb interactions on the EOS and the structure of non-relativistic white dwarfs. Auluck and Mathur (1959) extended this work by including exchange and correlation effects. Kirzhnits (1960) argued that the matter in white dwarfs should be in a 'condensed state', and Abrikosov (1961, 1962) computed the properties of this crystal lattice phase. A comprehensive discussion of the EOS, and the results for energy and pressure of a zero-temperature plasma were provided by Salpeter (1961). In the following sections we follow closely the approach by Van Horn and Liebert (1989), based on Salpeter's work.

The thermodynamic properties of matter are completely determined if the Helmholtz free energy $F(T, V, N)$ is known for the system. The pressure, for example, can be obtained from

$$P = \left(\frac{\partial F}{\partial V}\right)_{T,N} \tag{3.23}$$

Because $F = E - TS$, at zero temperature all that is needed is the energy $E(V, N)$, where $N$ stands collectively for the particle numbers. In the following we will discuss

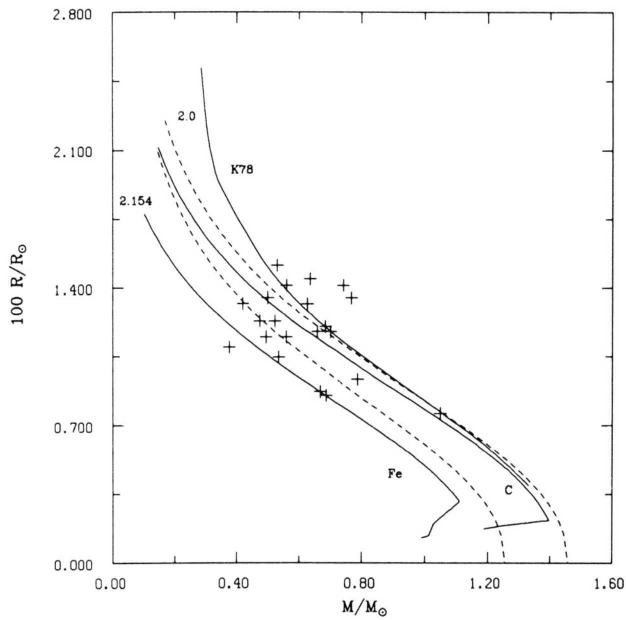


**Figure 2.** Mass–radius relations for white dwarfs. The two broken curves are the Chandrasekhar relations for $\mu_e = 2$ and $\mu_e = 2.15$. Continuous relations (C,Fe) are from Hamada and Salpeter (1961), and (K78) from Koester (1978) for a central temperature of $10^8$ K. The best observational data available are shown as crosses.

the various contributions to the energy, normalised to the total number of heavy ions. Some useful relations are:

$$n_e = N_e/V = ZN_i/V = Zn_i = \rho Z/AH \tag{3.24}$$

where the index i stands for positive ions, and e for electrons. We will consider only a single ionic species of charge $Z$ and atomic mass number $A$. Generalisations to a mixture of different ions by taking appropriate averages are straightforward but cumbersome. Instead of $n_i$ we will sometimes use a related variable $r_i$, defined through

$$\tfrac{4}{3}\pi r_i^3 n_i = 1 \tag{3.25}$$

which defines the radius of sphere containing on average one ion and $Z$ electrons.

*3.3.1. The kinetic energy of the electrons.* The average kinetic energy of the electrons is related to the Fermi energy, which in turn is completely determined by the matter density. In the non-relativistic limit, $\varepsilon_F = p_F^2/2m$, and (3.10) gives:

$$N_e/V = ZN_i/V = Z\rho/AH = \frac{8\pi}{3h^3}(2m\varepsilon_F)^{3/2}. \tag{3.26}$$

The kinetic energy in this case is:

$$E_e/V = \frac{4\pi}{5mh^3}\, p_F^5 = \frac{4\pi\; 2^{5/2} m^{3/2}}{5h^3}\, \varepsilon_F^{5/2}. \tag{3.27}$$

From the ratio the average electronic energy *per ion* can be derived:

$$E_e/N_i = \tfrac{3}{5} Z\varepsilon_F. \tag{3.28}$$

Similarly, in the extreme relativistic limit, $\varepsilon_F = p_F c$, and

$$N_e/V = \frac{8\pi}{3h^3c^3}\,\varepsilon_F^3 \qquad E_e/V = \frac{2\pi c}{h^3}\,p_F^4 = \frac{2\pi}{h^3c^3}\,\varepsilon_F^4 \qquad \text{and} \qquad E_e/N_i = \tfrac{3}{4} Z\varepsilon_F. \tag{3.29}$$

*3.3.2. The Coulomb interaction energy.* As discussed by Salpeter (1961) the electronic screening radius for an ion is of order $Z^{-1/3}a_0$, the Thomas–Fermi 'mean radius' of a free neutral atom; $a_0$ is the Bohr radius. If the typical separation between ions, $r_i$, is much less than this, as in typical white dwarfs, the electronic charge density is almost uniform and unaffected by the distribution and motion of the ions. The effective Coulomb repulsion between ions is essentially less than that for unshielded charges. The average repulsion energy between ions is a minimum for a regular lattice and, after including the attraction energy of the electrons, leads to a negative Coulomb energy $E_C$ per ion. Keeping the ions fixed near lattice sites leads necessarily to a positive kinetic energy, the zero-temperature vibrations of the lattice. This contribution is shown to be small compared to the Coulomb energy, and neglected by Salpeter (1961).

Under these conditions the Coulomb energy for regularly distributed ions in a homogenous negative charge distribution is given by

$$E_C/N_i = -S_M\frac{(Ze)^2}{r_i} \tag{3.30}$$

where the $S_M$ are the 'Madelung sums' and depend on the lattice geometry. Since the completely ionised ions have no preferred direction, the lowest energy lattice is a cubic lattice. The numerical values of $S_M$ for the three cubic lattice types are 0.880059442 (SC), 0.895929256 (FCC), and 0.895873615 (BCC). The most stable form at zero-temperature is the BCC lattice, although the differences are small. The numerical values are also very close to 9/10, the Coulomb energy of a 'Wigner–Seitz sphere' (Wigner and Seitz 1934), where each lattice site is replaced by a sphere consisting of a central positive ion of charge $+Ze$ embedded in a uniform density of neutralising electrons, and the interaction *between cells* is neglected. This is the model adopted by Salpeter (1961); it is an adequate approximation for zero-temperature models.

In order to estimate the importance of Coulomb effects we may compare $E_C$ with the electronic energy $E_e$. In the non-relativistic limit we have

$$E_C/E_e = -\frac{(9/10)Z^2e^2/r_i}{(3/5)Z\varepsilon_F}. \tag{3.31}$$

From (3.26), replacing the variable $n_i$ in $\varepsilon_F$ with $r_i$, we obtain

$$\varepsilon_F = \frac{9^{2/3}h^2}{2\;32^{2/3}\pi^{4/3}m}\;Z^{2/3}/r_i^2 \tag{3.32}$$

and

$$E_{\rm C}/E_{\rm e} = -\frac{32^{2/3}\,9^{1/3}}{12\pi^{2/3}}\,Z^{1/3}\,(r_{\rm i}/a_0) \simeq -0.8145\,Z^{1/3}\,(r_{\rm i}/a_0). \tag{3.33}$$

For realistic values of $Z$ (6 to 8) this is indeed much smaller than 1, although not negligible and by far the largest correction to the ideal Fermi gas EOS. In the relativistic limit

$$\varepsilon_{\rm F} = \left(\frac{9}{32\pi^2}\right)^{1/3} hc\,Z^{1/3}/r_{\rm i} \tag{3.34}$$

and

$$E_{\rm C}/E_{\rm e} = -\frac{(9/10)\,Z^2e^2/r_{\rm i}}{(3/4)\,Z\varepsilon_{\rm F}} = -\frac{6}{5}\left(\frac{4}{9\pi}\right)^{1/3}\alpha\,Z^{2/3} \simeq -0.6253\,\alpha\,Z^{2/3} \tag{3.35}$$

where $\alpha$ is the fine structure constant $2\pi e^2/hc$. This, too, is always a small correction.

Because the dominant contribution to the energy and pressure of the plasma is the kinetic energy of the electrons, the electron energy eigenstates are very nearly plane waves. An adequate model of the plasma therefore consists of a nearly uniform distribution of degenerate electrons with embedded classical ions. As pointed out by Salpeter (1961) the energy of the plasma can then be written as the sum of the electron kinetic energy plus correction terms, of which the Coulomb interaction is the largest:

$$E = E_{\rm e} + E_{\rm C} + E_{\rm ex} + E_{\rm TF} + E_{\rm corr}. \tag{3.36}$$

We now turn to a discussion of the remaining terms, following Salpeter (1961).

*3.3.3. Thomas–Fermi correction* $E_{\rm TF}$. In the evaluation of the Coulomb energy the distribution of the electrons was assumed to be homogeneous within each Wigner–Seitz 'cell'. The 'Thomas–Fermi' contribution accounts for the effect produced by the fact that the electron distribution is polarised by the positively charged ions. It could, in principle, be obtained from the equations of the Thomas–Fermi atomic model, but the solution would have to be carried through numerically for each value of $Z$ and $r_{\rm i}$. However, because $E_{\rm C}$ is much smaller than $E_{\rm e}$, the perturbation of the electron density from uniformity is small, and one finds

$$E_{\rm TF}/E_{\rm C} \sim E_{\rm C}/E_{\rm e}. \tag{3.37}$$

It is then possible to derive this term from an expansion of the energy of the Thomas–Fermi model, with $E_{\rm e}$ and $E_{\rm C}$ as the two leading terms. The third term in this expansion is given by Salpeter (1961) as

$$E_{\rm TF} = -\frac{162}{175}\left(\frac{4}{9\pi}\right)^{2/3}\alpha^2mc^2\,Z^{4/3}(1+x^2)^{1/2} \tag{3.38}$$

where $x = p_{\rm F}/mc$ is the relativity parameter. In the extreme non-relativistic regime ($x \to 0$) this term is independent of the density (or volume) and therefore does not contribute to the pressure.

*3.3.4. Exchange energy* $E_{ex}$ . This contribution arises from the fact that the electrons are indistinguishable in quantum mechanics. More precisely, one has to evaluate the interaction energy between pairs of electrons in first-order perturbation theory using antisymmetrised Dirac wave functions for the electrons. For general values of the relativity parameter this involves a complicated double integration over momentum space first carried out by Zapolski (1960), with the result

$$E_{ex} = -\left(\frac{3}{4\pi}\right) \alpha mc^2 Z\, x\, \Phi(x) \tag{3.39}$$

where

$$\Phi(x) = \frac{1}{4x^4}\left[\frac{9}{4} + 3(y^2 - y^{-2})\ln y - 6(\ln y)^2 - (y^2 + y^{-2}) - (y^4 + y^{-4})/8\right] \tag{3.40}$$

and $y$ is given by $y = x + (1 + x^2)^{1/2}$.

*3.3.5. Correlation energy* $E_{corr}$. This is the next term after $E_{ex}$ if the interaction energy between electrons is expanded in powers of $r_i$. It has been obtained by Gell-Mann and Brueckner (1957):

$$E_{corr} = (0.062\ \ln(Z^{1/3}r_i/a_0) - 0.096)\ \tfrac{1}{2}\alpha^2 mc^2 Z. \tag{3.41}$$

Salpeter (1961) discussed some other corrections — self-energy terms, the influence of internal electric fields, higher order terms in the various expansions — but generally found them to be negligible.

With the total energy thus determined (3.35), the pressure EOS is derived from

$$P = \left(\frac{\partial E}{\partial V}\right)_{N_i} = P_e + P_C + P_{TF} + P_{ex} + P_{corr}. \tag{3.42}$$

The derivative $\partial/\partial V$ can be replaced here with $\partial/\partial n_i$ or $\partial/\partial r_i$ as appropriate using (3.25) and (3.26). Explicit results for the pressure corrections are listed in Salpeter (1961) and Van Horn and Liebert (1989) and will not be repeated here.

### *3.4. Hamada–Salpeter zero-temperature models*

The EOS derived in the previous sections was used by Hamada and Salpeter (1961) to construct zero-temperature models for white dwarfs. Because the EOS now depends explicitly on $A$ and $Z$ instead of $\mu_e$ only, it is necessary to specify the chemical composition of the model. According to our present understanding of the previous history we expect an interior composition of carbon and oxygen for most white dwarfs. Irrespective of this knowledge, however, the possible composition of zero-temperature models is restricted by two physical processes.

*3.4.1. Pycnonuclear reactions.* At sufficiently high densities the ions in a lattice can undergo nuclear reactions even at zero-temperature as a result of the quantum zero-point motions about the equilibrium lattice sites and the screening of the positive charges by the electrons. This 'pycnonuclear' regime was studied originally by Wildhack (1940), a more recent reference is Salpeter and Van Horn (1969). Formulae for the reaction rates can be found in the latter paper, and a review is given by Shapiro and Teukolsky (1983).

The critical density for the transformation of hydrogen into helium is about $1 \times 10^6$ g cm$^{-3}$, for C $\longrightarrow$ Mg $\sim 10^{10}$; for He $\rightarrow$ C the rate is probably negligible (Shapiro and Teukolsky 1983). All these numbers are still very uncertain. For elements other than hydrogen, which is not expected in the interior of white dwarfs (Hamada and Salpeter 1961), a much faster instability occurs at about the same densities (see below). Pycnonuclear reactions are therefore not likely to be important in white dwarfs.

*3.4.2. Inverse β-decays.* At higher densities the Fermi energy of the electrons can exceed the mass difference between the nuclei $(A, Z)$ and $(A+1, Z-1)$. It then becomes energetically favourable to 'capture' an electron onto the nucleus, transforming a proton into a neutron. The neutrinos formed in this transition can easily escape from a white dwarf. For $C^{12}$ this instability occurs for $\rho = 3.9 \times 10^{10}$ g cm$^{-3}$ (Wapstra and Bos 1977) and the transformation then rapidly proceeds to $Ne^{24}$ because the intermediate nuclei have lower thresholds. Hamada and Salpeter (1961) have taken this into account by assuming that a carbon model actually consists of $Ne^{24}$ in those regions where the density exceeds the critical value. Similarly, at very high central densities, an iron model has a central core of $Cr^{56}$ for $\rho_c > 1.14 \times 10^9$ g cm$^{-3}$, and a $Ti^{56}$ core for $\rho_c > 2.45 \times 10^{10}$ g cm$^{-3}$. The results of their calculations for these two model sequences in the mass–radius diagram (note that Hamada and Salpeter (1961) used slightly different values for the critical densities) are given in figure 2.

The neutronisation at extremely high densities reduces the number of free electrons and increases the molecular weight per electron $\mu_e$, thus leading to kinks in the mass–radius relations at high masses. The Chandrasekhar limiting mass for infinite density is now replaced by a maximum mass at finite density and radius. For the 'carbon' sequence in Hamada and Salpeter's (1961) calculations this maximum mass is 1.381 $M_\odot$. It can be shown (e.g. Shapiro and Teukolsky 1983, Skilling 1968) that *homogeneous* models on the 'returning branch' with $M > M_{max}$ are dynamically unstable, although for an equilibrium equation of state maximum mass and onset of instability do not necessarily coincide exactly (Wheeler *et al* 1968, Chanmugam 1977).

*3.5. Influence of general relativity*

For the maximum mass carbon model, $GM/Rc^2 \simeq 1.37 \times 10^{-3}$. Even at this extreme point the effect of general relativity (GR) on the structure of white dwarfs and the shape of the mass–radius relation is very small. The mass–radius relation in GR was first derived by Kaplan (1949). GR has, however, a significant influence on the dynamical stability of high mass models. Chandrasekhar and Tooper (1964) discovered that GR leads to instability at a central density $\rho_c = 2.328 \times 10^{10}(\mu_e/2)^2$ g cm$^{-3}$; Shapiro and Teukolsky (1983) derive a critical value of $2.646 \times 10^{10}$ $(\mu_e/2)^2$ g cm$^{-3}$. The radius corresponding to these densities is still much larger than the Schwarzschild radius ($\simeq$ 3.6 km). For $Fe^{56}$ this is higher than the threshold for neutronisation, so GR is irrelevant for iron white dwarfs. For $C^{12}$, however, $\rho_c = 2.65 \times 10^{10}$ g cm$^{-3}$, lower than the neutronisation density ($3.9 \times 10^{10}$ g cm$^{-3}$), and in this case it is GR that determines the maximum stable mass. The problem of stability will be discussed further in the section on pulsations.

It should be noted, however, that single white dwarfs are not observed with masses higher than $\simeq 1.2\ M_\odot$— they are certainly extremely rare, if they exist at all — and iron is an unlikely composition. The refinements at the high end of the mass–radius relation discussed in the preceding paragraphs are therefore at present only of theoretical interest.

### 3.6. Corrections at finite temperature

Zero-temperature models provide an adequate description of the overall structure of white dwarfs. However, observed white dwarfs have surface temperatures ($T_{\rm eff}$) ranging from 5000 K to over 100 000 K, and they are still losing energy, implying a temperature gradient and even higher interior temperatures. Moreover, as we will see later, their evolution is governed by their *thermal* properties. A comparison of observations with models therefore requires the calculation of finite-temperature models.

Comprehensive discussions of the EOS at finite temperatures have been given by DeWitt (1969), Kovetz and Shaviv (1970), Shaviv and Kovetz (1972), and Lamb (1974), where a thorough discussion of earlier work and many references can be found. In our presentation we will follow closely the approach taken by Lamb.

The dominant contribution to the energy and pressure at all reasonable temperatures remains that of a non-interacting Fermi gas of the electrons. The general equations have already been given in equations (3.5) through (3.8). The Coulomb contributions can be most conveniently described by two dimensionless parameters:

$$\Gamma = \frac{(Ze)^2}{r_{\rm i}kT} \qquad \text{and} \qquad \theta = \frac{\hbar\Omega_{\rm P}}{kT} = \frac{\theta_{\rm D}}{T},$$

where

$$\Omega_{\rm P} = \sqrt{\frac{4\pi n_{\rm i} Z^2 e^2}{AH}} \tag{3.43}$$

is the ion plasma frequency, and $\theta_{\rm D}$ the Debye temperature. The parameter $\Gamma$ measures the Coulomb potential energy, while $\theta$ measures the importance of quantum effects. At very high temperatures the ions behave essentially as an ideal classical gas with small corrections arising from the Coulomb interactions, which for $\Gamma \lesssim 0.05$ can be calculated according to the Debye–Hückel theory. As the temperature falls, $\Gamma$ becomes much greater than 1 and the Coulomb interactions increasingly dominate the thermal properties of the ions, forming a 'Coulomb liquid'. Eventually the ions crystallise into a solid lattice and the star approaches the zero-temperature configuration. Quantum effects become important when $\theta > 1$, and at somewhat lower temperatures the thermal energy of the ions declines rapidly until they are effectively frozen into their zero-point oscillations.

The thermodynamics in the gas/liquid and solid phases are best described by modelling their Helmholtz free energy, with the ideal Fermi energy of the electrons as the leading term and the other contributions as additive corrections. The energy and pressure can be obtained in a thermodynamically consistent way through straightforward but tedious derivation. In the following we will consider only the most important terms, which are responsible for qualitatively new effects compared to the zero-temperature models. For a more detailed treatment, for example, the extension of the exchange and Thomas–Fermi contributions to finite temperatures, we refer the reader to Lamb (1974). The most important contributions to the free energy can be written as

$$F = F_e + F_C + F_i. \tag{3.44}$$

$F_e$ is the electronic contribution, $F_C$ the Coulomb potential energy, and $F_i$ the remaining contributions of the ions, which depend on the phase.

*3.6.1. Coulomb energy.* The Coulomb energy can generally be written as $E_C = N_i\, kT\, f(\Gamma)$ from which the free energy can be derived using the thermodynamic relation

$$F_C = -T \int^T (E_C/T^2)\, \mathrm{d}T. \tag{3.45}$$

For sufficiently large $\Gamma$ the Wigner–Seitz model gives $f(\Gamma) = -0.9\,\Gamma$. For very small $\Gamma$ ($< 0.05$) Debye–Hückel theory (e.g. Landau and Lifshitz 1969) gives

$$f(\Gamma) = -\frac{\sqrt{3}}{2}\left(\frac{Z+1}{Z}\right)^{3/2} \Gamma^{3/2}. \tag{3.46}$$

No rigorous treatment of the intermediate range exists as yet, and it is usually treated essentially using numerical methods. As we have noted before, the distribution of the electrons is very closely homogeneous. The ions therefore behave roughly as a one-component plasma (OCP) in a uniform negatively charged background. The physics of OCPs has been studied using Monte Carlo techniques by Brush *et al* (1966), Kovetz and Shaviv (1970), Hansen (1973) and Pollock and Hansen (1973). To our knowledge the most accurate calculation to date is that of Slattery *et al* (1980). The function $f(\Gamma)$ for the intermediate range can be taken directly from these numerical experiments or fitted with analytical approximations (see e.g. Lamb 1974). Slattery *et al* (1980) give the following fit for the free energy:
In the liquid phase

$$F_C(\Gamma) = N_i kT[-0.89752\Gamma + 3.78176\Gamma^{1/4} - 0.71816\Gamma^{-1/4} - 0.80049 \ln \Gamma - 2.58608] \tag{3.47}$$

while in the solid phase

$$F_C(\Gamma) = N_i kT[-0.895929\Gamma + 1.5 \ln \Gamma - 1.1703 - 1490/\Gamma^2] \tag{3.48}$$

*3.6.2. Other ionic contributions.* In the solid phase the temperature-dependent term in the free energy connected with the oscillations of the ion lattice, to which the zero-point contributions of Salpeter (1961) have to be added, is:

$$F_i = N_i kT \sum_\lambda \left\langle \frac{1}{2}\frac{\hbar\omega_{l\lambda}}{kT} + \ln\left[1 - \exp\left(-\frac{\hbar\omega_{l\lambda}}{kT}\right)\right]\right\rangle \tag{3.49}$$

where the $\omega_{l\lambda}$ are the frequencies of the phonon spectrum (depending on the wave vector $l$, tabulated for a BCC lattice by, for example, Carr (1961) and Kugler (1969)), the sum is taken over the excitation modes $\lambda$ and the average over the $l$ in the first Brillouin zone (Kovetz and Shaviv 1970, Lamb 1974). A slightly less accurate but computationally easier method uses the Debye model to describe the thermal properties of the lattice. The energy in this model is:

$$E_{\rm i} = 3N_{\rm i}kTD(\theta) \tag{3.50}$$

with the Debye function

$$D(\theta) = 3\theta^{-3}\int_0^\theta \frac{x^3\,\mathrm{d}x}{\exp(x)-1} \tag{3.51}$$

and $\theta = \theta_{\rm D}/T$. The free energy can be obtained as in (3.45):

$$F_{\rm i} = 3N_{\rm i}kT\int^{\theta_{\rm D}/T}\frac{D(\theta)}{\theta}\,\mathrm{d}\theta. \tag{3.52}$$

As is well known the energy approaches the classical limit $E_{\rm i} = 3N_{\rm i}kT$ for high temperatures, while for $T \rightarrow 0$ we obtain $E_{\rm i} \propto T^4$ and the thermal energy rapidly decreases for $T \ll \theta_{\rm D}$.

In the gas/liquid phase at high temperatures the dominant term in $F_{\rm i}$ is that for an ideal classical gas

$$F_{\rm i} = -N_{\rm i}kT\left[1+\frac{3}{2}\ln\left(\frac{2\pi AHkT}{h^2}\right)+\ln V-\ln N_{\rm i}\right]. \tag{3.53}$$

At high densities and low temperatures a correction for the quantum nature has also to be applied for the ions in the liquid phase, the so called 'ionic quantum correction'. It can be obtained as the first term in the Wigner (1932) expansion in terms of $h^2$; it is essentially an expansion in $\theta$ (Lamb 1974, Kovetz and Shaviv 1972):

$$F_{\rm iq} = \frac{1}{24}N_{\rm i}kT\theta^2 \tag{3.54}$$

which is a valid approximation for $\theta \ll 1$. Equations (3.43) through (3.54) constitute the model free energy, from which the energy, pressure and all other thermodynamic quantities can be derived.

*3.6.3. Phase transitions.* The transition between the liquid and the solid phase is expected to be a first-order phase transition (Lamb 1974) with an associated release of latent heat. Its location can, in principle, be obtained by equating the chemical potential (obtained from $\partial F/\partial N_{\rm i}$) of the two phases. The result, however, depends sensitively on the approximations used in the EOS and various authors have obtained values of $\Gamma_m$ (the Coulomb parameter at crystallisation) ranging from $\sim 60$ to $\sim 170$. Because of this uncertainty it has become customary to regard $\Gamma_m$ as a free parameter, but $\Gamma_m = 160$, close to the result obtained by Lamb (1974), seems to be the preferred choice at present. This is supported by the result of Slattery *et al* (1980), who obtained $168 \pm 4$ from their OCP calculations. The latent heat released can then be calculated from the entropy difference of the two phases at this point.

Cesare *et al* (1973) have suggested that a similar phase transition should exist at $\Gamma \approx 1$ between the gaseous and the liquid phase. The OCP calculations, however, which should give an excellent approximation in this range, show no indication of a first-order phase transition, and Lamb (1974) has given strong theoretical arguments that a smooth transition is expected from a region exhibiting no order to one exhibiting short-range order.

The different regions relevant for the EOS in a temperature–density diagram are displayed in figure 3 together with the approximate location of white dwarfs with $M = 0.6\ M_\odot$ and central temperatures of $10^7$ and $10^6$ K.

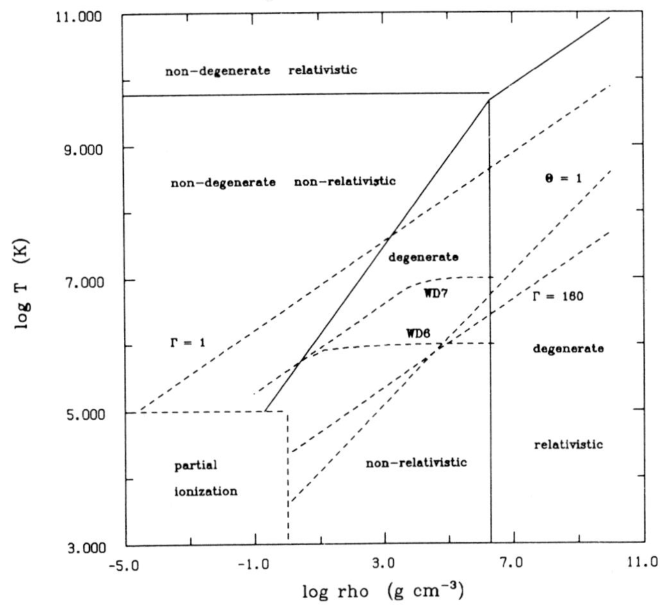


**Figure 3.** Temperature–density diagram for completely ionised carbon. The lower left corner, bounded approximately by the broken lines, is the region of incomplete ionisation and not considered here. The continuous straight lines are the boundaries between relativistic/non-relativistic, respectively degenerate/non-degenerate regions. Broken straight lines limit the regions where quantum ($\theta = 1$) or Coulomb ($\Gamma = 1, 160$) effects for the ions become important. The broken curved lines labelled WD6 and WD7 show the approximate structure for the interior 99% of the mass of 0.6 $M_\odot$ white dwarf models (the outer 1% is not shown) with central temperatures of $10^7$ and $10^6$ K. These calculations were done for the work in Koester (1978), but not published.

*3.6.4. Construction of models at finite temperature.* If the pressure is no longer a function of $\rho$ alone but also of $T$, the full set of four equations describing the structure of a star has to be used (e.g. Clayton 1968). In addition to (3.15) and (3.16) these are the equations for the energy transport mechanism and energy generation. The energy transport mechanism determines the temperature gradient:

$$\frac{\mathrm{d}T}{\mathrm{d}r} = \frac{3\kappa\rho}{16\sigma T^3}\frac{L_r}{4\pi r^2} \tag{3.55}$$

where $L_r$ is the total energy flow through the sphere with radius $r$, and $\kappa$ is the opacity coefficient. In the interior of a white dwarf heat conduction by the highly degenerate electrons is the dominant mechanism, leading to a very small temperature gradient and an almost isothermal interior, a fact exploited widely by early calculations of white dwarf evolution. Conductive opacities have been calculated by Marshak (1940), Mestel (1950), Lee (1950), Hubbard and Lampe (1969), Canuto (1970), Urpin and Yakovlev (1980), and Itoh *et al* (1983, 1984). If the electrons are not extremely degenerate, radiative energy transport becomes important. The calculation of all the relevant absorption coefficients is a formidable task that in the past has mainly been pursued by groups at the Los Alamos Laboratory (Cox and Stewart 1970, Huebner *et al* 1977).

In the outer layers of cool white dwarfs convective energy transport may become dominant and the temperature gradient has to be determined from the mixing length approximation (e.g. Böhm–Vitense 1958).

The boundary condition to (3.55) can be applied in several different forms, a typical one being

$$T(R) = T_{\mathrm{eff}} = \left(\frac{L}{4\pi\sigma R^2}\right)^{1/4} \tag{3.56}$$

at the surface.

The last equation to be used describes the energy generation in a spherical shell of inner and outer radii $r$, $r + \mathrm{d}r$:

$$\frac{\mathrm{d}L_r}{\mathrm{d}r} = \left(\varepsilon_{\mathrm{N}} - \varepsilon_\nu - T\frac{\Delta s}{\Delta t}\right) 4\pi r^2\rho \tag{3.57}$$

where $\varepsilon_{\mathrm{N}}$ is the energy generation rate due to nuclear processes and normally negligible in white dwarfs; $\varepsilon_\nu$ is the energy loss due to neutrinos leaving the star without being absorbed again. The quantity $T\Delta s/\Delta t$ is the change in heat content of the shell from a previous model at time $t - \Delta t$. It is the most important term for white dwarf evolution as we will see later. The boundary condition for this equation is $L_r(0) = 0$ at $r = 0$.

In addition to $M$ a second parameter, besides composition, is now necessary to specify a model, which is usually taken to be the total luminosity $L$. The most significant formal difference to zero-temperature models, however, is that a model now depends on the previous evolution through (3.57). An accurate treatment therefore requires the complete methods of stellar evolution calculations, starting at much earlier phases in the life of the star. Due to the heavy demands on computing power this has been attempted on a broader basis only during the last five years. In the previous twenty years it was customary to make simplifying assumptions in order to determine the temperature distribution within the model. One such possibility is to integrate equation (3.55) from the surface inwards, assuming $L_r = L =$ constant, until the temperature gradient becomes very small and can be set to zero until the centre is reached. The main emphasis in these calculations was on the evolution of white dwarfs. Koester (1978) studied the influence of finite temperatures on the mass–radius relation; his result for an interior temperature of $10^8$ K is presented in figure 2.

*3.6.5. Comparison with observed masses and radii.* Finally, figure 2 also shows a comparison with observed data. With the exception of very few objects in binary systems (the most massive object in figure 2 is Sirius B) the most reliable masses can now be obtained from gravitational redshifts. These redshifts give the ratio $M/R$, the radii are obtained from effective temperatures and distances using relation (2.1) as explained in section 5. The data in the figure are from Koester (1987b), Wegner *et al* (1989), and Wegner (1989b). While they definitely rule out a hydrogen interior (the relation would be far outside the range of the figure), any discrimination between the different relations (e.g. different compositions other than hydrogen) is lost in the scatter. This scatter is very probably due to observational errors: typical uncertainties are 15% for the radius and 20% for the mass.

## 4. Stellar evolution and white dwarfs

The theory of stellar evolution is an immense and very active field of study and we cannot hope to report even the major results adequately in this review. Instead we will try to give a very brief description of the prehistory of a typical white dwarf and refer the reader interested in more details to the excellent reviews of some of the leading experts (Iben 1967, 1974, Iben and Renzini 1983).

The main phases in the life of a star are best visualised in the Hertzsprung–Russell diagram of figure 4. The full lines in this diagram trace the position of a star (model) as a function of time and are called 'evolutionary tracks'. The present diagram is intended only for a qualitative description. The tracks are based on actual calculations (Mengel *et al* 1979, Sweigart and Gross 1978, Schönberner 1979, Koester and Schönberner 1986); they do not, however, use exactly the same chemical abundances and input physics in the various calculations.

### *4.1. Prehistory of a typical white dwarf*

After a star has completed its contraction from interstellar matter it enters the major phase of its life on the main sequence (MS). On the MS a star is chemically homogenous, at least at the very beginning, and derives its energy from the nuclear burning of hydrogen to helium at the centre. The part of the MS shown in the figure refers to the range of stellar masses between 0.7 $M_\odot$ (lower end) and 3.5 $M_\odot$ (high end). The evolutionary track shown is calculated for 1 $M_\odot$. Due to the high energy gained from the transformation of hydrogen to helium and the high abundance of hydrogen ($\simeq$ 90% by numbers) this is a very long lived phase and takes of the order of $8 \times 10^9$ yr for a solar mass star. During this phase the luminosity increases slowly.

When $\simeq$ 12% of the mass has been transformed to helium at the centre, a major structural change occurs. Because hydrogen is depleted at the centre, nuclear energy generation now takes place in a spherical shell surrounding the inert helium core. This change is accompanied by an expansion and cooling of the outer layers of the star: in the HRD the star moves to the right and upward to greater radii. This part of the track is therefore called the red giant branch (GB). At the top of the GB the star reaches at the centre the conditions necessary to ignite helium and start the next nuclear process, transforming helium into carbon and oxygen.

In a low mass star like our sun the density at the centre at this point is so high that the electrons are partially degenerate. In non-degenerate matter the increase in temperature caused by the new energy source would be accompanied by an increase

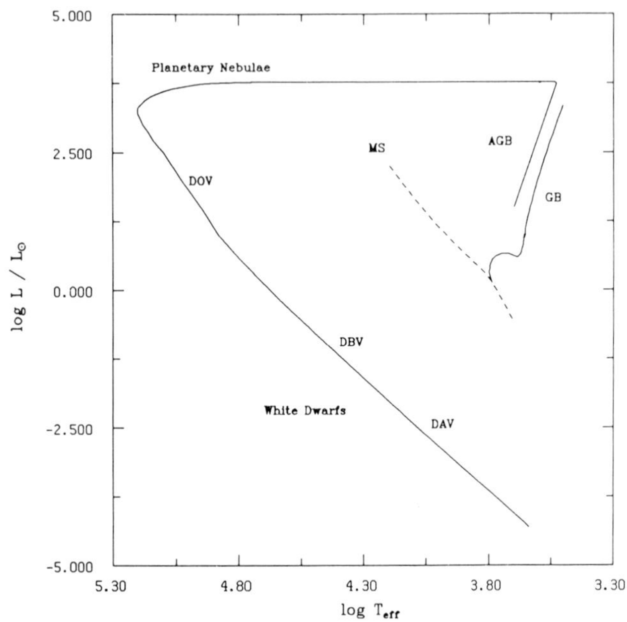


**Figure 4.** Qualitative description of the prehistory of a typical white dwarf.

in pressure, leading to expansion, cooling and stabilisation of the helium burning rate. However, due to the degeneracy, no significant change in pressure results; the increased temperature leads to further energy release and a thermal runaway develops. This phenomenon is called the 'helium flash'; it leads to a dynamic restructuring of the star after the degeneracy is lifted at high temperature.

When the star has adjusted to the new conditions it finds itself near the bottom of the sequence labelled AGB ('asymptotic giant branch'), quietly burning helium at the centre under non-degenerate conditions. Once helium is exhausted in the core, evolution follows a pattern similar to the departure from the main sequence: helium burns in a shell surrounding the inert carbon/oxygen core and the star moves again up and right on the AGB.

At the top of the AGB the star undergoes another instability, which leads to the expulsion of most of the remaining hydrogen-rich outer envelope ($\sim 0.2$–$0.3\ M_{\odot}$). We do not yet understand this process, which might be a continuous, very strong mass loss rate sustained over a few thousand years, or a singular event. In any case, when the remaining envelope mass fraction is reduced to $\sim 10^{-3}$ the remnant star moves rapidly to the left in the HRD, on the way exciting and ionising the previously expelled matter around it because of its high surface temperature. This becomes visible as a planetary nebula, a luminous cloud of gas, with the hot central star, the previous core of the red giant (not always visible), supplying the ionising photons. When the remaining hydrogen layer is decreased to $\sim 10^{-4}\ M_{\odot}$ nuclear energy generation comes to a halt. The luminosity decreases rapidly and the star enters the final phase of its life as a white dwarf. At about this time the nebula around it is expanded and diluted so much as to become unobservable.

The white dwarf is left with only gravitational and thermal energy sources available. Because the electrons are already degenerate in the interior at this stage, the radius is not far from the radius of the zero-temperature model and the remaining contraction is small. The star thus evolves roughly at a constant radius along the diagonal straight line given by equation (2.1) in the white dwarf region of the HRD.

Many important differences occur in physical details of the evolution of higher mass stars in the range $\sim$ 1–8 $M_\odot$, but the final outcome is qualitatively the same. Very high mass stars (say 10 $M_\odot$), however, have a different destiny. In these stars electron degeneracy never becomes important at the centre, the temperatures become high enough to ignite carbon and all the following nuclear burning phases up to iron, on rapidly decreasing time scales. The final fate of such a star is a supernova explosion leaving a neutron star, or possibly (in some cases) a black hole or nothing, if the star is totally disrupted.

### *4.2. Which stars become white dwarfs?*

The maximum mass of a carbon white dwarf is $\simeq 1.4\ M_\odot$, the low mass star described above will have no difficulty in reaching this final stage. There is, however, convincing evidence today that stars with initially (on the MS) much higher masses also manage to reach this stage. Direct evidence comes from the observation of stellar clusters, groups of stars with different masses, born at the same time. The luminosity on the MS increases much more steeply than the mass; a massive star therefore needs a shorter time to consume its fuel and start the evolution towards the giant branches and the final stages. The main sequences of clusters burn away from the top in the HRD, and the most massive stars still remaining on the MS ('turn-off point' in astronomical jargon) are an indication of the age of the cluster. In the famous Hyades cluster for example, stars with $M = 2\ M_\odot$ are still on the MS; nevertheless the cluster contains about a dozen white dwarfs! These white dwarfs have typical masses of 0.6 $M_\odot$, but on the MS they must have been more massive than the turn-off point, or they would not have evolved.

The solution to this puzzle is mass loss during the evolution. Even a normal star like our sun shows a continuous mass loss (solar wind); in the red giants the observed mass loss is many orders of magnitude larger (see, e.g., Reimers 1987). This mass loss obviously is large enough to bring fairly massive stars down to the white dwarf range. What then determines the final fate of a star?

Very massive stars end their lives as supernovae, whereas low mass stars certainly form white dwarfs. In a first approximation it is therefore reasonable to assume that the initial mass on the MS is the determining parameter, with all stars below a critical mass $M_{cr}$ becoming white dwarfs. One possible way to determine this critical parameter is to include mass loss (in a parametrised form, commonly that due to Reimers (1975, 1987) and Iben and Renzini (1983)) in stellar evolution calculations (the AGB branch in figure 4 includes mass loss, which reduces the original 1 $M_\odot$ to $\simeq$ 0.6 $M_\odot$ at the top of the AGB). Mass loss on the AGB was first considered by Mengel (1976) and Fusi-Pecci and Renzini (1976), and fully incorporated into the evolutionary calculations by Schönberner (1979) and many others since. A typical result for the critical mass is $\simeq 5\ M_\odot$. A difficulty of this approach is that while mass loss rates are fairly well known for the lower part of the giant branch, the extrapolation to the high luminosity region, where most mass is lost, is very uncertain.

A more empirical method compares the supernova rates (for type II supernovae, believed to originate from high mass stars) with the birth rate of massive stars. Unfortunately the supernova rate is very uncertain, and Tammann (1974) was only able to determine a range of 5 to 10 $M_\odot$ for $M_{cr}$. More recently, Kennicutt (1984) found $8 \pm 1$ $M_\odot$ for external galaxies (of type Sc).

The most direct empirical approach is based on the consideration of stellar clusters as previously mentioned. This method was first used by van den Heuvel (1975) for the Hyades cluster, and extended by Romanishin and Angel (1980). Since then Koester and Reimers have made a systematic search for white dwarfs in young clusters with as high turn-off masses as possible (Koester and Reimers 1981, 1985, Koester 1982, Reimers and Koester 1982, 1988a,b, 1989). Their best estimate for the value of the critical mass currently is 8 (+3,−2) $M_\odot$. In addition to the maximum mass for a star that can become a white dwarf, these observations can also help to establish a relation between the initial mass on the MS and the final mass of the remnant white dwarf, thus integrating the total mass loss during the stellar life. This relation has been studied extensively by Weidemann (Weidemann and Koester 1983); recent discussions, using not only white dwarf observations, are given in Weidemann (1987) and Weidemann and Yuan (1989). The scatter in this relation, which is certainly to a large degree observational, is considerable, but it nevertheless indicates, that the remnant of a massive star is a more massive white dwarf and vice versa. Typically a 8 $M_\odot$ star seems to lead to a remnant of $\sim 1.15$ $M_\odot$, a 1 $M_\odot$ star to one of $\simeq 0.5$ $M_\odot$, but the relation is probably not linear for the intermediate masses (Weidemann 1987).

In conclusion, it is generally accepted today that fairly massive stars, possibly up to 8 $M_\odot$, will finally end up as white dwarfs. Comparing this limit with the mass distribution on the MS at birth, this means that certainly more than 90% of all stars will become white dwarfs.

### *4.3. White dwarf evolution*

Following Mestel (1952), Mestel and Ruderman (1967), Van Horn (1971), Lamb and Van Horn (1975), Koester (1978) we will, in this section, study the energy balance of white dwarfs using the classical assumption that nuclear energy is no longer available. We will return later to a discussion of this point.

*4.3.1. Energy balance.* Using equations (3.16) and (3.57) we may write the luminosity equation:

$$\frac{\mathrm{d}L_r}{\mathrm{d}m} = -\varepsilon_\nu - T\frac{\Delta s}{\Delta t}. \tag{4.1}$$

Integrating over the whole star it follows that

$$\begin{aligned} -(L - L_\nu)\Delta t = \int_0^M T\Delta s \,\mathrm{d}m &= \int_0^M \Delta u \,\mathrm{d}m + \int_0^M P\Delta\rho^{-1}\,\mathrm{d}m \\ &= \Delta U + \int_0^M P\Delta\rho^{-1}\,\mathrm{d}m. \end{aligned} \tag{4.2}$$

Extensive thermodynamic quantities in lower case letters are defined per unit mass, those using capitals refer to the whole star, for example $U$ and $u$ are the internal energy. The limits in the following integrals are all 0 and $M$. The lefthand side must be equal to the total energy $\Delta E = \Delta U + \Delta\Omega$ lost during the time step $\Delta t$, where $\Omega$ is the gravitational energy, thus the second term above can be identified with $\Delta\Omega$. Now, assuming the internal energy to be a function of $\rho$ and $T$, we may write

$$\Delta U = \int \Delta u \mathrm{d}m = \int \left( \frac{\partial u}{\partial T}\Delta T + \frac{\partial u}{\partial \rho}\Delta\rho \right) \mathrm{d}m. \tag{4.3}$$

Using the thermodynamic relation $-\rho^2 \partial u/\partial\rho = T\partial P/\partial T - P$ we get

$$\begin{aligned} \Delta U &= \int \frac{\partial u}{\partial T}\Delta T \,\mathrm{d}m + \int T\frac{\partial P}{\partial T}\Delta\rho^{-1}\,\mathrm{d}m - \int P\Delta\rho^{-1}\,\mathrm{d}m \\ &= \Delta U_{\mathrm{th}} + \Delta E_{\mathrm{grav}} - \Delta\Omega. \end{aligned} \tag{4.4}$$

The total energy change is thus, somewhat artificially, divided into two parts stemming from the temperature and density changes respectively:

$$\Delta E = \Delta U_{\mathrm{th}} + \Delta E_{\mathrm{grav}}. \tag{4.5}$$

A further relation between the different energy forms can be obtained from the hydrostatic equations. Multiplying by $4\pi r^3 \mathrm{d}r$ and integrating over the whole star, using the appropriate boundary conditions, we get

$$3\int \frac{P}{\rho}\,\mathrm{d}m = -\Omega. \tag{4.6}$$

This is the virial theorem, applied to a star in hydrostatic equilibrium. In a normal star the EOS is well approximated by the ideal gas law $P \propto \rho T$, and $P/\rho = \frac{2}{3}u_{\mathrm{th}}$. In this case we have

$$\begin{aligned} \Delta E_{\mathrm{grav}} &= \int P\Delta\rho^{-1}\,\mathrm{d}m = \Delta\Omega \\ \Delta U_{\mathrm{th}} &= -\tfrac{1}{2}\Delta\Omega \\ \Delta E &= \tfrac{1}{2}\Delta\Omega = -\Delta U_{\mathrm{th}} \end{aligned} \tag{4.7}$$

a well known result: energy loss ($\Delta E < 0$) necessarily leads to gravitational contraction. One half of the released gravitational energy is radiated away, the other half increases the internal energy and temperature (because the internal energy is thermal energy in this case).

How does this picture change for white dwarfs? For simplicity we will consider here only the highly degenerate, moderately high $\Gamma$, and non-relativistic state of matter, typical of the bulk of the matter in a normal white dwarf. The ions are considered as an ideal gas and we include only the Coulomb corrections, therefore $P = P_{\mathrm{e}} + P_{\mathrm{i}} + P_{\mathrm{C}}$. From an expansion of (3.5) for $\eta \gg 1, \beta \ll 1$ we find for the electron contribution $P_{\mathrm{e}} = (2/5)Z\eta P_{\mathrm{i}}$, while at high $\Gamma$ the Coulomb contribution is $P_{\mathrm{C}} = -0.3\Gamma P_{\mathrm{i}}$. The corresponding energies are given by:

$$u_{\mathrm{e}} = \frac{3}{2}\frac{P_{\mathrm{e}}}{\rho} \qquad u_{\mathrm{i}} = \frac{3}{2}\frac{P_{\mathrm{i}}}{\rho} \qquad \text{and} \qquad u_{\mathrm{C}} = 3\frac{P_{\mathrm{C}}}{\rho}. \tag{4.8}$$

With these assumptions we find

$$\Delta E_{\rm grav} = \int T \left( \frac{\partial P_{\rm e}}{\partial T} + \frac{\partial P_{\rm i}}{\partial T} + \frac{\partial P_{\rm C}}{\partial T} \right) \Delta\rho^{-1}\, dm. \tag{4.9}$$

The last term in the bracket is 0, the first is of order $1/\eta^2$ and will be neglected, because the second is $O(1/\eta)$. Hence,

$$\begin{aligned} \Delta E_{\rm grav} &= \int P_{\rm i}\Delta\rho^{-1}\, dm = \int \frac{P_{\rm i}}{P_{\rm i}+P_{\rm e}+P_{\rm C}} P\Delta\rho^{-1}\, dm \\ &= \int \frac{P\Delta\rho^{-1}\, dm}{1-0.3\Gamma+\frac{2}{5}Z\eta} = \left\langle \frac{1}{1-0.3\Gamma+\frac{2}{5}Z\eta} \right\rangle \Delta\Omega. \end{aligned} \tag{4.10}$$

The average is defined by this equation. In a 0.6 $M_\odot$ white dwarf at a central temperature of $10^7$ K, $\Gamma \simeq 55$, $\eta \simeq 350$, and the quantity within the bracket is $\ll 1$.

How does this compare with the release of thermal energy? From the virial theorem, using our approximations for the EOS, we find

$$\Delta U_{\rm e} + \Delta U_{\rm i} + \tfrac{1}{2}\Delta U_{\rm C} = -\tfrac{1}{2}\Delta\Omega. \tag{4.11}$$

Using our knowledge of the EOS, we may also determine the changes directly. Because $\partial u_{\rm C}/\partial T = 0$, and $\partial u_{\rm i}/\partial \rho = 0$,

$$\int \Delta(u_{\rm e}+u_{\rm i})\, dm = \Delta U_{\rm th} + \int \frac{\partial u_{\rm e}}{\partial\rho}\Delta\rho\, dm. \tag{4.12}$$

Up to first order in $1/\eta$, $\partial u_{\rm e}/\partial\rho = P_{\rm e}/\rho^2$, and

$$\begin{aligned} \Delta U_{\rm e} + \Delta U_{\rm i} &= \Delta U_{\rm th} - \int P_{\rm e}\Delta\rho^{-1}\, dm \\ &= \Delta U_{\rm th} - \int P\Delta\rho^{-1}\, dm + \int \frac{P_{\rm i}+P_{\rm C}}{P_{\rm e}+P_{\rm i}+P_{\rm C}} P\Delta\rho^{-1}\, dm \\ &= \Delta U_{\rm th} - \left( 1 - \left\langle \frac{1-0.3\Gamma}{1-0.3\Gamma+\frac{2}{5}Z\eta} \right\rangle \right) \Delta\Omega. \end{aligned} \tag{4.13}$$

Similarly

$$\int \Delta u_{\rm C}\, dm = \int \frac{\partial u_{\rm C}}{\partial\rho}\Delta\rho\, dm = -\int P_{\rm C}\Delta\rho^{-1}\, dm = \left\langle \frac{0.3\Gamma}{1-0.3\Gamma+\frac{2}{5}Z\eta} \right\rangle \Delta\Omega. \tag{4.14}$$

Collecting the terms from the virial theorem we finally get

$$\Delta U_{\rm th} = \left( \frac{1}{2} - \left\langle \frac{1-0.3\Gamma}{1-0.3\Gamma+\frac{2}{5}Z\eta} \right\rangle - \frac{1}{2} \left\langle \frac{0.3\Gamma}{1-0.3\Gamma+\frac{2}{5}Z\eta} \right\rangle \right) \Delta\Omega. \tag{4.15}$$

Together with the equation for the total energy loss (4.5) and equation (4.10) this result can be interpreted in the following way:

(i) Energy loss is also accompanied by gravitational contraction in this case, and roughly one half of the gravitational energy supplies the luminosity (for $\Gamma = 0$ this would be exactly $\frac{1}{2}$).

(ii) The thermal energy and thus the temperature decrease. Because $E_{\rm grav}$ is small, equation (4.5) can be interpreted as showing that the white dwarf loses energy at the expense of its thermal energy. For this reason white dwarf evolution can be described as 'cooling'.

(iii) Unfortunately this terminology has led to much confusion in the literature and to statements of the kind: 'A white dwarf... cannot contract and release gravitational potential energy because of the great pressure of the degenerate-electron gas. Thus its only source of energy is the thermal energy ...' (Abell *et al* 1987). This is completely wrong, as (4.15) shows. The change in gravitational energy is comparable to the total energy loss, and only the fact that $\Delta\Omega \propto \Delta R/R^2$, and that $R$ is very small, makes the resulting radius change nearly unobservable.

The half of the gravitational energy that is not radiated away increases the internal energy as in normal stars. Very approximately, for high $\eta$, $\Gamma = 0$, we have:

$$\Delta U_{\rm i} \approx \tfrac{1}{2}\Delta\Omega \approx \Delta E \qquad \Delta E = \Delta U_{\rm i} + \Delta U_{\rm e} + \Delta\Omega \qquad \text{and} \qquad \Delta U_{\rm e} \sim -\Delta\Omega. \tag{4.16}$$

Thus, while the ionic energy decreases, the electronic (Fermi) energy increases by twice that amount and the total internal energy increases by $-\frac{1}{2}\Delta\Omega$.

*4.3.2. White dwarf cooling.* Equation (4.5) and the fact that $\Delta E_{\rm grav}$ is $O(1/\eta)$ suggest a very simple approach to the evolution of white dwarfs, proposed originally in a pioneering paper by Mestel (1952). If we neglect the gravitational energy release and the neutrino emission, and assume that the thermal heat content can be described as a function of $T_{\rm C}, M, A$, we have

$$L = -\frac{{\rm d}U_{\rm th}}{{\rm d}t} = -\frac{\partial U_{\rm th}}{\partial T_{\rm C}}\frac{{\rm d}T_{\rm C}}{{\rm d}t} = -\langle c_v\rangle M\frac{{\rm d}T_{\rm C}}{{\rm d}t}. \tag{4.17}$$

Here the average $\langle c_v\rangle$, as defined by the above equation, can be calculated from the temperature and density structure of a model.

To solve this equation we need another relation between $L$ and $T_{\rm C}$, which can be obtained from an integration of the structure equations as described in the previous section. Using an analytic approximation for the opacity (Kramer's law) Mestel (1952) found:

$$\frac{L}{M} = a\,T_{\rm C}^b \simeq 1.138\;10^{-27}\,T_{\rm C}^{3.5} \quad L_\odot/M_\odot. \tag{4.18}$$

Later papers using better opacity approximations and taking into account convection in the outer layers have usually obtained slightly smaller values for the exponent, e.g. 2.3 (Ostriker and Axel 1969), 2.7 (Van Horn 1968), 1.8–2.6 (Lamb and Van Horn 1975). Koester (1976) gives the following fit:

$$\frac{L}{M} = 9.743\;10^{-21}\,T_{\rm C}^{2.56} \quad L_\odot/M_\odot. \tag{4.19}$$

With such a relation at hand the simple differential equation (4.17) can be solved easily, assuming $\langle c_v\rangle$ = constant:

$$t - t_0 = \frac{\langle c_v\rangle M^{1-1/b}}{a^{1/b}}\left(L^{1/b-1} - L_0^{1/b-1}\right). \tag{4.20}$$

Here $t - t_0$ is the time needed to cool from luminosity $L_0$ to $L$, the so called 'cooling age'. Mestel's (1952) fundamental result can be derived from this with the following simplifications:

(i) the electronic contribution to $c_v$, which is $O(1/\eta)$ is neglected and the ions are treated as an ideal gas with $c_v = \frac{3}{2}k/AH$; and

(ii) the luminosity $L_0$ at $t_0 = 0$ is assumed much higher than $L$ such that the second term can be neglected.

With these assumptions the result is:

$$t = \text{constant} \times L^{-5/7}. \tag{4.21}$$

Using (4.18) and the specific heat for $A = 12$ one finds typical cooling ages for the faintest observed white dwarfs ($L \sim 10^{-4.5}\ L_\odot$) of $10^{10}$ yr. With similar assumptions, but $a$ and $b$ from Koester (1976), the cooling time is

$$t = \frac{10^8}{A}\left(\frac{M}{M_\odot}\right)^{0.609}\left(\frac{L}{L_\odot}\right)^{-0.609} \text{yr}. \tag{4.22}$$

The time spent in the white dwarf phase thus turns out to be a major phase in stellar life, longer than the MS life time for all but the smallest masses that have had time to evolve.

Mestel and Ruderman (1967) were the first to realise that the inclusion of Coulomb effects and crystallisation must have a significant effect on cooling times. Because in the lattice phase $c_v \simeq 3k/AH$, cooling ages at the same luminosity would be increased by a factor of two. On the other hand, if $T_C$ drops below the Debye temperature, $c_v \propto T^3$ and the decrease in heat capacity should lead to much more rapid (Debye) cooling. Similar calculations, following Mestel (1952) and Mestel and Ruderman (1967), have been made by a large number of authors, using increasingly sophisticated treatment of the outer layers important for the $L(T_C)$ relation and the interior EOS ($c_v$) (Van Horn 1968, Koester 1972, 1976, Sweeney 1976, Lamb and Van Horn 1975, D'Antona and Mazzitelli 1979). While these calculations greatly improved our understanding of white dwarf evolution, especially of the processes in the outer, non-degenerate layers, they suffer from several drawbacks, if the desired degree of reality is set ever higher. The problems with the simple approximations are:

(i) The release of gravitational energy depends on the previous history of the star, especially strongly in the very early phases at high luminosities. If these are to be described adequately a very realistic starting model is needed. This can only be obtained by following the complete evolutionary history, ideally from the MS, but at least from the bottom of the AGB.

(ii) The basic structure expected in a white dwarf consists of a carbon/oxygen core (the remains of helium burning) surrounded by a helium layer (left behind by the hydrogen burning shell) and an outer, largely unchanged hydrogen-rich envelope. The masses and chemical composition in these outer shells have a significant influence on the subsequent evolution, but again, they can only be determined from realistic stellar evolution calculations.

It thus became apparent in the last five years that progress can now only be made by using the full apparatus of stellar evolution theory and calculating the complete track through the AGB and the planetary nebula stage, including mass loss and a detailed EOS as learned over the past two decades. Such calculations were initiated by

Iben and Tutukov (1984), and followed by Iben and MacDonald (1985, 1986), Koester and Schönberner (1986), and Mazzitelli and D'Antona (1986). Two examples for cooling curves $L(t)$ taken from these papers, with the sources, are given in figure 5. All curves are for a 0.6 $M_\odot$ white dwarf and the times have been renormalised to give (arbitrarily) $10^6$ yr at $L = 10\ L_\odot$. At first sight it is truly remarkable how well the most modern calculations agree with the simple Mestel result. This agreement is, however, somewhat accidental because several of the improvements in theory have opposite effects on the cooling curves and tend to cancel when averaged over the whole evolutionary time span. Let us look somewhat more closely at the differences between Mestel (1952) and the modern results.

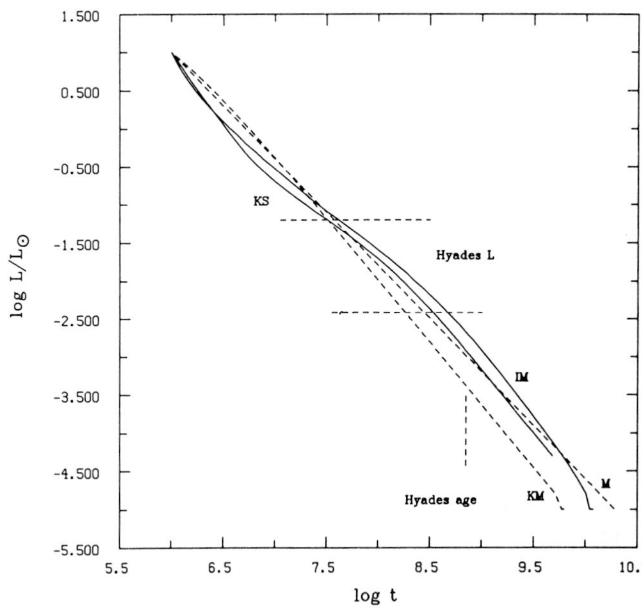


**Figure 5.** Cooling of white dwarfs with 0.6 $M_\odot$. Continuous lines: IM, Iben and MacDonald (1985); KS Koester and Schönberner (1986). Broken lines: simple models based on Mestel (1952) (M) and equation (4.20) with numbers $a$ and $b$ from Koester (1976). See text for further explanations.

At high luminosities and central temperatures energy is lost from the star not only in the form of photons, but also through the emission of neutrinos and for $\log L/L_\odot > -1$ this is the dominant mechanism. There are four processes that lead to the emission of neutrinos at high densities and temperatures: photo, pair, plasma and bremsstrahlung neutrino processes. Under the conditions prevailing in hot white dwarfs the plasma–neutrino process is usually dominant. In a plasma a photon may be considered, as a result of collective interactions with the plasma, to gain an effective mass ('plasmon') so that it decays into an $e^+e^-$ pair and subsequently into a $\nu_e\overline{\nu}_e$ pair. Itoh (1989) has recently discussed the energy loss rates in a review paper, which also contains many references and formulae.

Because neutrino emission provides an additional energy loss mechanism its effect is to accelerate the cooling and thus steepen the $L(t)$ relation. This effect is only partly compensated for by energy sources not taken into account by Mestel: electronic thermal energy and gravitational energy release $E_{\rm grav}$.

Below $\log L/L_{\odot} \sim -1$ neutrino losses become negligible and the cooling slows down, 'overtaking' the Mestel curve. For $\log(L/L_{\odot})$ between $-2$ and $-3$ Coulomb interactions become more and more important, increasing $c_v$, until below $\log L/L_{\odot} \sim -3$ the crystallisation starts at the centre, moving outward in a crystallisation front as the temperature drops. The final steep drop below $\log L/L_{\odot} \sim -5$ marks the onset of the 'Debye cooling' phase. From here on evolution would proceed rapidly, but it is doubtful whether any white dwarf has yet reached this stage.

*4.3.3. The role of nuclear burning.* Iben and Tutukov (1984) drew attention to the fact that the common neglect of nuclear energy generation might not be justified. The role of hydrogen burning as a major energy source ceases long before the beginning of the sequences in figure 5. However, it never stops completely and although the rate drops continuouslly, the other sources drop faster. Thus, in the phase between $10^8$ and $2 \times 10^9$ yr, hydrogen burning again becomes the dominant energy source (Iben and Tutukov 1984). Similar results were obtained by Mazzitelli and D'Antona (1986), while Koester and Schönberner (1986) found this contribution to be always below 12%.

These differences can be traced back to differences in the mass of the remaining hydrogen-rich layer. It is well known that the hydrogen-burning rate drops significantly once the hydrogen-mass has decreased to the order of $10^{-4}$ $M_{\odot}$(e.g. D'Antona and Mazzitelli 1979), because the density and temperature at the base of the hydrogen envelope become too low. The precise rate, however, depends strongly on the exact amount of hydrogen left. In Iben and Tutukov (1984) this is $1.5 \times 10^{-4}$ $M_{\odot}$, while Koester and Schönberner (1986) find $0.9 \times 10^{-4}$ $M_{\odot}$. This tiny difference is responsible for the different role of hydrogen burning.

The hydrogen-envelope mass thus turns out to be a very important parameter. Unfortunately it is not only changed by nuclear burning from the interior but also by mass loss — during the formation of the planetary nebula and in later phases. The formation of the nebula is very poorly understood and we do not know whether the modelling in stellar evolution calculations is very realistic. Some central stars of planetary nebula show very high mass-loss rates in the range $10^{-10}$–$10^{-7}$ $M_{\odot}$ yr (Hamann *et al* 1984, Perinotto 1987, Hutsemékers and Surdej 1989), and mass loss is also inferred from observations of hot white dwarfs (Bruhweiler and Kondo 1983) but the dependence on stellar parameters is not known. It is, therefore, doubtful, in our opinion, whether evolutionary calculations can predict the remaining envelope with any reliability.

On the other hand there are three independent arguments — pulsational stability, EUV observations of white dwarfs, and the number ratio of hydrogen versus helium dominated atmospheres — that all indicate hydrogen masses less than or equal to $10^{-7}$ $M_{\odot}$. These arguments will be discussed in later sections. If we accept these arguments as convincing, nuclear energy generation is negligible for white dwarfs with $L \leq 10$ $L_{\odot}$.

*4.3.4. Crystallising carbon/oxygen mixtures.* While our simplest models were constructed assuming a pure carbon interior, evolutionary calculations predict a mixture of carbon and oxygen, with the precise ratio depending mainly on the treatment of convection and the adopted rate of the $C^{12}(\alpha,\gamma)\,O^{16}$ reaction (Mazzitelli and D'Antona 1986) and rather uncertain at present. This ratio is important in view of another uncertainty concerning the behaviour of the mixture during crystallisation. The question is whether carbon and oxygen remain completely miscible in the solid state or whether oxygen, which freezes first, sinks towards the centre of the star. The resulting release of gravitational energy could lengthen the evolutionary time scale at low luminosities by up to $5 \times 10^9$ yr (Mochkovitch 1983, Garcia-Berro *et al* 1988a, b, Isern *et al* 1989). The phase diagram of a carbon/oxygen mixture has been studied by Stevenson (1980) and Barrat *et al* (1988) with ambiguous results. The latter study suggests that carbon and oxygen remain completely miscible, but that upon freezing a change in concentration occurs. The lengthening of the time scale is only moderate ($0.5 - 0.75 \times 10^9$ yr) in this case.

*4.3.5. Observational tests of cooling ages.* We have already noticed that the cooling ages of the faintest white dwarfs are of the order of $10^{10}$ yr; these are long but they are compatible with the age of our galaxy. It is possible, however, to make more detailed comparisons between theory and observations.

One relatively direct method is to compare the luminosities of white dwarfs in galactic clusters with the ages of the clusters. As an example we have in figure 5 indicated the range of luminosities for seven white dwarfs in the Hyades cluster. These values were calculated from the radii and effective temperatures given in Wegner *et al* (1989). The age of the cluster ($7 \times 10^8$ yr, Maeder and Mermilliod 1981) is also indicated by the vertical bar. All white dwarfs are younger than the cluster, but the faintest come close to the total age. Their MS life time cannot have been very long, which means they have had relatively high progenitor masses.

A more powerful, though less direct method, compares the distribution of observed white dwarfs over luminosity (in a normalised volume around the sun) with theoretical predictions. This distribution function, $n(\log L)$, is called the 'luminosity function', and $n(\log L)\mathrm{d}\log L$ gives the number of white dwarfs at $\log L$ in the interval $\mathrm{d}\log L$. Assuming a cooling curve $t = f(\log L)$, we can write the time interval spent by a cooling white dwarf in $\mathrm{d}\log L$ as:

$$\mathrm{d}t = \frac{\mathrm{d}f(\log L)}{\mathrm{d}\log L}\,\mathrm{d}\log L. \tag{4.23}$$

The predicted number of white dwarfs in $\mathrm{d}\log L$ is proportional to $\mathrm{d}t$, if the birthrate $\dot{n}$ is constant:

$$n(\log L)\,\mathrm{d}\log L = \dot{n}\,\mathrm{d}t = \frac{\mathrm{d}f}{\mathrm{d}\log L}\,\dot{n}\,\mathrm{d}\log L. \tag{4.24}$$

If the observed and theoretical distribution can be fitted, we will then automatically get the birthrate. We can also derive the total number of white dwarfs

$$N = \int_o^t \dot{n}\,\mathrm{d}t = \int_{L_{\max}}^{L_{\min}} n(\log L)\,\mathrm{d}\log L. \tag{4.25}$$

The actual comparison is slightly complicated by the fact that not all white dwarfs have the same mass, and that the birthrate cannot have been constant. But these

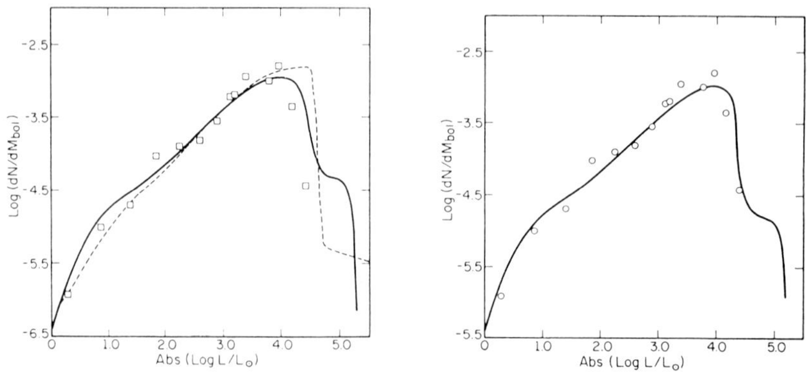


**Figure 6.** Observed and theoretical luminosity functions from Iben and Laughlin (1989, figure 14). Open symbols are the observed values after Liebert *et al* (1988), continuous curves are theoretical calculations by Iben and Laughlin (1989), while the broken curve is from Winget *et al* (1987a). The left panel is for a disk age of $9 \times 10^9$ yr, the right for $8 \times 10^9$ yr. The variable on the vertical axis in the figure is proportional to $\log(n)$ in our notation.

complications can be taken into account. This comparison has over the years been attempted by many authors, from the pioneering work by Weidemann (1967) to the most recent attempt by Iben and Laughlin (1989). Our figure 6 is taken from the latter paper and shows the observed luminosity function (Liebert *et al* 1988) together with some theoretical results.

The Mestel (1952) cooling function would correspond to a straight line in this diagram with a slope roughly equal to that in the central part of the figure. The deviation from the straight line at high $L$ is easily explained with the accelerated evolution due to neutrino losses. The deviations at the cool end have presented a more severe problem. The faintest observed white dwarfs have $\log L/L_\odot \simeq -4.6$ (Liebert *et al* 1988). The cooling times down to this luminosity are $\leq 10^{10}$ yr, less than the current best estimate of the age of the galaxy from globular clusters ($\simeq 1.5 \times 10^{10}$ yr, Iben and Renzini 1984). If white dwarf formation has occurred for more than $10^{10}$ yr, we would expect more and fainter white dwarfs ('Debye cooling', for some time suspected to be responsible for this 'red deficit' becomes important only at even lower luminosity). A possible effect that might theoretically prolong the life of white dwarfs are pycnonuclear reactions discussed in §3. With our current knowledge of white dwarf structure and masses this possibility can be ruled out: the densities in the interior are not high enough for the C $\rightarrow$ Mg transformation.

Extremely faint white dwarfs are difficult to detect and their numbers may be affected by severe selection effects. But continued effort (see, e.g., Liebert 1980, Sion 1986) has discovered only one new object (Ruiz *et al* 1986) and the falloff of the observed luminosity function is now considered to be real (Liebert *et al* 1988).

In a bold strike Winget *et al* (1987a) therefore reversed the arguments: if the oldest white dwarfs have cooling ages of $9 \pm 1.8 \times 10^9$ yr, as in their calculations, they conclude that star formation in the galactic disk did not occur much earlier. The correction for the MS life time of the progenitors adds only $0.3(+0.8, -0.1) \times 10^9$ yr to this number, thus the final conclusion is that the age of the galactic disk is $9.3 \pm 2 \times 10^9$

years. Iben and Laughlin (1989) in similar calculations improved the corrections due to the mass distribution of white dwarfs, but arrive at almost identical conclusions: the luminosity function of white dwarfs is best fitted with total cooling times for the faintest objects between 8 and $9 \times 10^9$ yr.

*4.3.6. **Reliability of theoretical ages.*** The conclusions discussed above depend heavily on the accuracy of theoretical ages at the low luminosity end. Unfortunately, a closer look at the results obtained by different groups within the last four years reveals large discrepancies, with the age at $L/L_\odot = -4.5$ ranging from 5 to $13 \times 10^9$ yr (Winget and Van Horn 1987). These authors have studied the differences in detail and found that they can all be attributed to differences in input physics and model parameters. The most important variables are the conductive opacities, carbon/oxygen ratio in the core, latent heat, hydrogen and helium layer thicknesses, and heavy element abundances in the outer layers. If they correct differentially for all these differences to bring all model calculations in accord with 'standard' assumptions they find that all ages agree at $8.2 \pm 0.5 \times 10^9$ yr. While this demonstrates that the differences between the groups can be understood it is not clear in all cases what the 'true' assumptions are. Clearly, the best available data for opacities etc should be used, but in other cases (carbon/oxygen ratio, carbon/oxygen crystallisation, remaining hydrogen and helium layers) theory does not predict unambigous results.

In conclusion: the general agreement between theoretical cooling curves and the observed luminosity function is satisfactory. It is highly probable that the drop-off in the luminosity function is real and connected to the beginning of star formation in the galactic disk $\sim 10^{10}$ yr ago. The actual numbers given for the age of the disk should, however, be regarded with some caution.

The total white dwarf number density calculated from the luminosity function is 0.0032 $pc^{-3}$ (Liebert *et al* 1988); considering the possible uncertainties (e.g. undetected white dwarfs obscured by a much brighter companion) it could be a factor of two higher (Liebert *et al* 1989). This is only 1% of the dynamically estimated total mass density in the solar neighbourhood (Bahcall 1984a,b). With a total cooling age for the faintest of $9 \times 10^9$ yr this corresponds to a white dwarf birthrate, time averaged over the history of the galactic disk, of $3.6 \times 10^{-13}$ $pc^{-3}$ $yr^{-1}$.

## 5. The visible surface: spectra and colours

Most of our knowledge about white dwarfs is derived from the light emitted at the surface. The spectral energy distribution is determined by the temperature structure (characterised by the effective temperature), the pressure stratification (determined by the surface gravity $g$) and the chemical composition of the atmospheric layers. It is the task of the theory of stellar atmospheres to construct theoretical models of the outer layers using the laws governing radiative and convective energy transfer, improve them until they fit the observations and then derive these parameters.

In this section we will give a short summary of these observations and their interpretation. This summary can by no means be complete, the main emphasis is on setting the stage for the next section. For more details we refer the reader to review papers (e.g. Liebert 1980) and the original literature.

The first step towards an understanding is, as with normal stars, to classify the observed spectra using empirical criteria. The classification scheme currently in use is that defined by Sion *et al* (1983). The major spectral types and their classification criteria are (McCook and Sion 1987; as a reminder: 'metals' are all elements heavier than helium):

DA: Only Balmer lines; no He I or metals present

DB: He I lines; no hydrogen or metals present

DC: Continuous spectrum; no lines deeper than 5% in any part of the electromagnetic spectrum

DO: He II strong; He I or hydrogen present

DZ: Metal lines only; no hydrogen or helium lines

DQ: Carbon features; either atomic or molecular in any part of the electromagnetic spectrum

In figure 7 we give a few example spectra. Historically, the letter 'D' in these designations stands for 'dwarf'. Some of the second letters have counterparts in the classification scheme of normal MS stars, although their meaning is quite different in the white dwarf case. In some cases the designations can be combined, for example 'DZA' stands for a spectrum dominated by metal lines but showing also weak hydrogen lines. If a temperature estimate is available a one digit temperature indicator (0...9) can be appended, being defined as the integer part of $50400/T$. The use of the whole electromagnetic spectrum in this classification reflects the tremendous observational progress in the last decade, mainly due to the IUE satellite, which made the whole range from 1200 to 3200 Å easily accessible.

### *5.1. Brief summary of spectroscopic results*

The basic observational facts are:

(i) No white dwarf has a chemical composition that even remotely resembles that of the sun. The differences are not factors of order unity, but many orders of magnitude.

(ii) White dwarf spectra generally fall into two well-defined groups: those with hydrogen-dominated atmospheres (spectral type DA) versus helium-dominated atmospheres (DO, DB, DC, DZ, DQ). Usually, though not in all cases, the dominating element is at least a factor of 1000 more abundant than all other elements together.

One of the most persistent problems in the field of white dwarfs is the distribution of these two basic types over the temperature range. At the highest temperatures we find the following:

DO: This class is characterised by helium-dominated atmospheres in the range 45 000 to over 100 000 K. Hydrogen is present with number abundances $\geq$ 1% (Wesemael *et al* 1985).

DA: This class is by far the largest and observed over almost the whole range of temperatures where the Balmer lines can be visible, from 6000 to $\simeq$ 70 000 K. Balmer lines should be visible at even higher temperatures, but no such DA has so far been observed. The DAO ($T_{\rm eff} \sim 60\,000$ K) has helium abundances of the order of 1% (Wesemael *et al* 1985). In all other cases it is probably much less abundant, although 1% is about the limit that can be determined from optical observations. Much lower fractions ($\sim 10^{-5}$) can be obtained from the analysis of EUV observations obtained with the Einstein and Exosat satellites (Kahn *et al* 1984, Petre *et al* 1986, Jordan *et al* 1987, Paerels and Heise 1989). Atmospheric parameters for large numbers of DAs have been determined by Shipman (1979), Koester *et al* (1979), Shipman and Sass (1980), Weidemann and Koester (1984), McMahan (1989) and especially for the hottest DA by Holberg *et al* (1986) and Finley *et al* (1989).

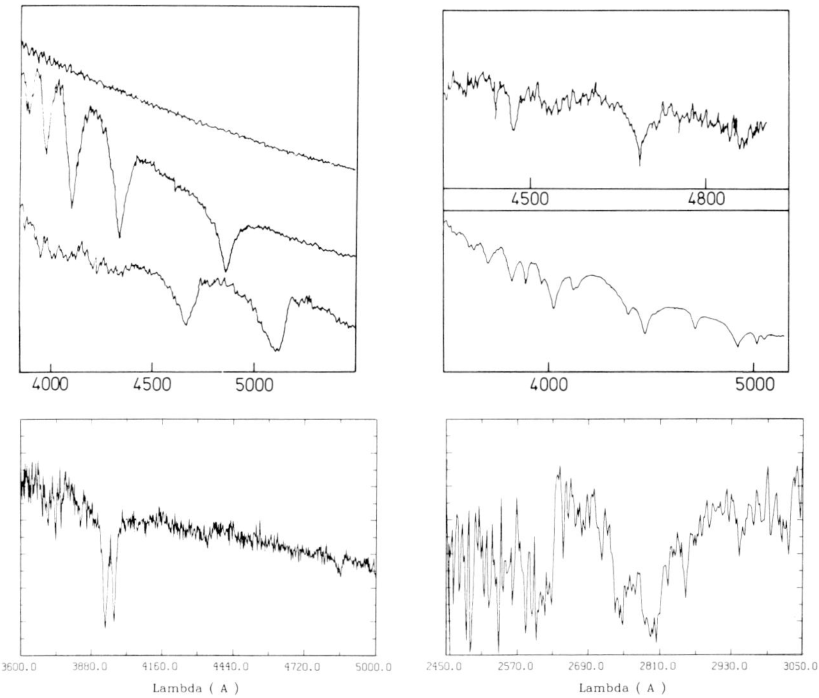


**Figure 7.** Example spectra showing the wide variety of spectral types. The two left panels show from top to bottom: L97-3 (DQ, former DC), LDS455A (DA), L870-30 (DQ), and K789-37 (DZA, from Koester *et al* 1989). The right panels show: HZ21 (DO, from Wesemael *et al* 1985), GD353 (a variable DB, from Koester *et al* 1985), and the ultraviolet spectrum of K789-37 as a demonstration of the large amount of additional information provided by UV observations (also from Koester *et al* 1989).

DB: This class shows only He I lines and is observed between 12 000 and 30 000 K, but mysteriously completely absent in the range 30 000 to 45 000 K, where the He I lines should be very strong, but no helium-rich object is observed. Improved observations have recently discovered a number of DBs showing weak hydrogen lines (DBA), with abundances of the order of $10^{-4}$, and Shipman *et al* (1987) estimate that these objects might make up as much as 20% of the DB class. Recent analyses of members of the DB class are given in the papers by Wickramasinghe and Reid (1983), Oke *et al* (1984), Liebert *et al* (1986), and Wegner and Nelan (1987).

The picture gets even more confusing at low temperatures ($<$ 12 000 K). Whereas for $T_{\rm eff}$ between 10 000 and 50 000 K the number ratio of hydrogen versus helium-dominated atmospheres is roughly 4 : 1, this ratio drops below one at the low temperature end (Fontaine and Wesemael 1987, Shipman 1989). The main helium-rich classes at low temperatures are:

DC: Historically most cool helium-rich spectra below the DB range ($<$ 12 000 K) were classified DC, because helium lines are no longer visible and only optical spectra were available. A real breakthrough was provided by IUE, which showed that all accessible DC white dwarfs (only very few unfortunately!) had strong CI resonance lines in their UV spectra and were thus reclassified DQ (Weidemann *et al* 1981, Vauclair *et al* 1981a, Wegner 1981, 1983). Whether this is true for all objects in this class is an open question.

DQ: Originally classified after the Swan bands of the $C_2$ molecule, all of the approximately twenty members of this class accessible to UV observations also show the CI resonance lines. Atmospheric parameters for many members of the DC and DQ classes have been derived by Koester *et al* (1982), Wegner and Yackovich (1984), and Wegner and Koester (1983), and references to the observations can be found in these papers. Typical carbon abundances in these objects range from $10^{-7}$ to $10^{-3}$ (Wegner and Yackovich 1984).

DZ: (approximately thirty members) As optical observations improved it turned out that several of the cooler non-DA stars formerly classified DC, DB or DBA showed CaII resonance lines (and in a few cases weak lines of Mg, Fe) and were given the designation DZ. Here again IUE observations made a breakthrough: some of the first of these (Cottrell and Greenstein 1980) revealed strong features of Mg, Fe and Si in the UV spectra. A number of observations were analysed by Zeidler-K.T. *et al* (1986); Liebert *et al* (1987a) used new and improved optical data to derive abundances. A summary of abundance determinations is given in the next section. The metal abundances, relative to helium, are always much lower than the solar values.

### *5.2. Photometry*

Obtaining spectra for faint white dwarfs is very time consuming and needs large telescopes, especially if large samples are desired. One way around this is to sacrifice high spectral resolution and observe the star through filter systems, with photomultiplier tubes, or more recently, CCDs as detectors. For historical reasons these measurements are called 'magnitudes' in astronomy, their relation to a more physical quantity such as energy flux $f(\lambda)$ (outside the earth's atmosphere, which has to have the appropriate correction) is, for example:

$$V = -2.5 \log \int_0^\infty S(\lambda) f(\lambda)\, \mathrm{d}\lambda + \text{constant} \tag{5.1}$$

where $V$ stands for visual magnitude and $S(\lambda)$ is the combined sensitivity function of the filter and the telescope optics. The constant in this equation is fixed by defining some well-observed stars as 'standards' for a particular system. The differences between magnitudes in different spectral regions are called 'colours', which are essentially intensity ratios. The most widely used system is the Johnson system UBV, with broad band filters of widths $\sim$ 1000 Å in the UV, blue and visual range of the spectrum. For white dwarfs the Strömgren system uvby with much smaller bandwidths has proven immensely useful. Observed colours in a two-colour diagram, e.g. u–b versus b–y, can be compared with theoretical predictions from model atmosphere calculations to determine the parameters $T_{\rm eff}$ and $\log g$ (an example is the work by Koester *et al* (1979) for DAs previously mentioned.).

### *5.3. Derivation of fundamental parameters*

The primary observational objective is to determine the spectral energy distribution for a star $f(\lambda)$, corrected for the influence of the earth's atmosphere. The theoretical analysis then gives, at least in principle, the atmospheric parameters $T_{\rm eff}$, $\log g$ and composition. The theoretical study of white dwarfs and their evolution on the other hand is concerned with their physical parameters mass, radius and luminosity. What is the connection between the two parameter sets?

*5.3.1. Luminosity.* The spectral energy distribution $f(\lambda)$ depends on the mean intensity $I(\lambda)$ of the stellar disk, which can never be resolved for white dwarfs, and the solid angle $\omega$ subtended by the star:

$$f(\lambda) = I(\lambda)\ \omega = I(\lambda)\ \frac{\pi R^2}{d^2} \tag{5.2}$$

assuming that no absorption occurs in the interstellar medium along the line of sight. Here $R$ and $d$ are the radius and distance of the star. The mean intensity is connected to the energy flux $F(\lambda)$ per unit area at the stellar surface by $I(\lambda) = F(\lambda)/\pi$ (e.g. Mihalas 1978). For the total energy flux, the luminosity, we then get

$$L = 4\pi R^2 \int_0^\infty F(\lambda)\, \mathrm{d}\lambda = 4\pi d^2 \int_0^\infty f(\lambda)\, \mathrm{d}\lambda. \tag{5.3}$$

The determination of luminosities thus requires two steps: the extrapolation of the energy distribution to longer and shorter wavelengths not covered by the observations; and the distance. The first step must use theoretical models, which are more accurate the broader the observed wavelength range. Fortunately most white dwarfs are relatively close to the sun and for $\sim$ 250 of them distances are known, mainly through trigonometric parallaxes (the apparent motion relative to distant stars caused by the yearly motion of the earth around the sun).

*5.3.2. Radii.* The most reliable method to determine radii utilises equation (2.1) with the effective temperature from an analysis of spectra or colours and the luminosity obtained as described above.

*5.3.3. Masses.* With the exception of seven objects (van Maanen 2, 40 Eri B, Stein 2051 B, Sirius B, Procyon B, G107-70B, BD+16°516) where the classical methods of binary orbit determination can be used, the determination of masses rests on the use of relations between mass and radius as obtained by some observational data or theory:

(i) The surface gravity as obtained from atmospheric analysis depends on mass and radius as $g = GM/R^2$.

(ii) The gravitational redshift as predicted by general relativity gives the relation $\Delta\lambda/\lambda = GM/c^2R$.

(iii) A third relation is provided by the theoretical mass–radius relation, either Hamada and Salpeter (1961), or one taking into account the effects of finite temperatures, if available.

Surface gravities can have large errors up to a factor of two and are useful mainly for statistical purposes. The most accurate determinations today use the gravitational redshift plus the theoretical relation, but are applicable only to a restricted number of bright white dwarfs in wide binary systems or clusters, because high resolution is necessary and the individual space motion has to be known from the companion or the cluster (Koester 1987b, Wegner 1989a). Any two of the above relations can be used to derive the radius in addition to the mass, even if the distance is not known.

A recent example is the study by McMahan (1989), who obtained masses and radii from the surface gravity and the mass–radius relation.

The mass distribution of DA white dwarfs is fairly well known today from statistical analyses of large samples by Koester *et al* (1979), Shipman and Sass (1980), Weidemann and Koester (1984) and McMahan (1989). This distribution is surprisingly narrow with two-thirds of all DA falling in an interval of 0.2 $M_\odot$. The mean mass is $\simeq 0.58 M_\odot$. McMahan (1989) finds a slightly lower average mass, but this is probably due to his use of surface gravities, which in our opinion are less accurate and may be affected by systematic errors, whereas the former authors preferred to use distances and derived radii plus the Hamada–Salpeter relation to define the average. For helium-rich objects the mass distribution is much less certain, but Oke *et al* (1984) found no significant difference from the DAs.

## 6. Physical processes in the non-degenerate envelope

The basic mechanism responsible for the pure, mono-elemental composition in the atmospheres of most white dwarfs was identified a long time ago by Schatzman (1949, 1958): gravitational separation of heavier and light elements in the extremely strong gravitational fields ($g \sim 10^8$ cm s$^{-2}$). The lightest element present floats to the top and forms the observable atmosphere, if no other competing mechanism exists to prevent this. The process leading to this distribution is diffusion, and ideally the diffusion equations should be solved simultaneously with the equations describing the evolution (cooling) of the white dwarfs. Such calculations have been started only recently by Iben and MacDonald (1985,1986), and with more emphasis on the chemical evolution of the outer layers by a group at Montreal (Pelletier *et al* 1986, 1989, Dupuis *et al* 1989, and references in these papers).

As we will see below the diffusion time scales are always much shorter than evolutionary time scales and the basic results for the outer layers can be obtained by neglecting the evolution as was done in the papers by Fontaine and Michaud (1979), Vauclair *et al* (1979), Alcock and Illarianov (1980a) and Muchmore (1984).

Diffusion in multi-component, non-uniform gases has been studied in detail by Chapman and Cowling (1970) and Burgers (1969) by solving the set of coupled Boltzmann equations for each species. An excellent review of the applications of this theory to element distributions in stars has been presented by Vauclair and Vauclair (1982). We will now derive the basic equation in a very simple, intuitive way and compare the final result with the more accurate derivation.

### 6.1. The diffusion equation

Let us assume a mixture of two species, distinguished by the indices 1 and 2, with partial pressures $P_i$, densities $\rho_i$, and masses $m_i$. If there is no interaction between the species, the equations of motion are:

$$\begin{aligned} \rho_1 \dot{v}_1 &= -\nabla P_1 + \rho_1 F_1 \\ \rho_2 \dot{v}_2 &= -\nabla P_2 + \rho_2 F_2 \end{aligned} \tag{6.1}$$

where $\nabla P_i$ stands for the radial gradient of the partial pressures $P_i$ and the $F_i$ are external forces (per mass) due to gravitational and electric fields (and possibly other effects).

If a relative motion exist between the species we expect interactions proportional to the relative velocity, which on dimensional grounds we write purely phenomenologically as a 'viscosity force'

$$F_v = \frac{\rho_1 \rho_2 (v_1 - v_2)}{\rho \tau} \tag{6.2}$$

with a typical collision time scale $\tau$. We then have

$$\begin{aligned} \rho_1 \dot{v}_1 &= -\nabla P_1 + \rho_1 F_1 - \frac{\rho_1 \rho_2}{\rho \tau}(v_1 - v_2) \\ \rho_2 \dot{v}_2 &= -\nabla P_2 + \rho_2 F_2 + \frac{\rho_1 \rho_2}{\rho \tau}(v_1 - v_2). \end{aligned} \tag{6.3}$$

Subtraction of the two equations leads to

$$(\dot{v}_1 - \dot{v}_2) + \frac{v_1 - v_2}{\tau} = \left(-\frac{1}{\rho_1}\nabla P_1 + F_1\right) - \left(-\frac{1}{\rho_2}\nabla P_2 + F_2\right). \tag{6.4}$$

Solutions of this type of differential equation show an initial phase with exponential increase of the relative velocity of duration $\sim \tau$. After this short time a stationary state is reached asymptotically with

$$v_1 - v_2 = \tau \left(-\frac{1}{\rho_1}\nabla P_1 + \frac{1}{\rho_2}\nabla P_2 + F_1 - F_2\right). \tag{6.5}$$

Using the hydrostatic equation this can be transformed into:

$$v_1 - v_2 = \frac{\tau k T}{\langle m \rangle}\left(-\frac{1}{c_2}\nabla \ln c_1 + \frac{m_1 F_1 - m_2 F_2}{kT}\right) \tag{6.6}$$

$$\text{with} \qquad \langle m \rangle = \frac{m_1 m_2}{c_1 m_1 + c_2 m_2} \qquad \text{and} \qquad c_i = n_i / n. \tag{6.7}$$

The exact solution using the Boltzmann equations (Chapman and Cowling 1970) can be brought into the form:

$$v_1 - v_2 = D_{12}\left(-\frac{1}{c_2}\nabla \ln c_1 + \frac{m_1 F_1 - m_2 F_2}{kT} + \alpha_T \nabla \ln T\right). \tag{6.8}$$

Here, $D_{12}$ is the binary diffusion coefficient; its order of magnitude can be estimated from a comparison of (6.8) and (6.6):

$$D_{12} \sim \tau k T / m \sim \tfrac{1}{3}\tau v_{\text{th}}^2 \sim \tfrac{1}{3} l v_{\text{th}} \tag{6.9}$$

where $v_{\text{th}}$ is the thermal velocity and $l$ the mean free path. The actual calculation of $D_{12}$ requires the calculation of a collision integral which depends on the nature of the particle interaction potential. The most accurate results available for stellar plasmas are currently those of Paquette *et al* (1986b). The additional term on the right hand side is the thermal diffusion due to a temperature gradient, with the thermal diffusion coefficient $D_{12}\alpha_T$. This term is taken into account in numerical computations, but is only of minor importance according to the results of Paquette *et al* (1986a) and will be neglected in our simple considerations.

In the equilibrium state the relative velocity vanishes and the concentration gradient is given by the right hand side of (6.8) . In order to understand the nature of the external forces let us, as an example, identify the species 1 and 2 with protons and electrons. Because the gravitational force is stronger on the protons gravity tends to separate electrons and protons, creating an electric field $E$ which counteracts the separation. As a starting point we assume that the matter remains very closely neutral ($n_\mathrm{p} = n_\mathrm{e}$). The concentration gradient therefore vanishes and in equilibrium we have:

$$m_\mathrm{p} g + eE = m_\mathrm{e} g - eE \tag{6.10}$$

$$E = -\frac{m_\mathrm{p} + m_\mathrm{e}}{2e}\, g \simeq -\frac{m_\mathrm{p}}{2e}\, g. \tag{6.11}$$

The electric force is thus of the same order of magnitude as the gravitational force. Is this consistent with our assumption about neutrality? From Poisson's equations we have:

$$\nabla \cdot \boldsymbol{E} = -4\pi e(n_\mathrm{p} - n_\mathrm{e}) = -\frac{m_\mathrm{p}}{2e}\, \nabla \cdot \boldsymbol{g} = \frac{m_\mathrm{p}}{2e}\, G\, 4\pi n_\mathrm{p} m_\mathrm{p} \tag{6.12}$$

or

$$-\frac{n_\mathrm{p} - n_\mathrm{e}}{n_\mathrm{p}} = \frac{G m_\mathrm{p}^2}{2e^2} \sim 4 \times 10^{-37}. \tag{6.13}$$

The charge separation is indeed extremely small, justifying our assumption (cf Rudkjøbing 1952). The total force on electrons and protons per unit volume is:

$$F = n_\mathrm{p}(m_\mathrm{p} g + eE) + n_\mathrm{e}(m_\mathrm{e} g - eE) = n\, \frac{m_\mathrm{p} + m_\mathrm{e}}{2}\, g \tag{6.14}$$

which means that we are justified in neglecting the electric field in the hydrostatic equation as long as we take the mean mass of a particle to be $(m_\mathrm{p} + m_\mathrm{e})/2$.

In the realistic white dwarf problem we are confronted not with binary, but at least ternary mixtures: two ionic species plus electrons. Because the velocity of the electrons is, however, much higher, it is a good approximation to assume that they have always reached their equilibrium distribution. The electric field can then be determined from

$$\frac{\partial P_\mathrm{e}}{\partial r} = -n_\mathrm{e}\,(m_\mathrm{e} g + eE) \tag{6.15}$$

and (6.8) describes the diffusion of the two ionic species in the neutralising background of the electrons. It should be clear from the preceding discussion that this diffusion must always occur in an initially homogeneous stratification because equilibrium between electric and gravitational forces cannot in general be obtained for two ions with different $(Z, A)$. The heavier ion always diffuses downwards and vice versa until equilibrium ($v_1 - v_2 = 0$) is achieved. In equilibrium the two species are completely separated at a transition zone, where the composition changes drastically over a few pressure scale heights $\sim 1/(\partial \ln P/\partial r)$. Only the presence of more effective mixing processes can prevent the separation. In order to clarify the importance of competing processes we need more quantitative estimates.

### 6.2. Diffusion velocities and time scales

Koester (1989a) has recently estimated the relative diffusion velocities in a hydrogen/helium mixture using the new diffusion coefficients (Paquette *et al* 1986b). Table 1 shows the results ($v_{12}$ in cm s$^{-1}$) at a depth below the surface corresponding to $\Delta M = 10^{-17}$ to $10^{-10}$ $M_\odot$ for white dwarfs with different effective temperatures. The second line for each mass entry gives a diffusion time scale $L/v_{12}$ in yr, where $L$ is the typical length scale of the outer layers. This time can therefore be interpreted as the time necessary to achieve complete hydrogen/helium separation. For comparison the last line gives approximate cooling ages at that effective temperature.

**Table 1.** Hydrogen/helium diffusion velocities in cm s$^{-1}$ (upper line for each entry) and time scales (lower line, in yr) evaluated at different depths below the surface (given as fractional mass).

| | $T_{eff}$ | | | | |
|---|---|---|---|---|---|
| $\log \Delta M/M_\odot$ | 60 000 | 50 000 | 40 000 | 30 000 | 20 000 |
| −10 | 1.6 x $10^{-5}$ | 1.4 x $10^{-5}$ | 1.2 x $10^{-5}$ | 9.8 x $10^{-6}$ | 7.3 x $10^{-6}$ |
| | 1387 | 1457 | 1538 | 1635 | 1798 |
| −11 | 4.9 x $10^{-5}$ | 4.2 x $10^{-5}$ | 3.6 x $10^{-5}$ | 2.9 x $10^{-5}$ | 2.2 x $10^{-5}$ |
| | 298 | 316 | 336 | 362 | 393 |
| −12 | 1.5 x $10^{-4}$ | 1.3 x $10^{-4}$ | 1.1 x $10^{-4}$ | 8.7 x $10^{-5}$ | 6.5 x $10^{-5}$ |
| | 63 | 68 | 73 | 79 | 87 |
| −13 | 4.7 x $10^{-4}$ | 4.0 x $10^{-4}$ | 3.4 x $10^{-4}$ | 2.7 x $10^{-4}$ | 2.0 x $10^{-4}$ |
| | 13 | 14 | 15 | 17 | 19 |
| −14 | 1.5 x $10^{-3}$ | 1.3 x $10^{-3}$ | 1.0 x $10^{-3}$ | 8.2 x $10^{-4}$ | 5.9 x $10^{-4}$ |
| | 2.8 | 3.0 | 3.3 | 3.6 | 4.1 |
| −15 | 4.7 x $10^{-3}$ | 4.0 x $10^{-3}$ | 3.3 x $10^{-3}$ | 2.6 x $10^{-3}$ | 1.8 x $10^{-3}$ |
| | 0.6 | 0.6 | 0.7 | 0.8 | 0.9 |
| −16 | 1.5 x $10^{-2}$ | 1.3 x $10^{-2}$ | 1.0 x $10^{-2}$ | 8.1 x $10^{-3}$ | 5.7 x $10^{-3}$ |
| | 0.1 | 0.1 | 0.1 | 0.2 | 0.2 |
| −17 | 4.9 x $10^{-2}$ | 4.1 x $10^{-2}$ | 3.3 x $10^{-2}$ | 2.6 x $10^{-2}$ | 1.8 x $10^{-2}$ |
| | 0.02 | 0.03 | 0.03 | 0.03 | 0.04 |
| Cooling age | 1.6 x $10^{6}$ | 2.7 x $10^{6}$ | 5.3 x $10^{6}$ | 1.5 x $10^{7}$ | 7.9 x $10^{7}$ |

The interpretation is the following: the diffusion velocities seem rather small (although they are much larger than in MS stars due to the high $g$). However, *the derived time scales are much shorter than the cooling ages.* The fundamental result is therefore that in the absence of competing processes we should expect the equilibrium situation with only the lightest element present (hydrogen in this case) visible at the surface. This is in agreement with the time-dependent studies of the diffusion process by Pelletier *et al* (1989) and also with the majority of observations.

At the high temperature end of the white dwarf sequence, however, some objects (DO, DAO) show both helium and hydrogen in their optical spectra. Moreover, the interpretation of Einstein and Exosat observations in the EUV region has demonstrated that even the DAs, with a pure hydrogen spectrum in the optical range, must have traces ($10^{-3}$ to $10^{-5}$) of helium in their atmospheres (see previous section). Why did the helium not disappear long ago?

### 6.3. Competing processes

Detailed calculations of diffusion in the presence of other processes (see below) have not yet been made. For the purpose of estimating their influence we may, however, estimate typical velocities associated with these processes and compare them with diffusion velocities. If they are much smaller the process can probably be neglected safely, whereas at the other extreme diffusion should be inhibited. Recent discussions under this aspect have been presented by Michaud (1987) and Koester (1989a).

*6.3.1. Meridional circulation.* It has long been known that the rotation of a star should lead to large-scale internal motions, the Eddington–Vogt meridional circulation (Eddington 1925, Vogt 1925). Tassoul and Tassoul (1983) found a stationary solution for the flow pattern in a cooling white dwarf with the assumption that viscous stresses exactly balance the transport of angular momentum in the surface boundary layer. They discussed a 0.8 $M_\odot$ white dwarf with $L = 10^{-2}$ $L_\odot$ and found time scales for mixing by circulation currents greater than the Hubble time. Since circulation velocities increase with luminosity and decrease with mass they could, however, be substantially larger for a 60 000 K (with $L \sim L_\odot$) white dwarf of 0.6 $M_\odot$. Also, Tassoul and Tassoul (1983) estimate that the time scale to reach the stationary solution is $\sim 10^{10}$ yr, much longer than the cooling ages of hot white dwarfs. It cannot, therefore, be completely ruled out at present that rotation might have an effect in the surface layers of very hot white dwarfs (Koester 1989a), considering the observed helium/hydrogen ratios in the hot DO and DAO objects. On the other hand, white dwarfs are observationally found to be very slow rotators (Greenstein and Peterson 1973, Pilachowski and Milkey 1984, 1987, Koester and Herrero 1988), which would tend to make the effects of rotation less important.

*6.3.2. Mass loss.* The possibility of mass loss in hot DAs has been inferred from the presence of shortward-shifted absorption components of highly ionised species in their UV spectra (Bruhweiler and Kondo 1983). The interaction of diffusion with mass loss has been discussed in detail by Michaud (1987). Because of the continuity of the flow, mass loss leads to a systematic velocity in the transition region which is superposed on the diffusion velocity. It is easy to estimate with the numbers in table 1 that these velocities are comparable if the mass loss is of the order of $10^{-15}$ $M_\odot$ yr$^{-1}$. With somewhat different assumptions about the original abundance distribution Michaud (1987) derived $10^{-14}$ $M_\odot$ yr$^{-1}$ as the critical rate at which diffusion becomes ineffective.

We have already mentioned that there are strong arguments that at least a large fraction of the DA white dwarfs have very thin outer hydrogen layers, probably $\leq 10^{-14}$ $M_\odot$. Such a thin layer would be stripped off by a mass loss of this magnitude in a very short time, which is not observed. It might well be, however, that mass loss in previous evolutionary phases has led to these thin hydrogen layers and stopped when it came close to the hydrogen/helium transition zone, although at the moment this is pure speculation. For the remainder therefore let us simply assume that mass loss, if present, is small enough not to interfere with diffusion.

*6.3.3. Accretion of interstellar matter.* Accretion is the flow of matter from outside the star onto the surface, and except for the sign of the flow the estimate for the critical rate is the same as in the case of mass loss. The accretion rate necessary ($\sim 10^{-15}$ $M_\odot$ yr$^{-1}$) to compete with diffusion is much higher than expected on theoretical grounds for an isolated hot white dwarf in the tenuous interstellar medium, although it could be much higher when the white dwarf traverses a cloud, a region of higher density. However, from the distribution of these clouds, the time between encounters is estimated as $\sim 5 \times 10^7$ yr, larger than the age of the hottest DAs and much larger than diffusion times. For the helium/hydrogen ratio in hot white dwarfs accretion therefore seems to be unimportant. This process plays, however, a crucial role in our current interpretation of the DZ objects at much lower temperature.

*6.3.4. Radiative levitation.* In our simple derivation of the diffusion equation we have neglected the radiation pressure on the ionic species in the consideration of external forces. Depending on the ionisation state and absorption cross sections the absorption of photons within the outer layers of the star, and the associated transfer of momentum to the absorbing particle, results in an outward force that decreases the effective gravitational acceleration felt by the ions. Because absorption efficiencies can vary over many orders of magnitude for different ions this force is highly selective and therefore can lead to relative diffusion.

In their pioneering paper Vauclair *et al* (1979) have shown that radiation forces in some cases can support carbon, nitrogen and oxygen ions in the atmosphere against the effect of gravitational separation. This concept has been extended in recent years in work by Morvan *et al* (1986), Vauclair (1989), and Chayer *et al* (1989), who demonstrated that the elements mentioned above (and silicon) can be supported in the form of 'metal clouds' in the outer layers, although the present attempts to interpret the silicon and carbon observed in three DAs in terms of this model are so far not completely successful (Koester 1989a).

Selective radiation pressure on helium had been invoked (Petre *et al* 1986, Shipman 1987) as an explanation for the helium seen in DAO and DA. However, Vennes *et al* (1989) have recently shown that this effect is not strong enough to explain the observed abundances.

*6.3.5. Convection.* Whenever the temperature gradient $\mathrm{d}\ln T/\mathrm{d}\ln P$ in a star becomes steeper than the adiabatic gradient $(\mathrm{d}\ln T/\mathrm{d}\ln P)_{\mathrm{ad}}$ the stratification becomes unstable (Schwarzschild criterion) and macroscopic motion or convection sets in. Because no convection theory exists that is applicable under the conditions in stars (extremely high Reynolds numbers), the energy transport is usually calculated in the 'mixing-length' approximation (Böhm-Vitense 1958).

The importance of convection in the envelopes of white dwarfs was first pointed out by Böhm (1968). The more efficient energy transport decreases the temperature gradient and changes the relation between the surface and interior temperatures. Far more important, however, is the role of convection for the interpretation of the element distribution in the outer layers and its possible change during the cooling history. The flow velocities in convection zones are much higher than diffusion velocities and the flow is highly turbulent. Within a convection zone, matter will therefore be completely mixed and no concentration gradient can develop (Vauclair and Vauclair 1982). If the concentration of a species is lower below the convection zone, diffusion can still occur at the bottom of the zone; in this case the whole mass in the zone acts as a 'reservoir' and the time scales for the disappearance of elements out of the atmosphere may become much longer.

### *6.4. Spectral types and spectral evolution*

The interpretation of spectral types and atmospheric abundances in white dwarfs is at the forefront of current research; even review papers appear at the rate of several per year (Vauclair 1989, Koester 1987a, Shipman 1989, Shipman 1987, Fontaine and

Wesemael 1987, Liebert *et al* 1987b, Sion 1986, Vauclair and Liebert 1987). It is not possible to report all speculations in this review, we will instead concentrate on a few major problems and our current understanding, following the papers by Fontaine and Wesemael (1987), Koester (1987a) and Shipman (1989).

*6.4.1. Hydrogen versus helium-dominated atmospheres.* The main observational facts that need explanation are:

(i) the absence of hydrogen atmospheres with $T_{\mathrm{eff}} \geq 70\,000$ K, whereas helium-rich objects are observed well over 100 000 K (the PG1159 objects, a subgroup of the DO);

(ii) the absence of DBs in the gap between 30 000 and 45 000 K; and

(iii) the change in the number ratio of hydrogen versus helium atmospheres from $\simeq 4:1$ at higher temperatures to $\leq 1$ below 10 000 K.

Because the change in luminosity and surface temperature of cooling white dwarfs is a continuous function of time, the above facts strongly suggest that white dwarfs can change their atmospheric composition and even spectral type during their evolution. This does not exclude, however, the possibility that different progenitor channels exist, with some stars entering the white dwarf phase without any hydrogen left, while others retain a small hydrogen envelope. This point of view was strongly emphasised by Shipman (1989). An argument in favour of this view is that observations of the central stars of planetary nebulae show a strong variation of helium/hydrogen ratios (Mendez *et al* 1986, 1988). On the theoretical side Iben (1984) has argued that it might depend on the exact phase, between helium-shell flashes, when a star leaves the AGB whether it retains a hydrogen envelope or not.

While the possibility of more than one birth channel cannot be excluded it cannot by itself explain the observed gaps. Fontaine and Wesemael (1987) have therefore taken the opposite view and tried to explain the relevant facts assuming a single channel. In their view, the extremely hot helium-rich PG1159 objects and the somewhat cooler DO are the progenitors of all white dwarfs. They argue that the mass loss on the AGB and in the planetary nebula phase strips the star of all matter outside the hydrogen-burning shell. It thus enters the white dwarf stage with a helium-rich atmosphere. However, a small amount of hydrogen is hidden in the deeper layers and floats to the top, until enough hydrogen ($\sim 10^{-16}\ M_\odot$) is accumulated that the star appears as a DA. This could explain the mixed abundances in the DO and DOA, and it could be responsible for the fact that all white dwarfs are DAs when reaching $T_{\mathrm{eff}} \simeq$ 45 000 K.

During further cooling a convection zone develops in the helium layer underneath the hydrogen, and around 30 000 K it can mix with the hydrogen and transform the star into a DB, if the hydrogen layer thickness is $\leq 10^{-15}\ M_\odot$. This explanation of the DB gap was first proposed by Liebert *et al* (1986, 1987b). If the hydrogen layer is larger the star remains a DA. This scenario requires that a substantial fraction of the hot DAs have extremely thin hydrogen layers (MacDonald 1989), in conflict with the results of traditional stellar evolution calculations.

Michaud *et al* (1984) recently suggested an interesting theoretical possibility. They assumed that the tail of the carbon distribution from the core and of the hydrogen distribution from the top meet at the intermediate layer, leading to quiet hydrogen burning, which could reduce the hydrogen mass to $10^{-7}\ M_\odot$. The detailed calculations by Iben and MacDonald (1985), however, did not support this conclusion.

Fortunately there are several observational clues that can give information about the deeper layers in DA white dwarfs. We have already mentioned that the EUV observations of $\sim$ 20 DA white dwarfs require the presence of an absorbing element below 200 Å, which is usually assumed to be helium. Most of the analyses so far have assumed a homogenous helium/hydrogen mixture in the atmosphere and determined helium abundances of the order of $10^{-3}$ to $10^{-5}$. If we believe instead that the DAs have very thin hydrogen layers we would have to use stratified atmosphere models, where the transition between hydrogen and helium is determined from the diffusion equilibrium and the total hydrogen mass present. The hydrogen opacity decreases strongly at shorter wavelength below 912 Å, and the emerging spectrum originates from deeper and deeper layers. The presence of helium can therefore become 'visible' in the EUV, although the star appears as a normal DA without any helium features in the optical spectrum.

Such atmospheric models and synthetic spectra were published by Jordan and Koester (1986) and applied to the interpretation of Exosat observations by Koester (1989a, 1989b) and Vennes *et al* (1989). These papers demonstrate that the EUV observations can indeed be interpreted with stratified models and the required hydrogen layers are very thin, ranging from $4 \times 10^{-14}$ to $4 \times 10^{-16} M_{\odot}$. The one channel/mixing scenario of Fontaine and Wesemael (1987) thus looks quite promising, but several severe problems remain:

(i) The PG1159 and DO groups can account for only $\sim$ 20% of all cooler white dwarfs, unless the assumed evolutionary time scales at high temperatures are completely wrong (Weidemann and Yuan 1989).

(ii) In his detailed study of stratified atmospheres Koester (1989a) was not able to find models that would look like a DA above 30 000 K and turn into a DB around this value. The model grid was relatively coarse, but if this transition is to occur at 30 000 K for about 20% of all DAs, it certainly requires an extreme fine-tuning of the mass in the hydrogen layer.

The situation seems to be much clearer at the low temperature end regarding the change in the number ratio of hydrogen versus helium atmospheres. At about 15 000 K a hydrogen atmosphere develops a convection zone below the surface, which increases dramatically in depth below 10 000 K. If this zone reaches into the helium zone, the much higher densities there lead to a complete dilution of the surface hydrogen and turn the star into a helium-dominated object. Following original suggestions by Strittmatter and Wickramasinghe (1971) and Baglin and Vauclair (1973) the evolution of convection zones and the mixing process has been studied in detail by Koester (1976) and Vauclair and Reisse (1977). Both groups found that mixing would occur below 10 000 K, if the hydrogen layer mass is less than $\sim 10^{-10}$ $M_{\odot}$. This process is now generally believed to be responsible for the changing number ratio, and the strong implications are that a large fraction of the DAs must indeed have thin hydrogen envelopes of the order of this value or even smaller. In §7 we will encounter yet another argument for the existence of very thin layers.

*6.4.2. The DBA white dwarfs.* With the discovery of six new DBs with traces of hydrogen in their spectra from the Palomar–Green survey (Shipman *et al* 1987) the DBA now constitute a well established subclass. Shipman *et al* (1987) estimate that about 25% of all DBs might show hydrogen abundances in the range $10^{-3}$ to $10^{-5}$. Koester (1987a) gives a table of recent abundance determinations for these objects. It is tempting to connect these observations with the above DB/DA scenario, assuming that the hydrogen is primordial and mixed within the helium convection zone around $T_{\rm eff}$ = 30 000 K. However, the mass in this convection zone is $\sim 10^{-6}$ $M_{\odot}$; with a typical observed hydrogen abundance of $10^{-4}$ this means a total hydrogen mass of $10^{-10}$! Such hydrogen layers (although low compared to 'evolutionary' masses) are much too thick to mix with the helium above $T_{\rm eff} \simeq 10\,000$ K — these objects would remain pure DAs throughout most of their life.

The only explanation we are left with seems to be accretion of interstellar matter onto stars with helium atmospheres. Since hydrogen is lighter than helium, it remains within the convection zone, whereas all heavier elements can leave this zone at the bottom by diffusion. Reasonable accretion rates in dense clouds are $10^{-15}$ $M_\odot$ yr$^{-1}$, a typical crossing time for a cloud $10^5$ to $10^6$ yr, which could easily produce the required amount of hydrogen.

*6.4.3. Metals in DZ white dwarfs.* This class of about thirty objects shows metal lines, predominantly CaII, in the optical range, in rare cases also magnesium and iron. Table 2 shows the present status of abundance determinations.

**Table 2.** Metal abundances in DZ, DBZ, DBAZ and DZA.

| | | Log($N_{\rm metals}/N_{\rm He}$) | | | | | |
|---|---|---|---|---|---|---|---|
| Object | $T_{\rm eff}$ | Ca | Mg | Si | Fe | H | Source |
| GD303 | 19 000 | −10.0 | | | | ≤−5.5 | SAK |
| GD378 | 16 000 | −10.0 | | | | −4.0 | SAK |
| GD61 | 15 000 | −10.0 | | | | −3.8 | SAK |
| GD85 | 15 000 | ≤−10.0 | | | | ≤−5.0 | SAK |
| G200-39 | 14 500 | −10.3 | | | | −4.0 | KSSA, KW |
| GD40 | 14 000 | −7.4 | −7.2 | < −6.7 | −7.2 | ≤−5.2 | SGB, S, SG |
| GR488 | 13 500 | −9.0 | | | | −4.0 | KSSA |
| GD408 | 13 000 | −10.0 | | | | ≤−5.0 | SAK |
| PG2322+119 | 12 000 | −8.7 | | | | −5.2 | LWG |
| CBS78 | 12 000 | −8.0 | | | | <−4.7 | SSWLS |
| GD124 | 12 000 | ≤−10.0 | | | | ≤−5.0 | SAK |
| K789-37 | 9 500 | −8.8 | −8.1 | <−9.0 | −8.0 | −3.8 | KWK |
| L119-34 | 9 300 | −9.1 | −7.7 | −9.2 | −8.1 | | ZWK |
| Ross640 | 8 800 | −9.1 | −7.2 | −7.3 | −8.3 | −3.5 | ZWK |
| GD401 | 8 800 | −8.0 | <−8.0 | <−8.5 | <−8.2 | <−3.4 | CG, L, K |
| G105-B2B | 8 800 | −9.5 | <−7.2 | <−7.5 | | | ZWK |
| GD95 | 8 200 | −9.6 | −7.2 | | −8.5 | | LWG |
| L745-46A | 7 800 | −10.4 | −9.0 | −9.5 | < −11.0 | −5.9 | ZWK, K |
| G165-7 | 7 500 | −7.2 | | | | <−4.0 | WL |
| G163-28 | 6 500 | −10.7 | <−8.0 | | | | ZWK |
| G157-35 | 6 000 | −10.5 | −8.7 | | −9.2 | | ZWK |
| vMa2 | 6 000 | −10.6 | −8.8 | | −8.8 | | LWG, ZWK |
| Solar ratios to hydrogen | | −5.70 | −4.58 | −4.48 | −4.40 | | |
| Solar ratios to calcium | | | 1.12 | 1.22 | 1.30 | | |

Sources for abundances: CG, Cottrell and Greenstein 1980; K, Koester 1987a KSSA, Kenyon *et al* 1988; KW, Koester and Weidemann 1989; KWK, Koester *et al* 1989; L, Liebert 1987 (H$\alpha <$ 2Å); LWG, Liebert *et al* 1987a; S, Shipman 1984; SAK, Sion *et al* 1988a; SG, Shipman and Greenstein 1983; SGB, Shipman *et al* 1977; SSWLS, Sion *et al* 1986; WL, Wehrse and Liebert 1980; ZWK, Zeidler-K T *et al* 1986.

What can we learn from these data? The abundances are always very low, much lower than the solar ones. That they can be detected at all these levels is due to the extreme transparency of helium at these temperatures. Clearly these elements cannot be primordial, because the diffusion times ($\sim 10^6$ yr) are orders of magnitudes smaller than the age of these white dwarfs ($>10^9$ yr). There is also no chance that these elements could have been mixed up again from deep below, the convection zones are not deep enough.

Again, the only solution seems to be accretion. In the basic scenario, the convection zone is regarded as a homogeneously mixed reservoir. Matter is added through accretion at the outer boundary, predominantly during a passage through an interstellar cloud, and diffuses out of this zone at the bottom. The time scales for this diffusion are defined by the mass contained in the convective zone divided by the diffusive flux; typical times are $10^5$ to $10^6$ yr. The typical accretion rates that are needed under these circumstances to produce the observed metal abundances are $10^{-16}$ to $10^{-14}$ $M_\odot$ yr$^{-1}$. With these simplifications, the equation governing the time evolution of abundances in the convection zone is extremely simple and leads immediately to the following conclusions:

(i) Hydrogen as the lightest element does not diffuse, but always stays in the convection zone. If the accreted matter has solar composition, this sets an upper limit to the total accreted mass (from the observed hydrogen abundances or upper limits).

(ii) If accretion times are very long, an equilibrium is reached, where the abundances are solely determined by the accretion rate and the diffusion time scales. If significant accretion, however, is possible only in dense interstellar clouds, this equilibrium stage would never be reached, because the crossing time for a cloud, of the order of $5 \times 10^5$ yr (Petre *et al* 1986), is not long compared to the diffusion time scales. We would then expect that the atmospheric composition shortly after the encounter should reflect the composition of the accreted matter (with the exception of helium).

(iii) When accretion stops, elements (except hydrogen) disappear with an e-folding time determined by their diffusion time scales. The ratio of, for example, calcium/hydrogen can thus only decrease from the original value.

(iv) Diffusion time scales for magnesium are always longer than those of calcium (Paquette *et al* 1986a), meaning that the abundance ratio of calcium/magnesium can only decrease from the value in the accreted matter.

(v) The diffusion time scale for iron in the relevant temperature range is only slightly longer than that of calcium, the ratio of iron/calcium should therefore roughly reflect the accretion abundances.

In interpreting the observations in view of this scenario we have to keep in mind that the determined abundances have uncertainties of typically a factor of two to three and the diffusion velocities of about a factor two. To facilitate the conclusions, table 2 also contains the solar ratios of the observed elements to hydrogen and to calcium.

If we assume a conservative upper limit of $10^{-4}$ for the hydrogen abundance in the six cooler objects where calcium is observed, but hydrogen is not, there are 21 cases with calcium/hydrogen determinations or lower limits. Taking the data at face value the ratio in 11 of them would be larger than that for solar abundances. This problem has been known for several years and the inevitable conclusion seems to be that matter cannot be accreted with solar abundances in these cases. There is no indication that the predictions of the diffusion theory can be that much in error, but since much less is known about the accretion process it seems easier to blame it and assume that the accretion of hydrogen is prevented.

There is indeed a process that might do just that: the so called 'propeller mechanism' proposed originally by Illarionov and Sunyaev (1975) and applied to white dwarfs by Wesemael and Truran (1982). In this mechanism, a rotating white dwarf with a small magnetic field might screen out protons and other light charged particles, while heavier elements might be accreted, especially if they are locked up in dust grains. In order for this mechanism to work the hydrogen near the star must be ionised, and this only happens if the effective temperature is high enough, $\geq 10\,000$ K. Liebert *et al* (1987a) were the first to point out that the slightly cooler objects K789-37 and Ross640, which in fact show about the solar calcium/hydrogen ratio, seem to provide empirical evidence for a critical temperature of 10 000 to 11 000 K for the accretion of hydrogen. This could be an acceptable explanation, however, the most likely explanation for the DBA phenomenon requires just the opposite, accretion of hydrogen, in exactly the same range of effective temperatures. Moreover, the even cooler objects GD401, L745-46A, and G165-7 also show a calcium/hydrogen ratio above the solar value although the 'propeller mechanism' is not supposed to work at these temperatures.

The evidence against solar accretion of hydrogen is weakened considerably if we allow an uncertainty of a factor of ten in the ratio. That would leave only four contradictory cases, with only one of them (G165-7) below 10 000 K. Sion *et al* (1988a) have used their results for the DBZ as a strong argument *in favour* of solar accretion.

For GD40, GD401 and K789-37 there is also a contradiction with the predictions for the magnesium/calcium ratio, which is a factor of three to ten smaller than the solar ratio, although, taking all possible uncertainties together, this does not seem to be as severe as in the hydrogen problem.

Summarising the discussion of metal abundances, we could state that the basic accretion-diffusion-mixing scenario is supported at least by part of the observations and is probably correct. The details, however, present some puzzles, which will only be solved if we understand the accretion mechanism, especially the accretion of dust grains and the possible deviations from solar abundances in the accreted matter (Alcock and Illarionov 1980b).

*6.4.4. Carbon in DQ.* What is different with carbon compared to other elements? If we believe in gravitational settling as the basic mechanism working in the outer layers of white dwarfs, we would, in helium-rich objects, expect that the thin helium layer floats on top of the next heavy element (carbon), built up by helium burning in previous stages of evolution. This provides the base for the current explanation of the DQs, first proposed by Koester *et al* (1982) and studied in much more detail by Fontaine *et al* (1984) and Pelletier *et al* (1986). We assume that there is a smooth transition between helium and carbon layers, with a structure given by the diffusion equilibrium. If the convection zone in the helium layer reaches down to the tail of the carbon distribution it can bring up traces of carbon to the surface without transforming the outer layers into almost pure carbon, as would be the case for an infinitely sharp transition layer. The detailed calculations predict a maximum carbon abundance in the range 11 000 to 12 000 K, due to a combined effect of extent of convection zones and the physics of pressure ionisation. This correlation is indeed present in the observations as was first noted by Wegner and Yackovich (1983). This theory even allows, at least in principle, an empirical determination of the thickness of the helium layer: log $M_{\rm He}/M$ must be between $-3.5$ and $-4.0$ in order to fit the observations (Pelletier *et al* 1986), a result, which again is in conflict with the current predictions of pre-white dwarf evolution (Mazzitelli and D'Antona 1986, Iben and Tutukov 1984, Iben and MacDonald 1985).

## 7. Pulsating white dwarfs

The luminosity of an isolated star may vary because the star is oscillating about its equilibrium state. Examples for this phenomenon are RR Lyrae and Cepheid variables, where the entire star expands and contracts radially with a well-defined period. The study of the pulsational properties of stars has a long history and excellent reviews may be found in Rosseland (1949), Ledoux (1958), Ledoux and Walraven (1958), Cox (1980) and Van Horn (1984).

The principal aim of stellar pulsation theory is to study the reaction of a stellar model in hydrostatic equilibrium to various types of perturbations. The simplest but as it turns out most important of these are small, adiabatic, radial perturbations, which determine whether the star is dynamically stable or not, on a dynamical time scale $\sim (G\rho)^{-1/2}$. This is the shortest of the various instability time scales in stellar models (Ledoux and Walraven 1958).

### *7.1. Radial oscillations*

The radial oscillations of white dwarfs were first studied by Sauvenier-Goffin (1949) in the Newtonian approximation. The differential equation describing such oscillations obtained by linearising the basic equations of momentum, energy and mass conservations with respect to small radial perturbations $\xi(r) = \delta r/r$ (Cox 1980) is:

$$r\ddot{\xi} = 4\pi r^2 \xi \frac{\mathrm{d}}{\mathrm{d}m}\left[(3\Gamma_1 - 4)\,P\right] + \frac{1}{r}\frac{\partial}{\partial m}\left(16\pi^2\Gamma_1 P\rho r^6 \frac{\partial \xi}{\partial m}\right) \tag{7.1}$$

where $r$, $P$, $m$, $\rho$, $\Gamma_1$ refer to the equilibrium solution and $\Gamma_1$ is the adiabatic index $(\mathrm{d}\ln P/\mathrm{d}\ln\rho)_{\mathrm{ad}}$. As a simple example consider a motion where $\xi$ and $\Gamma_1$ are spatially constant, i.e. independent of $m$ so that

$$\ddot{\xi} = -(3\Gamma_1 - 4)\frac{Gm}{r^3}\,\xi. \tag{7.2}$$

Hence, for solutions with $\xi \propto \exp(\mathrm{i}\sigma t)$,

$$\sigma^2 = (3\Gamma_1 - 4)\frac{Gm}{r^3} \tag{7.3}$$

and an oscillatory solution with a period $\Pi = 2\pi/\sigma$ exists if $\Gamma_1 > 4/3$. For $\Gamma_1 < 4/3$, we have $\sigma^2 < 0$, and the star is dynamically unstable on a time scale $\tau \sim (G\rho)^{-1/2}$.

It can be shown that these conclusions remain valid for a general stellar model with the changes that $\Gamma_1$ is replaced by an appropriate average over the star and the single frequency $\sigma$ is replaced by an eigenvalue spectrum, with the lowest frequency given by a relation similar to (7.3). The detailed numerical calculations of Sauvenier-Goffin (1949) confirmed the dimensional estimates derived from (7.3) that the fundamental pulsation periods of white dwarfs should be 10 s or somewhat less.

The dynamical stability of spherical stars within the framework of general relativity was studied by Chandrasekhar (1964), who found that the stability criterion is modified to

$$\langle \Gamma_1 \rangle > \frac{4}{3} + \frac{2KGM}{Rc^2} \tag{7.4}$$

where $K$ is positive and $\sim 1$. Hence general relativistic effects may be important for neutron stars, where $GM/Rc^2 \sim 0.1$. For the (ideal) degenerate electron gas in the interior of white dwarfs, $\Gamma_1 \simeq 5/3$ if the electrons are non-relativistic, and $\Gamma_1 \simeq 4/3$, if they are highly relativistic. Thus although $GM/Rc^2 \lesssim 10^{-3}$ for white dwarfs, general relativistic effects can play an important role for the highest density white dwarfs.

The dynamical stability of high mass white dwarfs had been of considerable interest about two decades ago. Finzi and Wolf (1967) had suggested that Type I supernovae may form because of the collapse of a white dwarf which contracts slowly due to electron capture reactions. Baglin (1968) argued that white dwarfs with reasonable compositions have a limiting mass of 1.2 $M_\odot$ corresponding to $\rho_C \simeq 7.8 \times 10^8$ g cm$^{-3}$ and that white dwarfs with $M > 1.2\ M_\odot$ would become dynamically unstable not because of electron capture but because of general relativistic effects.

On the other hand an analysis by Wheeler *et al* (1968) showed that white dwarfs with $1.8 \times 10^9 \lesssim \rho_C \lesssim 4 \times 10^{10}$ g cm$^{-3}$ could exist for various assumptions regarding $\mu_e$ and that for typical compositions with average atomic weights $A$ given by $12 \lesssim A \lesssim 40$ general relativistic effects are not the prime cause of the instability. The difference in the results arose from the Wheeler *et al* (1968) assumption that, because the dynamical time scale is shorter than the electron capture time scale, the composition of the star may not change during the oscillation. This leads to a larger value of $\Gamma_1$ (Meltzer and Thorne 1966, Chanmugam 1977).

The discovery of the first radio pulsars with a period of 1.3 s (see Smith 1977 and references therein) led to speculations that it might be a radially oscillating white dwarf star, since the shortest periods for such radial oscillations in the fundamental mode are $\simeq 2$ s (Wheeler *et al* 1968, Faulkner and Gribbin 1968, Skilling 1968, Chanmugam 1977) and occur for white dwarfs near the limiting mass with the highest central density. The discovery of the Crab pulsar with a period of 0.033 s, however, confirmed instead that pulsars are rotating neutron stars. At present, no white dwarfs with high central densities have yet been discovered, so that a Newtonian description is adequate.

Besides the determination of the stability and the frequency spectrum the aim of stellar pulsation theory is to study whether radial oscillations can be excited. For this purpose it is necessary to study the 'vibrational stability' of the star (Ledoux and Walraven 1958), which includes thermal effects and energy transport so that energy dissipation can be included in the analysis. In the fully non-adiabatic treatment $\sigma$ becomes complex, with the imaginary part proportional to the work done by pressure and gravitational forces during one cycle. The magnitude of this energy integral can be estimated in the linear, quasi-adiabatic approximation using the adiabatically derived eigenfunctions. In this case radial oscillations are excited if the energy integral $C < 0$, where

$$C = \int_0^M \left(\frac{\delta T}{T}\right)_{\rm ad} \delta\left(\varepsilon_{\rm N} - \frac{1}{\rho}\nabla \cdot \boldsymbol{F}\right) \, \mathrm{d}m \tag{7.5}$$

and $\varepsilon_{\rm N}$ is the energy generation rate due to nuclear burning, $\boldsymbol{F}$ the radiative flux and the suffix ad means that $(\delta T/T)$ is calculated in the adiabatic approximation.

In general the principal mechanisms that can excite oscillations are related to the two terms in brackets: the '$\varepsilon$-mechanism' caused by a certain temperature dependence of the energy generation, and, more importantly, variations in the energy transport caused by the temperature and density dependence of the stellar opacity ('$\kappa$-mechanism'). In this mechanism, radiation travelling through a partial ionisation zone is more easily absorbed when the star is compressed, because the opacity increases. This energy is then released when the star expands so that the pulsation is driven.

Vauclair (1971a, see also Ledoux and Sauvenier-Goffin 1950) has considered the effects of a hydrogen burning shell, the classical $\kappa$-mechanism in the He II ionisation zone, and the increase in the luminosity perturbations in the inner parts of the isothermal core due to opacity variations etc. He found that the $\kappa$-mechanism can drive the oscillations of both the fundamental and first overtone (Vauclair 1971b) for DA white dwarfs with $T_{\rm eff} \sim 10\,000$ K. This instability against radial oscillations was confirmed in all subsequent calculations (Cox *et al* 1980, 1987, Saio *et al* 1983, Starrfield *et al* 1983).

*7.1.1. Discovery of the first variable white dwarf.* The calculations by Sauvenier-Goffin (1949) had motivated searches for variable white dwarfs with periods of the order of 10 s, but these searches for over a decade failed to reveal any examples of variability (e.g. Lawrence *et al* 1967). Because white dwarfs had shown no evidence for variations they became useful examples of 'standard' stars whose luminosities do not vary with time. However, observations of one such white dwarf 'standard' (HL Tau 76) in December 1965 by Landolt (1968) revealed that its brightness varied with a period of 750 s. This is, as is so often the case in astronomy, another example of a major discovery, which was totally unexpected and serendipitous. This again stimulated searches for variability and soon other variable white dwarfs with periods between $\sim$ 100 and $\sim$ 1000 s were discovered (Lasker and Hesser 1969, 1971). These periods were puzzling, however, because they were substantially larger than the expected periods for the radial oscillations. In spite of continued searches no radial variability in the short period range has yet been discovered (Robinson 1984, and references therein).

### *7.2. Non-radial oscillations*

Extensive theoretical studies of non-radial oscillations in stars had been carried out a long time ago although, unlike the case of radial oscillations, there were no observational examples of stars exhibiting such oscillations. These studies showed that the possible pulsation frequencies split up broadly into three distinct classes of oscillations, the g- (or gravity) modes, the p- (or pressure) modes, and the f-mode which exists when the spherical harmonic index $l \geq 2$ (Cowling 1941). The p-modes correspond to oscillations where the restoring force is due to pressure so that they are essentially accoustic waves. They have periods which are $\leq (G\rho)^{-1/2}$, with the higher modes having shorter periods. Calculations of non-radial oscillations of highly luminous white dwarfs by Harper and Rose (1970) (see also Baglin and Schatzman 1969) showed that g-mode oscillations had periods of $\sim$ 50 s. On the other hand, these periods scale roughly $\propto 1/T$, where $T$ is the core temperature (Baglin and Heyvaerts). This led Chanmugam (1972), and Warner and Robinson (1972), to propose that if the variable white dwarfs were cooler they could exhibit g-mode non-radial oscillations consistent with the observed periods $\geq$ 100 s.

Consider a spherically symmetric Newtonian model of a star which is perturbed so that the Lagrangian perturbation $\delta \boldsymbol{r} = \delta r\, \boldsymbol{e}_r + r\delta\theta\, \boldsymbol{e}_\theta + r\sin\theta\, \delta\phi\, \boldsymbol{e}_\phi$, where $(r,\theta,\phi)$ are spherical polar coordinates and $\boldsymbol{e}_r$, $\boldsymbol{e}_\theta$, $\boldsymbol{e}_\phi$ are orthogonal unit vectors in the radial and transverse directions (see Cox 1980). If the perturbation has a time dependence $\propto \exp(i\sigma t)$ and the angular part is expanded in spherical harmonic functions, then the linearised mass, energy and momentum equations and Poisson's equation $\nabla^2\psi = 4\pi G\rho$ may be written as follows in the adiabatic approximation (Cox 1980):

$$
\begin{aligned}
u &= r^2 \delta r \\
y &= P'/\rho \\
\frac{\mathrm{d}y}{\mathrm{d}r} &= \frac{\sigma^2 + Ag}{r^2}\, u - Ay - \frac{\mathrm{d}\psi'}{\mathrm{d}r} \\
\frac{\mathrm{d}u}{\mathrm{d}r} &= \frac{\rho g}{\Gamma_1 P}\, u + \left(\frac{l(l+1)}{\sigma^2} - \frac{r^2\rho}{\Gamma_1 P}\right) y + \frac{l(l+1)}{\sigma^2}\, \psi' \\
\frac{1}{r^2}\frac{\mathrm{d}}{\mathrm{d}r}\left(r^2 \frac{\mathrm{d}\psi'}{\mathrm{d}r}\right) &= -\frac{4\pi G\rho A}{r^2}\, u + \frac{4\pi G\rho^2}{\Gamma_1 P}\, y + \frac{l(l+1)}{r^2}\, \psi'. \qquad (7.6)
\end{aligned}
$$

Here primes denote Eulerian perturbations and $u$, $y$ and $\psi'$ are functions only of $r$, $g$ is the local acceleration of gravity, $l$ is the spherical harmonic index, while

$$A = \frac{1}{\rho}\frac{\mathrm{d}\rho}{\mathrm{d}r} - \frac{1}{\Gamma_1 P}\frac{\mathrm{d}P}{\mathrm{d}r}. \qquad (7.7)$$

These equations may then be solved by using appropriate boundary conditions at the centre and the surface of the star. The eigenvalues corresponding to the angular oscillation frequencies are given by $\sigma_{nl}^2$ with $n = 1,2,3,\ldots$. The spectra of the oscillation frequencies show that the g-modes may be subdivided into two further classes called $g^+$- and $g^-$-modes. The $g^+$-modes are stable and correspond to $\sigma_{nl}^2 > 0$ while the $g^-$-modes are unstable and correspond to $\sigma_{nl}^2 < 0$. The f-mode and p-modes have $\sigma_{nl}^2 > 0$ and are stable. For $\sigma_{nl}^2 > 0$, $\sigma_{nl}^2$ increases with both $n$ and $l$. A schematic diagram, based on Smeyers (1967), of the different modes is shown in figure 8, taken from Cox (1980).

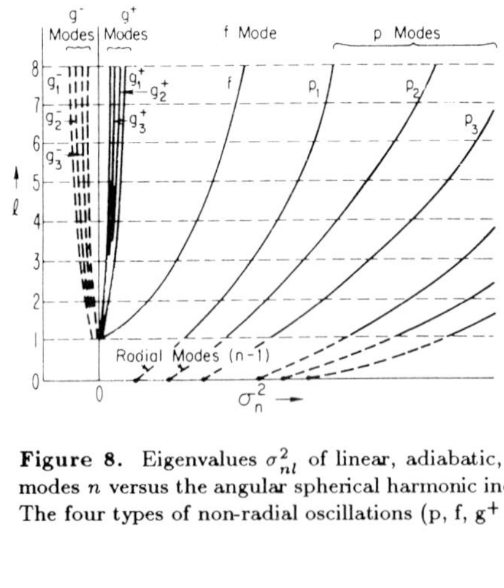


**Figure 8.** Eigenvalues $\sigma_{nl}^2$ of linear, adiabatic, non-radial oscillations for various modes $n$ versus the angular spherical harmonic index $l$ (schematic) from Cox (1980). The four types of non-radial oscillations (p, f, $g^+$ and $g^-$) are shown.

The quantity $A$ plays an important role in showing the connection between non-radial oscillations and convective instability. Consider a stratified fluid in which the physical quantities depend only on the radial coordinate $r$. If a fluid element is displaced from $r$ to $r + \delta r$ the buoyancy force per unit volume acting upwards on the element is $\rho g A \delta r$ (Cox 1980). The fluid element will tend to return to its original position if $A < 0$, in which case the fluid is said to be convectively stable and vice versa. If there is no friction or heat exchange with the fluid element and the fluid is convectively stable, it will undergo oscillations about the equilibrium position with an angular frequency $N$, known as the Brunt–Väisäila frequency, where $N^2 = -Ag$. If $|A|$ is sufficiently small it may be shown that for g-modes $\sigma_{nl}^2 \propto \langle -A \rangle$.

For a zero-temperature degenerate electron gas $A = 0$ so that there are no g-modes with non-zero frequencies. However, if the thermal corrections to the equation of state of a degenerate electron gas are included it may be shown that $A \propto T^2$ so that $\sigma_{nl} \propto T$ (Baglin and Heyvaerts 1969). Hence, simple dimensional analysis shows that the periods for g-mode non-radial oscillations of a white dwarf increase as the star cools (Chanmugam 1972).

Detailed numerical calculations of non-radial oscillations of white dwarfs composed of iron confirmed that the g-mode periods increase monotonically as the star cools and evolves (Osaki and Hansen 1973). Furthermore, they showed that a 0.398 $M_\odot$ white dwarf with a luminosity $L = 7.38 \times 10^{-5}\ L_\odot$ would have non-radial quadrupole oscillations ($l = 2$) for $n = 1$ and 2 of 209.8 and 291.1 s respectively, reasonably close to the basic periods of 212 and 273 s then reported for R548. Similar results were also obtained by Brickhill (1975).

The major unsolved problem at that time was to understand why the particular g-modes with low values of $l$ would be excited. The resolution of this problem took almost a decade and was aided by a number of observations of these variable white dwarfs.

### 7.3. *Observations of variable* DA *(*DAV*) white dwarfs*

Searches for low amplitude ($\lesssim 0.1$ magnitudes) luminosity variations in white dwarfs were significantly helped by the development of the two-star high-speed photometer by Nather (1973). This enabled high-speed photometric observations to be carried out on a star with simultaneous observations of a nearby 'standard' so that atmospheric fluctuations could be subtracted out and real variations in the luminosity more accurately determined. The use of the two-star photometer soon led to the discovery of additional variable DA white dwarfs, a class which became known as ZZ Ceti stars and is now called DAV. It also showed that some earlier reports of variable white dwarfs were spurious, while others were verified. Instead of recalling the whole history of variable white dwarf discoveries, we give in table 3 an up-to-date list of all presently known variables with references to the discovery papers.

In several cases multiple periods were observed. Thus, for example, R548 has two pairs of observed periods (table 3), which were interpreted (Stover *et al* 1980) as due to splitting by the rotation of the star with a period of 1.5 days. For a star oscillating with a frequency $\sigma_{nl}$ the $(2l + 1)$ fold degeneracy is lifted by rotation giving the $m$ modes:

$$\sigma_{nlm} = \sigma_{nl} \pm m(1 - c_l)\Omega, \tag{7.8}$$

with $m = -l, -l + 1, \ldots, l$, where $\Omega$ is the rotational frequency of the star and $c_l$ is a constant depending on the internal structure of the star (Ledoux 1958) with $c_l \simeq$

**Table 3.** Observations of variable white dwarfs.

| Object | Periods (s) | References |
|---|---|---|
| Variable PNN | | |
| K1-16 | 1520, 1730 | GB |
| Longmore 4 | 1800 | BC |
| Variable DO (DOV) | | |
| PG1159-035 | 390–833 | MSLG |
| PG1707+427 | 447.16, 448.95, 334.6 | BGGL |
| PG2131+066 | 450 | BGGL |
| PG0122+200 | 614, 467, 449, 435, 400, 380, 364, 336 | BG |
| Variable DB (DBV) | | |
| GD358 | 142.3–952 | WRNF |
| PG1654+160 | 148.5–851 | WRN |
| PG1351+489 | 489.5 | WNH |
| PG1115+158 | 1000 | WNH |
| PG1456+103 | 423–854 | GBGL |
| CBS114 | 650 | WC |
| KUV05134+2605 | 350–900 | GWGL |
| Variable DA (DAV) | | |
| HL Tau 76 | 746.2, 494.2 | L |
| R548 = ZZ Ceti | 212.77, 213.13, 274.25, 274.77 | LH |
| G29-38 | 613, 677, 931, 1016 | SK |
| G38-29 | 1024, 938 | MR75 |
| BPM31594 | 308, 617 | M76 |
| G117-B15A | 107.6, 119.8, 126.2, 215.2, 271.0, 304.4 | RU |
| Ross808 | 769, 1250 | MR76 |
| GD99 | ?? | MR76 |
| BPM30551 | 606.8, 682.7, 744.7, 840.2 | HLN |
| G207-9 | 739, 557, 318, 292 | RM |
| L19-2 | 192.4 | M77 |
| GD154 | 780, 1186 | RSNM |
| GD385 | 256.13, 256.33, 128.11 | FMCLPV |
| G191-16 | 865 | MFDGLS |
| G185-32 | 200 | MFDGLS |
| G255-2 | 685, 830 | VDC81 |
| G226-29 | 109.3 triplet | GM |
| GD66 | 197, 256, 273, 304 | DVC |
| G238-53 | 685, 830 | FW |
| PG2303+243 | 900.5, 794.5, 623.4, 397.3, 349.3 ... | VDC87 |

References: BC, Bond and Ciardullo 1989; BG, Bond and Grauer 1987; BGGL, Bond *et al* 1984; DVC, Dolez *et al* 1983; FMCLPV, Fontaine *et al* 1980; FW, Fontaine and Wesemael 1984; GB, Grauer and Bond 1984; GBGL, Grauer *et al* 1988; GM, Gustafson and McGraw 1980; GWGL, Grauer *et al* 1989; HLN, Hesser *et al* 1976; L, Landolt 1968; LH, Lasker and Hesser 1971; M76, McGraw 1976; M77, McGraw 1977; MFDGLS, McGraw *et al* 1981; MR75, McGraw and Robinson 1975; MR76, McGraw and Robinson 1976; MSLG, McGraw *et al* 1979; RM, Robinson and McGraw 1976; RSNM, Robinson *et al* 1978; RU, Richer and Ulrych 1974; SK, Shulov and Kopalskaya 1971; VDC81, Vauclair *et al* 1981b; VDC87, Vauclair *et al* 1987; WC, Winget and Claver 1989; WNH, Winget *et al* 1987b; WRN, Winget *et al* 1984; WRNF, Winget *et al* 1982b.

0.25 for white dwarfs (Hansen *et al* 1977). The recent observations of GD358 (a DBV), which was considered the most convincing case for rotational splitting, have, however, cast some doubt on this interpretation (Hill 1987). Hill showed that the splitting changed or even disappeared completely between two observations, which hardly can be reconciled with an interpretation due to rotation of the star.

The most important step towards the solution for the excitation mechanism was, however, the realisation that the structure of the outer layers as discussed in the previous sections plays a crucial role in it.

### *7.4. Excitation mechanisms*

In 1981 several groups independently proposed excitation mechanisms in white dwarfs involving ionisation zones in realistically stratified outer layers, corresponding to diffusion equilibrium. These calculations required the fully non-adiabatic solution of the non-radial pulsation equations. Dziembowski and Koester (1981) showed that the $\kappa$-mechanism operating in a $He^+$ ionisation zone could drive the oscillations. Calculations of white dwarf models with a helium layer of $10^{-4}$ $M_\odot$ showed that g-modes with periods $\sim$ 100 to 1000 s could be driven if the mass of the hydrogen layer is less than $1.2 \times 10^{-13} M_\odot$.

Dolez and Vauclair (1981) in similar calculations also found driving by the hydrogen ionisation zone for slightly more massive hydrogen layers up to $\sim 10^{-11}$ $M_\odot$. Winget (1981), and Winget *et al* (1981) carried out the most comprehensive study of the effects of hydrogen and helium ionisation zones in exciting non-radial oscillations in evolving white dwarfs with masses of 0.6 and 0.8 $M_\odot$. These authors and Starrfield *et al* (1982) showed that models with *relatively* thick surface layers $M_H$ in the range $10^{-14}$ to $10^{-8}$ $M_\odot$ could provide the appropriate periods in the observed DAV white dwarfs. Calculations of the radial part of the eigenfunctions $\delta r/r$ showed that in specific cases it is trapped in the hydrogen layer (see figure 9).

Thus, as may be seen in figure 9, the $g_{11}$ mode, which has relatively negligible amplitude in the inner parts, can be easily excited. These studies clearly showed that the manner in which the outer envelope is stratified plays a crucial role in selecting a particular mode of oscillations. An implication of this effect, on the other hand, is that the observations in combination with theoretical calculations might be used to empirically probe the stratifications of white dwarfs below the visible surface.

This success for the variable DAs led Winget *et al* (1982a) to suggest that DB white dwarfs may also exhibit non-radial oscillations driven by a partial helium ionisation zone.

### *7.5. DBV, DOV and PNN variables*

The prediction that the helium white dwarfs would also show g-mode oscillations was soon verified when Winget *et al* (1982b) made the exciting discovery of pulsations from the DB white dwarf GD358. Since then six more DB variables have been discovered (see table 3). Several of these variables show very complex frequency spectra displaying many periods, with GD358, the prototype, having over 20. In this and similar cases table 3 only gives the dominant periods or period ranges, because the periods are not always stable from one observing run to the next, probably due to beating between closely spaced and unresolved frequencies.

In addition to the DAV and DBV white dwarfs two other classes of variable white dwarfs (or pre-white dwarfs) have been reported. One of these is a subclass of the DO

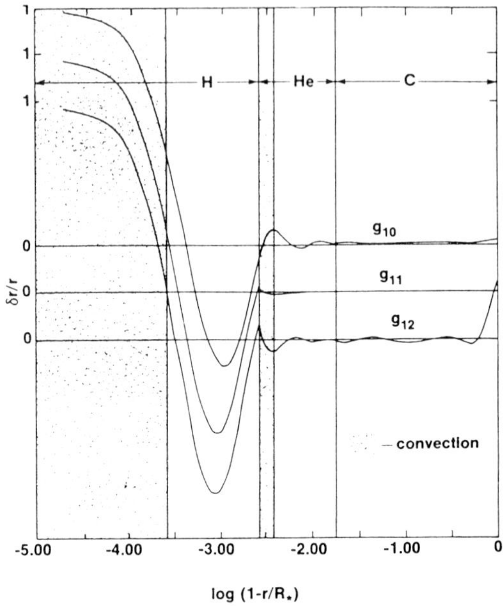


**Figure 9.** The radial part of the eigenfunctions $\delta r/r$ (as a function of the fractional radius $r/R_\star$ and normalised to unity at the surface) for the $g_{10}$, $g_{11}$, and $g_{12}$ modes for compositionally stratified white dwarf models. The scales are displaced for clarity. Note the trapping of the modes in the hydrogen layer. From Winget *et al* 1981.

stars and was originally named PG1159 variables after the first member (PG1159-035) to be discovered (McGraw *et al* 1979). The DOV stars (or GW Vir stars as they are also called) are extremely hot ($> 10^5$ K, Werner *et al* 1989), but the temperatures are not well known. The PNN class presently consists of two members that are central stars of planetary nebulae; the prototype K1-16 was discovered recently by Grauer and Bond (1984). They are less evolved and more luminous than the DOVs and their observed periods are slightly longer. The excitation mechanism for the two hot variable classes is poorly understood; the various possibilities are discussed in detail by Kawaler (1987).

### 7.6. *Instability strips, period changes and white dwarf 'seismology'*

The determination of the regions in the HRD where pulsational instability occurs, the 'instability strips', is hampered for the two hot classes by the lack of adequate spectral analyses and temperature determinations.

*7.6.1. Instability strips.* McGraw and Robinson (1976) made the important observation that the DAV stars have very similar effective temperatures between 10 000 and 15 000 K. The range of instability was subsequently studied by McGraw (1979), Greenstein (1982), Weidemann and Koester (1984), Fontaine *et al* (1985), Wesemael *et al* (1986) and Lamontagne *et al* (1987, 1989). The estimates seem at present to have converged on the values 11 500–13 000 K for the red and blue edges. Furthermore, it appears that essentially all white dwarfs in this temperature range are variable (Fontaine *et al* 1982). When comparing this observational fact to the theoretical predictions two main conclusions emerge (e.g. Winget *et al* 1981, 1982a, Bradley *et al* 1989):

(i) The thickness of the outer hydrogen layer must be in the range from $10^{-13}$ to $10^{-8}$ $M_\odot$. If the layer is thicker than $\sim 10^{-8}$ $M_\odot$ the blue edge of the instability strip drops rapidly to 8000 K, incompatible with the observations.

(ii) Energy transport in the convection zone must be much more effective than is usually assumed. In the mixing-length approximation different efficiencies are modelled by changes of the free parameter $\alpha$, the ratio of mixing-length to pressure scale height. Agreement with the observed instability strip is only obtained for $\alpha = 3$, whereas most stellar evolution calculations use values between 1 and 2.

The first conclusion, which is in sharp disagreement with predictions of stellar evolution calculations, has been challenged recently by Pesnell (1987) and Cox *et al* (1987). They proposed a new driving mechanism called 'convection blocking' based on a different treatment of the interaction of pulsations with convection and find pulsational instability even with thick hydrogen layers of the order of $10^{-4}$ $M_\odot$. The final decision about this mechanism may have to wait until a theory of time-dependent convection is available, although a comparison of convection and pulsation time scales casts some doubt on the realism of their assumptions (Koester 1987a). They were, however, unable to find unstable modes above $T_{\text{eff}} = 11\,500$ K, which is clearly not in agreement with the empirical instability strip.

The instability strip for the DB variables is less well known, because the determination of effective temperatures for hot DBs is extremely difficult in general. Koester *et al* (1985) found $T_{\text{eff}} = 24\,000 \pm 1000$ K for the prototype GD358 from an analysis of UV and very good optical data. The calculations of Winget *et al* (1983) had predicted instability in the range $16 \times 10^3$ to $19 \times 10^3$ K assuming the standard mixing-length theory, and up to $26 \times 10^3$ to $29 \times 10^3$ K with more efficient convection. Liebert *et al* (1986) found somewhat higher temperatures for the hot DBs from UV observations and determined $24 \times 10^3$–$28 \times 10^3$ K for the instability strip, with an uncertainty of 2000 K for the red and blue edges. In any case both determinations agree on the conclusion that convection must be very efficient if the theoretical calculations are correct. Contrary to the DAV case not all DBs in the instability strip seem to be variable. The location of the instability strip in the HRD are indicated in figure 4.

*7.6.2. Period changes.* In the cooler DBV and DAV white dwarfs cooling dominates the evolution. Because $\sigma_{nl} \propto T$ in this case (Baglin and Heyvaerts 1969) one can deduce that periods of g-mode oscillations increase with time roughly as

$$\frac{\dot{\Pi}}{\Pi} \propto -\frac{\dot{T}}{T} \propto \frac{1}{\tau_c} \tag{7.9}$$

where $\tau_c$ is the cooling time scale and $T$ the core temperature. Current predictions for $\dot{\Pi}$ from numerical models for the DAVs are in the range 2 to $7.5 \times 10^{-15}$ (Bradley *et al* 1989), while the best available observations are upper limits of $9.6 \times 10^{-15}$ for R548 (Tomaney 1987) and $1.25 \times 10^{-14}$ for G117-B15A (Kepler *et al* 1989). It will take a few more years of observations before meaningful comparisons become feasible.

Predicted period changes for the DBV white dwarfs are larger, $\sim 10^{-13}$ (Kawaler *et al* 1986). However, in spite of extensive observations of the best candidate

PG1351+489 the analysis so far has not given conclusive results. Kawaler and Hansen (1989) suggest that non-linear effects might be responsible, while Goupil *et al* (1988) claim that this object might be on the way to displaying chaotic behaviour.

Due to their much younger age and correspondingly faster evolutionary time scales the situation is more promising for the hot DOVs. Because these objects are not yet completely degenerate, their period changes should show the combined effect of contraction and cooling. While cooling *increases* the period, contraction *decreases* it. Theoretical computations (Kawaler *et al* 1985) indicated that cooling dominates and that the periods should increase by $\sim 10^{-11}$. From observations of PG1159-035 spanning five years Winget *et al* (1985) found a secular period change of the expected magnitude; however, the period was *decreasing* with time. The reason for this discrepancy is not understood and stellar rotation, non-adiabatic effects, magnetic fields, or excitation of modes with significant amplitudes in the core (where cooling dominates) have been proposed as possible solutions (Kawaler and Hansen 1989).

*7.6.3. Mass determinations from pulsation spectra.* The mean period spacing over several modes for successive values of $n$ is fairly constant, at least for higher values of $n$, with deviations less than 10% in the models (Kawaler 1987). This period spacing for g-modes in DOVs is very insensitive to the internal composition or the luminosity but depends very strongly on the total stellar mass. The identification of constant period spacing in two objects enabled Kawaler (1987) to determinate the masses of PG1159-035 to be 0.6 $M_{\odot}$ and that of PG0122+200 to be 0.73 $M_{\odot}$. This interpretation depends on the correct identification of modes but has the potential of becoming a very accurate 'seismological' tool.

## 8. Magnetic white dwarfs

The possible existence of strong magnetic fields ($\sim$ 1 MG) in white dwarfs was suggested by Blackett (1947), who argued that the magnetic moment of a star is proportional to its angular momentum and that angular momentum was conserved during stellar evolution, which is not the case (see below). Ginzburg (1964) and Woltjer (1964) suggested that if the surface magnetic flux ($\propto BR^2$), where $B$ is the surface magnetic field and $R$ the radius, is conserved during the lifetime and collapse of a normal (i.e. non-degenerate) star then the magnetic field of the white dwarf or neutron star produced would be very high. For typical initial values of $B \sim 100$ G and $R \sim 10^{11}$ cm for the normal star, a neutron star with a radius $\sim 10^6$ cm would then have a field $\sim 10^{12}$ G, while a white dwarf with $R \sim 10^9$ cm would have a field $\sim 10^6$ G. The discovery of pulsars, which are believed to be rotating neutron stars with magnetic fields $\sim 10^{12}$ G (Smith 1977) appeared, therefore, to confirm this suggestion and inspired searches, using Zeeman spectroscopy, for strong magnetic fields in white dwarfs (Preston 1970). This requires an understanding of the behaviour of atoms in strong magnetic fields, to which we now turn.

### *8.1. Spectroscopy in high magnetic fields*

The behaviour of atoms in strong magnetic fields has been discussed by a number of authors (see for example the review by Garstang 1977, and references therein). We briefly summarise here some of the essential features relevant for our discussion, following Garstang (1977).

The Hamiltonian for an atom in a magnetic field, neglecting terms involving the nuclear spin (i.e. hyperfine structure), is given by

$$H = H_{\rm KE} + H_{\rm C} + H_{\rm spin} + H_{\rm magI} + H_{\rm magII} \tag{8.1}$$

where $H_{\rm KE}$ denotes the kinetic energy of the electrons, $H_{\rm C}$ the Coulomb interaction between the electrons and the nucleus and between the electrons, $H_{\rm spin}$ the electron spin–orbit term, $H_{\rm magI}$ the term linear in the magnetic field $\boldsymbol{B}$ which gives rise to the linear Zeeman effect and $H_{\rm magII}$ the quadratic Zeeman term. For a uniform magnetic field with the vector potential $\boldsymbol{A} = (\boldsymbol{B} \times \boldsymbol{r})/2$ where $\boldsymbol{B} = \nabla \times \boldsymbol{A}$ and $\nabla \cdot \boldsymbol{A} = 0$, it follows that

$$H_{\rm magI} = e\boldsymbol{B} \cdot (\boldsymbol{L} + g_{\rm e}\boldsymbol{S})/2mc \tag{8.2}$$

where, for a many-electron atom, the total orbital ($\boldsymbol{L}$) and spin ($\boldsymbol{S}$) angular momenta are given by $\boldsymbol{L} = \sum \boldsymbol{l}_i$ and $\boldsymbol{S} = \sum \boldsymbol{s}_i$, and the sums are over the individual electrons $i$. Here $g_{\rm e} \simeq 2$ is the electronic g-factor.

If $\boldsymbol{B}$ is taken to be along the $\theta = 0$ axis in spherical polar coordinates $(r, \theta, \phi)$,

$$H_{\rm magII} = (e^2 B^2/8mc^2) \sum r_i^2 \sin^2 \theta_i \tag{8.3}$$

where the summation is over the atomic electrons $i$.

For sufficiently weak magnetic fields, $B \ll 5 \times 10^4$ G, except for light atoms, $H_{\rm magII}$ may be neglected and $H_{\rm magI} \ll H_{\rm spin}$ where the inequalities symbolically represent the corresponding typical matrix elements. Then the principal effects of the magnetic field arise through the linear Zeeman effect and the first-order effect on an atomic level characterised by $|SLJM\rangle$ is given by $\Delta E = \Delta E_0 + (e\hbar/2mc)Bg_J M$, where $g_J$ is the Landé g-factor (Garstang 1977) and $\Delta E_0$ is the transition energy in the absence of the field.

For stronger magnetic fields where $H_{\rm C} \gg H_{\rm magI} \gg H_{\rm spin} \gg H_{\rm magII}$, the magnetic interaction is large enough to break the spin–orbit coupling and the states are labelled by $|SM_S LM_L\rangle$. In this case the Paschen–Back effect is important and

$$\Delta E = \Delta E_0 + (e\hbar/2mc)B\Delta M_L \tag{8.4}$$

where, for electric dipole transitions, $\Delta M_L = 0, \pm 1$ (with $\Delta M_S = 0$).

For higher magnetic fields ($\geq 0.1$ MG), or at lower fields for large $n$, the quadratic Zeeman term $H_{\rm magII}$ could become more important. For a one-electron atom with the unperturbed state $|nlm_l m_s\rangle$ the first-order energy shift, assuming a nucleus of charge $Z$, is given by

$$\Delta E_{\rm Q} = \frac{e^2 B^2 a_0^2}{8mc^2 Z^2} n^2 \frac{[(5n^2 + 1 - 3l(l+1))]\,[l^2 + l - 1 + m_l^2]}{(2l-1)(2l+3)} \tag{8.5}$$

where $a_0$ is the Bohr radius. For transitions between two levels $\Delta l = 0, \pm 2$, while $\Delta n$ can take any value (see Hamada and Nakamura (1973) and Garstang and Kemic (1974) for improved calculations of the quadratic Zeeman effect using perturbation theory).

For magnetic fields in excess of 100 MG the magnetic terms in the Hamiltonian can become comparable to the Coulomb energy. This regime is intermediate to the

Paschen–Back (and quadratic Zeeman) regimes and the Landau regime where the magnetic field is so strong that the Coulomb interaction is small compared to the magnetic term (Garstang 1977). The intermediate field regime is very much more difficult to understand because it is not possible to use perturbation methods easily. By contrast the high field (Landau) regime is relatively easier to study (Yafet *et al* 1956). Variational calculations for the intermediate field have been carried out by a number of authors (e.g. Smith *et al* 1972), for $B$ less than about $7 \times 10^{11}$ G (see also Praddaude 1972). More recently, extensive calculations for the intermediate regime have been performed by research groups at Louisiana State University (LSU, Henry and O'Connell 1984, 1985) and at the University of Tübingen (Rösner *et al* 1984, Forster *et al* 1984, Wunner *et al* 1985, 1987). These calculations are in general agreement with one another in regions where they overlap.

### *8.2. Search for magnetic white dwarfs*

Because white dwarfs have in general very broad spectral features, or none at all, it is difficult to detect their magnetic fields using Zeeman spectroscopy. Preston (1970) attempted to detect strong magnetic fields ($\gg$ 1 MG) in white dwarfs by looking for the effects due to the quadratic Zeeman shift in the spectra of DA white dwarfs with strong Balmer lines. For the case of large $n$, neglecting terms of order $n^2$, the quadratic Zeeman term is given by

$$\Delta\lambda_{\mathrm{Q}} = -4.9654\ 10^{-23} Z^{-2}\ \lambda^2 B^2 n^4 (1 + m_l^2)\quad \text{Å} \tag{8.6}$$

for an s–$n$p transition. Note that $\Delta\lambda_{\mathrm{Q}} \propto \lambda^2 n^4$ and not $\lambda$ as in the case of the Doppler effect. This would allow the separation of the magnetic shift from that due to the space motion of the star along the line of sight. Furthermore, the shift in the $\sigma$ component ($m_l = \pm 1$) is twice that of the $\pi$ component ($m_l = 0$), so that the centroid of the triplet is shifted appropriately. Preston (1970) used this together with the radial velocities of several white dwarfs obtained by Greenstein and Trimble (1967) to place upper limits of about 0.5 MG for the average field over the stellar disk.

A novel approach for detecting magnetic fields in hot objects emitting an essentially featureless spectrum was proposed by Kemp (1970) with his magnetoemission model. He argued that the emitted radiation would nevertheless be circularly polarised and that its measurement could lead to the identification of the magnetic field. He noted that the Zeeman components of spectral lines have left and right circularly polarised components when viewed along the magnetic field. If the lines are significantly broadened so that they overlap and the spectrum becomes continuous, the radiation contains a hidden fractional circular polarisation. In order to show this he considered a simple model where the system consisted of three-dimensional harmonic oscillators distributed continuously with different natural angular frequency per unit angular frequency interval. Then the radiation at a frequency $\omega$ arises either from the transition from the $m = +1$, or the $m = -1$ state, to the ground state from two different oscillators, which have natural frequencies ($\omega \pm \omega_{\mathrm{c}}/2$) respectively, in the absence of a field. Here $\omega_{\mathrm{c}} = eB/mc$ is the cyclotron frequency. Since the magnetic field does not change the wave function to first order the transition probabilities are inversely proportional to ($\omega \pm \omega_{\mathrm{c}}/2$). Hence the intensities of the two components, which are left and right circularly polarised, are different and the radiation at the frequency $\omega$ is circularly polarised with fractional circular polarisation $V/I = -\omega_{\mathrm{c}}/2\omega$ for $\omega \gg \omega_{\mathrm{c}}$, where $I, Q, U, V$ are Stokes parameters. These results were experimentally confirmed, qualitatively, for thermal light sources in the laboratory (Kemp *et al* 1970a).

The first magnetic white dwarf to be discovered was Grw+70° 8247. It has an extraordinary spectrum whose main features defied identification for nearly half a century. Minkowski (1938) obtained several spectra of it which showed that it was not featureless as previously thought and that it had wide shallow absorption features at 4135 and 4475 Å. These together with another at 3910 Å discovered by Greenstein (1956) became known as Minkowski bands. Subsequent high resolution spectroscopy revealed a number of additional features (Wegner 1971).

In 1970, following the development of the magnetoemission model, Kemp discovered circularly polarised light from Grw+70° 8247 with a relatively small telescope, the 24 in of the University of Oregon. After observing the fractional circular polarisation as a function of wavelength Kemp *et al* (1970b) claimed a mean projected magnetic field of 10 MG for Grw+70° 8247. Because there were significant deviations from the simple linear relationship between circular polarisation and wavelength predicted by the magnetoemission model, and because the theoretical results were derived using a perturbation expansion in the magnetic field, there was some uncertainty in the interpretation. Nevertheless, the overall results were essentially unchanged, when the harmonic oscillator model was solved exactly (Chanmugam *et al* 1972).

After the discovery of the optical polarisation from Grw+70° 8247 attempts were made to understand the spectral features as those due to various atoms and molecules in strong magnetic fields (Mullan 1971, Angel 1972, Landstreet and Angel 1975, Wickramasinghe 1972, Greenstein 1970). Thus, for example, the discovery of a feature at 5835 Å raised the possibility that it was a Zeeman shifted He I 5876 Å line. However, Kemic (1974) presented calculations of the H$\alpha$–H$\delta$ and He I lines, and showed that none of the features could be explained with magnetic fields of less than 100 MG and hence Angel (1979) suggested that the features might be explained by hydrogen atoms in even higher magnetic fields.

Subsequent observations of Grw+70° 8247 with the IUE satellite showed a feature at 1347 Å in the ultraviolet which was interpreted by Greenstein (1984b) as due to the Lyman $\alpha$ line 1s0–1p1 in a magnetic field of 200–400 MG, retaining the labelling of energy levels by their zero field quantum numbers $nlm$. Calculations by Henry and O'Connell (1984) showed that this $\sigma$-transition corresponded to a maximum wavelength of 1342.6 Å for $B \simeq 500$ MG, while similar results were obtained by Rösner *et al* (1984) and Forster *et al* (1984), who found the maximum to occur at 1343 Å. The discrepancy with the observations of 4 Å is probably due to uncertainties of the IUE wavelength scale (Greenstein *et al* 1985).

A difficulty in determining the magnetic field spectroscopically is that the magnetic field varies across the surface of the white dwarf leading to different shifts for different parts of the stellar surface thus smearing out all spectral features. However, as conjectured from earlier calculations by Praddaude (1972), Greenstein (1984b) pointed out that the features may remain visible at those wavelengths, where the variation of the shift with field strengths becomes stationary. By searching for these stationary regions in the calculations of the Tübingen and LSU groups it became possible to identify the features in Grw+70° 8247 with transitions of hydrogen in a magnetic field of about 320 MG, which is much higher than the original estimate from the magnetoemission model (Greenstein *et al* 1985, Angel *et al* 1985, Wickramasinghe and Ferrario 1988). Observed spectra of Grw+70° 8247 and the position of spectral features in different fields are displayed in figure 10. However, the correct understanding of variations in the circular polarisation as a function of wavelength can only be obtained in detail if one takes into account the transfer of radiation through an atmosphere containing atoms in such strong magnetic fields (see below).

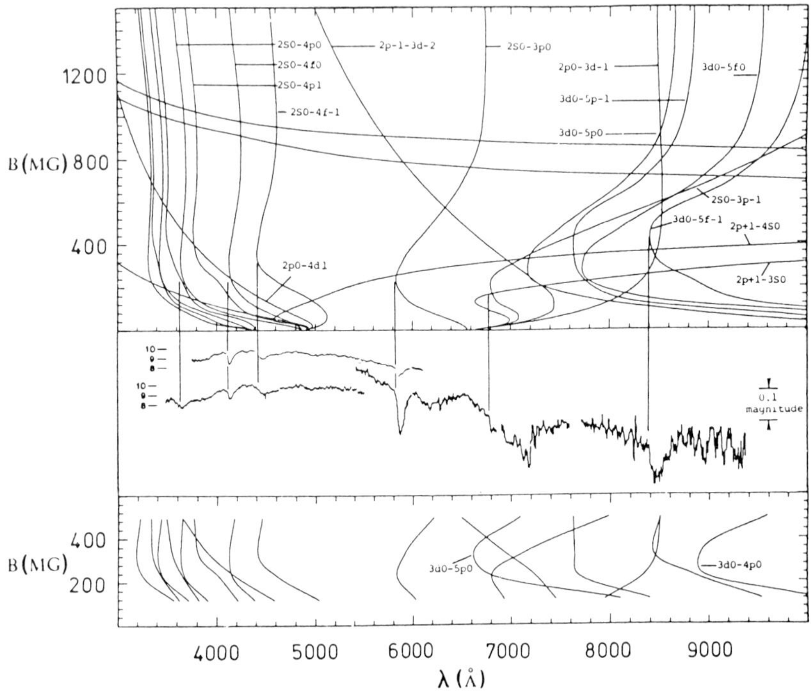


**Figure 10.** Observations of Grw+70° 8247 by Angel *et al* (1985) (centre) compared with the field–wavelength dependence of Zeeman components of hydrogen based on calculations by the Tübingen (upper) and LSU (lower) groups. The blue spectral region is on a flux scale ($F_\nu$), and the red spectral region is on a magnitude scale ($-2.5 \log F_\nu$ + constant). Reproduced from Wickramasinghe and Ferrario (1988).

The search for magnetic white dwarfs was continued after the discovery of polarisation in Grw+70° 8247, and 27 objects with magnetic fields, in most cases smaller than in Grw+70° 8247, have now been detected, either from polarisation measurements, or from the Zeeman splitting of spectral features. These results and references to the discoveries are collected in table 4.

For many white dwarfs, which showed no detectable fields, upper limits have been determined. A sensitive method was used by Angel and Landstreet (1970) who searched for polarisation in the wings of the H$\gamma$ line. If the mean longitudinal field over the stellar disk is $B_l$ and $I(\lambda)$ is the observed line intensity profile then the fractional circular polarisation is given by (Angel *et al* 1973):

$$V/I = 4.67\ 10^{-13} z (B_l/\lambda^2) \mathrm{d} \ln I / \mathrm{d} \ln \lambda \tag{8.7}$$

**Table 4.** Observations of magnetic fields in isolated white dwarfs. Adapted from Schmidt (1989).

| Object | $T$ | Spectral features | Period | Polarisation (%) linear | circular | Polar field (MG) | References |
|---|---|---|---|---|---|---|---|
| PG0136+251 | 45 000 | H | | | | 2 | LSGSM |
| PG1658+440 | 28 000 | H | | | | 3.5 | LSGSM, PMWG |
| G141-2 | 5600 | H | | | | 5 | G86 |
| GD90 | 9700 | H | | | <0.15 | 9 | ACSBH, WM, OD |
| G99-37 | 6200 | C2, CH | | <0.1 | 0.8 | 10 | AL, A |
| KUV03292+0035 | 15 000 | H | | | | 12 | WMB |
| GD356 | 7500 | H(em) | | | <0.05 | 15 | GM |
| HS1254+3430 | 15 000 | H | | | | 18 | HGEHTR, J |
| KPD0253+5052 | 17 600 | H | 4.1 h | | 0.2 | 18 | DM |
| WD1136-014 | | H | | | | 24 | FL |
| G99-47 | 5600 | H | | <0.1 | 0.4 | 25 | LAL, OD |
| KUV813-14 | 10 400 | H | | | 0.25 | 29 | SLSS, LSSSGB |
| PG1533-057 | 17 000 | H | | | | 31 | LSSSGB |
| Feige 7 | 17 800 | H, He | 2.2 h | | 0.3 | 35 | LASSB, MW |
| BPM25114 | 8600 | H | 2.8 d | | 1 | 36 | WB, WM |
| PG1313+095 | 15 000 | H | 5.4 h | | 1 | ≈50 | GSL |
| GD116 | 16 000 | H | | | | 65 | WSLS, SLWSS |
| GD229 | 16 000 | H | | 4 | <0.2 | 25–60 | LA, OSRW |
| ESO439-162 | 6300 | C2 | | | | ≈100 | RM |
| G195-19 | 8000 | none | 1.3 d | <0.13 | 0.8 | ≈100 | AIL |
| PG1015+015 | 9900 | H | 1.6 h | | 1 | 120 | A, WC |
| G227-35 | 7000 | ? | | 0.25 | 2 | ≈150 | HS, AHL |
| LP790-29 | 8600 | C2 | | 1 | 6 | 200 | LS |
| G240-72 | 6000 | ? | | 1 | 0.5 | ≈200 | AHSM |
| Grw+70°8247 | 12 000 | H | | 2 | 3 | 320 | KSLA, G84, GHO, ALS, WF |
| PG1031+234 | 15 000 | H | 3.4 h | 4 | 8 | 500 | SWLGS, LSG |

References: A, Angel 1978; AHL, Angel *et al* 1975; AHSM, Angel *et al* 1974b; AIL, Angel *et al* 1972; AL, Angel and Landstreet 1974; ALS, Angel *et al* 1985; ACSBH, Angel *et al* 1974a; DM, Downes and Margon 1983; FL, Foltz and Latter 1988; G84, Greenstein 1984b; G86, Greenstein 1986; GHO, Greenstein *et al* 1985; GM, Greenstein and McCarthy 1985; GSL, Green *et al* 1986; HGEHTR, Hagen *et al* 1987; HS, Hintzen and Strittmatter 1974; J, Jordan 1989; KSLA, Kemp *et al* 1970b; LA, Landstreet and Angel 1974; LAL, Liebert *et al* 1975; LASSB, Liebert *et al* 1977; LS, Liebert and Strittmatter 1977; LSG, Latter *et al* 1987; LSGSM, Liebert *et al* 1983; LSSSGB, Liebert *et al* 1985; MW, Martin and Wickramasinghe 1986; OD, O'Donoghue 1980; OSRW, Östreicher *et al* 1987; PMWG, Pravdo *et al* 1986; RM, Ruiz and Maza 1989; SLSS, Sion *et al* 1984; SLWSS, Saffer *et al* 1989; SWLGS, Schmidt *et al* 1986; WB, Wickramasinghe and Bessell 1976; WC, Wickramasinghe and Cropper 1988; WF, Wickramasinghe and Ferrario 1988; WM, Wickramasinghe and Martin 1979; WMB, Wegner *et al* 1987; WSLS, Wagner *et al* 1988. Most of the temperature determinations are from Sion *et al* (1988b).

where $\lambda$ is measured in Å and $z \simeq 1$ (Babcock 1958). This technique has been used to place upper limits typically of about 0.05 MG in several DA white dwarfs with the lowest upper limit being of the order of 1000 G for 40 Eri B (Angel *et al* 1981).

Even more sensitive is the search for Zeeman broadening in the narrow core of H$\alpha$ in DAs with high resolution spectroscopy, made possible by technological progress. This method has been used recently (Koester and Herrero 1988) to determine upper limits of typically 25 000 G for a number of brighter DA white dwarfs.

### 8.3. Theoretical modelling of magnetic atmospheres

The detection of circularly and linearly polarised optical radiation from white dwarfs confirms that they possess strong magnetic fields (Kemp *et al* 1970b). Qualitatively, the polarisation may be thought to arise as a result of circular dichroism or the existence of different opacities for the different modes of propagation of the observed radiation through the white dwarf's atmosphere (Angel *et al* 1981). For optically thin emission along the field at a frequency $\omega$ the fractional circular polarisation is given by

$$V/I = a\,(\omega_c/\omega) \tag{8.8}$$

where the constant $a = 2.0$ for the free–free opacity in a magnetic field (Kemp 1970, 1977) and bound–free transitions of hydrogen and helium in a magnetic field, with somewhat different values for $H^-$ and $He^-$ (Lamb and Sutherland 1974, Angel *et al* 1981). However, the observed circular polarisation does not increase linearly with wavelength as equation (8.8) predicts.

More extensive attempts have been made more recently to solve the complete transfer equations for the radiation from white dwarf atmospheres in the presence of a magnetic field by, for example, Martin and Wickramasinghe 1978, O'Donoghue 1980, Jordan 1989, and reviewed by Wickramasinghe (1987). In these calculations the effects of the Lorentz force on the atmospheric structure have been neglected, because the currents caused by the ohmic decay of the magnetic field (Landstreet 1986) are small (Jordan 1989) and hence the atmosphere is essentially force-free. The effects of the magnetic fields on the line and continuum opacities and on the equations of radiative transfer have been taken into account. In general the calculations have been performed for a plane parallel atmosphere, stratified in the z direction, which has a uniform magnetic field. The basic equations for radiative transfer are then given by:

$$\begin{aligned}
\mu\frac{\mathrm{d}I}{\mathrm{d}z} &= -\kappa_I(I-\mathcal{B}) - \kappa_Q Q - \kappa_V V \\
\mu\frac{\mathrm{d}Q}{\mathrm{d}z} &= -\kappa_Q(I-\mathcal{B}) - \kappa_I Q + r_R U \\
\mu\frac{\mathrm{d}U}{\mathrm{d}z} &= -r_R Q - \kappa_I U + r_W V \\
\mu\frac{\mathrm{d}V}{\mathrm{d}z} &= -\kappa_V(I-\mathcal{B}) - r_W U - r_I V.
\end{aligned} \tag{8.9}$$

Here, $\mathcal{B}$ is the Planck function for black body radiation, $\mu$ the cosine of the angle between the outward normal and the direction of propagation of the radiation and $r_R$ and $r_W$ the Faraday mixing coefficients (Pacholczyk 1976). These equations are essentially those due to Unno (1956) but include terms which allow for Faraday effects (Hardorp *et al* 1976, Wickramasinghe and Ferrario 1988). The absorption coefficients for the various Stokes parameters are given as follows in terms of $\kappa_\mathrm{L}$ and $\kappa_\mathrm{R}$, which correspond to the absorption coefficients for left and right circularly polarised radiation travelling parallel to the magnetic field, with $\kappa_\mathrm{P}$ the absorption coefficient for linearly polarised radiation travelling perpendicular to the field with the electric vector parallel to the magnetic field:

$$\begin{aligned}
\kappa_I &= \tfrac{1}{2}\kappa_{\rm P}\sin^2\xi + \tfrac{1}{4}(\kappa_{\rm L}+\kappa_{\rm R})(1+\cos^2\xi) \\
\kappa_Q &= \tfrac{1}{2}\kappa_{\rm P} - \tfrac{1}{4}(\kappa_{\rm L}+\kappa_{\rm R})\sin^2\xi \\
\kappa_V &= \tfrac{1}{2}(\kappa_{\rm R}-\kappa_{\rm L})\cos\xi.
\end{aligned} \tag{8.10}$$

Here $\xi$ is the angle between the magnetic field and the direction of propagation of the radiation. By a suitable choice of the coordinate system $\kappa_U$ is set equal to zero. The radiative transfer equations may be solved using various iterative techniques for an atmosphere in local thermodynamic equilibrium.

The continuum opacities in the presence of a strong magnetic field have been calculated by Lamb and Sutherland (1974), who showed that the bound–free opacity of hydrogen in a magnetic field can be deduced from the zero-field case because of the rigidity of the wave functions. They showed that these opacities can be expressed as

$$\kappa_q(\omega) = \frac{\omega}{\omega - q\omega_{\rm L}}\, g(\omega - q\omega_{\rm L}) \tag{8.11}$$

where $\omega_{\rm L} = \omega_{\rm c}/2$ is the Larmor frequency, $q = 0, +1, -1$ which correspond to $\kappa_{\rm P}$, $\kappa_{\rm R}$, $\kappa_{\rm L}$ respectively, and $g(\omega)$ the zero-field opacity.

Calculations of the circular polarisation and intensity as a function of wavelength have been performed for GD90 by Wickramasinghe and Martin (1979), see also O'Donoghue (1980), for centred and off-centred dipole magnetic fields and by Martin and Wickramasinghe (1984) for a quadrupole magnetic field with a polar field of 12 MG. Reasonably good fits to the data , with some discrepancies, have been obtained for the off-centred dipole and the quadrupole.

Wickramasinghe and Martin (1979) obtained good fits to the spectrum of BPM25114, which varies with a 2.8 d period, assuming a centred dipole field with a polar value of 36 MG. This value has been recently confirmed by Jordan (1989), who obtained a better fit by using the new line opacities obtained by the Tübingen group.

The magnetic white dwarf Feige 7 has both hydrogen and helium lines in its spectrum, which together with its polarisation varies with a period of 2.2 hr (Liebert *et al* 1977). Attempts to fit the observations with a centred dipole failed to produce the observed shifts of up to 10 Å in the field-dependent components with phase (Martin and Wickramasinghe 1936). Furthermore the period of the variations of the calculated circular polarisation is half the observed period. Off-centred dipoles also fail to fit the data and Wickramasinghe (1987) suggests that the star may contain a magnetic spot in addition to the centred dipole as suggested for PG1031+234 (Schmidt *et al* 1986).

More recently Wickramasinghe and Ferrario (1988) (see figure 10) have used the calculations of the spectra of hydrogen in magnetic fields greater than 100 MG by the LSU and Tübingen groups to model the spectrum of Grw+70° 8247. They show that the wavelengths, shapes and strengths of the features at 4135, 4480 and 5885 Å are in good agreement with a model in which the white dwarf has a centred dipole with a polar field of 320 ± 20 MG and a viewing angle with the field of 0 to 30 degrees and an effective atmospheric temperature of 14 000 K. The agreement is less satisfactory at long wavelengths (6200–9000 Å) and in the UV (2000–3000 Å), which they suggest may be due to departures from the assumed dipolar magnetic field. Their calculations confirm that the long unidentified Minkowski bands are due to hydrogen, the simplest and most abundant element in the universe, in a strong magnetic field of about 320 MG. Slightly better fits for Grw+70° 8247 have been obtained recently by Jordan (1989) using improved opacities.

The spectrum and polarisation of PG1031+234 have been analysed by Schmidt *et al* (1986) and Latter *et al* (1987) without solving the radiative transfer equations by assuming that the local field and geometry qualitatively determine them. They provide convincing evidence that the field strength in this white dwarf is of order 500 MG and that it may contain a magnetic spot with a field approaching $10^9$ G. This white dwarf has the strongest magnetic field of any white dwarf known and its field strength is comparable to that in the millisecond pulsars, which have fields ($\sim 10^8$ G) substantially weaker than those found in typical pulsars ($\sim 10^{12}$ G, White and Stella 1988).

Fits to the flux of PG1015+014, which shows variable circular polarisation with a period of 99 min, have been obtained by Wickramasinghe and Cropper (1988) assuming a polar field of 120 MG. Their model shows that, as in the case of Feige 7, the magnetic dipole axis is perpendicular to the rotation axis. Wickramasinghe (1987) speculates that magnetic white dwarfs prefer to be either perpendicular rotators or aligned rotators as in the cases of Feige 7 and PG1015+014. Then systems such as Grw+70° 8247, which show no evidence for polarisation variations may be parallel rotators. This implies that these fields must show a high degree of symmetry, which is difficult to justify.

Despite the significant progress and apparent success in fitting the observed spectra of Grw+70° 8247 it should be emphasised that the continuum opacities due to Lamb and Sutherland (1974), which have been used, are certainly not valid at such high field strengths above about 100 MG. More accurate calculations for this high field regime are badly needed.

### *8.4. Origin of magnetic fields*

Of all the isolated white dwarfs that have been surveyed only about 3–5% have observable magnetic fields (Angel *et al* 1981, Schmidt 1989). These have magnetic fields in the range of 1–500 MG (Schmidt 1989). Because it is not easy to detect magnetic fields weaker than about 1 MG and because complete surveys below this value have not been made, it is unclear, whether the fraction of white dwarfs with fields below 1 MG is the same as that above it.

The absence of detectable magnetic fields in the overwhelming majority of white dwarfs is puzzling in view of the arguments discussed at the beginning of this section, and the possibility that they are born with strong magnetic fields which subsequently decay has been explored. The time scale for the ohmic decay of the magnetic field of a conducting sphere of uniform electrical conductivity $\sigma$ and radius $R$ is given by $\tau \sim 4\sigma R^2/\pi c^2$, if the length scale of variation of the field is roughly $R$. By using this expression and a simple estimate for the electrical conductivity (Mestel 1965) found that $\tau \sim 10^{10}$ yr. The electrical conductivity, however, is both a function of the density and temperature in the star and hence varies both as a function of the radial distance $r$ from the centre of the star and also increases with time as the star cools. Detailed numerical calculations were performed by Chanmugam and Gabriel (1972), who used the electrical conductivity due to Canuto and Solinger (1970). They found that, for a symmetric dipolar magnetic field, the longest living mode decays on a time scale $\tau_0 \sim 10^{10}$ yr with the higher modes decaying increasingly faster.

More elaborate calculations were performed by Fontaine *et al* (1973), who used the Hubbard and Lampe (1969) values for the conductivity and found that $\tau_0 \simeq 3 \times 10^9$ yr. More recently these calculations have been further extended by Wendell *et al* (1987) who used the expressions for the conductivity due to Yakovlev and Urpin (1980) in the interior of the star. They found that the decay time exceeds the stellar age throughout the cooling sequence for a 0.6 $M_\odot$ carbon white dwarf and the field decays by a factor of only about 2 in 10 billion yr. These results are closer to those of Chanmugam and Gabriel (1972), than those of Fontaine *et al* (1973), and imply there is no justification for assuming that the longest living modes of white dwarfs would have decayed away.

The magnetic fields of white dwarfs may (or may not) be related to those of pulsars. An interesting feature of the white dwarfs is that their surface magnetic fluxes ($\sim BR^2 \sim 10^{24}$–$5 \times 10^{26}$ G cm$^2$ for white dwarfs of radius $10^9$ cm. By comparison the overwhelming majority of radio pulsars, which are believed to be rotating neutron stars with radii of $\sim 10^6$ cm, have magnetic fields $\sim 10^{11}$–$10^{13}$ G and hence have surface magnetic fluxes $\sim 10^{23}$–$10^{25}$ G cm$^2$, which is within an order of magnitude or so comparable to that of the magnetic white dwarfs. This has led to the suggestion that most pulsars and magnetic white dwarfs have a similar origin in the sense that they are descendants from red giant stars with similar magnetic fields produced during an earlier evolutionary phase. Magnetic flux is assumed to be conserved during the subsequent evolution, producing either a neutron star or a white dwarf (Ruderman and Sutherland 1973, Levy and Rose 1974). Note, however, that alternative origins for the magnetic fields of neutron stars have been proposed, where the fields are produced after the neutron star is formed by thermoelectric effects (Urpin and Yakovlev 1980, Blandford *et al* 1983). If such a model is correct then a separate origin for the white dwarf magnetic fields must be found since white dwarfs, unlike neutron stars, do not have hot crusts for a similar mechanism to work.

The rotation periods of seven isolated magnetic white dwarfs have been determined from the variations of their optical polarisation and range from 1.65 hr, for PG1015+015, to 2.8 days for BPM25114 (see table 4, and Schmidt 1989). In three cases (G240-72, GD229 and Grw+70° 8247) there is no evidence of variations in the position angle of the linear polarisation implying that either their rotation periods are greater than 100 yr or that the magnetic field is symmetric about the rotation axis. The magnetic white dwarfs appear to be slow rotators (cf Blackett 1947) in the sense that their rotational energy is much smaller than their gravitational energy and hence does not influence their structure significantly. On the other hand, since the moment of inertia $I \propto R^2$, their values are roughly $10^6$ times that of neutron stars and hence five of the magnetic white dwarfs with known rotation periods have angular momenta, which are comparable to pulsars with rotation periods of 7 ms to 0.3 s (Angel 1978, Brecher and Chanmugam 1978), suggesting that at least some magnetised white dwarfs and some pulsars may have had progenitors with similar angular momenta. There is no correlation between the rotation period and the surface temperature of the star (table 4) suggesting that they do not spin down with age (Schmidt 1987).

The rotation periods for five magnetic white dwarfs are a few hours, which corresponds to the orbital periods of cataclysmic binaries, which contain accreting white dwarfs. A subclass of the latter are known as AM Her binaries, which contain white dwarfs with magnetic fields of 20–50 MG rotating synchronously (Liebert and Stockman 1985). Binaries with fields larger than 50 MG are absent and Schmidt *et al* (1986) speculate that any such system would evolve relatively quickly so that the secondary is whittled down because of mass transfer to the white dwarf and mass ejection from the system so that it either does not exist anymore or is too faint to be seen. They suggest that the isolated magnetic white dwarfs may have unseen small companions. Hameury *et al* (1989) point out that for field strengths $\geq$ 40 MG such systems end up as detached binaries with low mass degenerate companions and orbital periods $\leq$ 7 hr. Thus such systems could provide an explanation for the short-period, apparently isolated, white dwarfs.

Greenstein and Oke (1982), from an accurate trigonometric parallax of Grw+70° 8247, have found its mass to be greater than 1 $M_\odot$. This is consistent with the evidence presented by Liebert (1988) that magnetic white dwarfs have masses which are on average slightly larger than non-magnetic white dwarfs, implying more massive and younger progenitors. This conclusion is supported by the space motions of the magnetic stars (Sion *et al* 1988b). Since pulsars have massive progenitors and similar surface magnetic fluxes (and in some cases angular momenta) as the magnetic white dwarfs it may well be that the fields of these degenerate stars are related by flux conservation to the fields of their progenitors. Thus, in the absence of a satisfactory theory for the origin of the magnetic fields of degenerate stars, flux conservation remains, because of its simplicity, an attractive hypothesis.

## 9. Summary of present knowledge and unsolved problems

Enormous progress has been made in the past two decades in our understanding of the structure of white dwarfs and their relationship to the evolution of stars in general. Thus, it had been a puzzle for a long time to understand, how stars with masses much larger than a typical white dwarf could evolve into one. We now realise that mass loss is extremely important and that stars with masses up to 8 $M_\odot$ can become white dwarfs. The evolution of white dwarfs is understood in principle, and it is plausible that in the near future calculated ages of the oldest visible white dwarfs will lead to consistent results, thus opening up the possibility to infer the age of the galactic disk with more certainty than is currently possible. Important questions regarding the masses of the hydrogen and helium layers in white dwarfs, the behaviour of a carbon/oxygen mixture in the crystalline phase, as well the precise abundance ratio of these elements, remain to be solved.

The relationship between the different spectral types, though studied very actively today, remains somewhat speculative. This is especially true for the basic dichotomy between hydrogen-rich and helium-rich spectral types. It is believed that the important processes, diffusion, convection, accretion and mass loss, are identified. How important mass loss really is, or what happens during the accretion of interstellar matter of solar composition, remains uncertain.

The variable white dwarfs are believed to be undergoing non-radial oscillations. The mechanisms exciting such oscillations in the DAVs and DBVs have been found, although it is controversial whether they require, in the case of the DAVs, an extremely thin hydrogen layer. The empirical location of the instability strip of the DBVS is uncertain by a few 1000 K, which is unfortunate because better knowledge could offer a means to calibrate the mixing-length approximation for convection. The red edge of the instability strip of the DAVs is probably caused by the interaction of pulsation with convection. The lack of a time-dependent theory of convection currently prevents a complete understanding.

The magnetic white dwarfs, with their extremely high fields, provide a test for the behaviour of hydrogen atoms in such fields. A severe problem in the theoretical modelling of spectra is the lack of accurate calculations of the bound–free opacity in strong fields. On the observational side, further efforts to detect field strengths below 1 MG and to search for a possible relationship between rotation and magnetic fields, are necessary.

Although white dwarfs are not very exotic like neutron stars and black holes, they nevertheless offer unique opportunities for studies with close connections to many branches of physics and astronomy.

## Acknowledgments

We have greatly benefited from being able to see parts of a preliminary version of the monograph currently being prepared by Van Horn and Liebert and want to thank Hugh Van Horn for this opportunity. We also gratefully acknowlege financial support through grants AST8814711 (DK) and AST8822954 (GC) from the National Science Foundation.

## Addendum

(i) Since the completion and initial publication of this article we have learnt that Slattery *et al* (1982) (Slattery W L, Doolen G D and DeWitt H E 1982 *Phys. Rev.* A **26** 2255) found slightly different constants in the approximation for the Coulomb Free Energy (equations (3.47) and (3.48)). The value for the phase transition changes to 178 with these results. The corrected equations are given in the text of the reprinted version here.

(ii) Schmidt and Norsworthy (1991) (Schmidt G D and Norsworthy J E 1991 *Astrophys. J.* **366** 270) discuss the evidence for rotation in magnetic white dwarfs. Nine objects now have measured periods ranging from 99 minutes to 18 days, while five others are apparently non-rotating (periods > 100 years!). There is no convincing evidence so far for a correlation between surface field strengths and rotation period, which would be expected, if the loss of angular momentum is due to magnetic braking. There is also no correlation between surface temperature (age) and rotation period of the white dwarf.

(iii) The only major conference dedicated to white dwarfs since the IAU Colloquium 114 in Hanover (1988) was the Seventh European Workshop on White Dwarfs, held in Toulouse (France) in September 1990. The major topics in a workshop, organized marvellously by G Vauclair, were: (*a*) evolution of white dwarfs and progenitors; (*b*) variability and period changes; and (*c*) white dwarfs in binary systems. The reader interested in the very recent developments in these fields is referred to the proceedings of this conference (*White Dwarfs (NATO ASI Series 336)* ed G Vauclair and E Sion (Dordrecht: Kluwer).

## References

Abell G O, Morrison D, and Wolff S C 1987 *Exploration of the Universe* (Philadelphia: CBS College Publishing) p 559

Abrikosov A A 1961 *Sov. Phys.-JETP* **12** 1254

—— 1962 *Sov. Phys.-JETP* **14** 408

Adams W S 1915 *Publ. Astron. Soc. Pacific* **27** 236
—— 1925 *Proc. Nat. Acad. Sci.* **11** 382
Alcock C and Illarionov A 1980a *Astrophys. J.* **235** 534
—— 1980b *Astrophys. J.* **235** 541
Anderson W 1929 *Z. Phys.* **54** 433
Angel J R P 1972 *Astrophys. J.* **171** L17
—— 1978 *Ann. Rev. Astron. Astrophys.* **16** 487
—— 1979 *White Dwarfs and Variable Degenerate Stars* IAU Coll. No. **53** ed H M Van Horn and V Weidemann (Rochester: University of Rochester Press) 313
Angel J R P, Borra E F and Landstreet J D 1981 *Astrophys. J. Suppl.* **45** 457
Angel J R P, Carswell R F, Strittmatter P A, Beaver P A and Harms R 1974a *Astrophys. J.* **194** L47
Angel J R P, Hintzen P and Landstreet J D 1975 *Astrophys. J.* **196** L27
Angel J R P, Hintzen P, Strittmatter P A and Martin P G 1974b *Astrophys. J.* **190** L71
Angel J R P, Illing R M E and Landstreet J D 1972 *Astrophys. J.* **175** L85
Angel J R P and Landstreet J D 1970 *Astrophys. J.* **160** L147
—— 1974 *Astrophys. J.* **191** 457
Angel J R P, Liebert J and Stockman H S 1985 *Astrophys. J.* **292** 260
Angel J R P, McGraw J T and Stockman H S 1973 *Astrophys. J.* **184** L79
Auluck F C and Mathur V S 1959 *Z. Astrophys.* **48** 28
Babcock H W 1958 *Astrophys. J.* **128** 228
Baglin A 1968 *Astrophys. Lett.* **1** 143
Baglin A and Heyvaerts J 1969 *Nature* **222** 1258
Baglin A and Schatzman E 1969 *Low Luminosity Stars* ed S S Kumar (New York: Gordon and Breach)
Baglin A and Vauclair G 1973 *Astron. Astrophys.* **27** 307
Bahcall J N 1984a *Astrophys. J.* **276** 156
—— 1984b *Astrophys. J.* **276** 169
Barrat J L, Hansen J P and Mochkovitch R 1988 *Astron. Astrophys.* **199** L15
Bath G T 1985 *Rep. Progr. Phys.* **48** 483
Blackett P M S 1947 *Nature* **159** 658
Blandford R, Applegate J and Hernquist L 1983 *Mon. Not. R. Astron. Soc.* **204** 1025
Böhm K H 1968 *Astrophys. Space Sci.* **2** 375
Böhm-Vitense E Z 1958 *Astrophysics* **46** 108
Bond H E, and Ciardullo R 1989 *White Dwarfs* IAU Coll. No. **114** ed G Wegner (Berlin: Springer) 473
Bond H E and Grauer A D 1987 *Astrophys. J.* **321** L123
Bond H E, Grauer A D, Green R F and Liebert J 1984 *Astrophys. J.* **279** 751
Boss L 1910 *Preliminary General Catalogue* (Washington: Carnegie Institution)
Bradley P A, Winget D E and Wood M A 1989 *White Dwarfs* IAU Coll. No. **114** ed G Wegner (Berlin: Springer) 286
Brecher K and Chanmugam G 1978 *Astrophys. J.* **221** 969
Brickhill A J 1975 *Mon. Not. R. Astron. Soc.* **170** 405
Bruhweiler F and Kondo Y 1983 *Astrophys. J.* **269** 657
Brush S G, Sahlin H L and Teller E J 1966 *Chem. Phys.* **45** 2102
Burgers J M 1969 *Flow Equations for Composite Gases* (New York: Academic)
Canuto V 1970 *Astrophys. J.* **159** 641
Canuto V and Solinger A B 1970 *Astrophys. Lett.* **6** 141
Carr W J Jr 1961 *Phys. Rev.* **122** 1437
Cesare L de, Forlani A and Platania G 1973 *Astrophys. Space Sci.* **21** 461
Chandrasekhar S 1931 *Astrophys. J.* **74** 81
—— 1935 *Mon. Not. R. Astron. Soc.* **95** 207
—— 1939 *An Introduction to the Study of Stellar Structure* (Cambridge: University of Cambridge Press)
—— 1964 *Astrophys. J.* **146** 417
Chandrasekhar S and Tooper R F 1964 *Astrophys. J.* **139** 1396
Chanmugam G 1972 *Nature* **236** 83
—— 1977 *Astrophys. J.* **217** 799

Chanmugam G and Gabriel M 1972 *Astron. Astrophys.* **16** 149

Chanmugam G, O'Connell R F and Rajagopal A K 1972 *Astrophys. J.* **175** 157

Chapman S and Cowling T 1970 *The Mathematical Theory of Non-Uniform Gases* (Cambridge: Cambridge University Press)

Chayer P, Fontaine G and Wesemael F 1989 *White Dwarfs* IAU Coll. No. **114** ed G Wegner (Berlin: Springer) 253

Clayton D D 1968 *Principles of Stellar Evolution and Nucleosynthesis* (Chicago: University of Chicago Press)

Cottrell P L and Greenstein J L 1980 *Astrophys. J.* **283** 941

Cowling T G 1941 *Mon. Not. R. Astron. Soc.* **101** 367

Cox A N 1986 *Highlights of Astronomy* **7** 229

Cox A N, Hodson S W and Starrfield S G 1980 *Non-radial and Non-linear Stellar Pulsations* ed H A Hill and W Dziembowski (Berlin: Springer) 453

Cox A N, Starrfield S G, Kidman R B and Pesnell W D 1987 *Astrophys. J.* **317** 303

Cox A N and Stewart J N 1970 *Astrophys. J. Suppl.* **19** 261

Cox J P 1980 *Theory of Stellar Pulsations* (Princeton: Princeton University Press)

Cox J P and Giuli R T 1968 *Principles of Stellar Structure* (New York: Gordon and Breach)

D'Antona F and Mazzitelli I 1979 *Astron. Astrophys.* **74** 161

—— 1990 *Ann. Rev. Astron. Astrophys.* in press

DeWitt H E 1969 *Low Luminosity Stars* ed S S Kumar (New York: Gordon and Breach)

Dirac P A M 1926 *Proc. R. Soc.* A **112** 660

Dolez N, Vauclair G and Chevreton M 1983 *Astron. Astrophys.* **121** L23

Dolez N and Vauclair G 1981 *Astron. Astrophys.* **102** 375

Downes R A and Margon B 1983 *Publ. Astron. Soc. Pacific* **95** 358

Dupuis J, Pelletier C, Fontaine G and Wesemael F 1989 *White Dwarfs* IAU Coll. No. **114** ed G Wegner (Berlin: Springer) 359

Dziembowski W and Koester D 1981 *Astron. Astrophys.* **97** 16

Eddington A S 1925 *Observatory* **48** 73

—— 1926 *Internal Constitution of the Stars* (Cambridge: Cambridge University Press)

Eggen O 1985 *Publ. Astron. Soc. Pacific* **97** 1029

Einstein A 1907 *Jahrbuch Radioakt.* **4** 411

Faulkner J and Gribbin J R 1968 *Nature* **218** 734

Fermi E 1926 *Z. Phys.* **36** 902

Finley D S, Basri G and Bowyer S 1989 *White Dwarfs* IAU Coll. No. **114** ed G Wegner (Berlin: Springer) 139

Finzi A and Wolf R A 1967 *Astrophys. J.* **150** 115

Foltz C B and Latter W B 1988 preprint referenced in Schmidt (1989)

Fontaine G, Bergeron P, Lacombe P, Lamontagne R and Talon A 1985 *Astron. J.* **90** 1094

Fontaine G, McGraw J T, Coleman L, Lacombe P, Patterson J and Vauclair G 1980 *Astrophys. J.* **239** 898

Fontaine G, McGraw J T, Dearborn D S P, Gustafson J and Lacombe P 1982 *Astrophys. J.* **258** 651

Fontaine G and Michaud G 1979 *Astrophys. J.* **231** 826

Fontaine G, Thomas J H and Van Horn H M 1973 *Astrophys. J.* **184** 911

Fontaine G, Villeneuve B, Wesemael F and Wegner G 1984 *Astrophys. J.* **277** L61

Fontaine G and Wesemael F 1984 *Astron. J.* **89** 1728

—— 1987 *The Second Conference on Faint Blue Stars* IAU Coll. No. **95** ed A G D Philip, D S Hayes and J Liebert (Schenectady: L Davis Press) 319

Forster H, Strupat W, Rösner W, Wunner G, Ruder H and Herold H 1984 *J. Phys.* B **17** 1301

Fowler R H 1926 *Mon. Not. R. Astron. Soc.* **87** 114

Fusi-Pecci F and Renzini A 1976 *Astron. Astrophys.* **46** 447

Garcia-Berro E, Hernanz M, Isern J and Mochkovitch R 1988a *Astron. Astrophys.* **193** 141

—— 1988b *Nature* **333** 644

Gatewood G D and Gatewood C V 1978 *Astrophys. J.* **225** 191

Garstang R H 1977 *Rep. Progr. Phys.* **40** 105

Garstang R H and Kemic S B 1974 *Astrophys. Space Sci.* **31** 103

Gell-Mann M and Brückner K A 1957 *Phys. Rev.* **106** 364

Giclas H L, Burnham R Jr and Thomas N G 1980 *Lowell Obs. Bull. No.* **166**

Ginzburg V L 1964 *Sov. Phys. Dokl.* **9** 329

Goupil M J, Auvergne M and Baglin A 1988 *Astron. Astrophys.* **196** L13
Grauer A D and Bond H E 1984 *Astrophys. J.* **277** 211
Grauer A D, Bond H E, Green R F and Liebert J 1988 *Astron. J.* **95** 879
Grauer A D, Wegner G, Green R F and Liebert J 1989 *preprint*
Green R F, Schmidt M and Liebert J 1986 *Astrophys. J. Suppl.* **61** 305
Greenstein J L 1956 *Proc. 3rd Berkeley Symp.* ed J Neyman (Berkeley: University of California Press)
—— 1970 *Astrophys. J.* **162** L55
—— 1982 *Astrophys. J.* **258** 661
—— 1984a *Astrophys. J.* **276** 602
—— 1984b *Astrophys. J.* **281** L47
—— 1986 *Astrophys. J.* **304** 334
Greenstein J L, Henry R W and O'Connell R F 1985 *Astrophys. J.* **289** L25
Greenstein J L and McCarthy J K 1985 *Astrophys. J.* **289** 732
Greenstein J L and Oke J B 1982 *Astrophys. J.* **252** 285
Greenstein J L, Oke J B and Shipman H L 1971 *Astrophys. J.* **169** 563
—— 1985 *Quart. J. R. Astron. Soc.* **26** 279
Greenstein J L and Peterson D M 1973 *Astron. Astrophys.* **25** 29
Greenstein J L and Trimble V 1967 *Astrophys. J.* **149** 283
Gustafson R G and McGraw J T 1980 *Bull. Am. Astron. Soc.* **12** 863
Hagen H J, Groote D, Engels D, Haug U, Toussaint F and Reimers D 1987 *Astron. Astrophys.* **183** L7
Hamada T and Nakamura Y 1973 *Publ. Astron. Soc. Japan* **25** 527
Hamada T and Salpeter E E 1961 *Astrophys. J.* **134** 683
Hamann W R, Kudritzki R-P, Mendez R H and Pottasch S R 1984 *Astron. Astrophys.* **139** 459
Hameury J M, King A R and Lasota J P 1989 *Mon. Not. R. Astron. Soc.* **237** 845
Hansen C J, Cox J P and Van Horn H M 1977 *Astrophys. J.* **217** 151
Hansen J P 1973 *Phys. Rev.* A **8** 3096
Hardorp J, Shore S N and Wittmann A 1976 *Physics of Ap Stars* IAU Coll. No. **32** ed W W Weiss, H Jenker and H J Wood p 419
Harper R van and Rose W K 1970 *Astrophys.J* **162** 963
Henry R J W and O'Connell R F 1984 *Astrophys. J.* **282** L97
—— 1985 *Publ. Astron. Soc. Pacific* **97** 333
Hesser J E, Lasker B M and Neupert H E 1976 *Astrophys. J.* **209** 853
Hill J A 1987 *The Second Conference on Faint Blue Stars* IAU Coll. No. **95** ed A G D Philip, D S Hayes and J Liebert (Schenectady: L Davis Press) p 681
Hintzen P and Strittmatter P A 1974 *Astrophys. J.* **193** L111
Holberg J B, Wesemael F and Basile J 1986 *Astrophys. J.* **306** 629
Hubbard W B and Lampe M 1969 *Astrophys. J. Suppl.* **18** 297
Huebner W F, Merts A L, Magee N H Jr and Argo M F 1977 *Astrophysical Opacity Library, UC-34b*
Humason M L and Zwicky F 1947 *Astrophys. J.* **105** 85
Hutsemékers D and Surdej J 1989 *Astron. Astrophys.* **219** 237
Iben I Jr 1967 *Ann. Rev. Astron. Astrophys.* **5** 571
—— 1974 *Ann. Rev. Astron. Astrophys.* **12** 215
—— 1984 *Astrophys. J.* **277** 333
Iben I Jr and Laughlin G 1989 *Astrophys. J.* **341** 312
Iben I Jr and MacDonald J 1985 *Astrophys. J.* **295** 540
—— 1986 *Astrophys. J.* **301** 164
Iben I Jr and Renzini A 1983 *Ann. Rev. Astron. Astrophys.* **21** 271
—— 1984 *Phys. Rep.* **105** 330
Iben I Jr and Tutukov A V 1984 *Astrophys. J.* **282** 615
Illarionov A and Sunyaev R 1975 *Astron. Astrophys.* **39** 185
Isern J, Garcia-Berro E, Hernanz M and Mochkovitch R 1989 *White Dwarfs* IAU Coll. No. **114** ed G Wegner (Berlin: Springer) p 278
Itoh N 1989 *White Dwarfs* IAU Coll. No. **114** ed G Wegner (Berlin: Springer) p 66
Itoh N, Mitake S, Iyetomi H and Ichimaru S 1983 *Astrophys. J.* **273** 774
Itoh N, Kohyama Y, Matsumoto J and Seki M 1984 *Astrophys. J.* **285** 758
Jordan S 1989 *White Dwarfs* IAU Coll. No. **114** ed G Wegner (Berlin : Springer) p 333
Jordan S and Koester D 1986 *Astron. Astrophys. Suppl.* **65** 367

Jordan S, Koester D, Wulf-Mathis C and Brunner H 1987 *Astron. Astrophys.* **185** 253
Kahn S, Wesemael F, Liebert J, Raymond J, Steiner J and Shipman H L 1984 *Astrophys. J.* **278** 255
Kaplan S A 1949 *Naukovy Zapiski (Sci. Notes Univ. Lwow)* **15** 109
Kawaler S D 1987 *The Second Conference on Faint Blue Stars* IAU Coll.No. **95** ed A G D Philip, D S Hayes and J Liebert (Schenectady: L Davis Press) p 297
Kawaler S D and Hansen C J 1989 *White Dwarfs* IAU Coll. No. **114** ed G Wegner (Berlin: Springer) p 97
Kawaler S D, Winget D E and Hansen C J 1985 *Astrophys. J.* **298** 752
Kawaler S D, Winget D E, Hansen C J and Iben I Jr 1986 *Astrophys. J.* **306** L41
Kemic S B 1974 *Astrophys. J.* **193** 213
Kemp J C 1970 *Astrophys. J.* **162** 169
—— 1977 *Astrophys. J.* **213** 794
Kemp J C, Swedlund J B and Evans B D 1970a *Phys. Rev. Lett.* **24** 1211
Kemp J C, Swedlund J B, Landstreet J D and Angel J R P 1970b *Astrophys. J.* **161** L77
Kennicutt R C 1984 *Astrophys. J.* **277** 361
Kenyon S J, Shipman H L, Sion E M and Aannestad P A 1988 *Astrophys. J.* **328** L65
Kepler S O, Vauclair G, Nather R E, Winget D E and Robinson E L 1989 *White Dwarfs* IAU Coll. No. **114** ed G Wegner (Berlin: Springer) p 341
Kirzhnits D A 1960 *Sov. Phys.-JETP* **11** 365
Koester D 1972 *Astron. Astrophys.* **16** 459
—— 1976 *Astron. Astrophys.* **52** 415
—— 1978 *Astron. Astrophys.* **64** 289
—— 1982 *The Messenger* **28** 25
—— 1987a *The Second Conference on Faint Blue Stars* IAU Coll. No. **95** ed A G D Philip, D S Hayes and J Liebert (Schenectady: L Davis Press) p 329
—— 1987b *Astrophys. J.* **322** 852
—— 1989a *White Dwarfs* IAU Coll. No. **114** ed G Wegner (Berlin: Springer) p 206
—— 1989b *Astrophys. J.* **342** 999
Koester D and Herrero A 1988 *Astrophys. J.* **332** 910
Koester D and Reimers D 1981 *Astron. Astrophys.* **99** L8
—— 1985 *Astron. Astrophys.* **153** 260
Koester D and Schönberner D 1986 *Astron. Astrophys.* **154** 125
Koester D, Schulz H and Weidemann V 1979 *Astron. Astrophys.* **76** 262
Koester D, Vauclair G, Dolez N, Oke J B, Greenstein J L and Weidemann V 1985 *Astron. Astrophys.* **149** 423
Koester D, Wegner G and Kilkenny D 1989 *Astrophys. J.* in press
Koester D and Weidemann V 1980 *Astron. Astrophys.* **81** 145
—— 1989 *Astron. Astrophys.* **219** 276
Koester D, Weidemann V and Zeidler-K T E M 1982 *Astron. Astrophys.* **116** 147
Kothari D S 1931 *Phil. Mag.* **12** 665
—— 1936 *Mon. Not. R. Astron. Soc.* **96** 833
—— 1938 *Proc. R. Soc. London* A **165** 486
Kovetz A and Shaviv G 1970 *Astron. Astrophys.* **8** 398
Kugler A A 1969 *Ann. Phys., NY* **53** 133
Lamb D Q 1974 *Thesis* University of Rochester, Rochester
Lamb D Q and Van Horn H M 1975 *Astrophys. J.* **200** 306
Lamb F K and Sutherland P G 1974 *Physics of Dense Matter* ed C J Hansen (Dordrecht: Reidel) p 265
Lamontagne R, Wesemael F and Fontaine G 1987 *The Second Conference on Faint Blue Stars* IAU Coll. No. **95** ed A G D Philip, D S Hayes and J Liebert (Schenectady: L Davis Press) p 677
Lamontagne R, Wesemael F, Fontaine G, Wegner G and Nelan E P 1989 *White Dwarfs* IAU Coll. No. **114** ed G Wegner (Berlin: Springer) p 240
Landau L D 1932 *Phys. Z. Sowjet.* **1** 285
Landau L D and Lifshitz E M 1969 *Statistical Physics* (London: Addison-Wesley)
Landolt A U 1968 *Astrophys. J.* **153** 151
Landstreet J D 1986 *Mon. Not. R. Astron. Soc.* **225** 437
Landstreet J D and Angel J R P 1974 *Astrophys. J.* **190** L25
—— 1975 *Astrophys. J.* **196** 819

Lang K R and Gingerich O (eds) 1979 *A Source Book in Astronomy and Astrophysics, 1900–1975* (Cambridge: Harvard University Press) p 215
Lasker B M and Hesser J E 1969 *Astrophys. J.* **158** L171
—— 1971 *Astrophys. J.* **163** L89
Latter W B, Schmidt G D and Green R F 1987 *Astrophys. J.* **320** 308
Lawrence G M, Ostriker J P and Hesser J E 1967 *Astrophys. J.* **148** L161
Ledoux P 1958 *Handbuch der Physik* ed S Flügge (Berlin: Springer) **51** 605
Ledoux P and Sauvenier-Goffin E 1950 *Astrophys. J.* **111** 611
Ledoux P and Walraven T 1958 *Handb. der Physik* ed S.Flügge (Berlin: Springer) **51** 353
Lee T D 1950 *Astrophys. J.* **111** 625
Levy E M and Rose W K 1974 *Nature* **250** 40
Liebert J 1980 *Ann. Rev. Astron. Astrophys.* **18** 363
—— 1987 private communication
—— 1988 *Publ. Astron. Soc. Pacific* **100** 1302
Liebert J, Angel J R P and Landstreet J D 1975 *Astrophys. J.* **202** L139
Liebert J, Angel J R P, Stockman H S, Spinrad H and Beaver P A 1977 *Astrophys. J.* **214** 457
Liebert J, Dahn C C and Monet D G 1988 *Astrophys. J.* **332** 891
—— 1989 *White Dwarfs* IAU Coll. No. **114** ed G Wegner (Berlin: Springer) p 15
Liebert J, Fontaine G and Wesemael F 1987b *Mem. Soc. Astron. Italiana* **58** 17
Liebert J, Schmidt G D, Green R F, Stockman H S and McGraw J T 1983 *Astrophys. J.* **264** 262
Liebert J, Schmidt G D, Sion E M, Starrfield S G, Green R F and Boroson T A 1985 *Publ. Astron. Soc. Pacific* **97** 158
Liebert J and Stockman H S 1985 *Cataclysmic Variables and Low Mass X-ray Binaries* ed D Q Lamb and J Patterson (Dordrecht: Reidel) p 151
Liebert J and Strittmatter P A 1977 *Astrophys. J.* **217** L59
Liebert J, Wehrse R and Green R F 1987a *Astron. Astrophys.* **175** 173
Liebert J, Wesemael F, Hansen C J, Fontaine F, Shipman H L, Sion E M, Winget D E and Green R F 1986 *Astrophys. J.* **309** 241
Luyten W J 1970 *White Dwarfs* (Minneapolis: University of Minnesota Press)
—— 1977 *White Dwarfs - II* (Minneapolis: University of Minnesota Press)
MacDonald J 1989 *White Dwarfs* IAU Coll. No. **114** ed G Wegner (Berlin: Springer) p 172
Maeder A and Mermilliod J C 1981 *Astron. Astrophys.* **93** 136
Marshak R E 1940 *Astrophys. J.* **92** 321
Martin B and Wickramasinghe D T 1978 *Mon. Not. R. Atron. Soc.* **183** 533
—— 1984 *Mon. Not. R. Astron. Soc.* **206** 407
—— 1986 *Astrophys. J.* **301** 177
Mauche C W 1990 (ed) *Proc. 11th North American Workshop on CVs and LMXRBS* (Cambridge: Cambridge University Press) in press
Mazzitelli I and D'Antona F 1986 *Astrophys. J.* **308** 706
McCook G P and Sion E M 1987 *Astrophys. J. Suppl.* **65** 603
McGraw J T 1976 *Astrophys. J.* **210** L35
—— 1977 *Astrophys. J.* **214** L123
—— 1979 *Astrophys. J.* **229** 203
McGraw J T, Fontaine G, Dearborn D S P, Gustafson J, Lacombe P and Starrfield S G 1981 *Astrophys. J.* **250** 349
McGraw J T and Robinson E L 1975 *Astrophys. J.* **200** L89
—— 1976 *Astrophys. J.* **205** L155
McGraw J T, Starrfield S G, Liebert J and Green R F 1979 *White Dwarfs and Variable Degenerate Stars* IAU Coll. No. **53** ed H M Van Horn and V Weidemann (Rochester: University of Rochester Press) p 377
McMahan R K 1989 *Astrophys. J.* **336** 409
Meltzer D W and Thorne K S 1966 *Astrophys. J.* **218** 514
Mendez R H, Miguel C H, Heber U and Kudritzki R P 1986 *Hydrogen Deficient Stars and Related Objects* IAU Coll. No. **87** ed K Hunger, D Schönberner and N K Rao (Dordrecht: Reidel) p 323
Mendez R H, Kudritzki R P, Herrero A, Husfeld D and Groth H G 1988 *Astron. Astrophys.* **190** 113
Mengel J G 1976 *Astron. Astrophys.* **48** 83
Mengel J G, Sweigart A V, Demarque P and Gross P G 1979 *Astrophys. J. Suppl.* **40** 733
Mestel L 1950 *Proc. Cambridge Phil. Soc.* **46** 331

—— 1952 *Mon. Not. R. Astron. Soc.* **112** 583
—— 1965 *Stellar Structure* ed L H Aller and D B McLaughlin (Chicago: University of Chicago Press) p 297
Mestel L and Ruderman M A 1967 *Mon. Not. R. Astron. Soc.* **136** 27
Michaud G 1987 *The Second Conference on Faint Blue Stars* IAU Coll. No. **95** ed A G D Philip, D S Hayes and J Liebert (Schenectady: L Davis Press) p 249
Michaud G, Fontaine G and Charland Y 1984 *Astrophys. J.* **280** 247
Mihalas D 1978 *Stellar Atmospheres* (San Francisco: Freeman)
Minkowski R 1938 *Ann. Rep. Mt. Wilson Obs.* **55** 44
Mochkovitch R 1983 *Astron. Astrophys.* **122** 212
Morvan E, Vauclair G and Vauclair S 1986 *Astron. Astrophys.* **163** 145
Muchmore D 1984 *Astrophys. J.* **278** 769
Mullan D J 1971 *Ir. Astron. J.* **10** 25
Nather R E 1973 *Vistas in Astronomy* **15** 91
O'Donoghue D E 1980 *Astrophys. Space Sci.* **68** 273
Oke J B, Weidemann V and Koester D 1984 *Astrophys. J.* **281** 276
Osaki Y and Hansen C J 1973 *Astrophys. J.* **185** 277
Östreicher R, Seifert W, Ruder H and Wunner G 1987 *Astron. Astrophys.* **173** L15
Ostriker J P and Axel L 1969 *Low Luminosity Stars* ed S Kumar (New York: Gordon and Breach) p 357
Pacholczyk A G 1976 *Radio Galaxies* (Oxford: Pergamon) ch 3
Paerels F B S and Heise J 1989 *Astrophys. J.* **339** 1000
Paquette C, Pelletier C, Fontaine G and Michaud G 1986a *Astrophys. J. Suppl.* **61** 197
—— 1986b *Astrophys. J. Suppl.* **61** 177
Patterson J 1984 *Astrophys. J. Suppl.* **54** 443
Pelletier C, Fontaine G and Wesemael F 1989 *White Dwarfs* IAU Coll. No. **114** ed G Wegner (Berlin: Springer-Verlag) p 249
Pelletier C, Fontaine G, Wesemael F, Michaud G and Wegner G 1986 *Astrophys. J.* **307** 242
Perinotto M 1987 *Planetary and Proto-Planetary Nebulae: From IRAS to ISO* ed A Preite Martinez (Dordrecht: Reidel) p 13
Pesnell W D 1987 *Astrophys. J.* **314** 598
Petre R, Shipman H L and Canizares C R 1986 *Astrophys. J.* **304** 356
Pilachowski C A and Milkey R W 1984 *Publ. Astron. Soc. Pacific* **96** 821
—— 1987 *Publ. Astron. Soc. Pacific* **99** 836
Pollock E L and Hansen J P 1973 *Phys. Rev.* A **8** 3110
Praddaude H C 1972 *Phys. Rev.* A **6** 1321
Pravdo S H, Marshall F E, White N E and Giommi P 1986 *Astrophys. J.* **300** 819
Preston G W 1970 *Astrophys. J.* **160** L143
Reimers D 1975 *Mem. Soc. R. Sci. Liège* **8** 369
—— 1987 *Circumstellar Matter* IAU Symp. **122** ed I Appenzeller and C Jordan (Dordrecht: Reidel) p 307
Reimers D and Koester D 1982 *Astron. Astrophys.* **116** 341
—— 1988a *Astron. Astrophys.* **202** 77
—— 1988b *The Messenger* **54** 47
—— 1989 *Astron. Astrophys.* **218** 118
Richer H B and Ulrych T J 1974 *Astrophys. J.* **192** 719
Robinson E L 1984 *Astron. J.* **89** 1732
Robinson R L and McGraw J T 1976 *Astrophys. J.* **207** L37
Robinson E L, Stover R J, Nather R E and McGraw J T 1978 *Astrophys. J.* **220** 614
Rösner W, Wunner G, Herold H and Ruder H 1984 *J. Phys. B: At. Mol. Phys.* **17** 29
Romanishin W and Angel J R P 1980 *Astrophys. J.* **235** 992
Rosseland S 1949 *The Pulsation Theory of Variable Stars* (Oxford: Clarendon)
Ruderman M A and Sutherland R G 1973 *Nature Phys. Sci.* **246** 93
Rudkjøbing M 1952 *Publ. Köbenhavns Obs. No.* **160**
Ruiz M T and Maza J 1989 *White Dwarfs* IAU Coll. No. **114** ed G Wegner (Berlin: Springer-Verlag) p 126
Ruiz M T, Maza J, Wischnjewski M and Gonzáles L E 1986 *Astrophys. J.* **304** L25
Russell H N 1914 *Popular Astronomy* **22** 275

Saffer R A, Liebert J, Wagner R M, Sion E M and Starrfield S G 1989 preprint
Saio H, Winget D E and Robinson E L 1983 *Astrophys. J.* **265** 982
Salpeter E E 1961 *Astrophys. J.* **134** 669
Salpeter E E and Van Horn H M 1969 *Astrophys. J.* **155** 183
Sauvenier-Goffin E 1949 *Ann. Astrophys.* **12** 39
Schatzman E 1949 *Publ. Köbenhavns Obs.No* **149**
—— 1958 *White Dwarfs* (Amsterdam: North-Holland)
Schmidt G D 1987 *The Second Conference on Faint Blue Stars* IAU Coll. No. **95** ed A G D Philip, D S Hayes and J Liebert (Schenectady: L Davis Press) p 377
—— 1989 *White Dwarfs* IAU Coll. No. **114** ed G Wegner (Berlin: Springer-Verlag) p 305
Schmidt G D, West S C, Liebert J, Green R F and Stockman H S 1986 *Astrophys. J.* **309** 218
Schönberner D 1979 *Astron. Astrophys.* **79** 108
Shapiro S L and Teukolsky S A 1983 *Black Holes, White Dwarfs and Neutron Stars. The Physics of Compact Objects* (New York: Wiley)
Shaviv G and Kovetz A 1972 *Astron. Astrophys.* **16** 72
Shipman H L 1979 *Astrophys. J.* **228** 240
—— 1984 *The Future of Ultraviolet Astronomy Based on Six Years of IUE Research* ed J M Mead, R D Chapman and Y Kondo, NASA Conf. Publ. **2349** 281
—— 1987 *The Second Conference on Faint Blue Stars* IAU Coll. No. **95** ed A G D Philip, D S Hayes and J Liebert (Schenectady: L Davis Press) 273
—— 1989 *White Dwarfs* IAU Coll. No. **114** ed G Wegner (Berlin: Springer-Verlag) p 220
Shipman H L and Greenstein J L 1983 *Astrophys. J.* **266** 761
Shipman H L, Greenstein J L and Boksenberg A 1977 *Astron. J.* **82** 480
Shipman H L, Liebert J and Green R F 1987 *Astrophys. J.* **315** 239
Shipman H L and Sass C A 1980 *Astrophys. J.* **235** 177
Shulov O S and Kopalskaya E N 1971 *Astrofizika* **10** 117
Sion E M 1986 *Publ. Astron. Soc. Pacific* **98** 821
Sion E M, Aannestad P A and Kenyon S J 1988a *Astrophys. J.* **330** L55
Sion E M, Fritz M L, McMullin J P and Lallo M D 1988b *Astron. J.* **96** 251
Sion E M, Greenstein J L, Landstreet J D, Liebert J, Shipman H L and Wegner G A 1983 *Astrophys. J.* **269** 253
Sion E M, Liebert J, Schmidt G D and Starrfield S G 1984 *Bull. Am. Astron. Soc.* **16** 724
Sion E M, Shipman H L, Wagner R M, Liebert J and Starrfield S 1986 *Astrophys. J.* **308** L67
Skilling J 1968 *Nature* **218** 923
Slattery W L, Doolen G D and DeWitt H E 1980 *Phys. Rev.* A **21** 2087
Smeyers P 1967 *Bull. Soc. R. Sci. Liège* **36** 357
Smith E R 1972 *Phys. Rev.* D **6** 3700
Smith E R, Henry R J W, Surmelian G L, O'Connell R F and Rajagopal A K 1972 *Phys. Rev.* D **6** 3700
Smith F G 1977 *Pulsars* (Cambridge: Cambridge University Press)
Starrfield S G, Cox A N, Hodson S W and Clancy S P 1983 *Astrophys. J.* **269** 645
Starrfield S G, Cox A N, Hodson S W and Pesnell W D 1982 *Pulsations in Classical and Cataclysmic Variable Stars* ed J P Cox and C J Hansen (Boulder: University of Colorado Press) p 78
Stevenson D G 1980 *J. Physique. Suppl. 3* **41** C2/61
Stoner E C 1930 *Phil. Mag.* **9** 944
Stover R J, Hesser J E, Lasker B M, Nather R E and Robinson E L 1980 *Astrophys. J.* **240** 865
Strittmatter P A and Wickramasinghe D T 1971 *Mon. Not. R. Astron. Soc.* **152** 47
Sweeney M A 1976 *Astron. Astrophys.* **49** 375
Sweigart A V and Gross P G 1978 *Astrophys. J. Suppl.* **36** 405
Tammann G A 1974 *Supernovae and Supernova Remnants* ed C B Cosmovici (Dordrecht: Reidel)
Tassoul M and Tassoul J-L 1983 *Astrophys. J.* **267** 334
Thejll P and Shipman H L 1986 *Publ. Astron. Soc. Pacific* **98** 922
Tomaney A B 1987 *The Second Conference on Faint Blue Stars* IAU Coll. No. **95** ed A G D Philip, D S Hayes and J Liebert (Schenectady: L Davis Press) p 673
Unno W 1956 *Publ. Astron. Soc. Japan* **8** 1956
Urpin V A and Yakovlev D G 1980 *Soviet Astron.* **24** 425
van den Heuvel E P J 1975 *Astrophys. J.* **196** L121
Van Horn H M 1968 *Astrophys. J.* **151** 227

—— 1971 *White Dwarfs* IAU Sympos. No. **42** ed W Luyten (Dordrecht: Reidel) p 87
—— 1984 *Theoretical Problems in Stellar Stability and Oscillations* ed A Noels and M Gabriel (Cointe-Ougree: Université de Liège)
Van Horn H M and Liebert J 1989 preprint
Van Maanen A 1913 *Publ. Astron. Soc. Pacific* **29** 258
Vauclair G 1971a *Astrophys. Lett.* **9** 161
—— 1971b *White Dwarfs* IAU Sympos. No. **42** ed R J Luyten (Dordrecht: Reidel) p 145
—— 1989 *White Dwarfs* IAU Coll. No. **114** ed G Wegner (Berlin: Springer-Verlag) p 176
Vauclair G, Dolez N and Chevreton M 1981b *Astron. Astrophys.* **103** L17
—— 1987 *Astron. Astrophys.* **175** L13
Vauclair G and Liebert J 1987 *Exploring the Universe with the IUE Satellite* ed Y Kondo (Dordrecht: Reidel) p 355
Vauclair G and Reisse C 1977 *Astron. Astrophys.* **61** 415
Vauclair S and Vauclair G 1982 *Ann. Rev. Astron. Astrophys.* **20** 37
Vauclair G, Vauclair S and Greenstein J L 1979 *Astron. Astrophys.* **80** 79
Vauclair G, Weidemann V and Koester D 1981a *Astron. Astrophys.* **100** 113
Vennes S, Fontaine G and Wesemael F 1989 *White Dwarfs* IAU Coll. No. **114** ed G Wegner (Berlin: Springer-Verlag) p 368
Vogt H 1925 *Astron. Nachrichten* **227** 325
Wagner R M, Sion E M, Liebert J and Starrfield S G 1988 *Astrophys. J.* **328** 213
Wapstra A H and Bos K 1976 *At. Data Nucl. Data Tables* **17** 474
Warner B and Robinson E L 1972 *Nature Phys. Sci.* **239** 2
Wegner G 1971 *Publ. Astron. Soc. Pacific* **83** 205
—— 1981 *Astrophys. J.* **245** L27
—— 1983 *Astrophys. J.* **268** 282
—— 1989a (ed) *White Dwarfs* IAU Coll.No. **114** (Berlin: Springer-Verlag)
—— 1989b *White Dwarfs* IAU Coll. No. **114** ed G Wegner (Berlin: Springer-Verlag) p 401
Wegner G and Koester D 1985 *Astrophys. J.* **288** 746
Wegner G, McMahan R K and Boley F I 1987 *Astron. J.* **94** 1271
Wegner G and Nelan E P 1987 *Astrophys. J.* **319** 916
Wegner G, Reid I N and McMahan R K 1989 *White Dwarfs* IAU Coll. No. **114** ed G Wegner (Berlin: Springer-Verlag) p 378
Wegner G and Yackovich F H 1983 *Astrophys. J.* **275** 240
—— 1984 *Astrophys. J.* **284** 257
Wehrse R and Liebert J 1980 *Astron. Astrophys.* **86** 139
Weidemann V 1967 *Z. Astrophys.* **67** 286
—— 1987 *Astron. Astrophys.* **188** 74
—— 1990 *Ann. Rev. Astron. Astrophys.* in press
Weidemann V, Koester D and Vauclair G 1981 *Astron. Astrophys.* **95** L9
Weidemann V and Koester D 1983 *Astron. Astrophys.* **121** 77
—— 1984 *Astron. Astrophys.* **132** 195
Weidemann V and Yuan J W 1989 *White Dwarfs* IAU Coll. No. **114** ed G Wegner (Berlin: Springer-Verlag) p 1
Wendell C E, Van Horn H M and Sargent D 1987 *Astrophys. J.* **313** 284
Werner K, Heber U and Hunger K 1989 *White Dwarfs* IAU Coll. No. **114** ed G Wegner (Berlin: Springer-Verlag) p 194
Wesemael F, Green R F and Liebert J 1985 *Astrophys. J. Suppl.* **58** 379
Wesemael F, Lamontagne R and Fontaine G 1986 *Astron. J.* **91** 1376
Wesemael F and Truran J W 1982 *Astrophys. J.* **260** 807
Wheeler J C, Hansen C J and Cox J P 1968 *Astrophys. Lett.* **2** 253
White N E and Stella L 1988 *Nature* **332** 416
Wickramasinghe D T 1972 *Mon. Not. R. Astron. Soc.* **76** 129
—— 1987 *The Second Conference on Faint Blue Stars* IAU Coll. No. **95** ed A G D Philip, D S Hayes and J Liebert (Schenectady: L Davis Press) p 389
Wickramasinghe D T and Bessell M S 1976 *Astrophys. J.* **203** L39
Wickramasinghe D T and Cropper M 1988 *Mon. Not. R. Astron. Soc.* **235** 1451
Wickramasinghe D T and Ferrario L 1988 *Astrophys. J.* **327** 222
Wickramasinghe D T and Martin B 1979 *Mon. Not. R. Astron. Soc.* **188** 165

Wickramasinghe D T and Reid N 1983 *Mon. Not. R. Astron. Soc.* **203** 887
Wigner E 1932 *Phys. Rev.* **40** 749
Wigner E and Seitz F 1934 *Phys. Rev.* **46** 523
Wildhack W A 1940 *Phys. Rev.* **57** 81
Winget D E 1981 *Thesis* University of Rochester
—— 1986 *Highlights of Astronomy* **7** 221
Winget D E and Claver C F 1989 *White Dwarfs* IAU Coll. No. **114** ed G Wegner (Berlin: Springer-Verlag) p 290
Winget D E, Kepler S O, Robinson E L, Nather R E and O'Donoghue D 1985 *Astrophys. J.* **292** 606
Winget D E, Nather R E and Hill J A 1987b *Astrophys. J.* **316** 305
Winget D E, Robinson E L and Nather R E 1984 *Astrophys. J.* **279** L15
Winget D E, Robinson E L, Nather R E and Fontaine G 1982b *Astrophys. J.* **262** L11
Winget D E and Van Horn H M 1987 *The Second Conference on Faint Blue Stars* IAU Coll. No. **95** ed A G D Philip, D S Hayes and J Liebert (Schenectady: L Davis Press) p 363
Winget D E, Van Horn H M and Hansen C J 1981 *Astrophys. J.* **245** L33
Winget D E, Van Horn H M, Tassoul M, Hansen C J and Fontaine G 1983 *Astrophys. J.* **268** L33
Winget D E, Van Horn H M, Tassoul M, Hansen C J, Fontaine G and Carroll B W 1982a *Astrophys. J.* **252** L65
Winget D E, Hansen C J, Liebert J, Van Horn H M, Fontaine G, Nather R E, Kepler S O and Lamb D Q 1987a *Astrophys. J.* **307** 659
Woltjer L 1964 *Astrophys. J.* **140** 1309
Wunner G, Rösner W, Herold H and Ruder H 1985 *Astron. Astrophys.* **149** 102
Wunner G, Geyer F and Ruder H 1987 *Astrophys. Space Sci.* **131** 595
Yafet Y, Keyes R W and Adams E N 1956 *J. Phys. Chem. Solid* **1** 137
Yakovlev D G and Urpin V A 1980 *Soviet Astron.* **24** 303
Zapolski H S 1960 *Newman Laboratory of Nuclear Studies, Cornell University* unpublished report
Zeidler-K.T. E M, Weidemann V and Koester D 1986 *Astron. Astrophys.* **155** 356

Rep. Prog. Phys. **53** (1990) 1467–1557. Printed in the UK

# Type I supernovae

J Craig Wheeler and Robert P Harkness

Department of Astronomy, University of Texas at Austin, Austin, TX 78712, USA

**Abstract**

The observations and theory of hydrogen-deficient Type I supernovae (SN Ia) are reviewed. Classical Type Ia supernovae are thought to arise in the thermonuclear explosions of degenerate stellar cores, but more recently defined classes, Types Ib and Ic (SN Ib, Ic), may result from core collapse in more massive stars.

Binary evolutionary schemes for the origin of SN Ia are described including transfer of hydrogen from a non-degenerate companion and the disruption of a degenerate companion. No scheme adequately accounts for all the observational constraints. The process of degenerate carbon ignition thought to power SN Ia still has many uncertainties including the role of the convective Urca process and questions involving the mechanism of explosion propagation. Subsonic deflagration processes match many of the observational constraints, including the light curve, early- and late-time spectra, and the properties of young extended remnants, but supersonic detonation seems to follow from the physics of the ignition process. Calculations of theoretical spectra of SN Ia at various stages in their evolution are described in some detail and the importance of SN Ia for nucleosynthesis and for determination of the cosmological distance scale is summarized.

Less is known about the details of the origin of the other classes of SN I. SN Ib are characterized by the appearance of helium lines near maximum and emission lines of oxygen and calcium at later epochs. Like SN Ib, SN Ic lack the evidence for Si near maximum that characterizes SN Ia, but they have only weak indications of He. They are indistinguishable from SN Ib in the later supernebular phases. There are indications from the light curves that SN Ic may differ in significant ways from SN Ib although the spectra suggest that they are generically related. Circumstantial evidence suggests that SN Ib and Ic arise in the hydrogen-denuded cores of massive stars which may have lost their hydrogen envelopes to a binary companion. The possible connection of SN Ib and Ic events to the 'transition' event, SN 1987K, is also discussed.

An appendix describes the operation of the supernova atmosphere code used to calculate early-time spectra of SN I.

This review was received in its present form in April 1990.

0034-4885/90/121467+89$14.00 

**Contents**

## 1. Introduction

The world has been made aware of the excitement surrounding supernova research with the discovery of supernova 1987A in the Large Magellanic Cloud. This highly publicized event, the brightest since the time of Johannes Kepler, has dominated supernova research for the past three years, and will continue to do so for some time to come. It has great intrinsic interest and much to teach us about supernovae in general. SN 1987A came on top of a rush of developments which had already changed our views of supernovae and heralded the establishment of a new category of supernovae. The purpose of this review is to summarize some of the most interesting new ideas of the last few years, the context in which SN 1987A arrived, and our growing understanding of its relation to other types of supernovae. We will concentrate on the hydrogen-deficient Type I supernovae where some of the most interesting developments have occurred, but point out implications for other types of supernovae and where old schemes of nomenclature are breaking down as physical understanding grows.

Their transient nature makes supernovae very difficult to study. Although several hundred extended supernova remnants and hundreds of pulsars have been discovered in our Galaxy, no contemporary supernova has been witnessed in our Galaxy in any wavelength band—radio, infrared, optical, x-ray—in at least three hundred years. Supernovae are discovered about twenty times a year in external galaxies. No specific supernova nor even the host galaxy can be anticipated and because of their large distances, typically greater than 10 Mpc, they soon fade below detectability. Statistically, many supernovae must go undetected, and too few are followed through with detailed, efficient study. Supernova 1987A is the dramatic exception that proves the rule.

Modern techniques for discovering and studying extragalactic supernovae were developed by Zwicky and Minkowski in the 1930s. The first few supernovae discovered by accident contained examples corresponding to the two types of event now recognized, but the first 36 Zwicky found himself were of only one type (Trimble 1982, see also Branch 1990 for an excellent review of the history of supernova spectral identifications). The spectra of this type were not immediately interpretable even though they were similar from event to event. They showed broad features which could not be readily associated with any known element. The light curve (luminosity as a function of time) had a predictable regularity. It rose to peak brightness in the space of about two weeks and then after declining for two more weeks, it faded in an almost precise exponential fashion with a half-life of about 50 days (Barbon *et al* 1973). Other observations identified another kind of explosive event. These were dimmer at maximum by a factor of a few in luminosity and had light curves which rose and faded in a manner that varied from supernova to supernova. The most important aspect of these events, which differentiated them as a separate class, was that the spectrum contained recognizable features—the common optical lines of hydrogen. These two different kinds of supernovae were differentiated by Minkowski (1941) as Type I and Type II, respectively. Zwicky (1965) added Types III, IV and V. Although interesting in their own right, these categories consisted only of single reasonably well studied events and the vast majority of extragalactic supernovae have fallen into one of these two broad categories—Type I (SN I) for those that displayed no evidence of hydrogen and Type II

(SN II) for those that showed spectral evidence of hydrogen in the ejecta. Figure 1 gives sample spectra of Type I and Type II supernovae. A more extended discussion of supernova spectral classifications is given in Branch (1990) and Harkness and Wheeler (1990).

Further study suggested that Type II events might themselves be differentiated by the shape of their light curves (Barbon *et al* 1979). Some had light curves for which the luminosity declined roughly in an exponential fashion (Doggett and Branch 1985, Young and Branch 1989), although with a different time constant than SN I. These came to be known as linear Type II supernovae (SN II-L). The other class tend to remain at peak light for one or two months before declining. These plateau Type II supernovae (SN II-P) tend to be less bright than SN I and the brightest SN II-L by a factor of three or four. The question of whether SN II-L and SN II-P are physically distinct or merely the extremes of some continuous distribution is debated (van den Bergh 1988, Young and Branch 1989). SN 1987A displayed strong evidence for hydrogen and hence falls in the class Type II (Arnett *et al* 1989). Its decline from maximum was parallel to the decline of some SN II-P from their plateaux, and hence it may have some direct connection to that class, although there is probably appreciable dispersion in the luminosity (Young and Branch 1989). Figure 2 presents schematic B-magnitude light curves for Type I and Type II events. The late-time behaviour is especially not to be taken too literally. The light curves of SN Ib and Ic are discussed in some detail in section 3.2.1. See Chevalier (1981b), Doggett and Branch (1985),

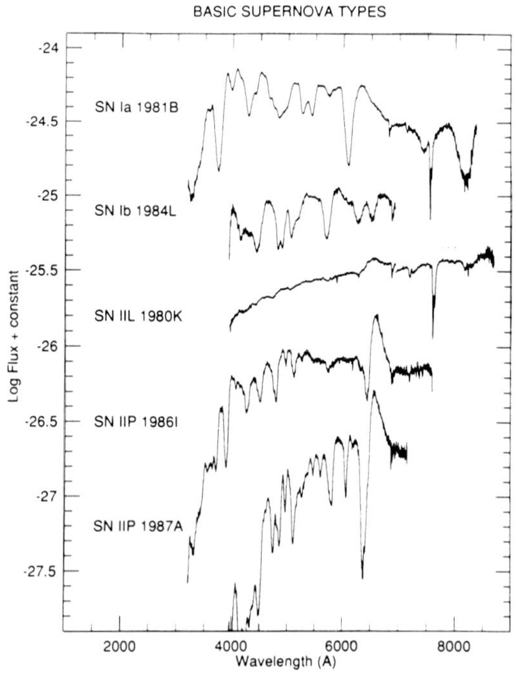


**Figure 1.** Representative spectra of supernovae of Types Ia, Ib, II-L, II-P and of SN 1987A several weeks after the explosion, near maximum light.

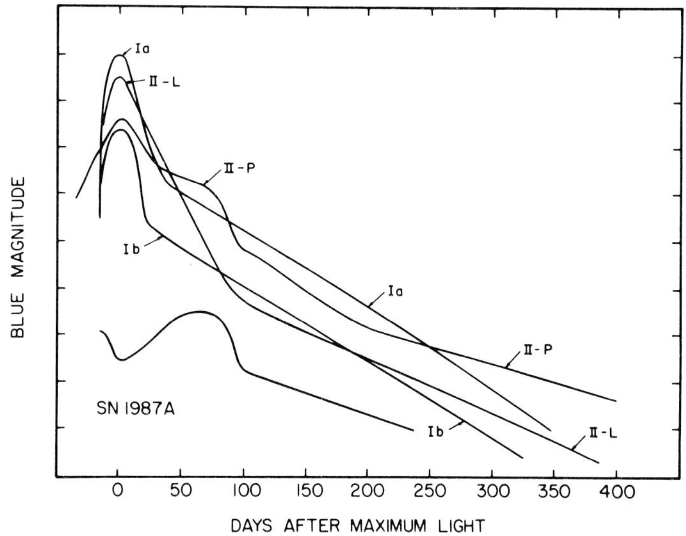


**Figure 2.** Schematic light curves for supernovae of Types Ia, Ib, II-L, II-P and SN 1987A.

Kirshner (1990) and Young and Branch (1989) for more extended discussions of supernova light curves.

Theoretical understanding of the processes leading to supernova explosions has proceeded in parallel with the observational study. Baade and Zwicky (1934a, b, c, d) perceived the potentially important role of neutron-star formation in triggering a supernova explosion, shortly after the discovery of the neutron. Hoyle and Fowler (1960) elucidated the two basic processes that are still at the heart of the subject 30 years later. One mechanism operates when the core of a star reaches conditions of high electron degeneracy for which thermonuclear burning is thermally and dynamically unstable. The star is consumed by the resulting supersonic shock wave (detonation) or by a rapid but subsonic combustion (deflagration) or a complex mix of the two. Such an explosion tears the star completely apart, leaving no compact remnant. The environment typically considered for such an explosion is in a dense evolved stellar core composed of carbon and oxygen in which the burning of carbon triggers the runaway. In particular, a star of less than about 8 $M_{\odot}$ is thought to form such a dense core. In a binary stellar system the hydrogen envelope can be lost by transfer to the companion star, or by loss of mass entirely from the system. If mass is added back onto the white dwarf when the second star evolves and swells, the white dwarf may grow in mass until it reaches a critical condition of density and temperature and explodes. Since the white dwarf may have very little hydrogen on its surface, this type of model has long been suggested as the origin of Type I supernovae.

A star in excess of about 8 $M_{\odot}$ will form a core that implodes rather than undergoing a thermonuclear explosion. The best studied cases involve stars which develop iron cores. Iron has the highest binding energy per nucleon of any common element and hence it can only undergo endoergic reactions that remove energy and pressure support from the star, leading to stellar collapse. In the collapse, the protons in the core capture electrons thereby forming neutrons and emitting neutrinos. The collapsing sphere of neutrons halts and bounces at nuclear densities. Although the details are not yet resolved, the process of neutron star formation liberates a great deal of energy, perhaps a hundred times more than is necessary to eject the outer layers of the star as a supernova. Most of the energy is lost in the neutrinos to which the outer stellar layers are transparent, but only one per cent needs to be deposited to cause a healthy explosion. In this process the neutron star is necessarily left behind. Core collapse is thought to occur in massive stars with extended hydrogen envelopes and has therefore been associated with Type II supernovae. SN 1987A was a star of about 15–20 $M_\odot$, within the mass range predicted to undergo core collapse. The detection of the burst of neutrinos from SN 1987A (Hirata *et al* 1987, Bionta *et al* 1987) represented the birth of extragalactic neutrino astronomy and gave a remarkable confirmation of the basic theoretical picture of core collapse (Arnett *et al* 1989 and references therein).

With the long-standing hypothesis of a thermonuclear explosion of a bare white dwarf making a Type I and the core collapse and explosion of a hydrogen-rich evolved red giant star making a Type II, there has been a tendency to equate 'thermonuclear explosion' with 'Type I' and 'core collapse' with 'Type II'. Progress in the last few years has eroded that scheme. An important step in that direction, unrecognized at the time, were the observations by Bertola of two supernovae, SN 1962L in the galaxy NGC 1073 and 1964L in NGC 3938. There was no spectral evidence for hydrogen, so these events fell in the broad category of hydrogen-deficient Type I supernovae. Classical SN I, however, show a strong absorption at 6150 Å attributed to the Doppler shifted line of once-ionized silicon with rest wavelength of 6355 Å. SN 1962L and 1964L did not show this feature. Bertola called them 'peculiar' Type I supernovae.

Little attention was paid to these peculiar Type I events until the discovery of SN 1983N in the M83 and 1984L in the galaxy NGC 991. Spectra of the 'peculiar' Type I supernovae 1983N and 1984L obtained near maximum light in the blue end of the optical wavelength range resembled spectra of classical SN I a month or so after maximum. This led to the suggestion that the 'peculiar' events were variants of, and generically related to, SN I, but evidence has accumulated that this is not the case. Further study showed that the spectra are distinctly different despite important qualitative similarities to the standard variety at early times. The distinct spectral differences have been formalized by denoting the classical Type I supernovae as SN Ia and the 'peculiar' Type I events as SN Ib (Elias *et al* 1985, Branch 1986, Harkness *et al* 1987, Porter and Filippenko 1987). The growing body of evidence suggests that the 'peculiar' Type I supernovae represent a completely different class of events with very different progenitor stars and explosion mechanism than the classical SN I of Minkowski. There is a growing belief that SN Ib result from massive stars so that their evolution and the mechanism of their explosion is quite distinct from the white dwarf evolution and thermonuclear explosion presumed to power SN Ia. It is plausible that these supernovae are driven by core collapse and hence are physically closely related to some Type II events and to SN 1987A.

Clues to the origin of supernovae are found in their rates of occurrence in various types of galaxies. Type Ia supernovae are traditionally associated with old stellar populations (but see below). Type II supernovae and Type Ib are associated with young stellar populations (Huang and Thaddeus 1986, Huang 1987, Harkness *et al* 1987). Estimates of rates are difficult to make because of the sparse sample of supernovae of known spectral type and selection effects such as the degree of obscuration in edge-on spiral disks. Table 1 gives estimates of SN rates in galaxies of various types from Tammann (1982), the recent redetermination of Cappellaro and Turatto (1988)

**Table 1.** Supernova rates per 100 years per $10^{10}$ $L_\odot$ (for $h = 0.5$) as a function of galaxy type.

| Galaxy type | Compilation† | Number | SN | SN I | SN II |
|---|---|---|---|---|---|
| E+SO | 1 | 19 | 0.34 | 0.34 | 0 |
| | 2 | 10 | | 0.17 | 0 |
| | 3 | 1 | | 0.08 | 0 |
| SOa, Sa | 1 | 9 | 0.28 | 0.28 | 0 |
| | 2 | 3 | | 0.27 | <0.12 |
| | 3 | 2 | | 0.15 | 0 |
| Sab, Sb | 1 | 38 | 0.69 | 0.37 | 0.32 |
| | 2 | 6 | | 0.13 | 0.14 |
| | 3 | 7 | | 0.23 | 0.45 |
| Sbc, Sc, Scd, Sd | 1 | 93 | 1.38 | 0.77 | 0.61 |
| | 2 | 31 | | 0.54 | 0.63 |
| | 3 | 5 | | 0.15 | 0.33 |
| Sdm, Sm, Im | 1 | 11 | 1.02 | 0.83 | 0.19 |
| | 2 | 1 | | <0.58 | 0.64 |
| | 3 | 0 | | ~0 | 0‡ |
| Milky Way (Sb-Sbc) | 1 | | 4.02 | 2.22 | 1.79 |
| (per 100 years, | 2 | | 2.04 | 1.14 | 0.92 |
| $L = 3.9 \times 10^{10}$ $L_\odot$) | 3 | | | 0.74 | 1.52 |

† 1 = Tammann (1982); 2 = Cappellaro and Turatto (1988); 3 = Evans *et al* (1989).
‡ Neglecting SN 1987A.

and from Evans *et al* (1989). The latter two groups accounted for the varying 'visibility' of events of different type due to different light curves. These rates are presented in terms of the rate per unit galactic luminosity. Although the rates do increase with parent galaxy luminosity in different galaxy classes (Cappellaro and Turatto 1988) there is little direct evidence that SN rates are linearly proportional to galactic luminosity except in face-on Sc galaxies. Jorgenson (1989) has suggested that far infrared luminosity might be a better measure of star formation rates in host galaxies. It is not clear why current star formation rates (measured by either blue or red light) should have any connection to old population SN Ia events. For more discussion of rates of SN Ia, see Wheeler (1990a).

In table 1, the rates from Evans *et al* (1989) are the sum of those for SN Ia and SN Ib (plus SN Ic). Cappellaro and Turatto give rates for individual morphological types, but their rates are averaged to correspond to the morphological groups adopted by Evans *et al* (1989). The lower part of table 1 gives an estimate of the rates in the Galaxy. Note that there are several serious uncertainties involved in this interpolation. The intrinsic rates per unit blue luminosity scale with $h^2$ where $h$ is the Hubble constant $H_0$ in units of 100 km s$^{-1}$ Mpc$^{-1}$. The blue luminosity of the Galaxy is not well known. The value adopted here from Cappellaro and Turatto is a factor $\gtrsim 2$ larger than other estimates (de Vaucouleurs and Pence 1978). Finally, the morphological type of the Galaxy (Sb-Sbc) is not well established, but the supernova rates per unit blue luminosity may be about a factor of two higher for Sbc than for Sb (Cappellaro and Turatto 1988). Cappellaro and Turatto adopted a mean between the rates for Sb and Sbc to estimate the Galactic rate and we have done a similar average between Sab, Sb and Sbc–Sd to estimate the Galactic rate based on the statistics of Evans *et al.*

Cappellaro and Turatto note, as have previous workers, e.g. Oemler and Tinsley (1979) (see also Caldwell and Oemler 1981, Stryker *et al* 1981), that the rates of SN

I increase as one goes to galaxies of earlier morphological type, in intuitive violation of the notion that SN Ia are associated with older stellar populations. Cappellaro and Turatto speculate that the increase may be due to the increasing fraction of SN Ib in the galaxies more dominated with early stellar populations. This may be true, but their analysis contains no SN Ib events and they have not considered quantitatively the fact that SN Ib events are intrinsically dimmer than SN Ia events and thus their probability of discovery will be lower. There seems a good probability that the raw rates, and hence the trend with galaxy type deduced for SN I are domimated by SN Ia, and that the mystery of the trend remains. All of these groups neglect to consider the question originally raised by Oemler and Tinsley (1979) of the apparently high rates of SN I in IO galaxies which are thought to be dominated by ongoing star formation. Tammann (1982), having addressed the question in earlier works, suggests that the obscuration is so great in such galaxies that no reliable rates are obtainable. See Caldwell and Phillips (1989) for a recent discussion of the star formation rate in NGC 5253 and of the two events (SN 1985B and 1972E) it has produced.

Table 2 gives the estimated rates for supernovae of Type Ia, Ib, Ic, II-L and II-P as estimated from the rates of the sample of the Shapley Ames catalogue as studied by Evans *et al* (1989). They give, as an average over the Shapley Ames sample of galaxy types, a rate for SN Ia of 0.28 $h^2$, for SN Ib of 0.27 $h^2$, and for SN II of 1.04 $h^2$, where the units are rates per 100 years per $10^{10} L_\odot$. Their sample included two 'helium-rich' SN Ib and two 'helium-poor' SN Ic events (see section 3.3). We have assumed that the SN II are divided 2/3 SN II-P and 1/3 SN II-L (Barbon *et al* 1979, 1989). The fraction of SN II-L may be considerably lower if supernova discoveries are strongly biased against dim SN II-P (Schmitz and Gaskell 1988, Gaskell 1990, Miller and Branch 1990).

**Table 2.** Rates of occurrence of supernovae of various types per 100 years per $10^{10}$ $L_\odot$ in the Shapley Ames sample of Evans *et al* (1989).

| SN Ia | SN Ib | SN Ic | SN II-L | SN II-P |
|---|---|---|---|---|
| $0.28 \pm 0.1\ h^2$ | $0.14 \pm 0.07\ h^2$ | $0.14 \pm 0.07\ h^2$ | $0.35 \pm 0.1\ h^2$ | $0.69 \pm 0.2\ h^2$ |

The similarity of these rates suggests that none of these supernovae come from radically different stellar populations, neither solely from stars of very low mass which occur in great abundance, nor solely from stars of very high mass which are rare. The small number statistics, and uncertainties in the observations and in the role of binary evolution make it difficult to interpret these rates directly in terms of birthrates and the stellar mass function in the Galaxy or in the solar neighbourhood. One could estimate to zeroth order that the relative rates of the various supernova types in the Galaxy reflected those derived (crudely!) for the Shapley Ames sample, but it should be borne in mind that the statistics reflected in table 2 may be very misleading. In general, the rates in table 2 correspond to explosions in stars of moderate mass $\sim$5–25 $M_\odot$. If one makes the simplifying assumption that a given class of supernova, for instance SN II-P, come from single massive stars then one may estimate the corresponding mass range from the stellar mass distribution, which itself is uncertain in the high mass range (Scalo 1986). For a typical mass function with slope $\mathrm{d}N(m)/\mathrm{d}m \propto m^{-3}$, a supernova rate of 0.01 per year in our Galaxy corresponds to the explosion of all stars above about 15 $M_\odot$.

Figure 3 shows the basic supernova classification scheme, traditionally based on the spectrum near maximum light with the primary differentiating property being whether or not the spectrum shows evidence of hydrogen. The events that fail to show the strong Si feature near maximum light can be further differentiated by the presence or absence of strong lines of He I (Wheeler *et al* 1987a). The events that show no Si near maximum light, but do show He I, especially He I $\lambda 5876$, are identified as Type Ib. There are other events that fail to show either H or Si near maximum light, and very little evidence for He. Wheeler and Harkness (1986) proposed the category SN Ic for these events. Wheeler *et al* (1987a) presented atmosphere models that suggested that SN Ib and the candidates for Type Ic might be similar except for the relative concentrations of He and O in the envelope. Nevertheless, these events can be differentiated from SN Ib events like SN 1983N and 1984L during the first few months. We will use the terminology Type Ic in the current review.

Recent years have seen a great deal of work and new insight into the nature of spectra in the later phases of supernovae roughly six months after maximum when the spectra become dominated by nebular emission lines. Most events preserve the distinction of SN I or SN II as defined by the early-time spectra. In the nebular phase, events with H lines near maximum light also show strong $H_\alpha$ line emission in the nebular phase. Some events show strong lines of [O I], particularly the $\lambda 6300$, 6364 doublet as well as $H_\alpha$ and permitted and forbidden Ca II lines. Others are dominated by the H and Ca emission with weaker evidence for O. Whether this represents a fundamental

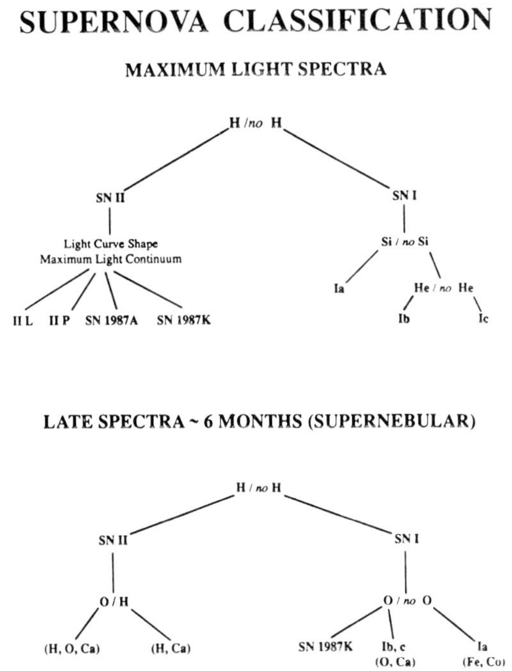


**Figure 3.** Classification scheme for supernovae based on the early- and late-time spectra and other features (from Harkness and Wheeler 1990).

difference in the physical properties and whether it is correlated with the SN II-L and II-P subclassification scheme is not clear at this time.

The Type I events have proven to show very distinct differences in their nebular spectra. In this phase, SN Ia show strong emission lines of [Fe II], [Fe III] and [Co III] with Ca H and K in absorption but no evidence for O (Axelrod 1980a, b). SN Ib events show strong [O I] λ6300, 6364 as well as Ca and Mg I] λ4562 emission (Gaskell *et al* 1986). It is not yet clear whether the SN Ic events show identifiable differences from the SN Ib in the nebular phase.

In the remainder of this review we will lay out the well studied subject of Type Ia supernovae and outline the problems that remain to be solved in terms of the progenitor evolution and the explosion mechanism. The subject of Type Ib (and Ic) supernovae is much less well developed. We will outline the recent progress and the prospects for future developments. Other reviews covering these and other topics associated with supernovae are Wheeler (1981), Trimble (1982, 1983), Woosley and Weaver (1986a, b), Wheeler (1990a, b) and the books edited by Petschek (1990), Woosley (1990), and Wheeler *et al* (1990).

## 2. Type Ia supernovae

### 2.1. Overview

The study of classical Type I supernovae has matured greatly in the last decade both observationally and theoretically. Axelrod (1980a, b) showed that the radioactive decay of $^{56}$Ni could account for the late-time spectra and light curves of SN Ia. Observation and analysis of the Type Ia SN 1981B (Branch *et al* 1982, 1983) found that the outer layers of the exploding material must be composed primarily of elements of intermediate mass: O, Mg, Si, S, Ca. This means that the outer portions of the progenitor cannot be processed to nuclear statistical equilibrium (NSE), a condition violated by many otherwise reasonable models. These constraints and many quantitative features of SN Ia are satisfied by models based on white dwarfs in which the explosion mechanism is the subsonic deflagration of degenerate carbon and oxygen (Nomoto *et al* 1976, Chevalier 1981, Müller and Arnett 1982, Sutherland and Wheeler 1984, Nomoto *et al* 1984b, Woosley and Weaver 1986b and references therein). Such a model gives a central portion burned to nuclear statistical equilibrium, primarily $^{56}$Ni, and an outer mantle of unburned or partially burned material. The radioactive decay of the nickel gives a reasonable representation of the light curve and the radiative transfer of the trapped decay energy near maximum light gives a satisfactory account of many details of the spectrum (Branch *et al* 1985, Harkness 1985, 1986), and the excitation of the optically thin central Co/Fe core by the radioactive decay gives a fair account of the spectrum at late times (Woosley *et al* 1984). The main aspects of SN Ia will be addressed in a chronological fashion: progenitor evolution, explosion, near maximum light, and later stages. Despite the overall success of this picture, there are some interesting unsolved problems.

### 2.2. Evolution

*2.2.1. Basic constraints.* Two long-standing features have guided research on SN Ia. One is their basic hydrogen deficiency. The other is the production of SN Ia, but not SN II, in elliptical galaxies. This is still regarded as evidence that some means of production of SN Ia require progenitors of long lifetime, since most elliptical galaxies are thought to have ceased star formation billions of years ago. As mentioned in the introduction, questions have been raised about this assumption, but a convincing case that *all* SN Ia come from relatively young progenitors has not been made. These two features continue to suggest that a white dwarf which has lost mass in a binary system and then is later brought to an explosive condition by mass transfer from a companion is the basic mechanism behind the production of SN Ia (Iben and Tutukov 1984, Webbink 1984). The more recent work on the analysis of light curves and spectra of SN Ia showing that carbon/oxygen white dwarf deflagration models provide a reasonable match to basic observations adds support to this picture.

*2.2.2. Problems with hydrogen transfer.* One of the principal problems regarding SN Ia remains the difficulty in determining the evolutionary processes leading to an appropriate explosion. For a variety of reasons the simplest scenario of transfer of hydrogen onto a white dwarf does not seem to be satisfactory to produce the majority of SN Ia events. Accretion of hydrogen onto a white dwarf produces different results for different rates, but there are objections to all rates from zero to infinity. The different possibilities as a function of the (assumed constant) accretion rate and the initial mass of the accreting C/O white dwarf are illustrated in figure 4.

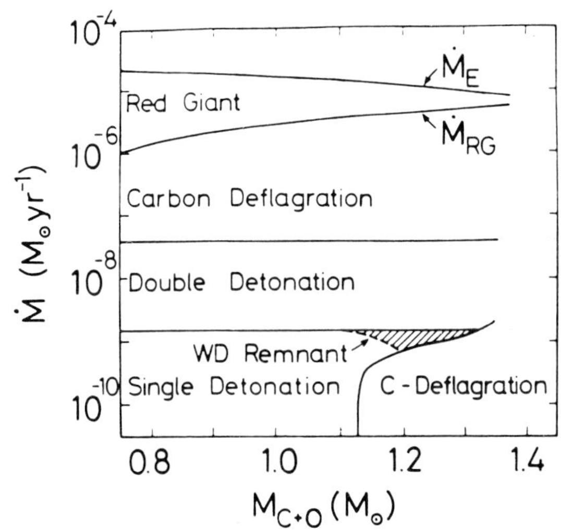


**Figure 4.** Possible evolutionary outcomes as a function of the accretion rate of hydrogen onto a carbon–oxygen white dwarf of given initia mass (from Nomoto 1985).

For $\dot{M} \leqslant 10^{-9}\ M_\odot\ \text{yr}^{-1}$ so much mass is expected to be ejected in nova explosions, that the time to bring the white dwarf to the Chandrasekhar limit will exceed the Hubble time or the companion must provide a reservoir of mass in excess of a solar mass (Fujimoto and Taam 1982). In addition, the number of progenitor systems required by this mechanism may exceed the number of cataclysmic variables, most of which have too small a mass anyway (Trimble 1982, Sutherland and Wheeler 1984, Iben and

Tutukov 1984, Webbink 1984, MacDonald 1984). At these low accretion rates, the hydrogen burns to helium during the accretion, and the helium builds up as a degenerate layer. If the initial white dwarf mass is less than 1.2 $M_\odot$, only the outer helium layer ignites, burning to nuclear statistical equilibrium and giving the wrong composition (Nomoto 1982b, Woosley *et al* 1986). If the white dwarf has cooled for $\gtrsim 10^8$ yr, the central density could rise to $\sim 10^{10}$ gm cm$^{-3}$ (Taam 1980, Nomoto 1982a). A deflagration at such densities is expected to produce an excess of neutron rich species (Sutherland and Wheeler 1984, Woosley *et al* 1984, Thielemann *et al* 1986).

For $\dot{M}$ between $10^{-9}$ and $5 \times 10^{-8}$ $M_\odot$ yr$^{-1}$ the accumulated helium triggers a detonation in both the helium envelope and the inner C/O core (Woosley *et al* 1986, Nomoto 1982b), giving a composition almost entirely in NSE and hence a result at odds with the observed composition. For $\dot{M}$ between $5 \times 10^{-8}$ and $10^{-6}$ $M_\odot$ yr$^{-1}$, helium detonation is avoided because the helium burns non-degenerately during accretion. The result, however, is an exceedingly bright progenitor system, and there are constraints on the number of such systems observed (Sutherland and Wheeler 1984, Paczyński 1985). Finally, for $\dot{M} \gtrsim 10^{-6}$ $M_\odot$ yr$^{-1}$, shell hydrogen burning cannot keep up with the accretion rate. The argument has been made that hydrogen must then accumulate in the system, in violation of the observed limits of hydrogen in the spectra (Nomoto *et al* 1976).

All these arguments need to be carefully re-examined. At the lowest rates, mass loss during nova explosions is one of the most severe constraints. Recent arguments have been made that novae may not accrete at a constant rate, and may 'hibernate' at low accretion rates with a rather long duty cycle (Shara *et al* 1986). This may mean that there are many more cataclysmic variables than previously estimated, and hence more potential progenitors of SN Ia. Ciardullo *et al* (1987) have argued that novae may be a bulge rather than a disk population. If SN Ia share this distribution they would be of an old population but might still be subject to an inclination effect as observed, thus encouraging the association with nova-like systems. See Wheeler (1990a) for a discussion of these and related issues.

The arguments against low and moderate accretion rates which lead to helium detonation are based on spherically symmetric models. Three-dimensional models would show ignition at a point, not simultaneously in a spherical shell, and might give a very different result. Two-dimensional models by Dgani and Livio (1989) do not, however, give radically different results. The bright progenitors which are required to burn both hydrogen and helium during accretion may produce most of their luminosity in the UV, where they may be unobserved. Another possibility is to seek a connection of such systems with the symbiotic variables.

The argument that the highest accretion rates will lead to a red giant envelope which will then contaminate the spectrum with hydrogen needs particular attention. Hachisu *et al* (1989) have proposed that the hydrogen transfer from a small mass helium core-burning red giant might be regulated in such a way as to provide stable accretion at the hydrogen shell-burning rate without the accumulation of excess hydrogen. They point out that the radius of the accreted matter surrounding a C/O white dwarf varies sensitively with the accretion rate. For mass accretion in excess of the hydrogen shell burning rate, the radius of accreted matter swells rapidly to fill the Roche lobe. Hachisu *et al* postulate that back pressure will staunch the mass transfer resulting in a smaller accretion rate, and hence a tendency for the radius of the accreted matter to decrease, decreasing the back pressure and allowing an increased mass transfer rate. They conclude that the result will be regulated transfer at the hydrogen shell-burning rate and argue that this condition can be satisfied for certain binary system parameters, whereas other conditions will result in slower accretion and nova explosions or in an overfilling of the Roche lobe and the spilling of matter into the circumstellar environment through the outer Lagrangian point, both of which they rule out as conditions for SN Ia. There are two problems with this suggestion. One is that it is not at all clear that the back pressure from a filled Roche lobe prevents further mass transfer. It seems equally likely that a common envelope would form with the matter spilling beyond the Roche lobes of both stellar cores. The other problem is observational. The configuration favoured by Hachisu *et al* demands a luminosity of about 50 000 $L_\odot$ from a volume the size of the Roche lobe. This yields effective temperatures corresponding to a B star. There must be several such systems within 1 kpc of the Sun to account for the rate of SN Ia, but no such systems are seen.

There might be ways to consider the option that Hachisu *et al* reject, namely that in which the transferred matter swells to form a common envelope. Barkat (private communication) argues that such envelopes would be expelled by the same mechanism that causes planetary nebulae to form as the core exceeds about 1.2 $M_\odot$, so that the simplest picture in which most of the mass is retained in a thick hydrogen common envelope surrounding the mass-losing star and the white dwarf cannot hold. Retention of even a small amount of hydrogen in the progenitor system opens some interesting possibilities, however. The limit on hydrogen is not currently all that tight, being $\lesssim 0.1\ M_\odot$ (Sutherland and Wheeler 1984, Branch *et al* 1982, Section 2.5.2). This means that enough hydrogen might be retained to mask the progenitor without showing up in the spectrum near maximum light. With planetary-nebula-type mass loss to prevent the hydrogen from accumulating, a steady state might develop in which the hydrogen flows from the companion into a common envelope at a rate necessary to balance shell burning and pulsational mass loss. If the envelope mass became much less than 0.01 $M_\odot$, it would collapse down onto the core. This may be prevented by the accretion. The steady state solution might then be an equilibrium with a low mass extended envelope. Such a progenitor might resemble a Mira variable from the outside, and be very difficult to identify prior to the explosion.

### *2.2.3. Binary white dwarf evolution*

*Basic properties.* A mechanism which may avoid the difficulties with hydrogen accretion is the possibility that some binary systems form two white dwarfs and that these stars evolve to produce a Type I supernova (Iben and Tutukov 1984, Webbink 1984, Iben and Tutukov 1985, Tornambé and Matteucci 1986). This process is illustrated in figure 5.

The basic idea is that two dwarfs form as a result of a common envelope phase of evolution and spiral together under the influence of gravitational radiation until the smaller mass dwarf with the larger radius fills its Roche lobe. For mass ratios near unity, the tendency for the lower mass dwarf to expand faster than the orbital separation as it loses mass will cause it to undergo catastrophic mass loss on a dynamical timescale. This situation is generally expected when the dwarfs have the same composition, either He–He or C/O–C/O (Iben and Tutukov 1984, Webbink 1984). Much of the matter of the disrupted dwarf may be trapped in the deep gravitational well of the companion and be retained. The precise behaviour of this matter which will have a large amount of angular momentum, is subject to debate. Tutukov and Yungleson (1979) (see also Iben and Tutukov 1984) have suggested that it will form a thick disk. This has been contested by Hachisu *et al* (1986) on energetic grounds, but supported by Benz *et al*

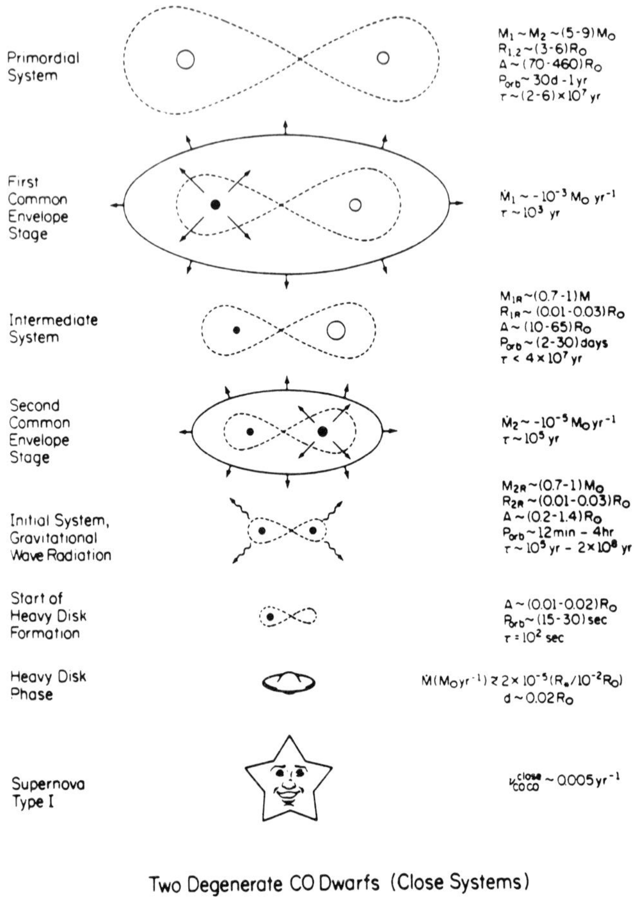


**Figure 5.** Schematic evolution from close binary to common envelope system to coalescing double degenerate system as a possible evolution scheme to produce a Type Ia supernova (from Iben and Tutukov 1984).

(1990a, b). Because of the angular momentum, the accreted matter will not necessarily be able to settle onto the companion on dynamical timescales.

*Accretion at constant rate.* At high accretion rates, matter accreted onto a degenerate companion is heated by gravitational compression and a burning shell may ignite. The existence and evolution of this burning shell is one of the key issues which determine whether the resulting configuration is a viable model for SN Ia. The simplified problem of accretion at constant rate has been investigated by various workers. Nomoto and Sugimoto (1977) calculated the accretion of He onto a He core of 0.4 $M_\odot$. For accretion rates of 1 and $2\times 10^{-8}$ $M_\odot$ $\mathrm{yr}^{-1}$, they found central helium ignition, a detonation, and total disruption of the model at 0.99 and 0.78 $M_\odot$, respectively (see also Mazurek

1973, 1980, Woosley *et al* 1986). For a larger accretion rate of $4 \times 10^{-8}$ $M_\odot$ $\mathrm{yr}^{-1}$, Nomoto and Sugimoto found that a helium burning shell ignited non-explosively off-centre when the model reached 0.66 $M_\odot$, but they did not follow the subsequent evolution of the star. For the high accretion rates expected as a binary dwarf is disrupted, shell ignition may be the rule, leading to some form of C/O mantle or perhaps one of O/Ne/Mg. The subsequent evolution is unclear. The burning shell might proceed to the centre, leaving a stable C/O dwarf. Alternatively, central ignition and detonation might occur at a later phase, particularly if the accretion rate declines as the envelope is consumed and the shell is extinguished, so that matter beyond the shell can cool and join the core.

The calculations of Nomoto and Iben (1985) assume accretion of C/O onto a C/O dwarf of 1 $M_\odot$ at constant rates approaching that which would generate an Eddington limit luminosity. For $\dot{M} = 2 \times 10^{-6}$ $M_\odot$ $\mathrm{yr}^{-1}$ (0.1 $\dot{M}_{\mathrm{Edd}}$), they find that central carbon ignition just barely manages to precede shell ignition because of adiabatic compression of the whole star. This presumably results in a thermonuclear explosion that completely disrupts the star. For $\dot{M} \geqslant 4 \times 10^{-6}$ $M_\odot$ $\mathrm{yr}^{-1}$ (0.2 $\dot{M}_{\mathrm{Edd}}$) they find that a carbon burning shell ignites off-centre before conditions are reached for central carbon ignition. Saio and Nomoto (1985) calculate the evolution of the burning shell in a model with $\dot{M} = 1 \times 10^{-5}$ $M_\odot$ $\mathrm{yr}^{-1}$ (0.5 $\dot{M}_{\mathrm{Edd}}$). The shell moves inward in a series of flashes and consumes the carbon in the core. This leaves an O/Ne/Mg dwarf, which is susceptible to core collapse and may play a role in the formation of compact x-ray sources (Nomoto 1984), but which seems unlikely to reproduce the observational properties of SN Ia.

These calculations assume a constant accretion rate for the time during which the burning shell eats into the core, of the order of $10^5$ yr. They do not allow for the structure, finite extent, and dynamics of the envelope of disrupted dwarf material, nor for its angular momentum. More realistic conditions might result in variable accretion rates and a very different outcome. The accretion rate might decrease with time, allowing the shell to go out and the matter beyond it to cool and settle onto the core as a mantle of degenerate O/Ne/Mg although Saio and Nomoto (1985) argue that once ignited, the burning shell does not require continued accretion for it to propagate. If the total mass of core and mantle exceeds the Chandrasekhar limit, central carbon ignition and deflagration or detonation would be expected. Between the extremes of central carbon ignition prior to shell ignition and complete carbon consumption by quasi-static shell burning, there may be a range of conditions for which the off-centre shell ignites, but for which adiabatic compression triggers central carbon ignition before the shell reaches the centre.

Wheeler *et al* (1987c) have calculated the dynamics of a number of configurations that correspond to possible outcomes of binary white dwarf evolution and disruption in which mantles of heavy elements presumed to result from accretion and shell burning overlay cores of lighter elements. Models based on central ignition of He cores within C/O mantles show once again that explosions based on degenerate helium configurations have great difficulty accounting for the observed features of SN Ia. In these models, C/O at densities too low to support a Chapman–Jouguet detonation is burned by the over-driven detonation proceeding from the helium core. The result is the production of insufficient non-iron peak matter, and what little is produced is ejected at excessive velocities. Models involving carbon-depleted mantles on top of C/O cores suggest that if a detonation forms in the core even a mantle of O/Ne/Mg is likely to be burned to NSE as well, yielding too little partially burned matter. If the explosion mechanism is a subsonic deflagration the higher ignition threshold of oxygen impedes the deflagration of the carbon-depleted mantle. The result is a minimum C/O core mass necessary to provide agreement with observed SN Ia. The core must be in excess of 0.5–0.9 $M_\odot$, depending on the deflagration model, to provide enough burning to account for SN Ia, a conclusion similar to that reached for pure C/O models by Sutherland and Wheeler (1984). The accretion-driven burning shell thus need not propagate all the way to the centre of the core to prevent the resulting configuration from being a reasonable SN Ia progenitor.

Wheeler *et al* (1987c) thus conclude that the coalescence of twin *helium* degenerates is unlikely to lead to an explosion like that observed in classical SN Ia. This process may occur occasionally, but it cannot be the principal production mechanism of SN Ia. The evolution of twin C/O degenerates need not produce an accretion rate so low as to lead to central ignition with *no* carbon shell burning, but the calculations show that the shell burning must not propagate appreciably into the core.

*Dynamical accretion.* A way out of this dilemma is suggested by Iben (1988), who points out that the smaller dwarf is disrupted dynamically upon filling its Roche lobe. He questions whether the calculation of Saio and Nomoto (1985) which spans a time of $10^5$ yr is relevant and argues instead that the disrupted mass may cascade onto the larger mass white dwarf on a dynamical timescale so that thermal, shell-burning timescales are irrelevant. The actual process may thus be more complex, and interesting, than indicated by simple calculations assuming constant accretion rate. Iben speculates that the sudden addition of a rather large quantity of mass would send a shock wave into the accreting white dwarf, igniting carbon and triggering the explosion even though the total mass involved did not equal or exceed the Chandrasekhar limit. This is a very interesting suggestion, because the dynamically rebounding star might quench any detonation that forms, and hence leave an outer layer of partially burned matter as demanded by the observations even though the explosion mechanism is a detonation, not a deflagration.

Preliminary models to explore this possibility have been calculated (by JCW). Initial white dwarf models of equal mass fractions of C and O were constructed in hydrostatic equilibrium with an isothermal temperature distribution of $10^7$ K. Dynamic accretion was then simulated by increasing the mass of the outer 10–15 zones (out of 60–80). A model with $M = 1.32\ M_\odot$ was augmented by 0.1 $M_\odot$ and the result was an inwardly propagating shock that triggered a detonation. Outer zones also ignited and generated a Rayleigh–Taylor instability as low density, high pressure material drove a shock inward as well as outward. The Rayleigh–Taylor instability caused sufficient mixing of heat to drive a deflagration into the star. The deflagration met the outgoing detonation and the result was the complete incineration of the model, leaving no intermediate mass elements which would be required for a viable model of a SN Ia. A model with 0.95 $M_\odot$ was augmented by 0.1 and 0.4 $M_\odot$. The addition of the smaller mass generated a weak shock that did not result in ignition. The larger mass accretion, bringing the total to near the Chandrasekhar limit did ignite a detonation in the centre (but no outer deflagration in the lower density outer portions), which consumed nearly all the matter of the star, leaving only a small unburned portion, $\simeq 0.08\ M_\odot$, travelling at $\gtrsim 16\,000$ km s$^{-1}$. These models have only a very simple burning prescription for carbon and oxygen and need to be repeated using a more complete reaction rate network to look more carefully at the burning on the outside where the detonation may die, and at the abundances left in the wake of the inwardly propagating deflagration. This type of model has some interesting possibilities, but evidently will require, at the very least, some tuning of the parameters to achieve a model that has the basic characteristics of SN Ia.

### 2.3. Observational constraints

Robinson and Shafter (1987) have used narrow band filters in the wings of lines to monitor Doppler shifts due to orbital motion induced by a close degenerate companion in a sample of white dwarfs. They find no new candidates for degenerate binaries with orbital periods between 30 seconds and 3 hours in a sample of 44 white dwarfs. They conclude that no more than 1 white dwarf in 20 in the local population could be a close binary in this period range, and argue that binary white dwarfs could produce SN I at the observed rate only if their upper limit to the space density of such systems is the actual value and that all such systems produce SN I events. The latter is very unlikely since most systems will have a total mass less than the Chandrasekhar limit, and will be unlikely to explode. Bragaglia *et al* (1988) and Foss *et al* (1990) have searched for double degenerates at longer orbital periods by means of variable radial velocities. Bragaglia *et al* report two candidates. Foss *et al* found no binaries with orbital periods between 3 and 10 hours.

The study of Robinson and Shafter was designed to detect very close degenerate companions with short orbital periods and large orbital velocities. Their analysis assumes that the double degenerate systems are born with larger orbital radii which decay under the dissipative action of gravitational radiation. Thus the results of Robinson and Shafter may not limit the rate of production of SN Ia by the coalescence of twin degenerate stars, but only the specific mechanism which depends on the slow spiral to small orbits due to gravitational radiation. Saffer *et al* (1988) have discovered a twin DA white dwarf system of large orbital radius (period 1.56 days) which suggests that such systems may be plentiful. They suggest that other avenues to form ultraclose systems should be explored. One possibility discussed by Saffer *et al* is that some double degenerates undergo coalescence while the stellar cores are still enmeshed in the common envelope, and hence never undergo the evolution tested by Robinson and Shafter. On the other hand, as pointed out by Robinson (private communication 1989), some systems are found at the short periods he investigated, so this phase cannot be skipped by all double degenerate systems.

In summary, we do not presently have a satisfactory picture for the progenitor evolution of SN Ia. The common idea that they are white dwarfs in some type of binary system is still regarded as the best active hypothesis, but no quantitative model has proved satisfactory, never mind compelling. The simple idea that SN Ia come from accretion onto a white dwarf from a red giant companion has not yet proved viable. The subject of double degenerate systems is certainly a rich new area of study for stellar evolution, but it, too, has failed to yet give a satisfactory account of what leads a star to become a SN Ia.

### 2.4. Ignition

*2.4.1. Degenerate carbon ignition.* The pioneering study of degenerate carbon ignition by Arnett (1969) raised questions that are still actively pursued today. Arnett studied the evolution of the core of a star of moderate mass ($\sim 5\ M_\odot$) to the point of carbon ignition. He found a brief convective burning phase controlled by neutrino losses, but eventual thermal runaway. The basic physical point is that the Fermi pressure of a degenerate gas is insensitive to the temperature. Thus as burning ensues and the temperature rises the pressure does not increase appreciably. With no increase in pressure, the star cannot respond mechanically to expand doing work against gravity, the mechanism that makes normal non-degenerate stars thermally stable. Without the mechanical response, a degenerate star heats when burning commences and the very temperature-sensitive nuclear reactions run away until sufficient thermal pressure is added to lift the degeneracy and initiate a mechanical response. For degenerate carbon ignition, the burning runs away on a dynamical timescale and the response is a violent explosion.

Arnett (1969) argued that the result of degenerate carbon ignition would be a supersonic detonation in which a shock front initiates the burning which drives the shock. The steady state version of this process is known as a Chapman-Jouguet detonation (see Mazurek and Wheeler 1980 and references therein for details). Two points in particular arose in Arnett's work which are still debated: one is the nature of the brief convective phase prior to thermal runaway and the other is whether the runaway leads to a supersonic detonation or to a subsonic deflagration process.

*2.4.2. Convective Urca process.* The Urca process is one of neutrino loss driven by a cyclic electron capture and nuclear transformation, e.g.

$$M(A, Z)+e^- \rightarrow M(A, Z-1)+\nu_e$$

followed by

$$M(A, Z-1)\rightarrow M(A, Z)+e^- +\bar{\nu}_e$$

(Gamow and Schoenberg 1941). The net result is no change in composition, but the generation and loss of a $\nu_e$, $\bar{\nu}_e$ pair in each cycle.

Paczyński (1973) first pointed out that the convection associated with degenerate carbon ignition could lead to just such a cyclic neutrino loss mechanism. Elements with appropriate electron capture thresholds could be swept into the centre of the star where the high electron Fermi energy would exceed the capture threshold and then carried to lower density regions where beta decay back to the original state would occur. Paczyński argued that this convective Urca process would regulate degenerate carbon burning and allow a degenerate C/O core to evolve to considerably higher densities before runaway occurred.

Bruenn (1973) added a complication by noting that electron capture would remove an electron with less than the Fermi energy on average leaving a 'hole' in the Fermi sea. An electron with the Fermi energy would then drop into the hole giving up its excess energy as heat. On the beta decay branch the emitted electron has an energy in excess of the Fermi energy and again the excess energy appears as heat. Bruenn thus argued that the convective Urca process could result in heating, not cooling, of the core.

Couch and Arnett (1975) argued that the mechanical energy of the circulation must also be part of the energy balance and that the net effect must be a sink for the nuclear energy. Iben (1982) has done the most careful numerical investigation of the convective Urca process to date. He finds that the central density at which thermal runaway occurs can be postponed from $\rho_c \sim 2\text{–}3\times10^9$ g cm$^{-3}$ to $\rho_c \gtrsim 4\times10^9$ g cm$^{-3}$ by the action of the convective Urca process. Such a change may not have a major effect on the subsequent dynamics, but it can substantially affect the rate of neutronization in a deflagration, as will be discussed below.

The physics of the convective Urca process has been re-analyzed by Barkat and Wheeler (1990). They argue that the central point is that the convection must rapidly lead to a steady state in which there is no net change in the composition with time. Rather, there must be a gradient in the composition of mother and daughter nuclei, *not* a uniform composition, as assumed implicitly or explicitly by other workers. The currents in these species implied by the gradients can be shown to exactly balance the heating terms so that in the steady state (e.g. on times long compared with a convective flow time, but short compared with an evolutionary time driven by accretion onto the core) *the heating is identically zero and the convective Urca process is a strong source of net cooling as originally proposed by Paczyński.* The growing convective core will overshoot an Urca shell by an amount which leads to a balance of the nuclear input by the regular and Urca neutrino losses.

A second conclusion is that the condition in which the convective Urca process controls the nuclear burning is thermally unstable. The Urca cooling is effectively a very steep function of the temperature ($\propto T^{170}$) since the process depends so sensitively on the extension of the convective core beyond the Urca shell at the critical threshold density, and the extension of the convective core in turn depends on the central temperature. A positive perturbation in the entropy of the convective core will lead to much stronger neutrino losses and to regulation of the nuclear burning ($\propto T^{20}$) and hence the temperature. A negative perturbation in the entropy of the convective core will, however, lead to a smaller convective core and Urca losses reduced to balance the decreased nuclear energy production. Such a state will be susceptible to negative entropy perturbations due to ordinary neutrino losses. Once a cool state is reached with small Urca losses, the cycle will probably begin again as nuclear burning dominates. Such thermal cycles were found by Iben (1982).

The subsequent evolution will be dictated by the slow growth of the core mass and central density due to accretion. Further modelling should show whether the result is collapse if the central density reaches the critical threshold for dynamical electron capture and general relativistic instability, or explosion due to off-centre carbon ignition in the hotter material which falls just beyond the Urca shell. If the core evolves to high density and collapse, it seems very unlikely that the result will resemble a SN Ia. Iben (1982) finds the formation of a new Urca core followed by uncontrolled runaway in an outer Urca shell. In any case, the work of Iben and Barkat has shown that the convective Urca process is one of dominant importance in the general problem of degenerate carbon ignition. It must be considered in thinking about the nature of any ultimate thermonuclear runaway.

*2.4.3. Effects of solidification.* A potentially important effect whose likelihood is still debated is the solidification and phase separation of the carbon and oxygen in the centre of a C/O white dwarf. Calculations by Stevenson (1980) suggested that carbon and oxygen would not be miscible in the solid phase that could occur if the white dwarf cooled for a sufficiently long time before accreting matter from a companion. Employing this concept Canal *et al* (1980; see also Hernanz *et al* 1988) suggested that oxygen would crystallize first and that oxygen 'snowflakes' would settle to the centre of the star. If this were to occur, the systematics of carbon ignition would be significantly altered. More recent calculations (Barrat *et al* 1986, Ichimaru *et al* 1987) have suggested that limited miscibility could occur in the solid phase, but the issue still depends on details of the ordering of the crystal lattice.

Canal *et al* (1988) suggest that immiscibility and separation of C and O will lead to off-centre ignition and models which can reproduce the Branch–Pskovskii correlation of peak luminosity with photospheric velocity (see section 2.1.). Canal *et al* argue that even with limited miscibility the relevant pycnonuclear reaction may be $^{12}C+^{16}O$, rather than $^{12}C+^{12}C$, and that the result will still tend to be off-centre ignition.

*2.4.4. Detonation versus deflagration.* Although deflagration models have received a great deal of attention because of their 'natural' production of unburned matter in quantities and with velocities that seem to match the observations, their physical basis is rather weak. The problem is that the deflagration process is intrinsically subsonic so that pressure waves from the burning matter precede the front, and the front is turbulent and therefore very difficult to treat realistically with current numerical techniques. In one-dimensional calculations, recourse has been made to a variety of rather simple prescriptions treating the front as convectively or Rayleigh–Taylor unstable. For the Rayleigh–Taylor instability only the pressure and density gradients are used to determine instability and the rate of mixing of turbulent heat, while the convective schemes involve the thermal contribution to the buoyancy (Nomoto *et al* 1976, Wheeler and Sutherland 1984, Wheeler *et al* 1987c, Nomoto *et al* 1984b, Woosley and Weaver 1986a, b). Both are limited by the result that the characteristic length scale, e.g. pressure scale height, changes rapidly through the front, making the results quantitatively unreliable. All deflagration prescriptions depend on parameters that are equivalent to specifying the propagation speed of the burning front. In this prescription, the amount of burning can then be adjusted to give the desired result, with a slow front speed ($\leqslant 0.1\ c_s$) giving little burning to NSE and little partial burning, intermediate values giving 'reasonable' results, and large front speeds ($\gtrsim 0.5\ c_s$) yielding complete incineration with virtually no intermediate elements. In addition, the simple one-dimensional deflagration calculations give a strictly stratified composition structure. Better agreement with observed spectra are obtained when an *ad hoc* mixing of the burning products is employed to spread the intermediate mass elements through a greater range in radius and hence velocity. This mixing may be the natural result of instabilities in the flow which are not modelled in the calculations, but may imply some deficiency in the basic ansatz of the deflagration models. Two-dimensional studies (Müller and Arnett 1982) can overcome some limitations of the 1D calculations, but are also limited by numerical resolution. Harkness (1990) has argued that a more thorough treatment of line opacity may make the *ad hoc* mixing unnecessary. A recent summary of the possible physical processes involved in degenerate C/O deflagration is given in Woosley (1990).

A potentially serious deficiency of deflagration models is that they expand relatively slowly compared to detonations. This gives more time for weak interaction processes to occur. Calculations show that excessive neutronization does occur in standard deflagration models, producing an excess of such species as $^{54}Fe$ and $^{54}Cr$ (Sutherland and Wheeler 1984, Woosley *et al* 1984). This problem will be discussed in some more detail below in the context of nucleosynthesis.

Woosley and Weaver (1986a; see also Woosley 1990) have proposed a variation on the deflagration process. In laboratory deflagrations the velocity of the combustion front tends to be proportional to the largest length scale of the physical configuration. Since the length scale is constrained to be $\leqslant r_F$ where $r_F$ is the radius of the front, Woosley and Weaver argue that the front velocity should vary in proportion to the front radius, yielding an accelerating front as the burning proceeds out through the star. In addition to being perhaps more physical than a constant front speed, this prescription should tend to reduce the excess neutronization. The reason for this is that as the deflagration front begins to propagate very slowly there is a rapid neutronization of the burned matter, but this can represent a very small mass fraction. In addition, the overpressure from the small inner burned region is communicated beyond the subsonic front to the outer unburned portion of the star. Thus the outer portions have expanded before the accelerating front reaches them, and the density is lower than would otherwise be the case, so weak interactions proceed at a negligible rate in the bulk of the ejecta. Woosley and Weaver (1986a) argue that for a reasonable choice of parameters, and ignoring the convective Urca process, a deflagration might reduce the neutronization to acceptable levels. In practice, the numerical models presented by Woosley and Weaver had a deflagration front speed that is essentially a constant fraction of the sound speed, as will be discussed below.

In their arguments in favour of the formation of deflagrations, Mazurek *et al* (1977) neglected the cumulative effects of the burning-induced shocks which amplify the shock from previously burned matter. Blinnikov and Kholkov (1986a; see also Kholkov 1989) pointed out that when care is taken to examine the process of carbon ignition and runaway at very fine mass scales, less than $10^{-5}\ M_\odot$, one finds that a region of spontaneous burning exists in which the timescale for adjacent regions to burn to nuclear statistical equilibrium is less than the sound crossing time. A small, but finite, amount of matter (which depends on the initial temperature distribution) thus burns at speeds for which the phase velocity of the 'front' is supersonic. Blinnikov and Kholkov concluded that this finite ignition region would lead to the formation of a detonation. Similar conclusions have been reached by Woosley and Weaver (1986a) who note, however, that these conclusions regarding detonation formation may not hold in three-dimensional calculations.

Barkat *et al* (1990; see preliminary discussion in Wheeler *et al* 1986) instituted such studies independently and found that Chapman–Jouguet pressure and density already hold at the boundary of the spontaneous burning region, but with no shock (see also Blinnikov and Kholkov 1986b). This conclusion seems to be independent of whether the ignition is central or somewhat off-centre (in the context of spherical models), and applies to all ignition temperatures in excess of $5 \times 10^8$ K. Whether there is an ignition temperature below which the temperature gradient is so flat that a detonation will not form is under investigation. This question is complicated by the increased length scale required to form a detonation as the ignition temperature is decreased. Woosley (1990) argues that the irregularity of the convective region where degenerate carbon runs away will ensure that there are regions of steep temperature gradient between convective elements that will damp any incipient detonation. It seems just as likely that the relatively small temperature gradients *within* the hotter convective elements will promote detonation. The size of the convective elements $M \sim \frac{4}{3}\pi\rho l^3$, where $l$ is the pressure scale height $\sim 10^7$ cm, is $M \sim 10^{-3}\ M_\odot$, which considerably exceeds the size of the region required for 'spontaneous' detonation formation, $M \sim 10^{-5}\ M_\odot$.

If these conclusions concerning the initiation of detonations following degenerate carbon ignition are accepted at face value, then a dilemma exists. The deflagration model matches the observations fairly well, but appears not to be self-consistent physically. Detonations may be more physically correct, and they reduce problems with overproduction of neutron rich species such as $^{54}$Fe (Sutherland and Wheeler 1984, Woosley *et al* 1984) because the expansion is intrinsically rapid. The simplest detonation models, however, lead to virtually complete incineration of the white dwarf, and hence do not correspond to the composition deduced from spectra.

The specific mode of explosion of SN Ia thus also remains ambiguous and various alternatives must be considered. Previous studies of the way in which a pre-existing detonation is damped in a degenerate carbon/oxygen white dwarf (e.g. Buchler *et al* 1971) assumed the detonation died when the burning timescale exceeded the dynamic time. Such a criterion is clearly uncertain by factors of two, and hence the critical density below which a detonation can be supported is also uncertain. A small expansion velocity due to burning prior to full thermal runaway could also affect the amount of mass that falls below the critical density. Gravitational diffusive separation is expected at the surfaces of white dwarfs. It is amplified by the large gravity, but inhibited by the rapidly increasing inward density profile. Diffusive separation in the outer layers of carbon and oxygen could lead to composition stratification with an oxygen-enriched layer underlying a carbon-enriched layer. Such an oxygen-rich layer with a higher ignition threshold might further dampen a dying detonation. The dynamical accretion model of Iben discussed above may also lead to quenching of a detonation. In addition, studies of laboratory and numerically simulated detonations show that detonation fronts consist of cellular patterns of interacting 'triple-points' of converging shock waves and the formation of unreacted pockets of material behind the detonation front. These complexities of the front structure are crucial in determining detonation propagation and failure (Oran *et al* 1982).

Finally, note that virtually all recent studies of carbon ignition and runaway have been based on the assumption of central carbon ignition at a density of about $2\text{–}3\times 10^{9}\,\text{gm cm}^{-3}$. This simple assumption ignores the strong influence of the convective Urca process to control central nuclear burning and hence to make the prospect of central thermonuclear runaway completely immaterial. Similar arguments hold if carbon and oxygen solidify and are immiscible.

### 2.5. *Explosion*

*2.5.1. Light curves.* The fact that light curves of SN Ia are more homogeneous than those of other types of supernovae has stood the test of time, but there is still a great deal of discussion of the true degree of homogeneity. The light curve contains important basic information about the physics of the explosion, and the question of homogeneity pertains directly to the issue of whether SN Ia represent standard candles that are useful as distance indicators.

*Light curve observations.* Barbon *et al* (1973) compiled a number of light curves for SN Ia and divided them into two basic classes, fast and slow, depending on the rate of decline from maximum, the contrast from peak to tail, and the rate of decline of the tail. Figures 6 and 7 show a compilation of blue and visual light curves for SN Ia. Barbon *et al* (1984) have compiled late-time light curves for SN Ia, as shown in figure 8, and argue that the mean decay rate corresponds to a half-life of $49.5\pm 0.8$ days.

Pskovskii (1977) and Branch (1981) argued that there is a continuum of light curve speeds for SN Ia. Branch also presented evidence for a correlation between light curve speed and of the velocity at the photosphere and peak magnitude in the sense that events with slower light curve speeds (broader peaks) have higher photospheric velocities and are brighter. This is counterintuitive, since more energy in a given mass would give a faster expansion and would represent more nuclear burning and hence more $^{56}$Ni to power a brighter light curve, but a faster decline. The case for a dispersion in the properties of the light curve has been disputed by Cadoneau *et al* (1985) who discriminated between photographic and photoelectric data and concluded that when

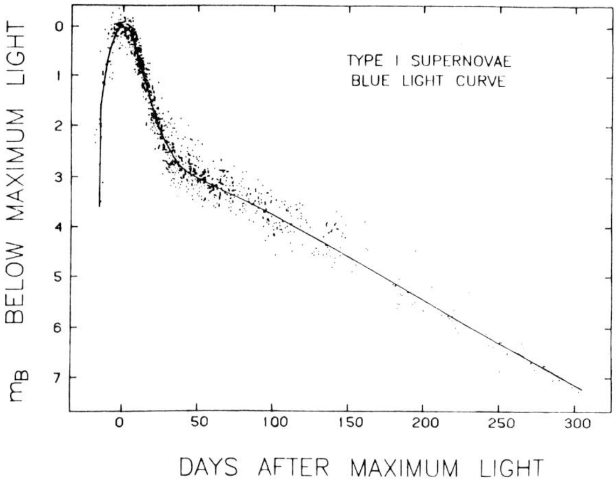


**Figure 6.** Composite B magnitude light curves for SN Ia (from Doggett and Branch 1985).

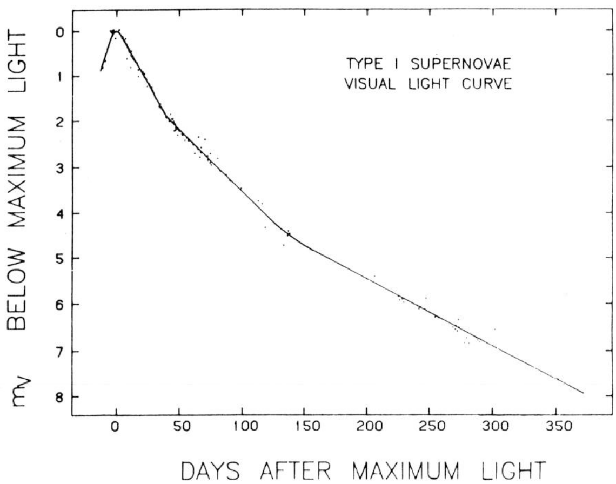


**Figure 7.** Composite V magnitude light curves for SN Ia (from Doggett and Branch 1985).

the higher accuracy photoelectric data were used alone, there was no evidence for a dispersion in light curve speeds (see also van den Bergh and Younger 1985). De Vaucouleurs (1988), noting that the light curves of SN Ia are more smoothly behaved in V than in B, fits V light curves with a single parameter from peak to tail, and concludes that there is a true dispersion, based on a rather small sample including S And, Tycho, and Kepler.

Broad-band infrared photometry of SN Ia has shown that SN Ia follow a predictable pattern in the infrared as well. Elias *et al* (1985) showed that SN Ia events have well defined infrared J, H, and K light curves characterized by a minimum ~20 days after maximum light, and a secondary peak ~30 days after maximum light, as illustrated in figure 9. The dip is thought to be due to some broad-band absorption. The secondary

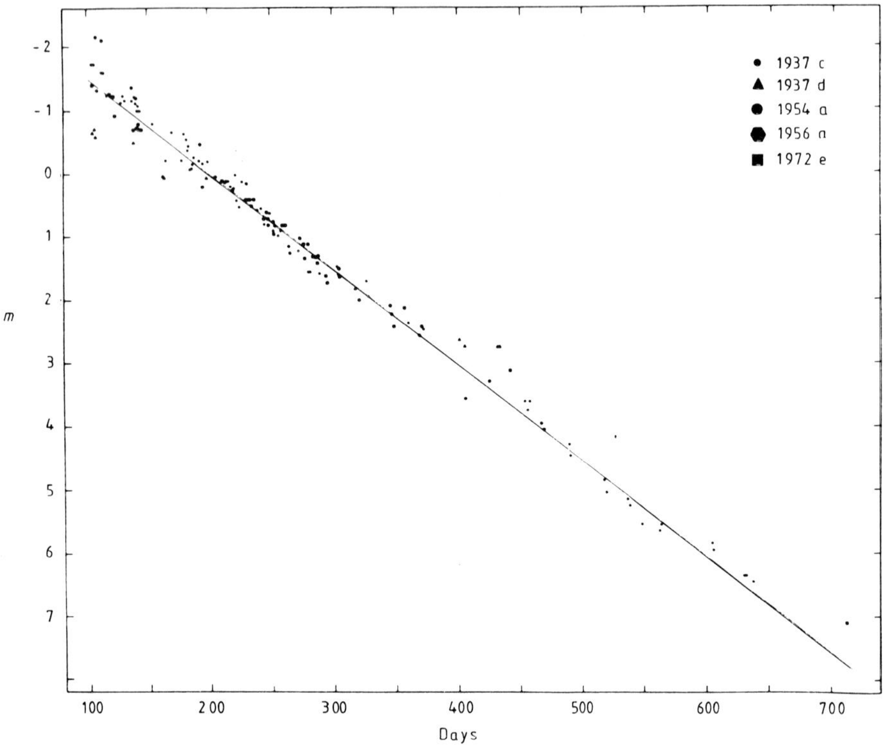


**Figure 8.** Composite late-time light curves (photographic and B) for SN Ia (from Barbon *et al* 1984).

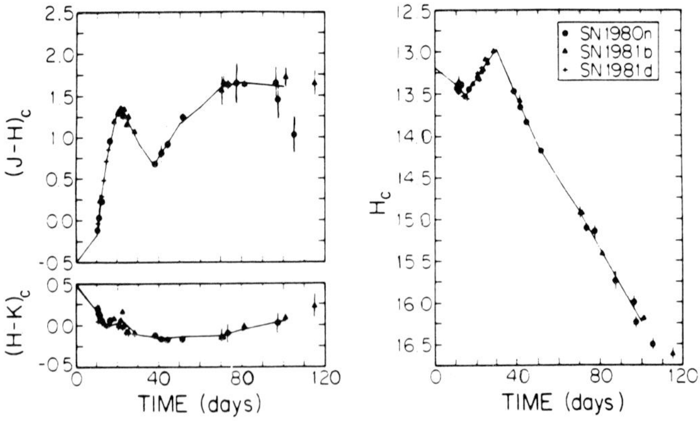


**Figure 9.** The H magnitude infrared light curves and H–K and J–H colour curves of three SN Ia 1980N, 1981B and 1981D showing the minimum at 20 days and the secondary maximum at 30 days after optical maximum (from Elias *et al* 1985).

maximum occurs as the optical light curve makes the transition from the initial peak to the exponential decline, but it is not known if this is significant. Elias *et al* argue that this minimum is due to an absorption that sets in at about 20 days. It cannot be caused by a single strong line, since it is broad band as shown by the spectra of SN 1986G (Frogel *et al* 1987). It might be due to blends of a single species. Graham (1986) has suggested that it might be caused by a dense array of lines of Si I, since that element is observed in the optical spectrum near maximum, and is not observed in SN Ib which also fail to show this dip in the IR light curve (see section 3.1).

*Light curve models.* There have been a large number of light curve calculations of SN Ia, both analytic and numerical. Arnett (1982) presented analytical light curves based on the explosion of a Chandrasekhar mass white dwarf and powered by radioactive decay. These models revealed that the bolometric luminosity at maximum was precisely equal to the amount of deposited decay energy at that instant, despite the hysteresis involved in trapping the energy of decay at earlier epochs and adiabatic cooling, rather independently of the exact opacity or density profile. This result has been confirmed by numerical studies of similar models, although it is not true in general, as models of SN 1987A have shown. The greater ejecta mass in this case traps more of the decay energy, and the peak luminosity is greater than the instantaneous deposition of decay energy (Woosley 1988).

Numerical models for the light curves of SN Ia have been given by Colgate *et al* (1980), Chevalier (1981), Arnett (1980), Schurmann (1983), Woosley *et al* (1984, 1986), Sutherland and Wheeler (1984), Nomoto (1982b). A key feature of these models is the nature of the gamma ray deposition function. There are nearly as many forms for this as there are authors. In the early phases, prior to maximum light, the precise prescription for the deposition function is probably not critical, since essentially all the gamma rays are deposited locally. At later times the various prescriptions seem to give relatively homogeneous results, but it is not clear why this should be so. For instance Woosley *et al* (1986) use a prescription in which the local deposition function is just given by $f_{\rm dep}=1-\exp(-\tau_\gamma)$ where $\tau_\gamma$ is the optical depth in gamma rays. This prescription neglects the deposition of gamma rays at points in the ejecta other than their point of origin. It is used by Graham (1987) to compute a case with an off-centre nickel shell even though it does not incorporate the deposition of the gamma rays from the nickel shell into the deeper mass fractions. The deposition function of Sutherland and Wheeler (1984) treats the gamma ray scattering as an absorptive process but accounts for the non-local deposition of energy by the gamma rays. This would seem to give a very different, and perhaps more correct, representation, but both prescriptions have been compared with Monte Carlo calculations with claims of good success, although in different models.

The various numerical light curve calculations show that the peak of the light curve can be reasonably well represented by the explosion of a C/O white dwarf with the production of $\sim$0.5–1.0 $M_\odot$ of $^{56}$Ni. The initial thermal energy deposited by the explosion is dissipated by adiabatic expansion of the compact progenitor, so aside from a very early prompt spike of hard radiation, the light curve is powered entirely by radioactive decay. Graham (1987) argues that the shape of the light curve peak and tail are sensitive to the distribution as well as the amount of $^{56}$Ni, since if the radioactive matter is closer to the surface, the luminosity will rise and decay more quickly for a given amount of nickel. Graham adopts analytic density distribution and gamma ray deposition functions and concludes that a configuration in which the inner core is stable $^{56}$Fe and the $^{56}$Ni lies in a concentric shell gives a good light curve. For the same density distribution, Graham finds that placing the same amount of $^{56}$Ni (0.6 $M_\odot$) in the centre gives too flat a light curve because the gamma rays are more efficiently trapped and thermalized for a longer time. It is not clear that Graham's results are quantitatively reliable because he uses an excessively large optical opacity ($\kappa$ = 0.3 cm$^2$ gm$^{-1}$), and density distribution and deposition functions that do not reflect the fact that in the deflagration models the unburned matter is compressed into a shell, resulting in a non-monotonic density distribution. Nevertheless, Graham's calculations illustrate that the light curve can, in principle, be used to constrain the distribution of the nickel.

Figure 10 shows the light curves that have been calculated for deflagration model W7 of Nomoto *et al* (1984b; see section 2.5.2. below) assuming different distributions of the nickel. The full curve corresponds to the original model with 0.58 $M_\odot$ of nickel distributed in an off-centre shell. The deposition function was originally calculated using a grey atmosphere code and the energy-dependent Klein–Nishina cross section for the gamma rays. A constant $\gamma$-ray opacity $\kappa_\gamma = 0.03$ gm cm$^{-3}$ reproduces the results well. The optical opacity was taken to be constant, $\kappa_{\rm opt} = 0.03$ gm cm$^{-2}$, a value obtained by adopting the constant value that reproduced the more detailed original model which computed the Rosseland mean corresponding to the actual composition. The broken curve corresponds to a model in which the nickel is arbitrarily redistributed to a spherical shell of pure nickel in the centre. With a judicious shift of amplitude and time of explosion, these curves both fit the bolometric light curves of SN 1972E and SN 1981B (Graham 1987) within the observational scatter although the curve with the central nickel gives a slightly flatter decline and does not fit as well. The curves in

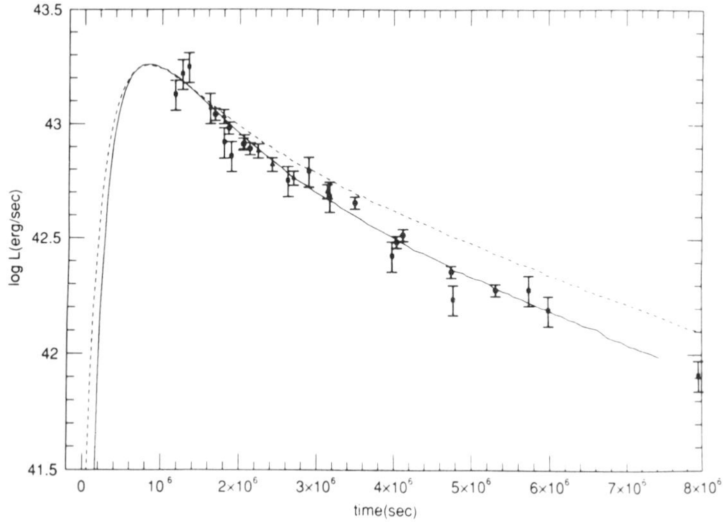


**Figure 10.** Two light curves are shown corresponding to the deflagration model W7. The full curve corresponds to the basic model, the broken curve to a model where all the radioactive nickel is placed in a spherical shell in the centre resulting in greater trapping of gamma rays and a slightly flatter light curve.

figure 10 are similar in post-maximum slope to those calculated by Nomoto and Shigeyama (1989) who also used $\kappa_\gamma = 0.03$ gm cm$^{-2}$, but they rise more quickly to somewhat brighter peaks, reflecting primarily a different assumption for the optical opacity by Nomoto and Shigeyama, $\kappa_{opt} = 0.1$ gm cm$^{-2}$. Graham's conclusion that the central-nickel case gives much too flat a light curve seems to be a result of his approximations. See Wheeler (1990c) for more discussion of these points.

After about 30 days, when the light curve makes its transition to the exponential tail, the models predict that most of the gamma ray deposition of energy is above the optical photosphere (see e.g. Sutherland and Wheeler 1984). The calculation of the optical light curve then becomes increasingly problematical. Approximations that have been employed include the assumption that all the deposited gamma rays are thermalized, that only the gamma rays that are deposited below the photosphere are thermalized, and that the photosphere assumes a constant temperature. None of these choices is particularly palatable. Branch *et al* (1983) show that the radius of the photosphere rises until about 30 days after maximum light in SN 1981B, and then begins to decline. Sutherland and Wheeler note that this behaviour cannot be reproduced in any model adopting a constant effective optical opacity. It remains to be seen whether evaluation of appropriate frequency-dependent opacity (section 2.5.2) will give this feature naturally.

A major unresolved point of discussion centres on the length of time that SN Ia are observed to continue their exponential decline. SN 1937A in IC 4182 (NGC 4157) and SN 1972E in NGC 5253 were observed to continue this decay to a rather high precision for upwards of 700 days (Minkowski 1939, Kirshner and Oke 1975). Axelrod (1980a, b) calculated the decreasing gamma ray deposition and heating of an expanding constant density shell of $^{56}$Ni with the assumption that the positrons, which represent about 4 per cent of the decay energy, are trapped locally, and showed that as the ejecta expand and cool, the matter cools precipitately and the bulk of the radiation is emitted by fine structure lines of Fe in the infrared. This shift of the emissivity from the optical to the infrared is referred to as the 'infrared catastrophe'. In many of Axelrod's models, he finds that the infrared catastrophe truncates the optical light curve at a time when some observations show the exponential decline in the optical continues, as illustrated in figure 11. Woosley *et al* (1984) show that proper treatment of density gradients in the structure allows an infrared catastrophe 'wave' to propagate out through the ejecta, rather than cooling it all at once. They argue that this gradual transition of infrared cooling coupled with observational uncertainty can plausibly account for the observed light curves.

Another point of view is expressed by Colgate *et al* (1980). They argue that the optical light curve can only be sustained if positrons as well as gamma rays leak progressively larger fractions of their energy out of the ejecta. Despite the theoretical attention to this subject, there has been little practical success in following the light curves of recent SN Ia with, e.g. CCD detectors, to very faint magnitudes to determine definite observational constraints and perhaps observe the infrared catastrophe. This remains a high priority for those interested in supernova observations.

### *2.5.2. Spectra*

*Observed spectra.* The fact that SN Ia follow a generally reproducible pattern in the evolution of their spectra has been known for a long time (Minkowski 1939, Kirshner and Oke 1975). The time evolution of the spectra of SN 1981B, a classical SN Ia

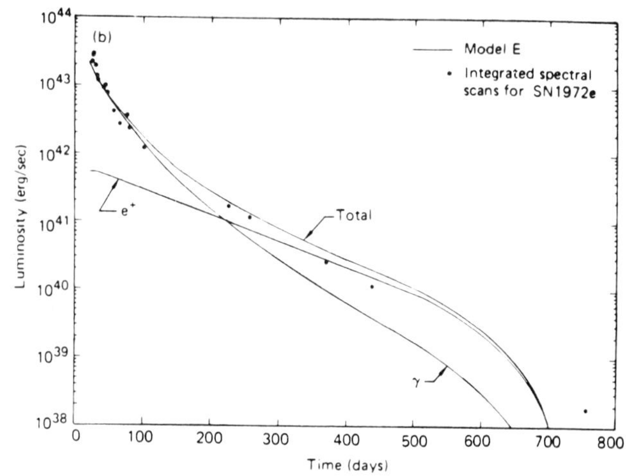


**Figure 11.** An example of a light curve of a detonation model which undergoes the 'infrared catastrophe' (from Weaver *et al* 1980).

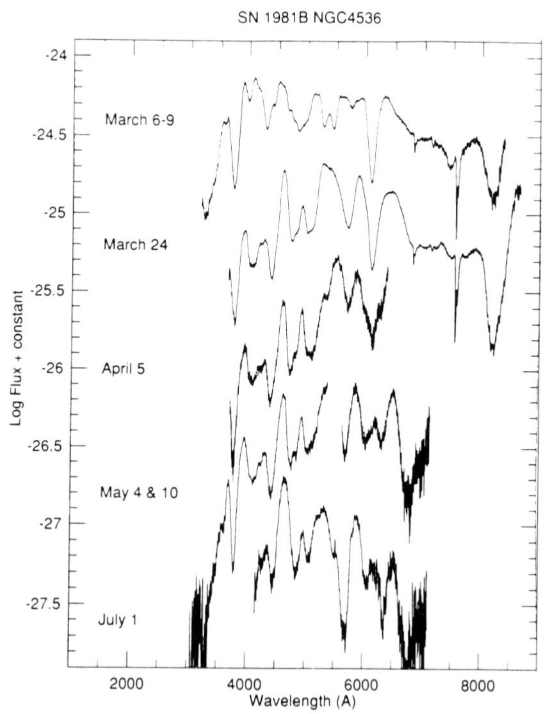


**Figure 12.** Time series of optical spectra of the SN Ia 1981b from near maximum light, to the early nebular phase about four months after maximum light. Note the transition in the spectrum from the first, obtained near maximum light, to the next two weeks later and the relative homogeneity in the spectral characteristics thereafter.

(Branch *et al* 1982, 1983) is given in figure 12. The optical spectrum at maximum light can be reproduced in terms of the Doppler-shifted resonance lines of intermediate to iron-peak heavy elements (Branch *et al* 1985; Harkness 1986, 1987, 1990). The observed spectrum makes a transition at some point so that by two weeks after maximum (the second spectrum in figure 12), the spectrum is dominated by Fe II. Models suggest that this transition could occur in a few days shortly after maximum light as the photosphere recedes into the iron-rich ejecta. There are little data near this transition time to elucidate its character. After this, the spectrum changes little even though after several months it is thought to become dominated by forbidden Fe II rather than lines of permitted Fe II, as it is at earlier times. The spectrum near maximum light shows a strong ultraviolet deficiency, as will be discussed in more detail below. There has been very little accurate spectral data in the IR of SN Ia (but see Phillips *et al* 1987, Frogel *et al* 1987).

Branch (1987) has argued that the spectrum of SN 1984A in NGC 4419 obtained prior to maximum light showed sufficiently high Doppler shift as to be inconsistent with other SN Ia. He thus argues that this is definite spectral evidence that SN Ia are not completely homogeneous. Pearce *et al* (1988) have responded that this higher than normal photospheric velocity may just be due to the early epoch of observation, but Branch *et al* (1988) argue against this possibility. The light curve for SN 1984A is apparently normal for an SN Ia when allowance is made for some discrepant data (Chevalier and Plait 1988, Graham 1988a). SN 1986G in Centaurus A (NGC 5128) was heavily reddened, $E(B-V)=0.9$, but well studied (Phillips *et al* 1987, Frogel *et al* 1987). The light curve dropped from maximum distinctly faster than that of SN 1981B. The spectrum showed subtle, but definable differences from SN 1981B, and resembled spectra of SN 1971I which also had a similar light curve (Barbon *et al* 1973). The photospheric velocity was always systematically slower than that of SN 1981B at a given phase. The question of whether or not there are significant differences in peak magnitude between SN 1986G and other SN Ia is obscured by observational uncertainties in distance and reddening. SN 1987N was another classic SN Ia discovered prior to maximum (Schneider *et al* 1988). This supernova showed distinctly faster photospheric velocities than SN 1981B. SN 1987A has taught us that the photospheric velocity can decline very rapidly in the early epochs of a compact progenitor, as Colgate *et al* point out, but the reality of genuine spectral and light curve differences among SN Ia seems to have been confirmed.

*Atmosphere calculations of the carbon deflagration model.* Prior to recent arguments summarized above that carbon ignition is likely to lead to detonation, not deflagration, the deflagration model received widespread acceptance as an explanation for SN Ia because of its success in explaining the spectra, particularly at early times. Most detailed spectral calculations for SN Ia are based on the models of carbon deflagration by Nomoto *et al* (1984b; see also Thielemann *et al* 1986). In these calculations, the mixing length in the deflagration front is a free parameter that determines the extent of the burning. Model W7 of Nomoto *et al* represents a core of 1.4 $M_\odot$ obtained from an evolutionary calculation. The deflagration was propagated with a mixing length of 0.7 times the pressure scale height. Slower deflagrations with smaller deflagration 'mixing length' of 0.6 produced a somewhat weaker explosion, and a faster deflagration with a mixing length of 0.8 consumed the whole star, leaving no intermediate mass elements. The Nomoto *et al* model W7, while not unique, has thus emerged as a popular model for spectral calculations.

Synthetic spectra for the W7 model were first computed by Branch *et al* (1985), for times of 15 and 32 days after the explosion. These calculations used the escape-probability approximation for resonance scattering lines formed above a blackbody photosphere. The temperature of the blackbody continuum (20 000 K) was derived from the optical colour temperature of SN 1981B at maximum light. Line optical depths were computed assuming LTE level populations at a separate, constant, excitation temperature which was used as a fitting parameter. These initial calculations demonstrated that the deflagration model could reproduce most of the optical features of SN 1981B at maximum light.

This conclusion is supported by more complex calculations of the emergent spectra of model atmospheres of the W7 explosion (Harkness 1985, 1986, 1987, 1990). The W7 deflagration model is expanded using a one-dimensional hydrodynamics code derived from the code used by Sutherland and Wheeler (1984), accounting for the time-dependent deposition of the radioactive decay energy. The deposition function is computed using the grey approximation to the spherical co-moving frame radiative transfer equation, with a fixed gamma ray opacity of 0.03 $cm^2\,gm^{-1}$. The hydrodynamical code uses flux-limited diffusion, with Rosseland opacities computed from the radially-dependent composition of the W7 model shown in figure 13.

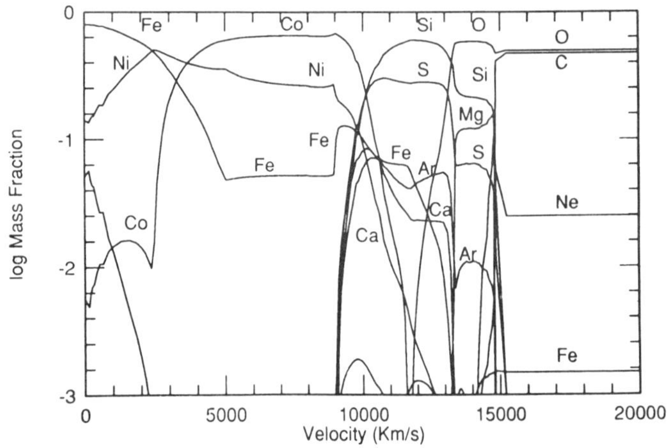


**Figure 13.** The composition of the W7 deflagration model as a function of expansion velocity as given by Thielemann *et al* (1986).

The emergent flux at a given time is computed from an atmosphere model, which uses the relativistic co-moving radiative transfer equation in spherical geometry, and the density structure, velocity profile, deposition function and radial abundance distribution from the hydrodynamical calculation. The model uses LTE ionization and level populations and assumes that lines are formed purely by resonance scattering, similar to the earlier calculations by Branch, but in this case incorporating the global effects of electron scattering and allowing for temperature gradients.

At maximum light the optical depth of the W7 model is about 20, so the entire star is included in the 'atmosphere'. This small optical depth alleviates the problems encountered with setting lower boundary conditions for very optically thick atmospheres such as SN 1987A (Harkness and Wheeler 1988). Furthermore, the errors involved in neglecting the explicit time-dependent terms in the transfer equation should be minimized because of the short diffusion time at and after maximum light. This is probably not the case for the first few days after the explosion.

A spectrum of the W7 deflagration model 14 days after the explosion is shown in figure 14, compared to the maximum light spectrum of SN 1981B obtained at McDonald Observatory and a heavily smoothed IUE spectrum from the same date. The observed optical and IUE spectra join at about 3200 Å, with no adjustment in the observed flux. Overall, the optical spectrum is a surprisingly good match, but the model fails to reproduce the observations in a narrow, but crucial, region between about 2700–3500 Å.

Branch *et al* (1985) found that their optical synthetic spectra were improved if the composition of the outer layers of W7 were mixed for matter expanding faster than 10 000 km s$^{-1}$. The middle spectrum shown in figure 14 is the spectrum of a W7

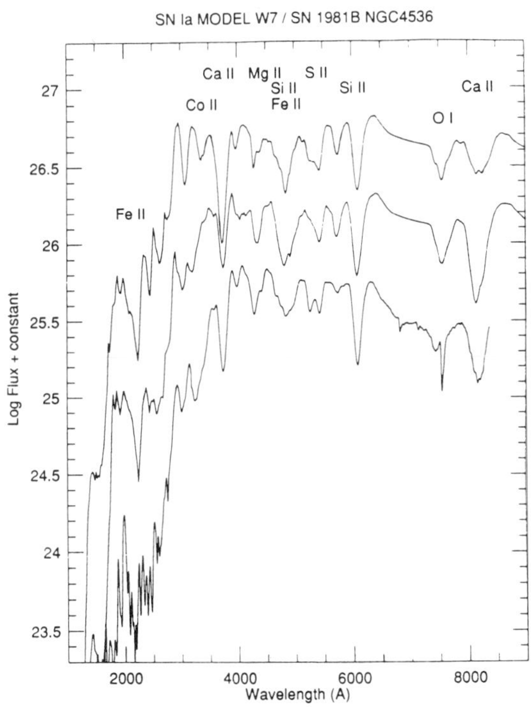


**Figure 14.** The observed optical and ultraviolet spectrum of SN Ia 1981b near maximum light (bottom) in comparison with two spectra based on the W7 deflagration model with the original stratified abundances (top) and with the composition homogenized above 11 000 km s$^{-1}$ (middle).

atmosphere calculation with mixing for matter expanding faster than 11 000 km s$^{-1}$, which represents the outer 0.4 $M_{\odot}$. In this case, there is some deterioration of the optical spectrum (note the 'W' feature of S II at about 5200 Å, and the Ca II H and K lines at 3700 Å), but substantial improvement in the broad complex of lines around 4800 Å and in the near-UV and also in the overall slope of the UV continuum between 1500 and 3500 Å. The changes are mainly due to an increase in the abundance of iron-peak elements in the high-velocity outer layers, particularly iron and cobalt. If too much iron-peak matter is mixed out to high velocity the strong lines of Fe II dominate the spectrum in disagreement with observation. In fact, the best results are obtained when the mixing is imposed for matter just above the point where NSE breaks down and the composition switches rapidly to intermediate mass elements. The mixing is thus presumed to be due to Rayleigh–Taylor instability in the hydrodynamical phase as the deflagration was dying away.

In both of the above cases the spectra were calculated by including only about 400 of the strongest resonance lines of neutral atoms and singly ionized ions. More recent calculations (Harkness 1990) indicate that the unmixed abundance profile gives superior results if tens of thousands of weaker lines are included. This is in contradiction to the results of Branch and earlier work by Harkness and Wheeler. The many weak lines do not contribute any clearly identifiable spectral features, but act more in the manner of a broad-band opacity. It now appears that these lines must be accounted for to obtain the correct flux levels in the ultraviolet.

The apparent absorption feature at 3200 Å is extremely important. In the mixed model this is due to Co II and it is considerably enhanced in mixed models. In unmixed models the shape of this feature is heavily influenced by the background opacity of the many weak lines, and is not well represented when only a few hundred lines are included in the models. Unfortunately, the Co lines occur just at the join between the long wavelength limit of IUE and the shortest wavelengths that can be observed from the ground. There is a possibility that the absolute flux depicted in figure 14 may be in error in precisely this region, but the observations should be qualitatively correct. More near-UV/optical observations of bright supernovae are urgently needed to confirm the presence of Co II at maximum light.

The carbon deflagration model is attractive because the subsonic burning allows the white dwarf to expand ahead of the burning front, so combustion occurs at lower densities and ultimately leads to a partially burned layer surrounding the NSE core. The W7 model is not unique in this respect. Harkness (1990) examined four carbon deflagration models which have appeared in the literature. At first glance, these four models appear to meet most of the observational criteria, but calculations of their emergent spectra at 14 days post-explosion clearly show that the abundance distribution as a function of expansion velocity determines whether a given model is viable. In fact, the requirements are quite strict. A good model must produce around 0.6 $M_{\odot}$ of $^{56}$Ni, and should not exceed 0.8 $M_{\odot}$ of $^{56}$Ni. The partial burning should proceed to the surface of the white dwarf, leaving less than about 0.1 $M_{\odot}$ of unburned C/O, and producing about 0.4 $M_{\odot}$ of intermediate mass elements which must be distributed over a large range of expansion velocity ($v \gtrsim 8$–$10 \times 10^3$ km s$^{-1}$).

Model W7 provides the closest match with observations, but it is important to remember that in all the presently published work the propagation of the deflagration wave is determined by an arbitrary prescription which is essentially just a convenient parametrization of the front velocity. Calculations of the emergent spectra place stringent requirements on the models, so we have a good idea of how the event must turn out. A quantitative understanding awaits the development of self-consistent deflagration or detonation models.

*The ultraviolet deficit.* Type II L supernovae such as SN 1979C and SN 1980K reveal a smooth continuum at maximum light which extends to ultraviolet wavelengths. In contrast, Type I supernovae have long been known to exhibit a severe 'ultraviolet deficit', where the observed flux at 2000 Å may be as much as 100 times less than the flux one would expect from a black body radiating at a temperature equal to the observed *optical* colour temperature (Holm *et al* 1974, Branch *et al* 1983).

This result was first discovered for the 'classical' Type Ia supernovae, but more recent observations have shown that SN Ib have similar UV spectra (Panagia 1985). The same is true for SN 1987A. IUE spectra of SN 1987A taken between about two days and the time the short wavelength flux fell below the level of detectability resemble those of SN Ia and Ib much more than those of other SN II.

The ultraviolet deficit arises quite naturally for expanding atmospheres characterized by moderately steep ($\rho \propto r^{-7}$ to $r^{-10}$) density profiles where the principal source of opacity is electron scattering and resonance lines of once-ionized metals (Harkness 1986, 1987, 1990, Harkness and Wheeler 1990, Wheeler and Harkness 1986, Wheeler *et al* 1987b). The optical colour temperature is characteristic of the temperature at the thermalization depth for optical photons. In typical maximum light conditions the ratio of optical pure absorption to scattering opacities can be in excess of 1 : 100, so the radiation is thermalized at an electron scattering optical depth of around 20. This is extremely close to the actual electron scattering optical depth of the entire atmosphere at maximum light for an SN Ia. In the ultraviolet the continuous absorption due to photoionization becomes dominant at wavelengths less than 1500–2000 Å; however, there are a colossal number of spectral lines due to singly ionized atoms at wavelengths less than about 3000 Å, which collectively provide the opacity to account for the major part of the deficit (see figure 15). In most SN Ia calculations to date, several hundred

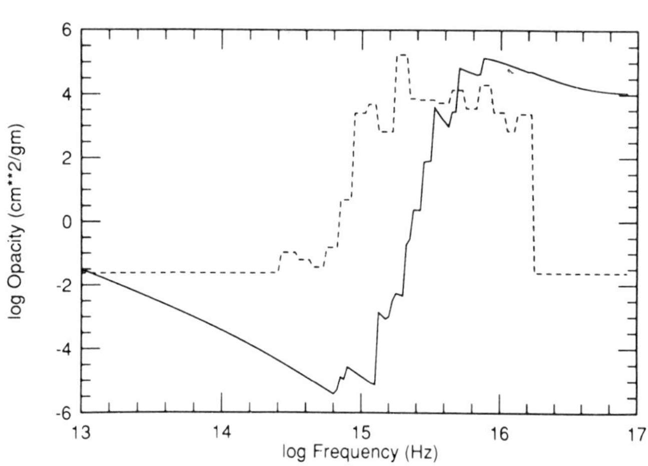


**Figure 15.** The opacity is given as a function of frequency for a composition representative of the inner regions of a deflagration. The full curve represents all relevant continuum opacities except electron scattering. The broken curve shows the added effect of electron scattering plus about 250 000 scattering lines averaged over wavelength bins.

of the strongest resonance lines are included explicitly with the radiative transfer computed on characteristics in the co-moving frame and then transformed to impact parameters in the stationary frame before summing to obtain the final spectrum (see figure 16 and the appendix). The great majority of these lines are due to Fe II, Co II and Ti II. The effective opacity of the lines is enhanced by the differential expansion of the atmosphere. In co-moving frame formulations the various contributions to the total opacity are accounted for simply, in the same manner as for static atmospheres. In the Eulerian frame of reference the opacity is frame-dependent. The effective opacity in this case was first considered by Karp *et al* (1977), who showed that it was probably necessary to include hundreds of thousands of weak spectral lines to obtain accurate results.

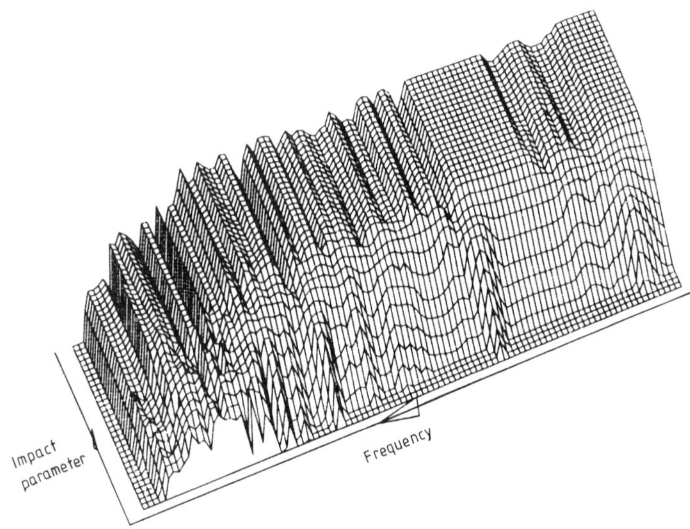


**Figure 16.** The emitted flux in the observer frame corresponding to a carbon deflagration model for a SN Ia near maximum light is given as a function of frequency and the impact parameter through the ejecta.

In the published deflagration models the highest expansion velocities are around 25 000 km $s^{-1}$ because of the relatively coarse zoning used in the outer layers. Fine zoning of the surface layers results in much higher expansion velocities for small mass fractions, up to a considerable fraction of the speed of light when the shock accelerates matter in the density gradient typical of white dwarfs and evolved cores. In model W7, some of the strongest resonance lines, such as Mg II 2797 Å, will have considerable optical depth at very high velocities, perhaps in excess of 100 000 km $s^{-1}$, with major effects on the ultraviolet spectrum. Calculations with expansion velocities limited to 25 000 km $s^{-1}$ give good agreement with observations, however, suggesting that perhaps the expansion velocity is limited by collision with circumstellar matter. Since SN Ia have not been detected in the radio, the amount of circumstellar matter must be small, but its composition may be of considerable interest.

If some circumstellar matter were left from some formative stage of a common envelope, it would be expected to be mainly hydrogen. In conditions of high accretion rate one might also expect some hydrogen to be spilled into the circumstellar environment, and some to be associated with the surface layers of the exploding star. No hydrogen has ever been detected in the spectra of SN Ia and currently there are no tight theoretical limits on the possible abundance of hydrogen in SN Ia, despite the considerable influence this would have on progenitor modelling.

*Late-time spectra.* As we have just seen, at maximum light the spectrum of an SN Ia is dominated by intermediate mass elements. About a week after maximum the spectrum becomes dominated by permitted lines of Fe II. At very late times the spectrum is dominated by forbidden emission lines of Fe II and Fe III. Surprisingly, the shift from the photospheric phase to the 'supernebular' phase is subtle, and the spectrum does not change dramatically from a week or so after maximum to several months. The transition from the optically thick, photospheric, permitted line phase to the optically thin, forbidden line, nebular phase has not been calculated yet, principally because the two regimes are amenable to different physical assumptions and computational techniques and the transition regime requires a difficult general analysis.

The pioneering calculations of the late-time nebula spectra of SN Ia were done by Meyerott (1980) and Axelrod (1980a, b; see also Woosley *et al* 1984). As mentioned previously, Axelrod tackled the problem of an expanding constant density shell of $^{56}$Ni as it decayed to $^{56}$Co and finally $^{56}$Fe. He followed the heating by gamma rays and positrons (assumed to be trapped) and the associated ionization, excitation and line emission. Axelrod's calculations provided a quantitative reproduction of the spectral evolution of SN 1972E, as illustrated in figure 17. He showed that the major features at ~4600 Å were blends of Fe II and Fe III forbidden lines. Moreover, he could account for a feature at 5900 Å as a blend of lines of Co III. In his model, this Co arose directly from the radioactive decay of $^{56}$Ni and the predicted feature decayed in strength as the $^{56}$Co decayed to $^{56}$Fe in just the way the spectral feature was observed to shift. This was the first direct evidence that radioactive decay of $^{56}$Ni was the power source of the long-term light curve.

In a more realistic case, the supernova ejecta will have gradients in the density, and the transition to the nebular phase will be gradual. Figure 18 shows the late-time spectra of detonation and deflagration models in comparison with spectra of SN 1972E. In the detonation model, rapid expansion results in fewer free electrons at a given phase and hence less efficient cooling. This gives rise to lines of highly ionized Fe which are not seen in the observations. The fact that the outer layers are burned to NSE also gives an excess of high velocity iron and line widths that are broader than observed. The deflagration gives less highly ionized Fe and Co which come only from the inner, more slowly moving core so that the lines are narrower. The result is a fair representation of the observed spectrum. The model does not include the Ca H and K absorption feature at log $\nu = 14.90$ ($\lambda = 3900$ Å).

Spectra of SN 1937C showed 'relatively narrow, bright lines' of [O I] $\lambda\lambda$ 6300, 6364 by day 275 (McLaughlin 1960). Whether this is supernebular emission from the supernova ejecta or emission from a shocked or irradiated circumstellar medium is not clear. This emission should be modelled and sought observationally in other SN Ia.

### 2.6. *Young supernova remnants*

After supernovae have ploughed up a mass comparable with their ejecta, information about the nature of the explosive event tends to be lost. In the youngest remnants information about the nature of the explosion may be preserved and hence they are worthy of study for clues to the nature of the progenitor star and the explosion mechanism.

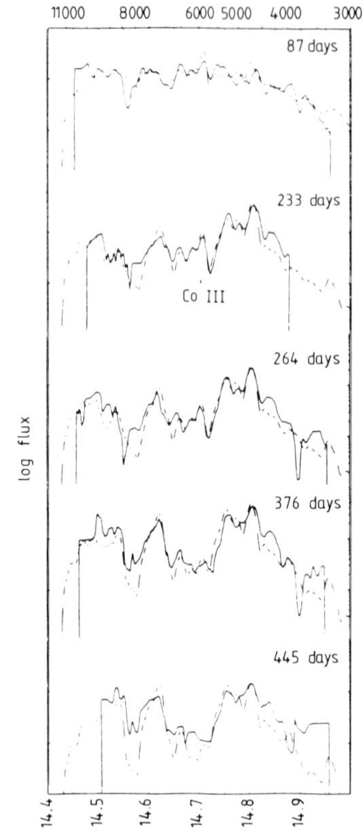


**Figure 17.** The time evolution of the spectrum of an expanding shell of initially pure $^{56}Ni$ is shown in comparison with the corresponding spectra of SN 1972E (from Axelrod 1980a). The feature at about 5900 Å (log $\nu \sim 14.72$) is due to decaying radioactive $^{56}Co$.

There are three historical galactic supernova remnants which have been associated with SN Ia events: SN 1006, SN 1572 (Tycho), and SN 1604 (Kepler) (Clark and Stephenson 1977). There is not a good light curve for SN 1006 but it was particularly bright, suggesting a SN Ia. Tycho and Kepler obtained visual light curves which are consistent with SN Ia, although Doggett and Branch (1985) have argued they cannot really be distinguished from bright SN II-L events, and from their shape alone, they also are very similar to SN Ib. Strom (1988) has re-examined the observations of historical remnants and attempted to assign *relative* distances, which may be done more easily than absolute distances. He concludes that Tycho may have been distinctly underluminous compared to Kepler and SN 1006, and suggests that it may have been a SN Ib. Whereas SN 1006 and Tycho seem to be interacting with the interstellar medium (Chevalier 1982), Kepler may be interacting with circumstellar matter. Bandiera (1987) has argued that Kepler may represent a runaway star and an SN Ib. Young and Branch (1989) and Gaskell (1990) argue that most SN II-L are also homogeneous and distinctly less luminous than SN Ia or the singularly bright SN II-L 1979C. Thus the possibility that Tycho was a SN II of some variety must still be

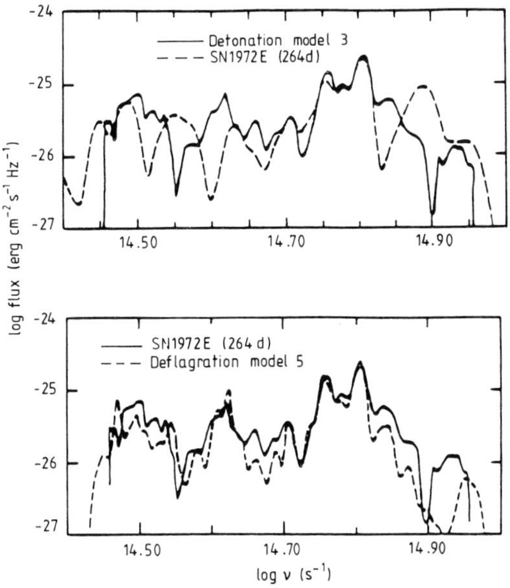


**Figure 18.** The spectrum 264 days after maximum of SN 1972E is given in comparison with two theoretical spectra corresponding to a detonation model (top) and a deflagration (bottom). The narrower lines and lower mean ionization in the deflagration model give a better agreement with observations (from Woosley *et al* 1984).

considered as well. Van den Bergh (1988) has discussed the possible connection of different types of supernova with different types of supernova remnants.

The connection of these historical remnants with SN Ia has been bolstered by studies of the X-ray emission which gives information about the mass and composition of the ejecta, and by the serendipitous alignment of an OB subdwarf star along a line of sight passing near the centre of SN 1006 against which absorption line studies may be done. If Tycho were indeed a different type than Kepler and SN 1006 then differences in its remnants, compact and extended, must be investigated carefully.

There has been some controversy about the mass and composition of the ejecta of SN 1006, Tycho and Kepler. A crucial point is that the ejecta are expected to be pure heavy elements so that estimates based on solar composition will underestimate the emissivity and hence overestimate the mass required to produce a given x-ray line spectrum. In addition, SN 1006 shows a more nearly line-free power law continuum from around 1–10 keV whereas Tycho and Kepler show strong lines in this range.

Hamilton *et al* 1985 (see also Hamilton *et al* 1986) have analysed the x-ray spectra of a reverse shock propagating into heavy-element dominated supernova ejecta due to collision with a uniform density interstellar medium with allowance for the effects of non-ionization equilibrium. Hamilton *et al* treat the ejecta as having uniform density and conclude that a thermal bremsstrahlung continuum will provide a power law spectrum with slope $\mathrm{d}L/\mathrm{dln}\ \nu \propto \nu^{-1}$ as compared with the observed value $\nu^{-1.2}$ (note this continuum spectrum slope has been contested by Tenma observations of Tycho (Tsunemi *et al* 1986) and SN 1006 (Koyama *et al* 1987)). Hamilton *et al* argue that the outermost shocked matter will dominate since the inner shocked material will be at lower densities, since the density will decline like $\rho \propto t^{-3}$ before the reverse shock arrives. The innermost regions will be cold and unshocked and therefore could be very rich in iron, for instance, without contributing to the x-ray spectrum. The model of uniform density ejecta is clearly oversimplified since the ejecta will contain a strong density gradient which will present the reverse shock with an increasing density which will offset the $t^{-3}$ effect of homologous expansion. In addition, deflagration or damped-detonation models will tend to have the unburned matter compressed into a shell of somewhat higher density than the Fe peak matter immediately beneath it.

With their uniform ejecta model, Hamilton *et al* (1985) argue that both SN 1006 and Tycho can be explained with a model based on a total ejecta of 1.4 $M_\odot$ with an outer 0.4 $M_\odot$ of carbon-rich matter, an intermediate region of 0.4 $M_\odot$ of intermediate mass elements and an inner core of Fe of 0.6 $M_\odot$ of which 0.3 to 0.4 $M_\odot$ is unshocked. This conclusion should be re-examined with the possibility in mind that they were two different kinds of explosion. They account for the difference between the x-ray spectra of Tycho and Kepler and that of SN 1006 as being due to the denser ISM surrounding Tycho as deduced from the model, 0.3 $\mathrm{cm}^{-3}$, versus that attributed to SN 1006, 0.04 $\mathrm{cm}^{-3}$, which is plausibly attributed to the higher distance of SN 1006 above the Galactic plane (~400 pc versus 70 pc).

Hamilton *et al* (1985) find that their model requires a rather low oxygen abundance in the outermost layer in contrast to the deflagration models and the observational constraint of an appreciable oxygen abundance observed in the optical spectra of SN Ia at maximum light. They note, however, that a more realistic model in which the outer C/O layer falls in a compressed shell might lead to more complete ionization of the oxygen in the higher density environment which would reduce the oxygen line emission relative to the continuum, thus allowing a larger oxygen abundance.

Further insight into the structure and composition of SN 1006 derives from the fact that it falls along the line of sight to an OB subdwarf, the Schweizer–Middleditch star (Schweizer and Middleditch 1980). Wu *et al* (1983) used the SM star as a probe to study UV absorption within SN 1006 and discovered broad (±5000 km $\mathrm{s}^{-1}$) centrally symmetric UV Fe II absorption features. This observational work was followed up in more detail by Fesen *et al* (1988) and an associated theoretical analysis by Hamilton and Fesen (1988).

Fesen *et al* (1988) derive a mass of Fe II of ~0.015 $M_\odot$. Hamilton and Fesen (1988) compute detailed reverse-shock models for the ionization and excitation of Fe in the context of two deflagration models. They argue that the most plausible mechanism for ionizing the Fe is photoionization of the inner iron-rich regions by UV and x-ray emission from reverse-shocked ejecta.

Hamilton and Fesen (1988) argue that model W7 of Nomoto *et al* (1984) gives too large a central density concentration in the inner iron core and hence 10 times more Fe II at $v \leqslant 5000$ km $\mathrm{s}^{-1}$ (as compared, for instance, with the Fe II 2600 Å feature) than the observations allow. Model W7 was based on a convection-driven deflagration with a front speed which was very slow at first, $v_f \sim 0.08\ c_s$, and then accelerated to approximately $v_f \sim 0.3\ c_s$. Hamilton and Fesen also did a calculation based on a model similar to that in Woosley and Weaver (1986a) in which the front velocity is approximately a constant fraction of the sound speed. For reasons that are not clear, such a model results in a smaller central density contrast than a model like W7, with the central density at a given epoch being only 1/3 that of W7. The comparison of the Fe density versus velocity derived from the observations with that implied by the constant speed deflagration model is shown in figure 19. Hamilton and Fesen describe the agreement between this model and the observed Fe II density as 'jolly good'. As described in section 2.5.2 this model did not account for the optical spectrum nearly as well as W7.

Itoh *et al* (1988) have analysed the Tenma x-ray spectral observations of Tycho in terms of the hydrodynamic deflagration model W7. They thus incorporate both the composition and density structure of the numerical model and the non-ionization equilibrium interaction with a uniform density interstellar medium. By matching the angular size and the known age, they obtain a density of the ambient ISM of about $1\,cm^{-3}$, somewhat larger than that of Hamilton *et al* (1985). There are two immediate problems with the resulting model. One is that the thickness of the blast-shocked ISM (0.8 pc) is about three times larger than the observed value. The other is that the blast wave has not reached the inner iron-rich regions, and the Fe K$\alpha$ line blend is too weak, in contrast to naive expectation based on the iron-rich inner core. Itoh *et al* experiment with models in which the inner and outer portions of the stratified ejecta are mixed by *ad hoc* homogenization. Mixing the outer layers, above $10^4\,km\,s^{-1}$ does not produce an adequate fit to the x-ray spectrum, but mixing some matter from the inner iron-rich regions at $\leqslant 8000\,km\,s^{-1}$ to about $13\,000\,km\,s^{-1}$ provides reasonable agreement. The Fe K$\alpha$ line strength is rather sensitive to this outer velocity which, coincidentally or not, is about the outer velocity of the partially burned region in the original W7 model. The problem with this solution to the nature of the x-ray spectrum is that such mixing between 8000 and $13\,000\,km\,s^{-1}$ would yield a spectrum at maximum light that is far too rich in iron to match the optical spectrum. Itoh *et al* raise the

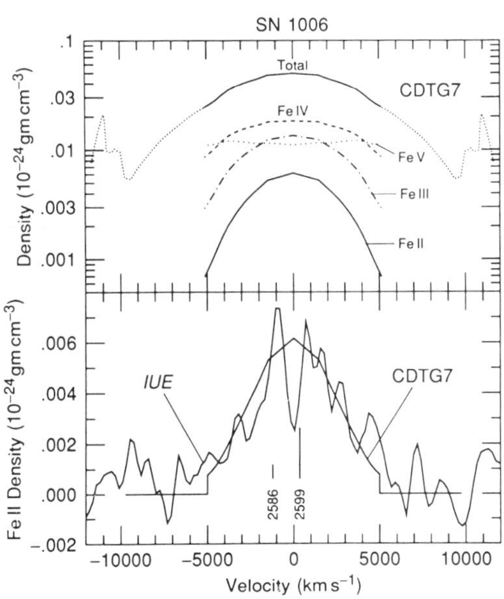


**Figure 19.** The total iron density (right-hand scale) against velocity deduced from observations of ultraviolet absorption lines of Fe II (left-hand scale) in SN 1006 (Fesen *et al* 1988) are compared with the iron density profile deduced from a model based on an accelerating deflagration calculation of Woosley and Weaver (1986a) (from Hamilton and Fesen 1988).

possibility that the inner mixing might occur after the phase of maximum light, but it seems rather unlikely that the mixing timescale at that epoch of homologous expansion is fast enough. Given the trouble Itoh *et al* have matching the thickness of the shocked ISM and the x-ray spectrum of Tycho with model W7, it would be interesting to examine other deflagration models. It is also important to bear in mind in that context that Tycho may not be the remnant of a SN Ia, and hence to explore other physical models, such as those for SN Ib events.

It would be very interesting to investigate the classes of models, if any, that simultaneously satisfy the constraint on the density distribution imposed by the predicted $^{54}Fe/^{56}Fe$ ratio (Sutherland and Wheeler 1984, Woosley *et al* 1984, Thielemann *et al* 1986), by the light curve in terms of the nickel distribution, that obtained by Hamilton and Fesen (1988) from the Fe II absorption in SN 1006, and that obtained from the x-ray emission of the young remnants of SN 1006, Tycho, and Kepler.

Our calculations of the light curve of the W7 carbon deflagration model using appropriate Rosseland opacities (figure 10) give a light curve that reproduces the observations fairly well. Note that the light curve after about 20 days after explosion is nearly independent of the optical opacity because the diffusion time is short. Model W7 overproduces $^{54}Fe/^{56}Fe$. The constant deflagration speed model of Woosley and Weaver (1986a) tends to have a core of $^{56}Fe$ with a very narrow transition region to $^{56}Ni$ so that little excess neutronization occurs. Such a model gives a fairly good light curve (Woosley and Weaver 1986a, Woosley 1990). The relatively shallow density profile also seems to give a good fit to the UV Fe II absorption in SN 1006, whereas the W7 model has a density distribution which is too centrally concentrated. The constant deflagration speed model of Woosley and Weaver produces significantly smaller quantities of Si group elements and hence produces a poor optical spectrum. It is not known what sort of remnant x-ray spectra it would produce.

An important question which has not been adequately explored is the degree to which a damped detonation model might be constrained by these various observations. Failing a physical understanding of why a detonation might decay, the 'incinerated' models of Sutherland and Wheeler (1984) provide a first-order approximation to such a configuration. Such models burn their fuel quickly and expand rapidly, so excessive neutronization is reduced. In this type of model, the $^{56}Ni$ is produced and remains in the central high density portions of the model. The resulting light curve seems to give as good agreement with the bolometric light curve as Graham's 'W7' model or the accelerating-deflagration model of Woosley and Weaver (1986a). The density distributions of Sutherland and Wheeler's incinerated models B and C give a central density contrast which is even somewhat less than that of the Woosley–Weaver deflagration model. Whether that then produces too little photoionized Fe II in the context of the models for SN 1006 would be very interesting to check.

### *2.7. Nucleosynthesis*

If the evidence from the spectra is taken at face value, then SN Ia can be substantial contributors to the synthesis of a variety of heavy elements: C and O from unburned white dwarf matter, Ne, Na, Mg, Si, S, Ar, Ca from partially burned matter, $^{56}Fe$ from the decay of $^{56}Ni$, and a distribution of other iron peak elements. If SN Ia come predominantly from long-lived systems, then their nucleosynthetic products could be ejected into the interstellar medium at a different rate than those of SN II. The latter presumably arise from short-lived massive stars which begin to contribute to nucleosynthesis very early in the history of the Galaxy. The contributions of SN Ia to nucleosynthesis have been discussed by Nomoto *et al* (1984a, b), Thielemann *et al* (1986), Tornambé and Matteuci (1986), Woosley and Weaver (1986b) and references therein. Ever since the calculations of Arnett (1969) on carbon detonation, there has been some concern that such a mechanism could produce too much iron compared to Galactic abundances (Tinsley 1980, Ostriker 1982). Twarog and Wheeler (1982) were the first to point out that the carbon deflagration model had the potential to contribute significantly to the production of carbon and oxygen as well.

Models by Nomoto *et al* (1984b), Thielemann *et al* (1986), and Woosley *et al* (1984) showed that the subsonic burning of a deflagration model for SN Ia gives a different distribution of silicon peak elements than is derived in massive star models in which a shock wave from core collapse is presumed to impinge on a heavy element mantle. The distribution of various elements in the W7 deflagration model from Thielemann *et al* (1986) is shown in figure 20. The deflagration model tends to fill in the relative paucity of silicon peak elements which come from massive star models but the details of the deflagration models depend on the assumed rate of propagation of the turbulent front, and those of the massive star models on the assumed rate for the $^{12}C(\alpha, \gamma)^{16}O$ reaction. Compared with Fe, Thielemann *et al* (1986) find the ratio of Si, S, Ar, and Ca to be about one-half the solar value from model W7, whereas Woosley *et al* (1984) find about half as much from a roughly comparable model.

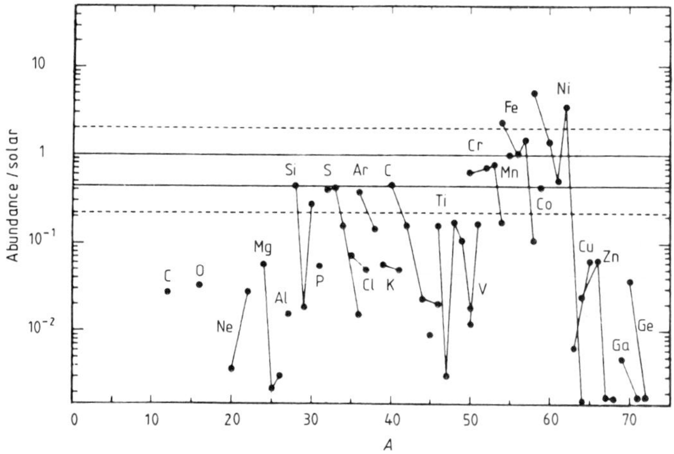


**Figure 20.** The abundances of stable isotopes relative to their solar values produced by the W7 deflagration model are shown normalized to $^{56}$Fe. Note the overabundances of $^{58}$Ni, $^{62}$Ni, and $^{54}$Fe (from Thielemann *et al* 1986).

Nomoto *et al* (1986a) combined the nucleosynthetic products of model W7 with that of a 25 $M_\odot$ model from Woosley and Weaver (1982). They find that if objects like W7 and the 25 $M_\odot$ model explode at comparable rates the distribution of elements from O to Fe is reasonably well reproduced. If the rate of explosion of SN Ia is ten times that of the massive stars, too much Ca as well as Fe is produced, compared with O. One would expect that the rate of explosion of massive stars of order 25 $M_\odot$ would be less than the mean rate of SN II because of the steep decline in the mass function. To require SN Ia to explode only as often as such massive stars is equivalent to requiring a rather low rate of explosion of SN Ia, or at least a low rate of injection of Fe into the ISM. The possibility that the rates are lower than estimated or that appreciable quantities of iron are ejected from the Galaxy abetted by the large scale height of the progenitors of SN Ia has been discussed.

Other elements are also produced in potentially excessive quantities in the deflagration models. A major problem is that the slow propagation of a deflagration allows more time for weak interactions and hence more time to produce neutron-rich species (Sutherland and Wheeler 1984, Woosley *et al* 1984, Nomoto *et al* 1984b, Thielemann *et al* 1986). In particular, $^{54}$Fe is over produced by a factor of 5 in model W7 compared with $^{56}$Fe (see figure 20). Nomoto *et al* (1984b) argue that there may be uncertainties in SN rates and the weak interaction rates that will circumvent this objection, but in fact Nomoto *et al* may have underestimated the $^{54}$Fe ratio. Model W7 ignited carbon at $2\times10^9$ gm cm$^{-3}$, but this ignition density ignores the effects of the convective Urca process which might increase the ignition density significantly. This would considerably exacerbate the problem. Nomoto *et al* (1984a) point out that the ignition density might have to be less than the threshold density for the $^{25}$Mg–$^{25}$Ne Urca shell at $1.3\times10^9$ gm cm$^{-3}$ in order to avoid the jump in ignition density. Such a low ignition density might only be achievable in extreme conditions as might be found in dynamical double degenerate merging. Woosley and Weaver (1986a, b) suggest that an initial slow burning might lead to an expansion of the star, a decrease in the central density and a decrease in the neutronization. The suggestion that carbon ignition leads to detonation is difficult to reconcile with such a slow burning phase. On the other hand, one advantage of detonation models is that they expand rapidly and neutronization processes do not have time to proceed on the subsequent dynamical timescale, although such models will also have trouble if the ignition density is increased significantly because of the convective Urca process. Woosley (1990) notes that in some circumstances detonation models can produce more $^{54}$Fe than a deflagration model.

In summary, the deflagration models do have some interesting nucleosynthetic properties and tend to fill in the gaps at the silicon peak left by massive stars. They tend to produce too much $^{54}$Fe compared with $^{56}$Fe and too much $^{56}$Fe compared with Galactic abundances. A detonation model might reduce the $^{54}$Fe problem, but tend to exacerbate the $^{56}$Fe problem.

### 2.8. *SN 1885A (S Andromeda)*

The first moderately well studied supernova in modern times, SN 1885A, remains somewhat of an enigma. De Vaucouleurs and Corwin (1985) have done a Herculean job of assembling and reanalysing the primitive photometric and spectroscopic data on this event. This event was atypically red at maximum and the light curve both rose and declined much faster than a typical SN Ia.

Graham (1988a) suggests that the 'high-velocity' SN Ia 1984A had a similar rapidly declining light curve to SN 1885A. The data of Kimeridze and Tsvetkov (1986) contradict the visual magnitude estimates of Verdenet (1984) quoted by Graham, however, and SN 1984A seems to have a normal light curve for an SN Ia (Chevalier and Plait 1988). This leaves SN 1885A as uniquely rapid in its decline.

The spectral information on SN 1885A is tantalizing but incomplete. De Vaucouleurs and Corwin (1985) argue that essentially every feature can be identified with one in the spectra of an SN Ia. This statement needs to be regarded with some caution, however. De Vaucouleurs and Corwin did not discriminate the spectrum as a function of epoch and the features in both SN 1885A and in SN Ia evolve with time. It is important not to simply match spectral features, but to match them at the appropriate epoch.

De Vaucouleurs and Corwin (1985) allude to the possibility that SN 1885A could be an SN Ib. Graham (1988a) argues against this based on both the light curves (which are too slow in decline) and the spectra. Graham argues in favour of an agreement with the spectra of SN Ia and identifies a strong peak in the spectrum of SN 1885A at 6568 Å as the emission associated with Si II $\lambda$6355 which characterizes SN Ia near maximum light. This argument ignores several factors. One is that the Si II feature disappears by two weeks after maximum in SN Ia (figure 12) and the feature persists in SN 1885A. Another is that the observed feature in SN 1885A is significantly to the red of the peak in SN Ia and this is difficult to explain since the peak in SN Ia is the P-Cygni emission peak of the Si II line which should be at the rest wavelength, certainly not red-shifted appreciably. SN Ia do develop a strong Fe II feature at about 6500 Å a few weeks after maximum. This could be the feature in SN 1885A, but Fe II appears in all types of supernova so its appearance in SN 1885A is certainly not a unique clue to its being an SN Ia.

We have independently examined the spectra of SN 1885A in some detail. The feature at 6568 Å could be H$\alpha$, but there is no reason to associate it with Si II $\lambda$6355. The Fe II blend in SN Ia forms a little after maximum and slowly shifts to the red. There is excellent agreement between the relevant feature in SN 1885A and that in the 4–10 May spectra of SN 1981B in figure 12. In SN Ib 1984L the minimum due to He I $\lambda$6678 falls right on this peak in SN 1885A discouraging any identification of SN 1885A with SN Ib events, nor is there any significant feature in SN Ic events with which to associate it. SN 1885A has several maxima in the range 5300–5600 Å. Some of these could be identified with features at maximum light in SN 1981B (figure 12), but they persist for 20–40 days past maximum whereas the spectrum of SN 1981B changes significantly in that epoch. These features correspond roughly to the Fe II features in the spectrum of SN 1981B on March 31, about three weeks after maximum. SN Ib and SN Ic also show Fe II features in this range, testimony to the ubiquity of Fe II in supernova spectra, but not to the identity of the progenitor of SN 1885A.

All this suggests that the spectrum of SN 1885A might be consistent with that of iron and little else, but there is a troubling exception in the blue. SN 1885A shows peaks at 4773 Å 20–30 days after maximum and at 4890 Å about 40 days after maximum. The latter nearly coincides with the small but distinct Fe II feature in SN 1981B at ~4900 Å which also appears in SN Ic 1983V (figure 28 below). It corresponds to the minimum of He I $\lambda$4771 in the spectra of SN 1984L (figure 26 below) again suggesting that SN 1885A did not contain distinct He I features and was not an SN Ib. The peak at 4773 Å, however, falls at a minimum in all spectra of SN Ia at all epochs. This is also true of SN Ic and there is certainly no strong identification with features in SN Ib events.

Chevalier and Plait (1988) argue that a model in which helium detonates on top of a degenerate, but unburned C/O core as computed by Woosley *et al* (1986) provides a good fit to the light curve of SN 1885A. They conclude that SN 1885A was perhaps a physical variation on, but was distinctly not a SN Ia. The analysis of the spectra just presented is in agreement with this conclusion. The most likely intepretation of the spectral information is that it was consistent with iron (Fe II) and little else. This interpretation is in accord with the helium single detonation model. The major disagreement is the emission peak at ~4773 Å in SN 1885A which seems to have no counterpart in the iron-rich spectra of known supernovae.

The potential for further fruitful study of SN 1885A comes with the discovery of the remnant by Fesen *et al* (1989). Fesen *et al* discovered the remnant with a narrow passband filter centred on Fe I $\lambda 3860$. They suggest that the remnant is thus probably rich in iron and hence perhaps consistent with an SN Ia event, although it would be difficult at this stage to differentiate such a model from that advocated by Chevalier and Plait (1988) on this basis alone. Fesen *et al* argue that any central source of ionizing radiation such as a neutron star or white dwarf must be less than $\sim 10^{35}$ erg s$^{-1}$ to preserve the Fe I. They also argue that ambient starlight from the bulge of M31 could destroy unshielded Fe I in ~5 yr. It will certainly be interesting to see whether future observations and modelling can discriminate between models which disrupt completely, producing an inner, expanding core of iron, from those that eject a shell of iron and leave a white dwarf remnant.

### 2.9. *SN Ia and the distance scale*

The deflagration model has interesting implications for the cosmic distance scale as pointed out by Sutherland and Wheeler (1984) and by Arnett *et al* (1985). For a recent discussion of this and other constraints on the distance scale based on supernovae, see Branch (1988b). Sutherland and Wheeler noted that if too little of the star burned, the expansion velocities would be lower than observed. If all the star burned, the spectral constraints requiring some unburned matter would be violated. Sutherland and Wheeler thus argued that a deflagration model must consume between 0.8 and 1.4 $M_\odot$ of degenerate C/O in the deflagration model. Arnett *et al* noted that about half the matter consumed in burning goes into $^{56}$Ni, the other half going to intermediate mass elements and other iron peak elements. Using Arnett's (1982) demonstration that the peak luminosity of a deflagration model was virtually identical to the rate of deposition of radioactive decay energy thus gives a direct determination of the peak luminosity, once the amount of $^{56}$Ni produced is known. Making standard assumptions about the relation between blue and bolometric magnitudes and about the epoch of maximum light, Arnett *et al* thus deduced that if the deflagration model is a reasonable representation of a SN Ia then the nickel mass ejected must be $0.4 \leqslant M/M_\odot \leqslant 1.4$, that the bolometric luminosity must be $0.87 \times 10^{43} < L(\text{erg/s}) < 3.05 \times 10^{43}$ and the blue magnitude $-20.4 < M_B < -19.1$. The latter can be used to calibrate the Hubble diagram for SN Ia in extinction-free ellipticals beyond the Virgo cluster, to avoid Virgo flow problems. The result is $0.39 < h < 0.73$. For deflagration model W7, 0.6 $M_\odot$ of $^{56}$Ni is produced. This corresponds to $L = 1.3 \times 10^{43}$ erg s$^{-1}$, $M_B = -19.5$ and $h = 0.59$.

These results do not depend in detail on the deflagration model, only on the production of a finite amount of $^{56}$Ni in the explosion of a Chandrasekhar mass white dwarf. They thus apply to detonation models in the limit that the whole star is incinerated, or in 'damped-detonation' models in which some matter emerges only partially burned. These results might not apply quantitatively to the dynamical coalescence of binary white dwarfs since the total mass might conceivably be less than the Chandrasekhar mass and hence the kinematics would be different for a given amount of $^{56}$Ni produced.

Taken at face value, the deflagration (detonation) model for SN Ia is compatible with a range in values of $H_0$, but not with $H_0 = 100$ km s$^{-1}$ Mpc$^{-1}$, a value which receives support on various astronomical grounds. To be consistent with this large value of $H_0$, a SN Ia would have to eject $\leqslant 0.2\ M_\odot$ of $^{56}$Ni and the resulting explosion would be too feeble unless another mechanism, e.g. core collapse, were available to provide the kinetic energy. On the other hand, we can argue that the lower limit quoted by Arnett *et al*, $h_0 = 0.39$, is also too extreme. That corresponds to producing a full Chandrasekhar mass of $^{56}$Ni in each event. Since a finite amount of intermediate mass elements must be ejected, then that limit is incompatible with the observations.

Fowler (1986) has used the results pertaining to the deflagration model to make a novel argument linking SN Ia to the properties of an inflationary Universe. An inflationary Universe has exactly the critical density at the current epoch $\Omega = \rho/\rho_{\rm crit} = 1$. In this case the ratio of the age of the Universe is exactly 2/3 the Hubble time $H_0^{-1}$. Fowler notes that his favourite age of the Universe derived from nucleocosmochronology, $12 \times 10^9$ yr, is 2/3 of the Hubble time $H_0^{-1} = 17.6 \times 10^9$ yr derived from the 'favoured' deflagration models $H_0 = 59$ km s$^{-1}$ Mpc$^{-1}$, within one standard deviation. If one is willing to consider three standard deviations then considerable latitude is still allowed in the choice of both the age of the Universe and $H_0$.

### *2.10. Summary*

Table 3 gives a summary of the properties of the competing models for SN Ia events. It is clear from inspection of this table that no model fully satisfies all the constraints. The deflagration model of Woosley and Weaver (1986a) with front speed a constant fraction of the sound speed accounts for many of the observed features, but not the optical spectrum near maximum and can be argued to be incompatible with the physics of ignition, recourse to unknown multidimensional affects aside. A carefully concocted 'damped-detonation' model would by definition be nearly indistinguishable from a deflagration model in that it would leave partially unburned material, and hence satisfy the spectral constraints. There may, nevertheless, be detectable differences between such detonation and deflagration models in terms of the final density distribution. Unfortunately, no such 'damped-detonation' models exist.

There are thus two major challenges remaining in the study of SN Ia, despite all the progress that has been made. One is the evolution that leads to ignition, and the other is the nature of the burning and runaway. The latter cannot be properly addressed without adequate attention to the convective Urca process.

**Table 3.** Observational constraints on models for SN Ia.

| | Full detonation | Damped detonation | Deflagration $v_f \propto r$ | Deflagration $v_f \sim$ constant |
|---|---|---|---|---|
| Physics of ignition | √ | √ | ×? | ×? |
| Physics of quenching | √ | × | √ | √ |
| Max-light spectra | × | ? | √ | × |
| Late-time spectra | × | ? | √ | √ |
| Light curve shape | √ | √ | √ | √ |
| $^{54}$Fe | √? | √? | × | × |
| SN 1006 UV absorption | ? | ? | × | √ |
| Remnant X-ray spectra | × | ? | ×? | ? |

## 3. Type Ib supernovae

### *3.1. Overview*

The modern history of SN Ib began in 1983. After a long hiatus, that year produced three 'SN Ib' events, SN 1983I, 1983V, and 1983N. By far the most influential was SN 1983N. The optical spectra of SN 1983N had a certain resemblance to Bertola's of SN 1962L and 1964L, but were obtained with modern electronic rather than photographic spectrographs. Thus while details could not be compared, SN 1983N, like SN 1962L and 1964L showed no evidence for the strong absorption at 6150 Å attributed to the Si II $\lambda$6355 line, making it very different than classical SN Ia events. A large amount of data, IR, optical, UV, spectral and photometric, was obtained on this event over the next year. A compendium of these observations has been circulated in preprint form by Panagia *et al* (1986). Various aspects of this modern prototypical event have been addressed by Richter and Sadler (1983), Panagia (1985), Uomoto and Kirshner (1985), Harkness *et al* (1987), Graham *et al* (1986) and Gaskell *et al* (1986).

The next important SN Ib event was SN 1984L. Wheeler and Levreault (1985) argued that the optical spectrum of SN 1984L near maximum light was very similar to that of SN 1983N (Panagia 1985) and distinctly different from classical SN Ia. Study of the spectra of SN 1983N and 1984L near maximum light resulted in the discovery of strong evidence for appreciable amounts of helium in the ejecta (Harkness *et al* 1987, Wheeler and Harkness 1986, Wheeler *et al* 1986). Gaskell *et al* (1986) discovered that SN 1983N made a transition to a 'supernebular' phase dominated by [O I] $\lambda\lambda$6300, 6364 and identified SN 1985F (Filippenko and Sargent 1985, 1986) as a member of the class SN Ib at late times on the basis of its similar spectrum. A late-time spectrum of SN1984L showed the same transition (Schlegel and Kirshner 1989). The optical light curve of SN 1983N was similar in shape to that of a typical SN Ia (Panagia *et al* 1986) but SN 1983N and 1984L were dimmer at maximum by about 1.5 mag (Wheeler and Levreault 1985, Uomoto and Kirshner 1985, Branch 1986, Panagia *et al* 1986).

Graham *et al* (1986) used infrared spectra to detect a feature they identify with the 1.644$\mu$ line of [Fe II] in SN 1983N about eight months after maximum light (in the 'supernebular' phase, see below). They estimated that the total amount of iron was about 0.3 $M_\odot$ and argued that such a quantity of iron was probably freshly synthesized in the explosion. The natural mechanism for producing the iron is to first make radioactive $^{56}$Ni which decays to $^{56}$Co and thence to stable $^{56}$Fe. This decay scheme provides the observed luminosity in the thermonuclear explosion model for SN Ia, as discussed earlier. The quantity of iron estimated for SN 1983N is on the small side for the amount expected from a SN Ia, and somewhat large for the particular thermonuclear explosion or gravitational core collapse models that have been proposed to account for the lower luminosity of SN Ib events. There are, however, uncertainties in the estimated iron mass due to lack of precise knowledge of the distance to M83, the host galaxy, and of the composition of the radiating matter which is more complex than the iron plasma assumed by Graham *et al.* The presence of a rather large quantity of iron in SN 1983N would strongly suggest that, as for SN Ia, radioactive decay plays an important role in driving the light curves of SN Ib.

Studies of supernovae in the infrared have also proved useful in categorizing SN Ib events. Elias *et al* (1985) showed that three 'SN Ib' events also followed well defined infrared light curves in which the post-maximum dip shown by SN Ia was conspicuous by its absence. The absence of this secondary minimum in the SN Ib presumably indicates a fundamental difference in the composition of the ejecta between the two classes. Elias identified SN 1983I (see below) as a Type Ib on the basis of the infrared light curve alone. We now realize that SN 1983I is a candidate SN Ic, based on its optical spectrum.

Oliva (1987) and Fransson and Chevalier (1989) argue that the $1.645\mu$ line of [Si I] is an equally likely identification for the IR feature identified as the $1.644\mu$ line of [Fe II] by Graham *et al* (1986) and that such a feature could plausibly be expected at the observed strength. Fransson and Chevalier estimate the fraction of Si I to be about 0.5 and the total mass of Si to be $\sim 7 \times 10^{-2}\ M_{\odot}$. This amount of silicon at late times suggests that the Si is concentrated in the core. Its detection does not necessarily contradict the hypothesis of Graham (1986) that the absence of the IR absorption dip at 20 days in SN Ib suggests a true paucity of Si in the outer layers observed near maximum light compared with the case of SN Ia.

There are other circumstantial ways to discriminate SN Ib events from those of SN Ia. SN Ib are apparently correlated with large patches of ionized hydrogen (H II regions) which represent the birthplace of young stars (Wheeler and Levreault 1985, Uomoto and Kirshner 1985, Harkness *et al* 1987). The line of sight to essentially all the SN Ib fall on or near to such a region, in strong distinction to SN Ia which tend to avoid spiral arms in general and such very young regions in particular. Type II supernovae do share this feature of apparent correlation with H II regions to some degree (Huang 1987). This suggests that both SN Ib and SN II arise from similar populations of rather short-lived massive stars.

No SN Ia has been detected in the radio. Sramek and co-workers (Sramek *et al* 1984, Panagia *et al* 1986, Sramek and Weiler 1990) observed radio emission from both SN 1983N and SN 1984L, and obtained radio 'light curves' in different frequencies. SN 1985F was not observed (Sramek 1989) at a level and phase where SN 1983N and SN 1984L were detected, indicating a real difference in the properties of SN 1985F in this regard. Chevalier (1984) has presented a plausible explanation for the radio emission in which the shock wave from the supernova interacts with the circumstellar 'wind' blown off the star prior to its explosion. The high luminosity of massive stars naturally expels such a wind with radiation pressure, but the wind could also arise from the highly evolved companion of a lower mass star. The radio emission gives a constraint only on the density of the circumstellar nebulae and the question of whether it could arise from the low velocity, low mass loss of a red star or the high mass loss, high velocity of a blue star is debated (Panagia *et al* 1986, Harkness *et al* 1987). In either case this evidence also suggests an origin in moderately massive stars.

Wheeler and Levreault (1985) argued for an origin of SN Ib in massive stars, but suggested that the relatively narrow light curve peak restricted the exploding core mass to less than 3-4 $M_{\odot}$ and hence the main sequence mass to $\sim 20\ M_{\odot}$. The estimated rates of explosions, which could be as high as those of SN Ia and SN II-L when selection effects are considered (Branch 1986, Evans *et al* 1989, table 2 of this paper), also argue against progenitors only of very large mass. Emission line strengths in the supernebular phase tend to give higher masses for the ejecta ($\gtrsim 5\ M_{\odot}$, Begelman and Sarazin 1986, Gaskell *et al* 1986) but such estimates may not take complete account of the relevant physics (section 3.3.4 below). Wheeler and Harkness (1986) noted that SN 1962L, 1964L, 1983I and 1983V seem to lack not only the 6150 Å line of SN Ia, but also the deep absorption at 5700 Å that characterizes SN 1983N and 1984L, which is identified as Doppler-shifted He I $\lambda$5876 (Harkness *et al* 1987). Wheeler and Harkness noted the similarity between the spectra of SN 1983I and SN 1983V and suggested that they belong to yet a third spectrally distinct class of hydrogen deficient supernovae, Type Ic. Wheeler *et al* (1987a) presented models to argue that, rather than a distinct spectral class, SN 1983I and 1983V, along with SN 1962L and 1964L, and perhaps 1975B and 1982R, are members of the general class of SN Ib, but with markedly less helium and more carbon and oxygen in the ejecta than SN 1983N and 1984L. Harkness and Wheeler (1990) have retreated to a more conservative point of view by acknowledging that there are clear empirical spectral differences between the events like SN 1983I and V and those like SN 1983N and 1984L, and hence that the SN Ic category should be empirically maintained until there is more substantial physical evidence to link or separate SN Ib from SN Ic. Other possible hints at distinction between the two classes will be discussed below.

Table 4 gives a compendium of information of the current candidates for SN Ib and SN Ic. The suggestion of SN 1966J comes from Barbon *et al* (1989). SN 1975B is rather problematical since the spectrum is marred by plate defects so that it is not clear whether the 6150 Å feature is present or not (Kirshner *et al* 1976). SN 1982R is included on the basis of a verbal description of the spectrum (Dennefeld (1982). SN 1989E is suggested as a candidate by Gaskell and Koratkar (1989) and SN 1990B by Filippenko (1990) and Benetti *et al* (1990). Some spectra are too noisy to classify and, of course, many supernovae have no spectral data at all. Given the steady rate of discovery of SN Ib/c since 1983 and the suggestion that they occur at rates comparable to SN Ia and SN II in spiral galaxies (Evans *et al* 1989, section 1 of this paper), it is curious that more SN Ib are not present among the 200+ older supernova events. The answer presumably lies in selection effects. SN Ib are dimmer than SN Ia (but comparable to SN II-P) and occur near H II regions, the surface brightness of which may mask any event not caught near maximum light. SN 1985F provides a case in point. It was discovered ~8 months after peak light (Filippenko and Sargent 1985, 1986) but

**Table 4.** Summary of candidate Type Ib and Ic events.

| SN | Type | Galaxy | $V_H^{1,2}$ | $\mu^2$ | Galaxy type[3] |
|---|---|---|---|---|---|
| 1954A | Ib | NGC4214 | 288 | 27.71 | I(s)m |
| 1962L | Ic | NGC1073 | 1212 | 30.91 | SB(rs)c |
| 1964L | Ic | NGC3938 | 805 | 31.15 | SA(rs)bc |
| 1966J | Ib | NGC3198 | 660 | 30.17 | SB(rs)c |
| 1975B | ? | anon. | — | — | |
| 1982R | Ib? | NGC1187 | 1394 | 30.05 | SB(r)c |
| 1983I | Ic | NGC4051 | 710 | 31.15 | SAB(rs)bc |
| 1983N | Ib | NGC5236 | 518 | 28.35 | SAB(s)c |
| 1983V | Ic | NGC1365 | 1639 | 31.14 | SB(s)b |
| 1984L | Ib | NGC991 | 1538 | 31.37 | SAB(rs)c |
| 1985F | Ib | NGC4618 | 546 | 29.31 | SB(rs)m |
| 1986M | Ib | NGC7499 | 11 595 | — | SAO(s) |
| 1987K | ? | NGC4651 | 800 | 31.13 | SA(rs)c |
| 1987M | Ic | NGC2715 | 1126 | 31.55 | SAB(rs)c |
| 1988L | Ic | NGC5480 | 1907 | 32.52 | SA(s)c |
| 1989E | Ib? | MCG+5-32-45 | | | |
| 1990B | Ic | NGC4568 | 2247 | 31.13 | SA(rs)bc |

[1] Corrected for solar motion.
[2] From Tully (1988).
[3] From de Vaucouleurs *et al* (1976).

was present in prediscovery plates at $V \sim 12$ (Tsvetkov 1986) at the magnitude 'predicted' for a SN Ib (Gaskell *et al* 1986). At the distance of the Virgo Cluster SN Ib should peak at about $V \sim 13$–14, and hence should be eminently 'discoverable'. Presumably an appropriate fraction of untyped supernovae were SN Ib/c, and the twenty-year hiatus in discovery between the early 1960s and the early 1980s just poor luck and insufficient attention.

The topic of SN Ib events has not matured to the extent that SN Ia has and there is still controversy over the fundamental nature of the progenitor, and hence its evolution. The subject of SN Ib will therefore be approached in a more empirical way in the following sections, where the topics outlined in this overview will be addressed in more detail.

### *3.2. Light curves*

#### *3.2.1. Observed light curves*

*Optical photometry.* Photometric data provide useful information on the various SN Ib events. A reanalysis of SN Ib and Ic light curves is in preparation by Miller and Wheeler (1990). Other compilations of light curves for SN Ib and c events are given by Porter and Filippenko (1987), Ensman and Woosley (1988), Nomoto *et al* (1988) and Schlegel and Kirshner (1989).

Figure 21 gives normalized B or photographic light curves for four strong helium line, SN Ib events (SN 1954A, 1983N, 1984L and 1985F) and three weak helium line, SN Ic events (SN 1962L, 1964L, and 1983I), along with the mean light curve of SN Ia (figure 6). The curves are adjusted in time to a common estimated maximum B magnitude normalized in apparent magnitude. The light curves of SN 1983N, SN 1984L and SN 1985F are nearly identical in the region of overlap. The long term light curves of SN 1984L and SN 1985F are consistent with the decay of $^{56}$Co, with efficient trapping of the gamma rays (Filippenko and Sargent 1986, Filippenko *et al* 1986, Schlegel and Kirshner 1989). The photographic light curve of SN 1954A falls somewhat more steeply. Pskovskii (1977) assigns this light curve the speed class 13, comparable to a rather fast SN Ia. The B light curves of the SN Ib and SN Ic events are similar around peak, but the curves for the Ic events tend to fall below those of SN 1984L and SN 1985F after about 10 days past maximum. The light curve for SN 1964L, for which there is the most extensive data, tends to flatten after 100 days, but there is a question as to whether galaxy contamination could be an increasingly important effect at this phase.

The failure to obtain any spectra near maximum light for SN 1985F leaves its subclass open to argument. Given the example of SN 1987K that switched from a SN II to something indistinguishable from a SN Ib (section 3.3 below), it is not even clear that SN 1985F could not have been a SN II at maximum. The similar light curves for SN 1984L and SN 1985F suggest, however, that SN 1985F was a SN Ib, as originally suggested by Gaskell *et al* (1986).

B light curves are popular because a considerable amount of older data is photographic, roughly equivalent to photoelectric B. Since the spectra of both SN Ia and Ib/c events have strong ultraviolet deficits, it is not clear that B light curves are appropriate to analyse for physical parameters. Figure 22 shows the V magnitude light curves for two SN Ib events (SN 1983N and 1984L) and two SN Ic events (SN 1962L and 1983I), and the bolometric light curve for SN 1983N (Panagia 1985). These light curves show that the V magnitudes all match around the peak (the two low points

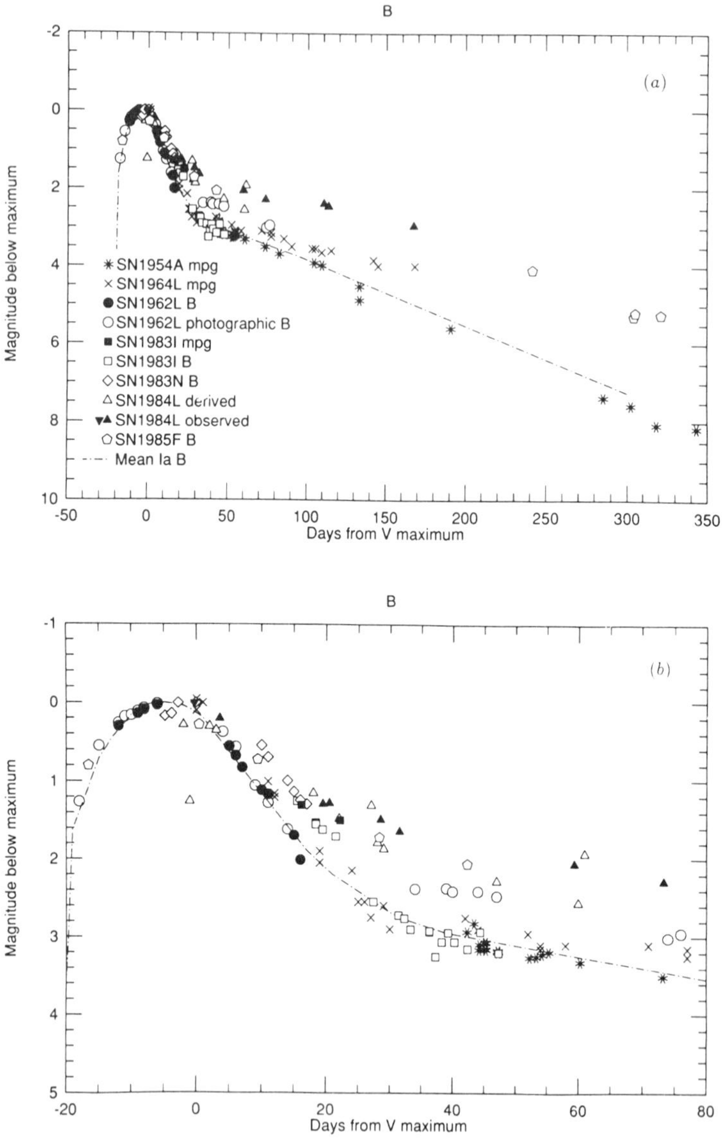


**Figure 21.** (*a*) The B (or photographic) magnitude against time is shown for four Type Ib (SN 1954A, 1983N, 1984L, 1985F) and three Type Ic (SN 1962L, 1964L, 1983I) events. Also shown for comparison is the mean B light curve for SN Ia (Doggett and Branch 1985). (*b*) Same on an expanded scale.

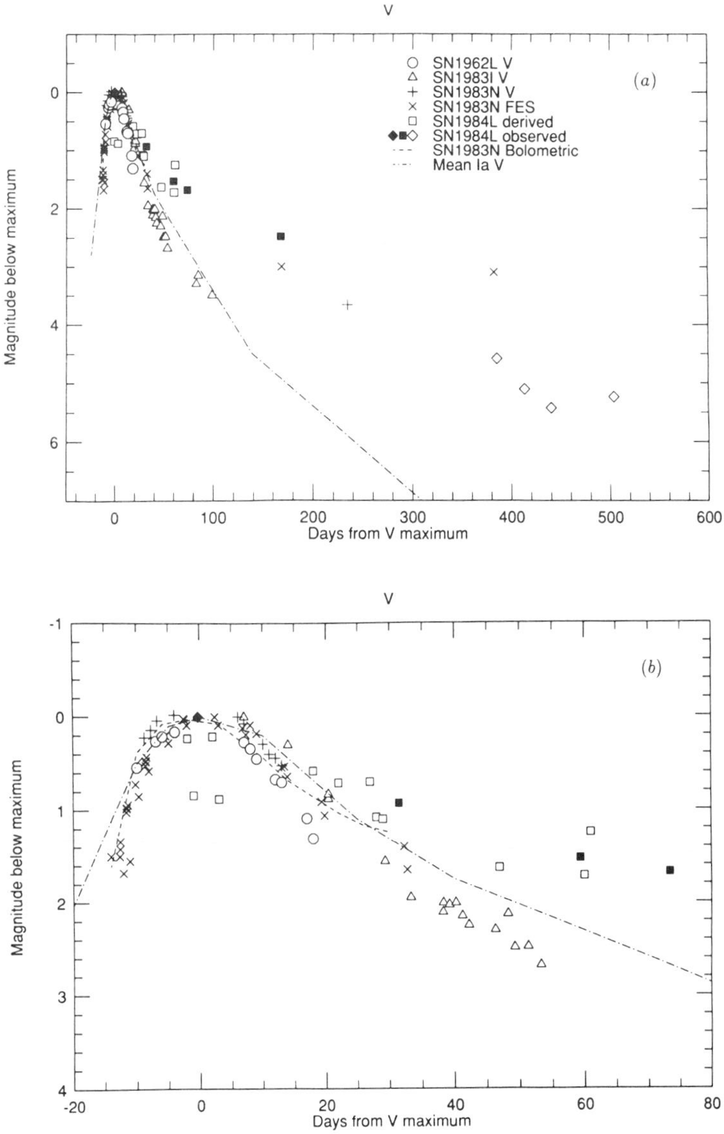


**Figure 22.** (*a*) The V magnitude against time is shown for two Type Ib (SN 1983N, 1984L) and two Type Ic (SN 1962L, 1983I) events. Also shown for comparison are the bolometric light curve for SN 1983N (Panagia 1985; broken curve) and the mean V light curve for SN Ia (Doggett and Branch 1985; chain curve). (*b*) Same as (*a*), but on an expanded scale to emphasize the relation of the data to the bolometric curve and SN Ia curves.

near peak for SN 1984L are derived from incomplete spectral scans in Harkness *et al* 1987), and that the V magnitude curves agree reasonably well with the bolometric curve for SN 1983N. This is presumably because the V filter samples the spectrum in a region where the continuum is approximately Planckian so that it is directly related to the bolometric flux.

Between 50 and 150 days past maximum, the V light curve of SN 1984L falls somewhat more slowly than that of SN 1983N. This may indicate a longer diffusion time and hence a larger mass ejected for SN 1984L. The point for SN 1983N at 380 days from Panagia *et al* (1986) may be subject to galaxy contamination. The (rather uncertain) point at 235 days from Gaskell *et al* (1986) defines a late time curve for SN 1983N which is similar to that of SN 1984L. For the first twenty days after maximum, the V light curves of candidate SN Ic events 1962L and 1983I fall more rapidly than SN 1984L and SN 1983N. SN 1983I continues to decline rather steeply after 30 days when SN 1984L flattens appreciably. This suggests that SN Ic events may differ in some physical way, perhaps by having less ejecta mass than SN Ib events. This may be a clue that there are real differences in the physical makeup of SN Ib and SN Ic events. It is very important to establish the light curves for recent SN Ic events such as SN 1987M and SN 1988L to see if this trend of steeper light curves (and hence, perhaps, smaller ejecta mass) holds true. Recall that SN 1954A, apparently a SN Ib with a steep light curve, contradicts this trend. As noted below, this event has a different colour curve, and so may have to be considered a 'special case'.

Figure 23 gives the B–V colour curves for a number of SN Ib and c events and the mean colour curve for SN Ia events. The individual colour curves have been corrected for Galactic extinction (Miller and Wheeler 1990) but not for absorption in the host galaxy which is probably the more dominant affect. The locus until about 40 days past

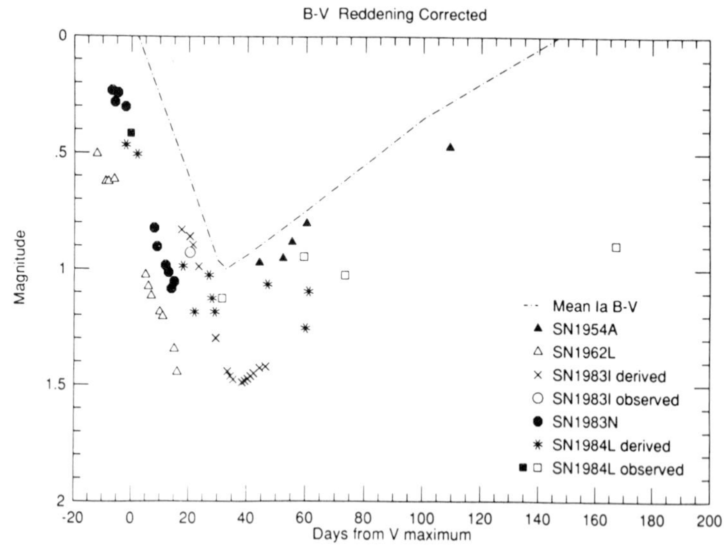


**Figure 23.** The B–V colour curve is shown for three Type Ib (SN 1954A, 1983N, 1984L) and two Type Ic (SN 1962L, 1983I) events. Also shown for comparison is the mean B–V curve for SN Ia.

maximum is roughly parallel to that for the mean SN Ia, but redder by $\sim 0^m.8$. There is some concern that in using SN I for distance calibrations one could confuse SN Ia and SN Ib since both are hydrogen deficient in the spectrum and the shape, but not the amplitude, of the light curves is similar. Miller and Wheeler conclude that maximum light colours can be used to distinguish the two types of event in a manner that is relatively free of internal absorption. An alternative point of view is that the colour curves of SN Ib and Ic are similar to SN Ia and that figure 23 could be used to determine the internal absorption of the SN Ib and Ic events.

The colour curve synthesized for SN 1984L does not show the well known blueward trend of SN Ia beyond +30 days. SN 1954A is tentatively designated a SN Ib because of the relatively strong lines of He I in the photographic spectra. Photographic B–V colours suggest that it evolved blueward more rapidly than SN 1984L. Clearly there is a need for more photometry of SN Ib events to determine whether there is a standard colour evolution or whether large variations are the rule. The interpolated colours (B and V were not obtained on the same nights) of SN 1983I give a well defined minimum at B–V = 1.5 at about 37 days after maximum. Allowing for the sparse data, it is possible that SN 1984L may have had a similar minimum. Unfortunately, there are no later data on SN 1983I to determine whether it stayed relatively red or evolved back to blue.

*Infrared photometry.* Elias *et al* (1985) have established that SN 1983N, SN 1984L and SN 1983I follow common J, H and K infrared light curves which are distinctly different than those of SN Ia, as illustrated in figure 24. Figure 24 also gives the mean H light curve for SN Ia and illustrates that SN 1983N and SN 1984L seem not to have the minima and secondary maxima which characterize SN Ia (figure 9). They identified SN 1983I as a SN Ib on this basis alone, without the support of spectral data. They note that despite the similarity of infrared light curves, SN 1983I may be about $1^m$ dimmer at H in absolute magnitude than SN 1983N and SN 1984L. This distance assignment depends on their adopted Virgo infall model. They put SN 1984L 1.7 times more distant than SN 1983I, whereas de Vaucouleurs (1979) assigns virtually equal

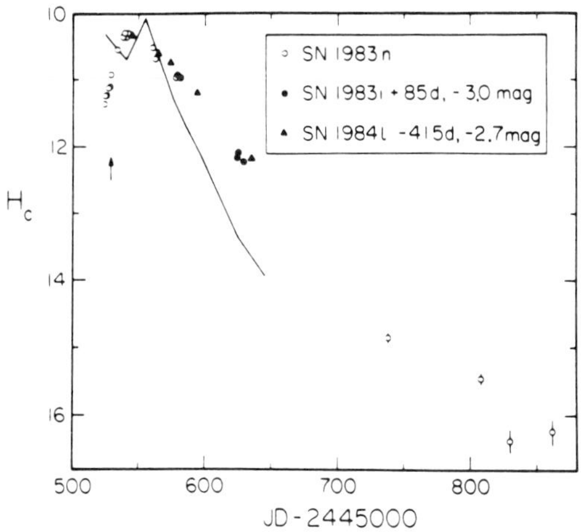


**Figure 24.** Infrared (H) light curves for two Type Ib (SN 1983N, 1984L) events and one Type Ic (SN 1983I) event are shown in comparison with the standard light curve derived for SN Ia (see figure 9; from Elias *et al* 1985).

distances despite the discrepant radial velocities. For equal distances, SN 1983I would be dimmer than SN 1984L by only $0^m.3$ at H. On the other hand, SN 1983I is now recognized to be a candidate for the class SN Ic, and hence the possibility that it could have real IR photometric differences from SN 1983N and SN 1984L should be considered.

Whitelock (1983 and private communication) gives IR photometry for SN 1983V obtained by Whitelock on 12 December, by Feast on 17, 18 December, and by Laney on 26 December, 1983. Unfortunately, due to a transcription error, some of these data were misreported in Wheeler *et al* (1987a). The 18 December values quoted there for J and H are actually H and K. The correct data reveals that the J light curve agrees with the 'standard curve' established by Elias *et al* (1985). The H and K light curves for SN 1983V are also in fair agreement with the standard curves of Elias *et al* when allowance is made for the fact that larger apertures were used for the data of 17, 18 December. Overall consistency suggests that the V maximum of SN 1983V was about 1 December 1983.

*Constraints from the photometry.* The optical and IR light and colour curves are useful to put constraints on the similarity of the candidate SN Ib and Ic events and to determine the spectral phase. Miller and Wheeler (1990) have re-examined the data and attempted to assign phases. They note that the phase assigned for SN 1983I by Tsvetkov (1985) is based on mean light curves for SN Ia which are not quantitatively relevant to this SN Ic event. They also note that, as shown in figure 24, the IR data for SN 1983I fall on the tail of the standard SN 'Ib' curve established by SN 1983N and SN 1984L by Elias *et al* (1985) where the slope is relatively constant, ~0.025 magnitude day$^{-1}$. SN 1983I can thus be assigned a different phase if the normalizing magnitude is adjusted appropriately.

Table 5 gives estimates of the phases of various SN Ib and SN Ic events. Maximum in SN 1962L seems to have been several days earlier than assigned originally by Bertola (1964). The estimate for the date of the H maximum in SN 1983I by Elias *et al* (1985) is earlier than, and hence incommensurate with, the phase obtained from comparison with the optical light curves of SN 1983N and 1962L. The data suggest that the maximum in H for SN 1983I was somewhat later than estimated by Elias *et al*, by perhaps 10 days. Conflicting data on SN 1983V (Wheeler *et al* 1987a) seems to be most consistent with a phase such that maximum V was about 1 December 1983. This assignment of phase is different than the conclusion reached by Wheeler *et al* (1987a) but consistent with the estimate of Branch (1988a).

**Table 5.** Phases of Type Ib and Ic supernovae.

| Event | Discovery date | Date of maximum B | Date of maximum V |
|---|---|---|---|
| SN 1954A | 30 May 1954 | 18 Apr 1954 ± 3 | — |
| SN 1962L | 23 Nov 1962 | 5 Dec 1962 ± 2 | 8 Dec 1962 ± 2 |
| SN 1964L | 12 Dec 1964 | 12 Dec 1964 | — |
| SN 1983I | 12 May 1983 | 27 Apr 1983 ± 2 | 2 May 1983 ± 4 |
| SN 1983N | 4 Jul 1983 | 16 Jul 1983 ± 2 | 19 Jul 1983 ± 2 |
| SN 1983V | 25 Nov 1983 | | 1 Dec 1983 ± 4 |
| SN 1984L | 29 Aug 1984 | 20 Aug 1984 ± 2 | 1 Sep 1984 ± 2 |
| SN 1985F | 28 Feb 1985 | 12 Jun 1984 ± 4 | |
| SN 1987M | 21 Sep 1987 | | 21 Sep 1987 ± 4 |

*3.2.2. Theoretical light curves.* Wheeler and Levreault (1985) argued on the basis of the width of the peak of the light curves of SN Ib that the mass of the ejecta might be somewhat in excess of that of SN Ia, but that the ejecta mass probably did not exceed about 2 $M_\odot$. Assuming the light curve to be powered primarily by radioactive decay of $^{56}$Ni, they estimated the mass of the nickel to be about 0.2 $M_\odot$. They concluded that these two features could be consistent with the explosion of the hydrogen-stripped core of 3-4 $M_\odot$ arising from a star of main sequence mass of 10-20 $M_\odot$. Wheeler and Levreault noted that the hydrogen deficiency implied by the spectra of SN Ib events was reminiscent of Wolf-Rayet stars, but that observed Wolf-Rayet stars were so massive ($\gtrsim 10\ M_\odot$) that they could be expected to result in light curves that were too broad to correspond to observed SN Ib light curves. Subsequent more detailed work on light curves has verified these basic points, and raised some associated questions.

Cahen *et al* (1986) adopted Arnett's (1980, 1982) analytic models to show that exploding oxygen-rich cores could be powered at a dim level by the energy of recombination of oxygen. Schaeffer *et al* (1987) added the energy of radioactive decay of a few tenths of a solar mass of $^{56}$Ni and concluded that the core mass associated with the explosion of a SN Ib could be consistent with a core mass of about 8 $M_\odot$. Schaeffer *et al* take their core structure from a 40 $M_\odot$ model of Maeder, and may have removed both the hydrogen and the helium layers to reduce the core to 8 $M_\odot$. They include 0.3 $M_\odot$ of $^{56}$Ni. Schaeffer *et al* point out that in their models the gamma rays are trapped for a considerable time so that the light curve does not begin to rise rapidly to the peak until 30 days after the explosion, after the oxygen recombines reducing the electron-scattering opacity. Peak light is thus at about 50 days after the explosion. Whereas Wheeler and Levreault assumed that the time of maximum and the width of the peak would be comparable, Schaeffer *et al* find that the peak, when it occurs, is only about 25 days wide, full width, one-half magnitude below the peak in their calculated V magnitude. Schaeffer *et al* calculate a colour near maximum of $B-V \sim 1$, typical of SN Ib. They also find a secondary hump in the V and B light curves that does not correspond to any observed feature. In these massive star models, the gamma rays are trapped for a considerable time, and the light curves tend to follow the natural decay time of the radioactive decay. Schaeffer *et al* find a decay time on the tail of ~0.012 magnitude day$^{-1}$ between about 100 and 200 days after the explosion which is perhaps rather fast for SN Ib.

Schaeffer *et al* note that Maeder and Lequeux (1982) have estimated that about 15-30% of all supernovae could be from Wolf-Rayet stars and that Panagia (1985) estimates that 20-40% of all supernovae are SN Ib. From table 2, van den Bergh *et al* estimate that SN Ib and Ic events represent about 20% of the supernovae in the ensemble of galaxies represented by the Shapley Ames catalogue. The rate for SN Ib and Ic events is about that for SN Ia. From table 1, the rate of explosion of SN Ib and Ic events might be as small as one per 200 years in our Galaxy. With the (uncertain) mass function of Miller and Scalo (1979), this corresponds to the explosion of all single stars with mass in excess of about 20 $M_\odot$. For a higher rate, or a higher fraction of SN Ib/c to SN Ia in the Galaxy as compared with the Shapley Ames sample (which includes early-type galaxies biased toward SN Ia), or explosions only up to a finite mass (~30 $M_\odot$, see Twarog and Wheeler 1986), the lower limit to the mass would be lower. The rate of explosion of all stars in excess of about 40 $M_\odot$ is about one per 2000 years in the Galaxy for the Miller-Scalo mass function. These numbers generally make it difficult to associate SN Ib with stars of typical progenitor mass of 40 $M_\odot$ as adopted by Schaeffer *et al* (see also van den Bergh 1988).

Ensman and Woosley (1988) have calculated evolutionary and hydrodynamic models of SN Ib events also assuming they arise in the cores of massive stars. The models of Ensman and Woosley retain the helium layers determined by the evolutionary calculations, but remove the hydrogen envelope from their otherwise mass-conservative calculations. They consider cores of mass 4–14 $M_{\odot}$, arguing that cores of less mass eject negligible amounts of nickel. The nickel mass in their models varies with the model, but is $\sim$0.1 $M_{\odot}$. In the models of Ensman and Woosley it is the recombination of He that lowers the opacity, releases the previously trapped radioactive decay energy, and initiates the rapid rise to maximum light. The oxygen in the inner layers remains predominantly ionized through maximum light, in contrast with the models of Schaeffer *et al.* A sample of light curves calculated by Ensman and Woosley is given in figure 25. They compare their calculated bolometric luminosities to the bolometric light curve of SN 1983N (FWHM of 25 days) and conclude that cores of mass much in excess of 6 $M_{\odot}$ would give light curves which are too broad unless the kinetic energy were unrealistically large. Allowing for the mass of a collapsed remnant, they thus deduce that the ejecta mass is less than $\sim$5 $M_{\odot}$. Considering the shape of the bolometric peak, they prefer a model with a core mass of about 4 $M_{\odot}$, and hence an ejecta mass of about 2–3 $M_{\odot}$, although the luminosity of such a model is rather low. The model bolometric light curves tend to show less contrast between the peak and the tail than the observed bolometric light curve for SN 1983N, but Ensman and Woosley argue that their theoretical light curves might be a better fit to the flatter curve of SN 1985F. We note from the shape of the V magnitude curve in figure 22 that a similar statement might be made for SN 1984L. Ensman and Woosley interpret the lack of the observed peak/tail contrast compared with SN 1983N as a potentially serious problem for the massive core models of SN Ib. Unfortunately, the late-time bolometric light curve is not known for other SN Ib events. Observations of SN 1983N show that the V light curve follows the bolometric curve for SN 1983N rather closely over the peak. The V

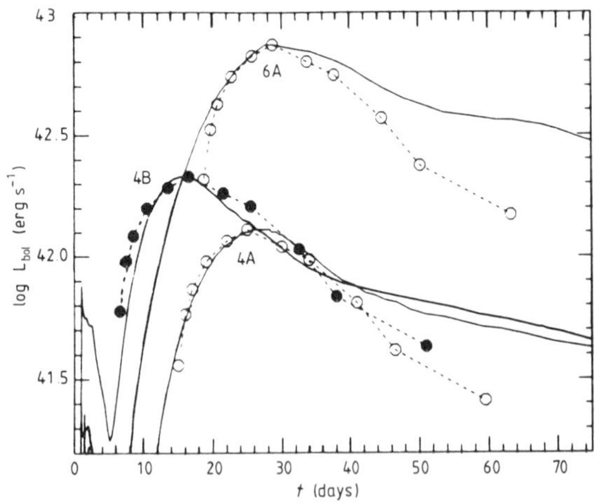


**Figure 25.** Light curves for selected models of Ensman and Woosley (1988) in comparison with the bolometric light curve of SN 1983N. Models 4A and 4B correspond to 4 $M_{\odot}$ helium cores exploded with 0.5 and $3.2 \times 10^{51}$ ergs, respectively, and model 6A corresponds to a 6 $M_{\odot}$ helium core exploded with $1.4 \times 10^{51}$ ergs.

light curves do fall more slowly than the B light curves (figures 21 and 22) and have less contrast from peak to tail. Failing the availability of complete bolometric light curves, it probably makes more sense to model the V light curves, and this may alleviate some of the problems noted by Ensman and Woosley.

Light curves for SN Ib events have also been modelled by Nomoto *et al* (1988a) and Nomoto *et al* (1988b). These works compare explosions in cores of massive stars with the visual light curve of SN 1984L, the bolometric light curve of SN 1983N, and the blue light curves of SN 1962L and 1964L. The latter may enhance the apparent dispersion in the light curve properties. Nomoto *et al* (1988a, b) consider the possibility of spreading the radioactive matter through the ejecta, as suggested by studies of SN 1987A. They find that they can fit the rather flat early decline of the visual light curve of SN 1984L with a model with 2.2 $M_\odot$ of ejecta and 0.07 $M_\odot$ of $^{56}$Ni confined to the inner regions, or a model with 4.4 $M_\odot$ of ejecta and the Ni mixed throughout the mass, both with $10^{51}$ergs of energy. The former model is the C/O core of a 6 $M_\odot$ He core, and it is not clear that it would have the requisite He for the spectrum if taken literally. They fit the bolometric curve for SN 1983N with a model with an ejecta mass of 2.6 $M_\odot$ with the Ni mixed out to 2.2 $M_\odot$ (the initial rise is too rapid if the Ni is fully mixed). These model light curves which agree moderately well with the decline from peak tend to fall too rapidly on the late-time tail. This is an example of the difficulty in reproducing the peak/tail contrast pointed out by Ensman and Woosley. Nomoto *et al* fit the rapid decline from peak of the blue light curves of SN 1962L and 1964L with a variation in which all the nickel is confined to the outer 0.2 $M_\odot$ of the model in a manner intended to be reminiscent of the exploding white dwarf model of Branch and Nomoto (1986). Nomoto *et al* (1988a, b) note that these two events may represent SN Ic, and so it is legitimate to consider a rather different physical model. Such a model does give a steep decline from the peak, but also gives much too steep a decline on the late-time tail, after about 50 days, the extreme version of the Ensman-Woosley problem. Once again it may be misleading to try to deduce physical properties of the explosion by comparing bolometric theoretical light curves with observed blue light curves.

Clearly the problem of the nature of light curves of SN Ib and Ic events has not been solved. These studies have used Rosseland opacities which may underestimate the true opacity or constant opacities, e.g. $\kappa = 0.1$ gm cm$^{-2}$, which is probably too high and does not adequately reflect the variation of opacity with composition, depth, and time that is expected to occur. It is important to compile a uniform set of data on the observed light curves, and to perhaps focus on the V rather than B light curves when bolometric light curves are not available.

### *3.3. Spectra*

#### *3.3.1. Observed early-time spectra*

*Type Ib.* The time evolution of the prototypical SN Ib event SN 1984L for the first two months after maximum is given in figure 26 along with the spectrum of SN 1983N about 8 months after maximum. Note the strengthening of the He I features in the SN 1984L spectra that make the identification unambiguous one or two weeks after maximum. The striking agreement with observed minima corresponding to the rest wavelength of He I lines at 4471, 5876, 6678 and 7065 Å, particularly in the second spectrum, is convincing evidence that He is a significant constituent of the ejecta of SN Ib as observed at this phase, the first two months. The third spectrum in figure 26 shows evidence for emission of [O I] $\lambda$5577 and [O I] $\lambda$6300, 6364 that signifies the

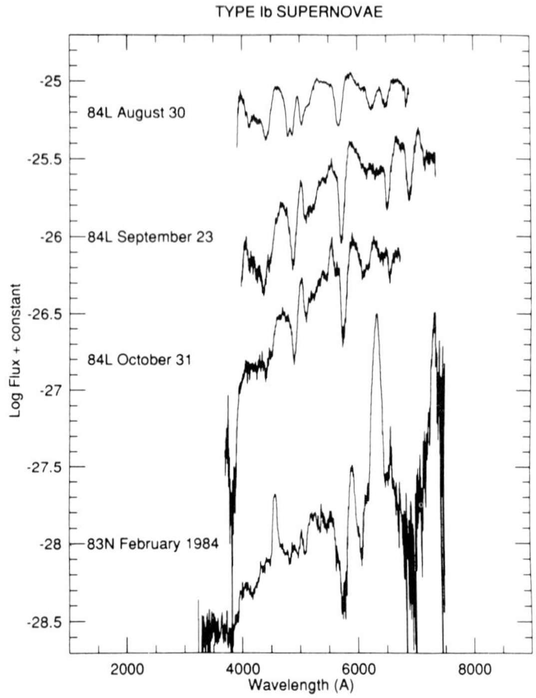


**Figure 26.** Time series of spectra for the first two months after maximum of SN 1984L plus the spectrum of SN 1983N at about 8 months after maximum.

onset of the nebular phase illustrated by the late-time spectrum of SN 1983N (see section 3.3.2 below).

An interesting feature of the spectra of SN 1984L in figure 26 is their relative homogeneity. The evolution of the spectrum is much less distinct than that displayed for SN Ia (figure 12). Between 3 September and 23 September, the absorption feature at 6300 Å disappears, but all the other major features are reproduced. During a similar two-week period after maximum light, SN 1981B (figure 12) underwent significant spectral changes, although the subsequent evolution in the spectrum was quite small.

The pre-maximum spectra of SN 1983N by Richtler and Sadler (1983) show a strong deceleration of the photospheric velocity. Richtler and Sadler were handicapped in their analysis because it was not generally recognized at the time they obtained their unique data and wrote their paper that SN 1983N was a SN Ib, not a SN Ia. They did establish that the absorption features in the spectra shifted by about 4000 km $s^{-1}$ from 3 July, about 14 days before V maximum to 13 July, about 4 days before V maximum. Recognizing SN 1983N as a SN Ib, we can now pick out the broad, blue-shifted He I $\lambda$5876 feature in the data of Richtler and Sadler. In the early spectra there is a small bump in the minimum that confuses the true minimum of the absorption, but one can estimate the minimum to be at about 5520 Å corresponding to a velocity of 18 200 km $s^{-1}$ in the composite 3–6 July data and a minimum at 5620 Å corresponding to a velocity of 13 100 km $s^{-1}$ in the composite 12–14 July spectra. The velocity at maximum luminosity is about 10 000 km $s^{-1}$ (Harkness *et al* 1987). The apparent deceleration caused by the retreat of the photosphere into the slower moving layers corresponds to about 500 km $s^{-1}$ $day^{-1}$. This velocity evolution has not been carefully modelled, but provides a potentially important constraint on the progenitor structure. Richtler and Sadler also note a narrow static emission feature in the earliest spectra which they identify as He I $\lambda$4471. They point out that this feature may represent a circumstellar nebula, possibly helium rich, surrounding the progenitor which could also be an important clue to the progenitor evolution.

Figure 27 shows a comparison of spectra of the Type Ia event SN 1981b with those of SN 1984L at roughly comparable epochs, near maximum light and 20 days later. Although there is some similarity between SN 1981B 6–9 March and SN 1984L 30 August in the wavelength range 3500–4600 Å, there are major differences throughout the longer wavelength range. Neither of these spectra shows evidence of H. The first spectrum shows the strong absorption at 6150 Å attributed to Si II $\lambda$6355 which is absent in the second. The second shows the strong absorption at 5700 Å attributed to He I $\lambda$5876 which in turn is absent in the first. A comparison of the *maximum* light spectra of SN 1984L in figure 27 with the three-week post-maximum spectrum of SN 1981B on 27 March shows a much closer resemblance at all wavelengths less than 6000 Å than does the comparison at maximum light. This agreement in the blue is what suggested that SN Ib near maximum resemble more mature SN Ia. The reason

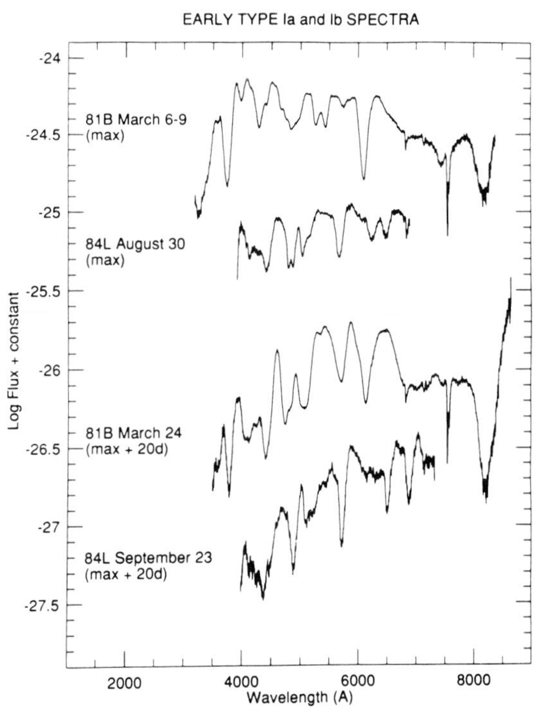


**Figure 27.** Spectra of the SN Ia 1981b are compared with those of the SN Ib 1984L near maximum and then about 20 days after maximum.

is that the blue ends of the SN Ia spectra are dominated by features of Fe II several weeks after maximum and some of these Fe II features are already manifested at maximum light in SN Ib. This is due to the fact that the photospheres in the SN Ib are somewhat cooler at maximum than those of SN Ia, consistent with the redder colours of the SN Ib events (section 3.2.1). While the blue spectra of SN Ia three weeks past maximum are probably due essentially entirely to Fe II, even the qualitatively similar features in the SN Ib are shaped by the added presence of He, Mg, and other species. The absorption minimum at ~5800 Å in the 27 March spectrum of SN 1981B has a qualitative similarity to the 5700 Å minimum in SN 1983N and 1984L near maximum, but we now see that the latter is probably primarily due to He I $\lambda$5876 and the former to Fe II or perhaps Na I (Branch *et al* 1983), so the resemblance is mostly a coincidence. The spectra are still quite different at wavelengths greater than 6000 Å. As the spectra evolve, the differences between SN Ia and SN Ib get more pronounced, even in the blue. The temporary similarity in the blue is thus not grounds for presuming a physical link between SN Ia and SN Ib events beyond the ubiquitous existence of Fe II which is a strong feature of SN II at appropriate phases as well.

*Type Ic.* The similarity of the spectrum of SN 1983I 5 days after discovery to that of SN 1983V 5 and 13 days after discovery in the common wavelength range (4700–6500 Å), as shown in figure 28, argues that these two events had a similar physical basis and that these spectra were obtained at roughly comparable phase. Figure 28 also illustrates the distinct absence of both the strong 6150 and 5700 Å features, respectively, in the spectra of SN 1983I and V. The spectra of SN 1983V on 30 November and 8 December are from about maximum and a week after maximum, respectively, and the spectrum of SN 1983I is from about two weeks after maximum (section 3.2.1). These spectra may show weak evidence for He I $\lambda$5876, and hence have a spectral connection to SN Ib, but Ca II, Fe II and O I are the only clear line identifications. This makes these spectra distinctly different from post-maximum spectra of SN Ib, since the helium lines grow significantly in strength in SN Ib events, as illustrated in figure 27. It is not clear which type, helium-rich SN Ib or helium-poor SN Ic, is in the majority from the present poor statistics. Over the last few years, they have been discovered in roughly equal numbers (section 1).

The presumption that SN Ib and Ic are associated with massive stars suggests that carbon and oxygen represent the bulk of the non-helium matter. The specific manifestation of oxygen in the calculated spectra (see section 3.3.3 below) is the presence of O I $\lambda$7773. This feature is predicted to occur in the helium-rich SN 1983N and SN 1984L (see section 3.3.3), but observations at the appropriate wavelength tend to be contaminated by terrestrial lines. More recent examples of SN Ic, SN 1987M and SN 1988L both clearly show the O I line (Filippenko 1988). SN 1988L also shows a more distinct absorption at about 5700 Å than other SN Ic. This could be stronger He I $\lambda$5876, but there is no evidence for stronger He I 6678 or 7026, so it probably represents enhanced Na I. Filippenko (1988) has suggested that the bump at 6600 Å surmounted by the H$\alpha$ emission from the host H II region in SN 1983I (figure 28) and SN 1988L could be H$\alpha$ in the supernova which would imply some generic link to SN II. Figure 28 also gives a spectrum of SN 1987K which bears an interesting similarity to those of the SN Ic events with the addition of the H$\alpha$ feature but without the O I $\lambda$7773 absorption (see section 3.3.2).

SN 1987M is, to date, the only SN Ic event that has been followed in its spectral evolution from the early to late phases (Filippenko 1988b). The spectrum shows a

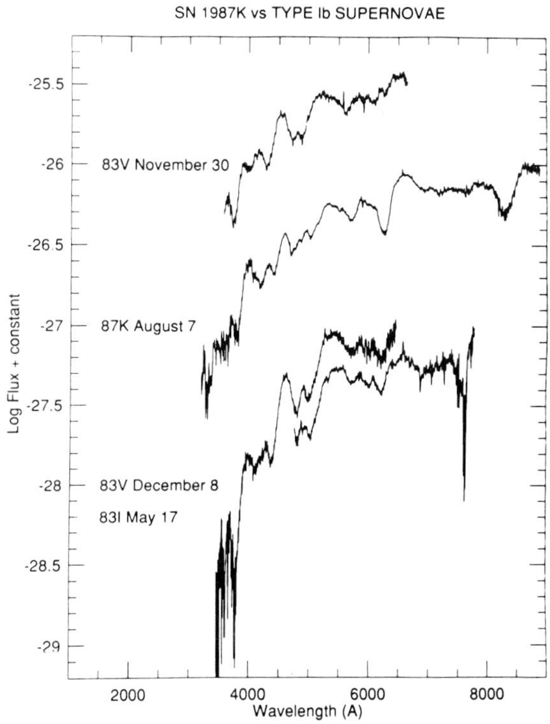


**Figure 28.** Spectra of SN Ic events 1983I and V from near maximum (top), and one and two weeks after maximum (bottom pair) are compared. Also shown is the spectrum of SN 1987K about a week after maximum light (see figure 29) which shows distinct similarities to the SN Ic except for the feature at 6200 Å attributed to $H_\alpha$.

definite evolution as the atmosphere cools and Fe II lines become more prominent (figure 29). SN 1987M was discovered on 21 September 1987 and the first spectrum was obtained by Filippenko on 28 September. This spectrum is a model for the type of data needed to analyse supernovae. It covers the wavelength range from 3100 to 10 000 Å. This spectrum resembles those of the SN Ic events SN 1983I on 17 May, 5 days after discovery, and SN 1983V from 8 December, 13 days after discovery, thus establishing this event as a SN Ic. This first spectrum of SN 1987M is also special because it shows a small minimum at about 3200 Å (figure 29) which results from Co II in atmosphere models of SN Ia events, and for which there is some evidence in the SN Ia events SN 1981B and SN 1989B. Confirmation of this feature and identification of it in the spectrum would be an important clue to the nature of the explosion. It could imply some form of thermonuclear explosion as adopted for SN Ia events, or it could mean that Ni produced by a shock wave emerging from core collapse is mixed by some as yet unknown process into the outer core, as is envisaged in SN 1987A to account for the early release of x- and gamma rays (Arnett *et al* 1989) and the late-time spectrum (Swartz *et al* 1989).

On closer inspection, the McDonald spectra of SN 1983I and V can be discriminated from that of SN 1987M on 28 September by the presence of the Fe II feature that

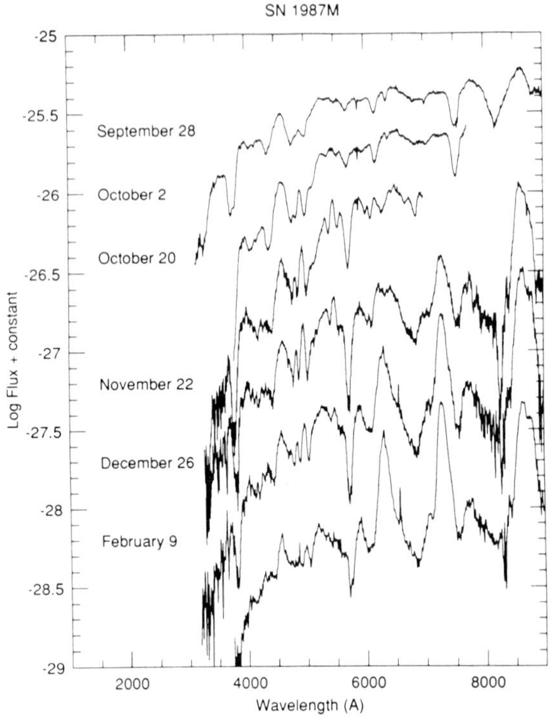


**Figure 29.** Time series of spectra of SN Ic 1987M showing the lack of evidence for helium in the first few months and the nebular phase spectrum which is very similar to that of SN Ib events. In the early phases the spectra evolve in an identifiable fashion unique to SN Ic events. (Courtesy A Filippenko, private communication.)

begins to come in as a peak at about 4900 Å which is indistinct in the 28 September spectrum of SN 1987M, moderate on 2 October, as it is in the McDonald spectra of SN 1983I and V, and relatively strong in the 20 October spectrum of SN 1987M. The spectra of SN 1983I and V clearly resemble one another and the spectra of SN 1987M on 2 October more than they do the next spectrum of SN 1987M obtained on 20 October which has developed a cluster of narrow Fe II lines around 5500 Å. From optical and IR photometry (section 3.2.1), the phase of the spectrum of SN 1983I is about $15 \pm 4$ days after V maximum and that of SN 1983V about $7 \pm 4$ days after V maximum. The spectral agreement suggests that 2 October 1988 was 7–15 days after V maximum for SN 1987M, which would have been in the range 17–25 September and hence that it was discovered at about maximum V magnitude.

By similar arguments, one may conclude that the spectrum of the SN Ic 1988L from 11 May 1988 given in Filippenko (1988) most closely resembles the spectrum of SN 1987M on 20 October. At earlier epochs in SN 1987M, the absorption at 5700 Å is indistinct, but by 20 October it is fairly strong, and probably due to Na, as remarked above for SN 1988L. The agreement of the spectrum of SN 1988L from 11 May 1988 with the 20 October 1987 spectrum of SN 1987M suggests that the spectrum of SN

1988L was obtained about 30 days after V maximum which would thus have been on about 11 April 1988.

It has long been known that the spectral evolution of SN Ia is sufficiently systematic that the phase can be determined rather accurately from a single spectrum. Although it was not obvious in the first meagre data, this discussion shows that the same seems to be true for SN Ic. Once again this emphasizes that SN Ic are distinctly and definably different from both SN Ia and SN Ib near maximum light.

This attempt at setting the phases of these SN Ic events was predicated on the assumption that SN 1987M represents a template for the other events. There are some interesting differences in the spectra, however, that give caution until they are better understood. Although the spectrum of SN 1988L from 11 May 1988 resembles that of SN 1987M on 20 October 1987, it has a distinct minimum between 4200 and 4400 Å adjacent to a peak at 4000 Å, presumably the P-Cygni peak of Ca H and K. SN 1987M has a secondary peak in this wavelength range. The origin of this difference is not immediately apparent. This secondary peak is present in the spectrum of SN 1983V on 8 December, but anonymous spectra of SN 1983V on 16, 22 and 30 December show that there is a more pronounced minimum in this wavelength range. It may thus be that in this detail SN 1983V evolves more like SN 1988L than SN 1987M. If SN 1983V was at V maximum on about 1 December 1983, then it does not show this dip 7 days after maximum V, but does show it about 15, 21 and 29 days after maximum. This is consistent with the existence of this dip in the spectrum of SN 1988L at about 30 days after maximum. The McDonald spectrum of SN 1983I goes insufficiently far to the blue to show this feature.

In addition to the narrow well defined Fe II feature that develops in SN 1987M at 4900 Å which we remarked on above, it also develops a somewhat smaller narrow feature, probably also Fe II, at 4800 Å. This feature is not present in the 28 September spectrum; it is beginning to show in the 2 October spectrum, and it is moderately pronounced in the 20 October spectrum when other Fe II features are strong throughout the spectrum. It is not clear whether this feature is present in the 11 May 1988 spectrum of SN 1988L because of noise in the data. The same is true of the spectrum of SN 1983I on 17 May 1983. The spectrum of SN 1983V on 27 November shows only one peak in this wavelength range, and the spectrum is similar, within the uncertainty of the noise, to that of SN 1987M on 28 September 1987. The spectrum of SN 1983V from 8 December shows only a single, moderately well defined peak at about 4900 Å. The spectra on 16, 22 and 30 December continue to show only a single peak at 4900 Å. None of the spectra of SN 1983V shows the cluster of Fe II lines at about 5500 Å that characterize the 20 October spectrum of SN 1987M and the 11 May spectrum of SN 1988L.

SN 1983V is of particular interest because the spectrum obtained by Axon *et al* on 30 November, a few days before maximum light, seems to show significantly higher velocities than the spectrum obtained at McDonald eight days later (figure 28). Branch (1988a) argues that the photospheric velocity at the earlier epoch is about 18 000 km $s^{-1}$. Wheeler *et al* (1987a) obtain a photospheric velocity of about 7500 km $s^{-1}$ for the spectrum of 8 December. This apparent deceleration of the photosphere corresponding to about 1200 km $s^{-1}$ $day^{-1}$ is reminiscent of, but considerably larger than, that observed in SN 1983N (Richtler and Sadler 1983, section 3.3.1 of this paper). The high velocity has been used to suggest (Branch 1988a) that the progenitor must be compact like a white dwarf, and not the core of a massive star. SN 1987A and models thereof have illustrated, however, that the progenitor only has to be as compact as a blue supergiant to eject a detectable amount of matter at rather high velocities. The photosphere can recede rapidly through this particularly high velocity matter yielding only moderate Doppler shift in the spectra after a few days. SN 1987A showed an apparent deceleration of about 700 km $s^{-1}$ $day^{-1}$ in the first ten days after discovery, and then about 200 km $s^{-1}$ $day^{-1}$ for the next 20 days. The suggestion is that any progenitor with an initial structure approximated by a polytrope of $n = 3$, a white dwarf, a Wolf–Rayet star, or a blue supergiant, may give a brief phase of particularly high velocity. The exact evolution of the spectrum should be a powerful probe of the original and ejected density distribution and clearly needs to be studied carefully in the case of SN 1983V. It remains true that the apparent deceleration in SN 1983V is abnormally large. A tantalizing prospect is that this velocity evolution is related to the fact that SN 1983V is a SN Ic, and hence a further clue that SN Ic are physically distinct from SN Ib. Since the November spectrum was obtained under poor conditions and is perhaps photometrically unreliable, there is some question as to whether the velocity calibration could also be in error. Both the original reduction and the independent reduction by Graham show the same velocity, however, and the supernova features are definitely shifted with respect to the night sky lines by the appropriate amount. There is thus no obvious means by which to impeach the deduced photospheric velocity.

It is very important, however, to attempt to verify this velocity evolution and other features of the spectral evolution in SN 1983V. Other spectroscopic data exist on SN 1983V and we are trying to assemble them to initiate a more detailed study of the explosion. P O Lindblad (private communication) has a spectrum obtained at nearly the same time as that of Cannon, and M A Dopita (private communication) has one obtained a day or so earlier. It would also be beneficial to obtain the digitized data of the early spectral phases of SN 1983N, and to model and compare the evolution of the early time spectra of these SN Ib and SN Ic events to seek clues to the nature of the progenitor structure.

The Fe II features at about 5500 Å in SN 1987M are supplanted in the 22 November spectrum by an emission feature at about 5600 Å. This feature which is present about 60 days after V maximum is probably [O I] $\lambda 5577$, indicating the onset of the nebular phase. There is no discernible sign of this feature in the 20 October spectrum of SN 1987M, about 30 days after maximum. The same feature appears in the spectra of SN 1984L about 20 days after maximum and it is prominent at 60 days. This feature appears in the spectra of SN 1983V by 22 December, about 20 days after maximum.

SN Ic cannot simply be placed in a single spectral category of 'peculiar Type I supernovae' along with SN 1983N and SN 1984L. The question is whether SN Ic are a variation of the spectral type represented by SN Ib, or are they the prototypes for a separate spectral classification that perhaps signifies an appreciably different physical basis? The spectra of SN Ib and Ic are clearly a fertile area, calling for careful physical modelling.

### *3.3.2. Observed late-time spectra*

*Type Ib.* Spectra of SN 1983N obtained 8 months after maximum light (Gaskell *et al* 1986) and of SN 1984L 14 months after maximum (Schlegel and Kirshner 1989) are very different than spectra near maximum light, and very different than those of SN Ia at a comparable phase. Figure 30 gives the nebular spectrum of SN Ia 1981B about 4 months after maximum when it is dominated by [Fe II], [Fe III] and [Co III]. Figure 30 also shows the nebular spectrum of the first SN Ib for which spectra were obtained

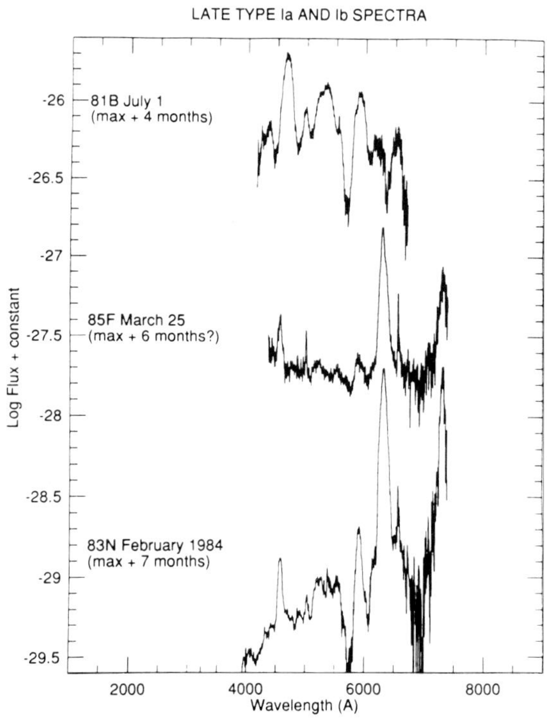


**Figure 30.** Late-time 'supernebular' spectra are shown for SN Ia 1981B, SN Ib 1983N, and probable SN Ib 1985F.

in that phase, SN 1983N. As illustrated, the later spectra of SN 1983N display strong broad lines of [O I] $\lambda\lambda$6300, 6364, Mg I] $\lambda$4571, Na I D $\lambda\lambda$5890, 5896 and [Ca II] $\lambda\lambda$7291, 7324.

Figure 30 also presents the spectrum of SN 1985F (Filippenko and Sargent 1985, 1986). Although only discovered of order 8 months after maximum, spectra of SN 1985F are very similar to late-time spectra of SN 1983N and SN1984L. There is thus a strong presumption that SN 1985F was a SN Ib in the 'supernebular' phase (cf Wheeler *et al* 1986, Gaskell *et al* 1986, Porter and Filippenko 1987), but with no spectra at early times, the question of whether SN 1985F was a SN Ic or yet something else cannot be ruled out. The relatively slow decline of the optical light curve, comparable to that of SN 1984L, suggests that the SN Ib identification is correct but, as discussed in section 3.2, the light curves of SN I b and c are not physically well understood as yet.

*Type Ic.* As we have seen, SN Ic events are easily distinguishable from SN Ib events for the first month or so because they never show the strong absorption of He I $\lambda$5876. Harkness *et al* (1987) hypothesize that this is because SN Ic events are helium poor, and hence probably carbon and oxygen rich compared with SN Ib events. Given the evidence from the dispersion in light curve speeds, the possible fast velocity evolution of SN 1983V, and this spectral difference at early times, it is possible that SN Ib and SN Ic events could represent rather different physical processes which could lead to very different late-time spectra, as is the case for SN Ia and SN Ib.

Late-time SN Ib events are characterized by strong oxygen and calcium emission lines. Emission lines of [O I] $\lambda$5577 began to be apparent in SN 1983N and 1984L from 10 to 20 days after maximum light, and [O I] $\lambda\lambda$6300, 6364 is seen 60 days after maximum in SN 1984L. No such emission lines are apparent in the spectra of SN 1983I and V 7–15 days after maximum, but [O I] $\lambda$5577 begins to be apparent in SN 1983V about 20 days and in SN 1987M about 60 days after maximum. Thus while this spectral evolution is not incommensurate with SN Ib events, there is no particular evidence that SN Ic are especially oxygen-rich. If the interpretation of SN Ib as the cores of massive stars is correct, then perhaps either a greater portion of the non-helium abundance is C rather than O, or the structure is sufficiently different to suppress the oxygen forbidden line emission.

In any case it is very important to obtain spectral data on more extreme late-time epochs to see the nature of the spectra of SN Ic at this phase. Unfortunately, there is no late-time spectral data on SN 1983I and V to ascertain whether or not they made the transition to the oxygen-rich supernebular phase that so characterizes SN 1983N, SN 1984L and SN 1985F. SN 1987M is the only SN Ic to have been followed systematically into the late-time nebular phase (figure 29). Filippenko (1988b) has shown convincingly that there are no major differences between SN Ib and SN Ic in the later phases exceeding about three months after maximum light. In SN 1987M, [O I] $\lambda\lambda$6300, 6364 begins to be visible at about 60 days after maximum and is very strong by about 90 days after maximum. The lines of [Ca II] $\lambda\lambda$7291, 7324 and the Ca II infrared triplet are also strong in emission with detectable contributions from Mg I] $\lambda$4571, Na I D $\lambda\lambda$5890, 5896, and O I $\lambda$7773. All these features also characterize SN Ib in the nebular phase. The emission line of [Ca II] $\lambda\lambda$7291, 7324 is moderately stronger than that of [O I] $\lambda\lambda$6300, 6364 in SN 1987M, whereas they were roughly equal in SN 1983N, and the [O I] line was moderately stronger in SN 1985F. It is not clear whether this characteristic is typical of SN Ic events. The data for SN 1987M supports the notion that SN Ic are closely related to SN Ib events.

*SN 1987K.* Filippenko (1988a) has challenged the standard classification scheme of supernovae with his data on SN 1987K. As illustrated in figure 31, SN 1987K showed a strong H$\alpha$ P-Cygni feature at around 6300 Å, so the event was identified as a SN II. In the first couple of spectra obtained, on 31 July and 1 August 1987, presumably near maximum light, an absorption feature at 4700 could be H$\beta$. These spectra are very similar to a SN Ic spectrum in the sense that there is no strong He I $\lambda$5876 nor Si II $\lambda$6347, but with an additional strong line at 6300 Å and minus the O I $\lambda$7773 line in absorption which appears in SN 1983I, SN 1987M and SN 1988L (see figure 28). Four months later SN 1987K was in the nebular phase, and there was no evidence of H. The late-time spectra are very similar to those of SN Ib or Ic events. The emission line of [Ca II] $\lambda\lambda$7291, 7324 is moderately stronger than that of [O I] $\lambda\lambda$6300, 6364, as it was in SN 1987M.

This transition between a SN II with a family resemblance to SN Ic near maximum and then a late-time spectrum indistinguishable from a SN Ib or Ic also tends to support the notion that SN Ib and Ic are more closely related to SN II than to SN Ia. The suggestion has been made (Filippenko 1988) that SN 1987K had only a small mass hydrogen envelope which contributed to the formation of the spectrum near maximum

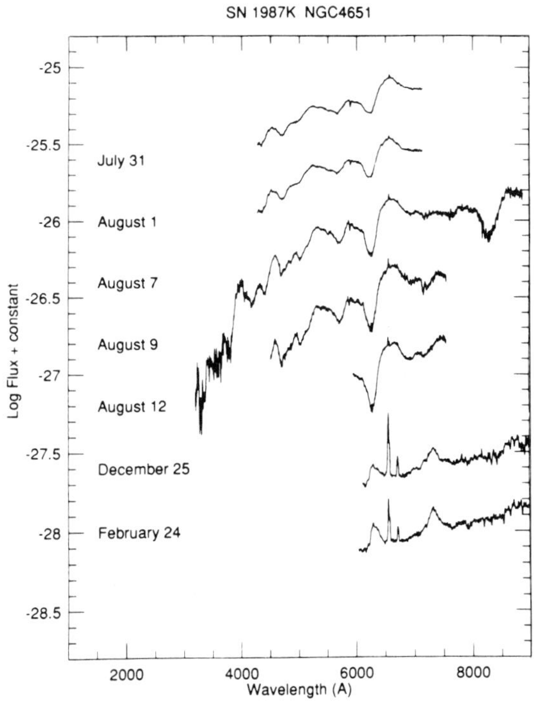


**Figure 31.** The time evolution of spectra is given for SN 1987K (Filippenko 1988).

light, but was too dilute to contribute later, a hypothesis that will be explored in section 3.3.5. The behaviour of SN 1987K thus puts yet another slant on the question of the spectral behaviour and type of SN 1985F near maximum light. With no spectra near maximum light, it is not possible to say that the spectrum of SN 1985F near maximum did not resemble SN 1987K. The fragmentary light curve of SN 1987K was not distinguishable from that of SN 1983N, so the light curve does not allow one to easily identify this class of event.

*3.3.3. Theoretical early-time spectra.* Model atmospheres for SN Ib and Ic events based on evolutionary models have not yet been computed. Synthetic spectra have been computed by Branch and Venkatakrishna (1986), Branch and Nomoto (1986) and Branch (1988a). These calculations were aimed at exploring possible white dwarf progenitors and will be explored below (section 3.3.5). Model atmospheres based on power law density profiles with uniform compositions rich in helium, carbon, and oxygen but with solar abundances of heavier elements have been presented for SN Ib by Harkness *et al* (1987) and for SN Ic by Wheeler *et al* (1987). These models assume that the density scales like $\rho \propto r^{-n}$ with $v \propto r$. The radii are chosen to correspond to expansion of the base of the atmosphere for 20 days with a fiducial velocity of 5000 km s$^{-1}$, giving $r_{\rm inner} = 8.6 \times 10^{14}$ cm. The outer radius was chosen to correspond to a density $10^5$ times less than that at the inner boundary. The temperature was chosen to be $T_{\rm inner} = 11\ 700$ K and to decrease in the atmosphere according to $T \propto r^{-1/2}$, as if the luminosity were constant in a stationary atmosphere. This fixed temperature distribution was used to isolate the effects of variation in composition and *ad hoc* deviations from Boltzmann excitation equilibrium. The luminosity varies with the opacity and hence the composition, but is $\sim 2.5 \times 10^{42}$ ergs s$^{-1}$. The total mass of the models is formally very large, $\sim 30\ M_\odot$ but more realistic models would have flatter density profiles at smaller radii, and hence less mass. Two homogeneous composition mixes were considered, a 'helium-rich' mix and an 'oxygen-rich' mix.

These atmosphere calculations reveal that even with a large helium abundance ($\sim$90 per cent) the helium lines are not as strong as observed. At the temperatures and densities typical of supernovae near maximum light, it proves to be exceptionally difficult to excite any of the visible lines of helium. In LTE atmospheres for main sequence stars, He I lines are only visible in B stars where the matter is very hot and the helium is already appreciably ionized. In SN Ib, the temperature is simply too low to account for the observed spectral lines of He I if the level populations are in LTE. The helium lines can be made apparent, however, by scaling the Boltzmann level populations by an *ad hoc* departure coefficient, *b*. In these exploratory calculations, *b* is taken to be the same for all levels and independent of radius.

*Type Ib.* Figure 32 presents theoretical spectra with departure coefficients $b = 1$, $10^2$, $10^3$ and $10^4$ for 'helium-rich' models with $X_{\rm He} = 0.9$, $X_{\rm C} = 0.01$, and $X_{\rm O} = 0.09$ and a density profile with $\rho \propto r^{-7}$. Figure 33 gives the model with $b = 10^2$ along with maximum light spectrum of SN 1984L and an identical model with $b = 10^4$ in comparison with the three-week post maximum spectrum of SN 1984L. Atmosphere calculations for different density profiles, $\rho \propto r^{-5}$, $r^{-10}$, gave much less satisfactory results. The shallower density profile puts more mass at large radii where the temperature ($\propto r^{-1/2}$) is low. This results in strong lines of Fe II and other species being formed at high velocity and the resulting spectral features are far too broad. The steeper profile produces an atmosphere which is more compact. The bulk of the matter is then at lower velocities, but the lines appear to be too narrow. These constraints on the density profile depend, of course, on the adopted temperature distribution.

For the 'helium' atmosphere a factor of $b = 100$ suffices to match the observed line strengths near maximum light. Although the requisite departure coefficient to match the observed spectra decreases as the He/O ratio is increased, suggesting some preference for a larger helium abundance, a pure helium composition can be ruled out. The ionization structure depends on the free electron density, and hence on the composition, for a given density and temperature structure. For the conditions assumed, a pure helium atmosphere would be essentially neutral and optically thin to electron scattering. The resulting spectrum would bear no relation to the observations. Some element of significantly lower ionization potential than helium, e.g. carbon or oxygen, must exist in moderate abundance to provide the requisite supply of free electrons so that the continuum is optically thick near maximum light.

Evidence for oxygen appears in the theoretical models only in the absorption feature at $\sim$7500 Å, which is due to O I 7773 Å. This is also present in some SN Ic. It is contaminated by terrestrial lines in SN 1983N and not obviously present, although at the predicted strength and breadth some evidence might be expected anyway. Calcium H and K are apparent in some of the spectra of SN 1984L, while the Ca II $\lambda$8446 triplet was observed in SN 1983N about 10 days prior to maximum (Harkness *et al*

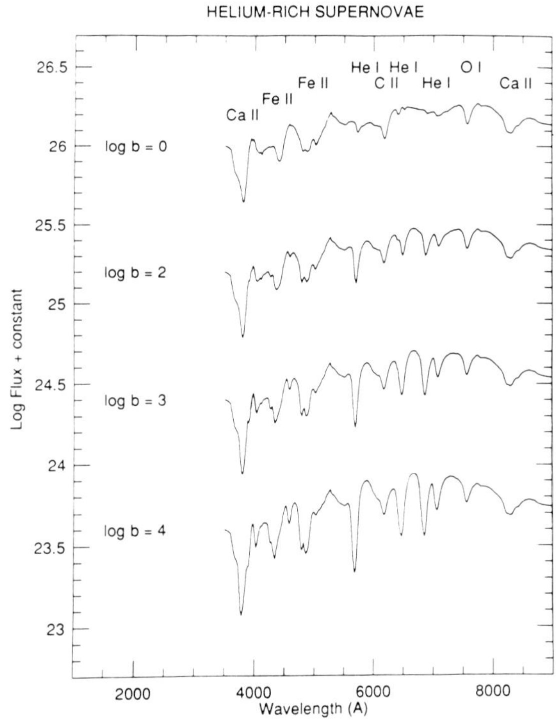


**Figure 32.** Theoretical spectra for 'helium-rich' models with $X_{He} = 0.9$, $X_C = 0.01$, and $X_O = 0.09$ and a density profile $\rho \propto r^{-7}$ are shown for departure coefficients $b = 1$, $10^2$, $10^3$ and $10^4$.

1987). Both observed features are somewhat stronger than the calcium lines in the theoretical models.

In the blue, the emergent spectrum is due mainly to blends of Fe II, Mg II with some contribution from neutral helium. Note that, unlike SN Ia which seem to require an appreciable enhancement of the iron-peak abundances in the matter visible near optical maximum (Branch *et al* 1985, Wheeler and Harkness 1986, Wheeler *et al* 1986), the spectra of SN Ib can be accounted for with essentially solar abundances of elements heavier than oxygen. This is apparently due to a combination of factors, including the lower velocity of expansion and the greater total mass.

One notable feature of SN Ib which remains in doubt is the line appearing at around 6300 Å. This line is discernible in the SN 1984L spectrum of 30 August, much less so in the spectrum of 4 days later and not seen thereafter, while it is most apparent in the 19 July spectrum of SN 1983N. In all of the model spectra shown here the closest plausible line is C II $\lambda$6580. The disappearance of this feature can be understood qualitatively as a decrease in the excitation of C II, but since substantial departures from LTE level populations are required in order to explain the helium lines the above identification remains in some doubt. For instance, Ne I $\lambda$6402 could contribute if similarly enhanced by non-LTE effects.

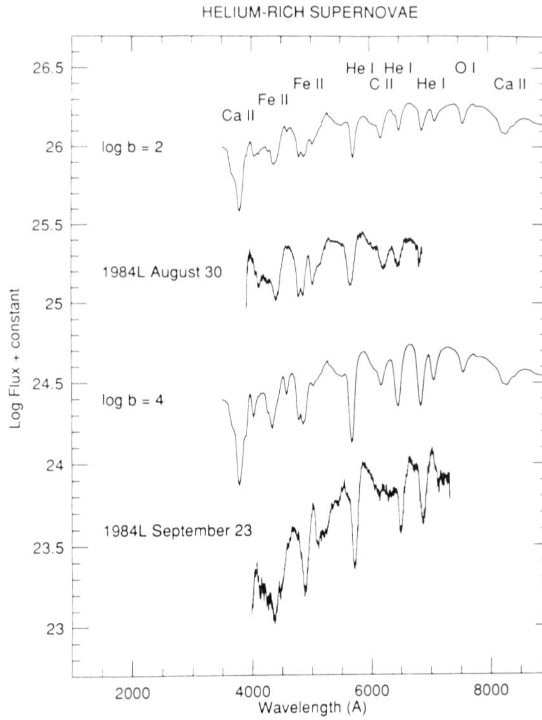


**Figure 33.** The model from figure 30 with $b = 10^2$ is shown in comparison with the maximum light spectrum of SN 1984L and the model from figure 30 with $b = 10^4$ is shown in comparison with the three-week post maximum spectrum of SN 1984L.

The simple model spectra reproduce the observations of SN Ib in the early phases remarkably well, confirming the identification of He, but raising the challenge of understanding the physical origin of the implied strong departure of He from LTE in SN Ib events. We note that He I is likely to be susceptible to departures from LTE because of its high excitation potential, ~20 eV. For instance, at a temperature of 10 000 K only one neutral He atom in $10^{10}$ is in the excited state and an enhancement in the number of excited atoms by a factor of $10^2$ or $10^4$ still leaves very few atoms in that state. In contrast, for Fe II, with an excitation of about 3 eV, ~3 per cent of the atoms would be in the excited state at 10 000 K and an enhancement of the excited population by a factor of $10^2$ would represent a major, and very unlikely, perturbation. Once a neutral helium atom is excited then only relatively low-level excitations from the 2p state are necessary to account for the observed features.

The strong suspicion is that radioactive decay plays a major role in driving the He out of LTE. Just such a process has been suggested for SN 1987A. In that event, He I $\lambda$5876 was observed in the first few days when the photosphere was hot. Two months later when the photosphere was far too cool to excite the upper levels in He, the line of He I $\lambda$10830 grew in strength. Graham (1988b) suggests that x-rays representing Compton down-scattered gamma rays (Xu *et al* 1988, Pinto and Woosley 1988) from the radioactive decay of $^{56}$Ni and $^{56}$Co could provide the necessary excitation mechanism. In this interpretation, the appearance of He I $\lambda$10830 represents the harbinger of the x-rays and gamma rays which were directly observed a month or so later and thus presents an important probe of the physical condition of the ejecta. These arguments need to be quantified for the case of SN 1987A and applied to SN Ib events.

The power-law atmosphere models discussed here are formally based on rather large masses, $\sim$30 $M_\odot$, but only the outer portions contribute significantly to the formation of spectral features since the deeper matter radiates only a thermal continuum. The atmosphere calculations suggest that there is a minimum ejecta mass of order 5–10 $M_\odot$. Below this mass the ejecta tend to turn optically thin, and the character of the spectrum changes drastically. These minimum mass estimates are at odds with those based on the light curve width. Calculations near these minimum masses are complicated by effects of the inner boundary condition. It will be important to determine more rigorous minimum ejecta masses from the atmosphere calculations based on more realistic structural models.

The properties of these SN Ib model atmospheres differ in several ways when compared to those for SN Ia. SN Ia models with iron-rich cores and O/Si envelopes are nearly optically thin near maximum light because the optical continuum opacity is due almost entirely to electron scattering. The ratio of absorption to scattering is typically 1/50 at optical frequencies, hence the true photosphere is not coincident with the surface of last scattering and a range of temperature and density conditions are reflected in the overall emergent flux. The opacity of the iron-rich core is lower than that of the mixture of intermediate mass elements in the envelope. When the latter becomes optically thin, there is a rapid transition to the iron-dominated phase, as pointed out in section 2.5.2.

In SN Ib an appreciably greater mass is involved. On the basis of the late-time observations, one expects a core consisting predominantly of oxygen and other intermediate mass elements. The envelope is presumed to contain a large mass fraction of helium, which contributes negligible continuous absorption at low temperatures. Any assumed departure from LTE, however, also enhances the bound–free opacity due to He I. When the departure from LTE is large, the absorptive opacity may rival electron scattering, giving rise to an emergent spectrum more characteristic of local conditions near the photosphere.

The atmosphere calculations confirm beyond any reasonable doubt that He exists in substantial amounts in the portions of the ejecta that are observable up to two months after maximum light in SN 1983N and SN 1984L. An important prediction that follows from this conclusion is that the 10830 Å line of He I should be a very strong infrared feature if the level populations are at all similar to those assumed here. The helium abundance probably cannot be as low as 0.1 by mass without requiring an implausibly large departure from LTE. The other principal constituent is probably oxygen, given the large oxygen abundance observed at late times and the need for a plentiful supply of free electrons beyond those available from helium alone. Spectra of SN Ib extending to 8000 Å are needed to determine the existence of the oxygen in the early phases. Carbon is also identified in the early phases and probably contributes substantially to the mass fraction of the ejecta. Iron features are seen, but the iron abundance of the matter observed near maximum light need not exceed the solar ratio. The composition used to reproduce the observed spectra, predominantly helium and oxygen, with a solar mix of heavier elements is representative of the composition of models of massive stars and hence evidence that SN Ib have their origin in the cores of massive stars.

*The ultraviolet deficit.* The observed and theoretical spectra of SN Ib are very different from those of the SN Ia for wavelengths longer than 6000 Å. As discussed by Harkness *et al* (1987), some of the qualitatively similar features in the blue have somewhat different physical origins due to wavelength coincidences between He I and Fe II, with the former contributing significantly to SN Ib. In the ultraviolet, the observed spectra of SN Ia and Ib are rather similar including both the slope of the UV continuum and the wavelength and the shape of individual features (Panagia 1985). The minimum at 3200 Å in the spectrum of SN 1981B is attributed to a blend of Co II lines in the models, as discussed in section 2.5.2. There is a hint of this feature in spectra of SN Ic 1987M. The general similarity of the UV spectra of SN 1983N and SN 1981B has been used by Branch and Venkatakrishna (1986) to argue that there may be a similarity in the physical origins of the two classes of event.

The basic model suggested here for SN Ib, rather large amounts of He, C and O as would be expected from the core of a massive star, also gives a natural explanation for the UV spectrum. Figure 34 gives the theoretical spectrum of figure 32, extended into the UV, together with the observed optical and UV spectra of SN Ib 1983N. The result is an ultraviolet spectrum that is very similar to that of the theoretical deflagration model for SN Ia (figure 14) despite the rather large differences in the underlying models and the distinct differences in the optical portion of the spectra. The theoretical spectrum

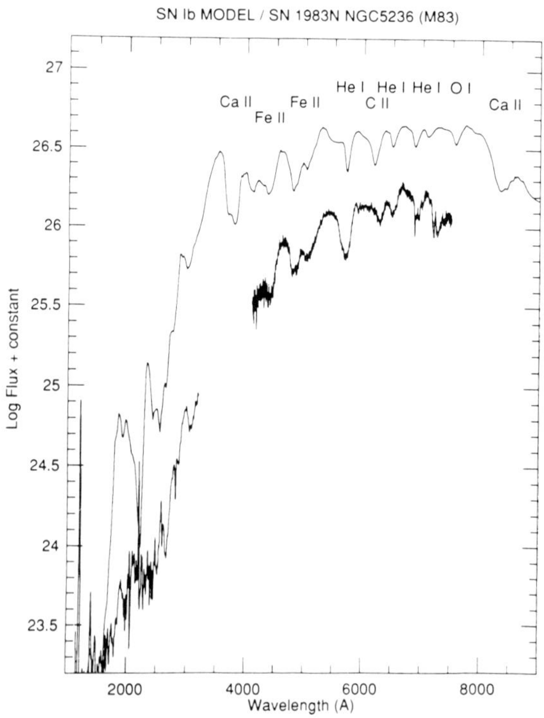


**Figure 34.** The spectrum from figure 30 with $b = 10^2$ is shown in comparison with the optical and UV spectra of SN Ib 1983N near maximum light.

also resembles observed spectra of both SN Ia and SN Ib. In particular, the SN Ib atmosphere model shows the same UV deficit due to the presence of strong Fe II lines, and the peak at about 2000 Å due to their absence. In the case of the Ib models, the Fe that gives rise to the UV features is just that due to the solar abundance of Fe in the helium-rich envelope.

*Type Ic.* Harkness *et al* (1987) and Wheeler *et al* (1987b) present atmosphere models with a larger abundance of oxygen and a smaller abundance of helium and argue that some versions of these 'oxygen-rich' models provide a reasonable match to the observed spectra of the SN Ic events SN 1983V and SN 1983I. For the 'oxygen' atmosphere a much larger value of the departure coefficient ($b \gtrsim 10^4$) is required to get any sign of helium in the spectrum. This is again due to the need to have a large contrast of line over electron scattering opacity. The electron scattering opacity is considerably enhanced in the oxygen atmosphere relative to the helium atmosphere for the same temperature and density and so, for a given mass, the oxygen atmosphere has a far greater electron scattering optical depth. Since lines are diluted by larger electron scattering opacity, the features due to C II λ6580 and O I λ7773 are not significantly stronger in the spectra with larger oxygen and carbon abundances. Although the 'helium' and 'oxygen' atmospheres and resulting spectra are similar for appropriate choices of *b*, the features fall more to the blue in the latter case. This is again a result of the larger electron scattering optical depth so that lines tend to form farther out in higher velocity material.

Figure 35 gives theoretical spectra with a composition of $X_{He} = 0.1$, $X_C = 0.1$, and $X_O = 0.8$ and departure coefficients of $b = 1$, $10^4$, $10^6$ and $10^7$. Figure 36 gives the model with $b = 10^4$ in comparison with spectra of SN 1983I and V obtained about one and two weeks after maximum, respectively. The data from SN 1983I and V are artifically combined to extend the wavelength range. The resulting spectrum is similar to the spectra of SN 1987M obtained by Filippenko (1989) on 28 September and 2 October 1987. The models did not include lines at shorter wavelength than 3000 Å, but show the absorption due to Ca H and K which appears in the observed spectra with a minimum at about 3900 Å. The rather sharp minimum at 4900 Å which is produced by the 4921 and 5015 Å lines of He I in the model for SN 1984L (figure 33) is muted in accordance with the smaller helium abundance and may be due entirely to Fe II lines at 4923 and 5018 Å. Toward the red there is a shallow absorption at 5700 Å, the remaining effect of He I λ5876 in the model, and maybe a suggestion of He I λ6678 at 6450 Å and λ7065 at 6850 Å. Filippenko's spectra of SN 1987M on 28 September and 2 October 1987 show a small minimum at about 6400 Å which might be attributed to He I λ6678, but there is no discernible evidence for He I λ7065. A theoretical spectrum similar to that shown but with $b = 1$ is basically unchanged except for the failure to show any helium line minima.

The minimum at 7500 Å, clearly shown in the Filippenko spectra of SN 1987M, and noisily in the McDonald spectra of SN 1983I is provided by O I λ7773 in the theoretical calculation. The atmosphere model also gives a fairly good representation of the rather flat sided absorption minimum at 8200 Å which is due to the calcium infrared triplet. The minimum observed at 6200 Å in SN 1983I and V and in SN 1987M is possibly associated with C II λ6580. The carbon line is present in the theoretical spectra, but is not appreciably stronger in those models with larger carbon abundance than in those for SN Ib events with smaller carbon abundance due to the higher electron scattering opacity as remarked above.

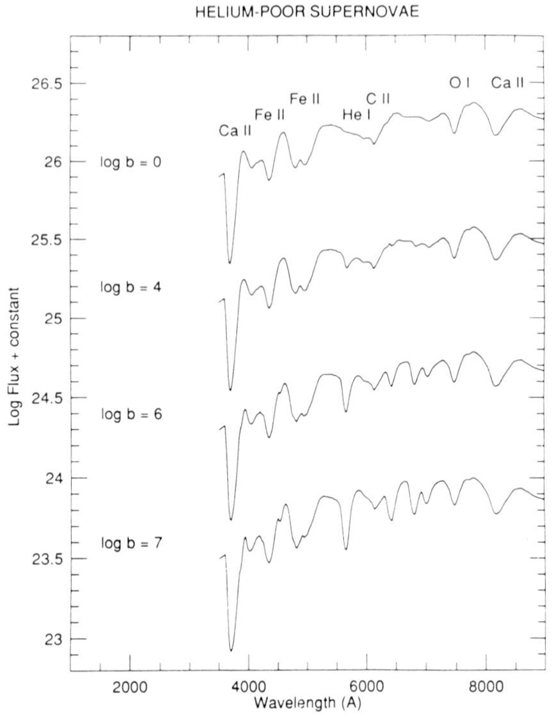


**Figure 35.** Theoretical spectra for 'helium-poor' models with $X_{He} = 0.1$, $X_C = 0.1$, and $X_O = 0.8$ and a density profile $\rho \propto r^{-7}$ are shown for departure coefficients $b = 1$, $10^4$, $10^6$, and $10^7$.

As just stated, the absorption feature at about 5700 Å in the spectra of SN Ic events is represented in the theoretical model by a weak feature due to He I $\lambda$5876. There may also be even weaker evidence for He I $\lambda$6678. By the same epoch, two weeks or so after maximum, the helium lines in SN Ib events are among the strongest features. The identification of helium in the spectra is thus considerably more marginal in SN Ic events than in SN Ib events. The helium abundance cannot be specified in either type of event without a better understanding of the departures from LTE but the calculation corresponding to figure 35 suggests that events such as SN 1983I and V are relatively helium deficient compared to SN 1983N and 1984L.

These atmosphere models coupled with the similar late-time spectra encourage the view that there may be a close generic relation between SN Ib and SN Ic events, namely a smaller He abundance in the outer envelope of the latter, despite the distinct spectral differences at early times and the possible differences in light curves. The non-LTE nature of the helium features makes a specific estimate of the helium abundance impossible at this time.

*3.3.4. Theoretical late-time spectra.* About a month after maximum light, supernovae evolve from the early optically thick phase dominated by permitted lines to an optically thin nebular phase dominated by forbidden lines. Efforts are now underway to analyse

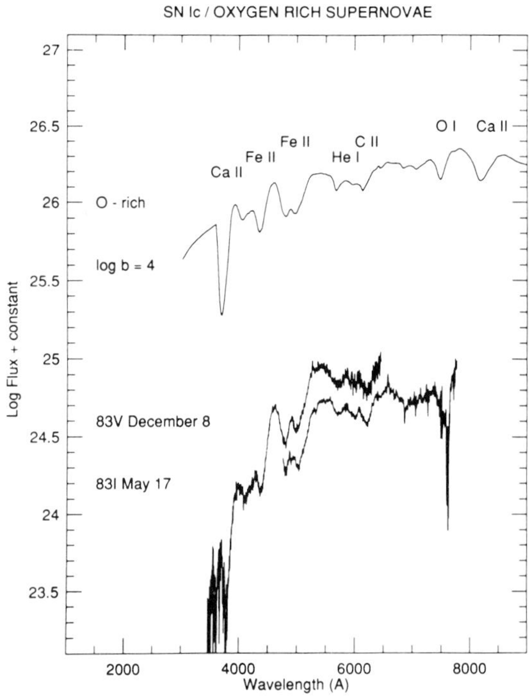


**Figure 36.** The model from figure 33 with $b = 10^4$ is shown in comparison with spectra of SN 1983V and I obtained about one and two weeks after maximum, respectively.

the supernebular phase of SN Ib. Study of this phase represents a crucial complement to attempts to reproduce the maximum light spectra because the nebula is presumed to be virtually optically thin so that much of the ejecta can be sampled at once and because the detected lines form in, and hence sample, the ejecta in different ways than the permitted lines that characterize the earlier phase.

The pioneering 'supernebular' calculations were done by Axelrod (1980; see also Woosley *et al* 1984) for SN Ia. Recently, SN Ib (and SN 1987A) have been the focus of attention. Begelman and Sarazin (1986) estimated from the line strengths of spectra of SN 1985F that that event ejected more than 5 $M_\odot$ of oxygen (depending on the distance). Pinto (1988) and Fransson and Chevalier (1989) point out that gradients in composition and deposition of radioactive decay energy are important. Pinto argues that the minimum emitting mass could be as small as 0.2 $M_\odot$ and suggests that a feature at 5900 Å may be Co, rather than Na as preliminary suggestions held. Fransson (1988) and Fransson and Chevalier (1989) show that the line profiles can be used, in principle, to deduce the density distribution of the ejecta, and stress that [O I] and [O II] may arise in differential spatial locations, introducing inhomogeneities neglected in the simplest one-zone analyses.

Fransson (1988), Axelrod (1988) and Fransson and Chevalier (1989) have discussed the physics of the formation of the late-time emission line spectrum of SN Ib. They adopt models based on the cores of massive stars. Fransson and Chevalier (1989) examine models from Woosley (1988) with helium core masses of 4, 6 and 8 $M_\odot$, corresponding to main sequence masses of 15, 20 and 25 $M_\odot$. The amount of $^{56}$Ni is 0.1, 0.2 and 0.3 $M_\odot$, respectively. They emphasize that the emission line flux from various layers is sensitive not only to the mass of the elements, but to the density structure imposed by the dynamics. Fransson and Chevalier point out that the line strengths are sensitive to such factors as the degree of intermixing of O and Ca. Both of these are prime coolants, but the Ca is more efficient so some admixture of Ca with O strengthens the Ca lines considerably at the expense of O. The possibility of intermixing, in the evolutionary or dynamical phases, makes it difficult to determine relative abundances solely from line strengths. The information contained in the line profiles is crucial to deconvolve the composition structure.

Fransson and Chevalier (1989) point out that more Ca is intermixed with the O in the models with 4 $M_\odot$ cores than in those with 8 $M_\odot$ cores. This makes the Ca/O emission line ratios much too large in the lower mass models. The higher mass cores are thus favoured in this regard, contradicting the conclusion from light curve studies (Ensman and Woosley 1988). A further problem is that the density distribution in the models does not give sufficiently wide line profiles to match the observations of SN 1985F. Fransson and Chevalier (1989) discuss the possibility that this problem is resolved by mixing instabilities which move some of the heavy elements to outer, higher velocity portions of the ejecta. They find microscopic mixing to once again excessively increase the strength of Ca at the expense of O. They point out that macroscopic mixing of clumps in radius-velocity space without homogenizing the composition microscopically might provide a better solution.

Although they address some questions of mixing in the ejecta, Fransson and Chevalier (1989) do not account for a number of other questions raised by studies of SN 1987A. In particular, they assume that the radioactive matter is confined in the innermost zones. Models of the light curves and optical, x-ray, and gamma ray spectra of SN 1987A suggest that the composition is strongly mixed and that the radioactive matter may reside rather far out in the hydrogen-rich envelope. If this has happened in a star which otherwise conformed to our expectations for the evolution and explosion of a massive star, then we must also ask whether such phenomena are within the range of possibility for SN Ib events, the progenitors of which are still debated. In addition, Fransson and Chevalier assume that the strong O and Ca lines arise in the heavy element core. Swartz *et al* (1989) show that the small abundances of these elements in the outer envelope can be sufficient to account for the strong lines in the spectrum of SN 1987A. The question then also must be addressed as to whether the strong O and Ca lines in SN Ib can arise in the small amount of O and Ca in the helium envelope suspected for SN Ib events, even if heavy elements from the core are not mixed up into the envelope. The models of Fransson and Chevalier show that although the helium mantle absorbs some energy, it is emitted as [C I] $\lambda$8727 and Ca II] $\lambda$7300, with no trace of helium in the spectrum. It would be interesting to see how this result would change with the admixture of some $^{56}$Co or other elements from the inner core. The estimate of Begelman and Sarazin (1986) that SN 1985F contained a minimum of 5 $M_\odot$ of O (see also Gaskell *et al* 1986) certainly needs to be revised with models which account for the stratification of composition and ionization structure and variations in the gamma ray deposition.

*3.3.5. White dwarf models.* Branch and Venkatakrishna (1986) have modelled the portion of the spectrum of the SN Ia 1981B between 2800 and 3400 Å with blends of

Fe II and Co II lines with $v \sim 12\,000$ km s$^{-1}$, consistent with the expected composition of an exploding white dwarf. They argue that the pre-maximum ultraviolet spectrum ($\lambda \gtrsim 3100$ Å) of the SN Ib 1983N is very similar to that of SN 1981B (Panagia 1985) and hence that the composition and the underlying cause of the explosion are likely to be closely related. They point out that a Wolf–Rayet star would not be expected to eject iron and cobalt at high velocities. This motivation led Branch and Nomoto (1986) to present a model for the explosion of SN Ib events based on the detonation of a layer of degenerate helium surrounding a degenerate C/O core. An immediate problem with this model is that, unlike observed SN Ib, the light curve of this detonation model evolves more rapidly than models of SN Ia assuming the same opacity ($k \sim 0.1$ cm$^2$ g$^{-1}$, (Woosley *et al* 1986)).

Atmosphere calculations add another argument against an exploding white dwarf model for SN Ib. Branch and Venkatakrishna argue that the similarity of the blue spectra of SN Ia and SN Ib require Co in the spectra. The combined optical and ultraviolet spectra for SN 1983N (Panagia *et al* 1986) do not, however, show strong evidence for the absorption dip at 3200 Å (primarily due to lack of relevant data) that seems to occur in SN 1981B and which is attributed to Co II (Branch *et al* 1983). On the other hand, there is a hint of this feature in the first spectrum of the SN Ic 1987M obtained by Filippenko (1988b). This does not necessarily mean that the explosion of SN Ic events is more closely related to SN Ia than to SN II. Observations of SN 1987A have strongly suggested that some form of mixing occurred (Arnett *et al* 1989 and references therein) in or after the explosion which brought radioactive matter, $^{56}$Ni and $^{56}$Co, into the outer hydrogen envelope. A similar process in the hydrogen-denuded core we envisage for SN Ib or SN Ic events could thus very well have left some Co in the oxygen and helium layers. The atmosphere models presented by Harkness *et al* (1987) show that the blue portion of the spectrum near maximum light can be reproduced by Fe II in both SN Ia and SN Ib, but with a large abundance of iron for SN Ia ($\sim$0.03 by mass) and a small abundance ($\sim$0.0018) for SN Ib. This difference in iron abundance for models of SN Ia and SN Ib shows that the blue spectra can be similar even though the underlying physical models and associated compositions are quite different. The atmosphere models for SN Ib argue that the iron mass fraction cannot be as large as 0.1 in SN Ib, a statement consistent with a massive star model in which the deep inner layers where Fe and Co may reside in large concentration have not yet been exposed, and inconsistent with the off-centre detonation model of Branch and Nomoto (1986).

The model for SN Ib suggested by Branch and Nomoto (1986) is based on the detonation of a layer of degenerate helium surrounding a degenerate C/O core. After freezeout from nuclear statistical equilibrium, the $\sim$0.4 $M_\odot$ of detonated He consists of 90 per cent $^{56}$Ni and 5 per cent He (Woosley *et al* 1986). In this model, the C/O core expands unburned, the potential origin of the oxygen enrichment in the super-nebular phase. Branch and Nomoto show that for an appropriate choice of the optical depths of Fe II and He I, a synthetic spectrum can reproduce the observed spectrum of SN 1984L (30 August). This calculation does not, however, specifically constrain the abundances of Fe and He.

The atmosphere models may be more nearly physically self-consistent in that, while a departure coefficient is assumed for He, the effect of Fe is computed rigorously for the assumed structure using LTE level populations. The two techniques each make an assumption for the temperature distribution. The synthesis technique assumes a constant ionization and excitation temperature above the photosphere which is independent of the blackbody temperature assumed for the photospheric temperature, whereas the atmosphere calculations for SN Ib and Ic events assumed $T \propto r^{-1/2}$. Figures 32 and 33 show that the blue end of the spectrum is adequately represented by Fe in only solar proportions. A spectrum was calculated for a model consisting of the incineration of a degenerate helium white dwarf of $\sim 0.6\ M_\odot$. This model is in the spirit of the suggestion of Branch and Nomoto, although it does not contain the lagging C/O core. The composition was taken to be an appropriate mix of Ni, Co and Fe, corresponding to the epoch of maximum light. The expected small helium abundance after freeze-out was omitted. The spectrum fails to resemble either a SN Ia or Ib in any significant way. At even 10 per cent by mass, the Fe II lines are far too strong not only in the blue, but throughout the optical portion of the spectrum (Harkness *et al* 1987). These atmosphere calculations seem to be substantial evidence that models based on detonations in white dwarfs, which demand a large abundance of Fe and a large Fe/He ratio, are not viable models for the origin of SN Ib. Even if non-LTE effects altered the iron spectrum in a major way, the helium in such detonation models is at too high a velocity to correspond to the observed spectra.

Fransson (1988) has also objected to the lagging-core model of Branch and Nomoto on the grounds that the oxygen contained therein cannot be adequately excited by the radioactive decay from the outer rapidly expanding incinerated layers. Fransson and Chevalier (1989) have explored other aspects of the late-time emissivity of such models, with generally negative results.

Sramek *et al* (1984) and Chevalier (1984) discuss the implications of the radio observations of SN 1983N for the progenitor. They note that the optical light curve precludes an explosion within a supergiant envelope and propose a symbiotic-type binary system with the supernova occurring in a white dwarf which is in turn immersed in the circumstellar nebula shed by an asymptotic giant branch companion. The objections to a white dwarf progenitor for SN Ib just presented apply to this version as well. Furthermore, the lack of evidence for hydrogen in either the early or later supernebular phases may constrain this picture. The models for radio emission determine that $\dot{M}/v_w \sim 3\times10^{-7}\ M_\odot\ \mathrm{yr}^{-1}/\mathrm{km\ s}^{-1}$ where $\dot{M}$ is the mass loss rate and $v_w$ the wind velocity. The symbiotic star model is based on the assumption that $v_w \sim 10\ \mathrm{km\ s}^{-1}$ and hence $\dot{M} \sim 3\times10^{-6}\ M_\odot\ \mathrm{yr}^{-1}$, consistent with a red supergiant. Alternatively, the radio emission might be explained with $\dot{M} \sim 3\times10^{-4}\ M_\odot\ \mathrm{yr}^{-1}$ and $v_w \sim 10^4\ \mathrm{km\ s}^{-1}$. This phase may have been brief, and does not preclude an earlier phase of slower mass loss, but could still create a high density fast wind sufficient to create the observed radio emission. The suggestion of a helium-rich nebula from the early spectra of SN 1983N by Richtler and Sadler (1983) encouraged the notion that SN 1983N was not simply surrounded by a circumstellar nebula that originated in a red giant wind alone. Chevalier assumed a density profile in the ejecta of $\rho \propto r^{-7}$, patterned after an exploding white dwarf. The present spectral calculations give some independent justification for such a choice of density profile. This does not necessarily imply a white dwarf progenitor, however. The $n=3$ polytrope structure shared by a white dwarf and a radiation-pressure dominated core are expected to give rise to similar density profiles in the ejecta. The preponderance of the evidence still seems to suggest that there is no commanding reason to adopt a white dwarf model for either SN Ib or SN Ic events, and that there are some arguments against such models.

*3.3.6. SN 1987K.* It remains an interesting question to determine where SN 1987K fits in the classification and physical scheme of supernovae. It was apparently a Type II near maximum based on the evidence for H in the spectrum. It was similar to a Type I b (or c) at later times and definitely neither a SN II nor a SN Ia. The late-time spectrum suggests an origin in a massive star, given the preponderance of circumstantial evidence that SN Ib events arise in the cores of massive stars. The primary evidence for that is the association with H II regions and the mass of the ejecta as determined from the light curve to be several solar masses (Wheeler and Levreault 1985, Schaeffer *et al* 1987, Ensman and Woosley 1988, Nomoto 1988a).

The appearance of H near maximum light and its subsequent disappearance at later times could require that the amount of hydrogen was small and that it became too dispersed to be detectable at later times. Both SN II-L and SN II-P show appreciable evidence for H in their late phases, so they must be substantially different than SN 1987K. The question of the appearance and then disappearance of H must be approached carefully. Studies of SN 1987A have shown that to see the hydrogen at late times there must be substantial radioactive heating by direct mixing of $^{56}$Co into the envelope or by inhomogeneities in the envelope which allow greater deposition of gamma rays from the interior than given by spherically symmetric models (Swartz *et al* 1989). The failure to detect hydrogen may then be determined by the deposition of energy in the envelope, and not only by the amount of H.

To explore such a question, and the related issue of the minimum mass allowed in SN Ib atmospheres, we have constructed a model with a hydrogen envelope of $\sim 0.1\ M_\odot$ on top of a core of about 5.3 $M_\odot$ of helium and heavier elements. This model is similar to those put forward for SN 1987A in which strong mass loss from a star of $\sim 20\ M_\odot$ leaves a small amount of hydrogen at the time of the explosion (Wood and Faulkner 1987). Such models do not work for SN 1987A because they give insufficient tamping of the inner core, but they are not astrophysically unreasonable. The inner 1.4 $M_\odot$ is assumed to remain behind as a neutron star, and the outer 4 $M_\odot$ is ejected by imposing a high velocity in the zones beyond the core to simulate a shock of chosen energy of $\sim 10^{51}$ ergs. The outer radius of the star was $\sim 3\times 10^{12}$ cm, appropriate to SN 1987A, but not necessarily to SN 1987K or to a SN Ib.

The light curve of this model was calculated for various values of the energy. Some models included a careful treatment of the recombination, although this did not seem to have a significant qualitative effect. A model with $2\times 10^{51}$ ergs fits the peak of the bolometric light curve of SN 1983N fairly well and the light curve of SN 1987K (Filippenko 1988a) within the rather large uncertainty. A model with an energy of $4\times 10^{51}$ ergs gives too narrow a light curve, and one with $10^{51}$ ergs, a light curve that is too broad. These conclusions are similar to those of Schaeffer *et al* (1987), Ensman and Woosley (1988) and Nomoto *et al* (1988a) for models of SN Ib which omit the thin outer layer of H.

The spectrum was computed at 20 days, corresponding to the epoch of maximum light in the theoretical light curve. The scattering photosphere is in the oxygen-rich layers at the base of the helium layer. The Balmer series is completely obscured in the theoretical spectrum by strong metal lines formed by the solar abundance of heavy elements in the hydrogen and helium layers. Homogenizing the composition and thus mixing the hydrogen with the heavier elements from the inner core resulted in little change. The conditions are too cool and too low in density and the resulting spectrum bears little resemblance to that observed.

Thus the most obvious explanation for the SN 1987K phenomenon, a thin outer envelope of hydrogen, does not work in its first manifestation. Based on these preliminary results, it seems that SN 1987K must have ejected a substantial amount of hydrogen. In this case the task of hiding the hydrogen at later epochs may fall to an appropriate distribution of the radioactive species. Why this event should differ in this regard from other SN II events is not clear. Only after the maximum light spectrum is reproduced will we have a better view of where SN 1987K fits into the classification scheme of supernovae. The related question of how much hydrogen could be present in SN Ia and SN Ib events is still open, and very important.

### *3.4. Discussion and conclusion*

The evidence for Population I environments, circumstellar nebulae, and estimates of the ejecta composition and mass based on light curves and the early photospheric and later supernebular phase spectra all imply that SN Ib and SN Ic have their origin in the hydrogen-denuded cores of massive stars. If one accepts a massive star origin for these events, then there are many interesting questions concerning their origin which may be posed.

Panagia and Laidler (1989) argue that SN Ib and Ic events are less tightly correlated with H II regions than SN II (Huang 1987) and hence that SN Ib, c must come, on average, from less massive progenitors than SN II. Van den Bergh (1988) argues that oxygen-rich, hydrogen-poor Cas A-like supernova remnants are probably remnants of SN Ib events and that these remnants share the same correlation with H II regions as O5 to O9.5 stars of mass $\gtrsim 16\ M_\odot$, suggesting SN Ib come from more massive progenitors than SN II. In practice, neither of these arguments is very rigorous in the face of uncertainties and small number statistics. SN 1987A was only loosely correlated with 30 Doradus, but nevertheless arose in a progenitor of $\sim 20\ M_\odot$, so SN Ib events could also come from such a mass range. On the other hand, given the expectation of oxygen from SN II and the uncertainty over how much hydrogen a SN Ib or SN 1987K harbours, it may be premature to identify the oxygen-rich remnants with SN Ib. In addition, van den Bergh's argument does not account for the degree of association with the H II regions raised by Panagia and Laidler.

Although it is tempting to associate SN Ib with Wolf–Rayet star progenitors, there are problems with this suggestion. Wolf–Rayet stars as an observed class are generally too massive ($\gtrsim 10\ M_\odot$; Massey 1982) to satisfy the observational constraints of the SN Ib. On the other hand Wolf–Rayet stars seem to lose mass at a rate $\dot{M} \sim 2\times 10^{-5}\ M_\odot\ \mathrm{yr}^{-1}$, independent of spectral type (Conti 1982) and Moffat (Moffat *et al* 1986, 1990 and references therein) has proposed that this high mass loss may systematically reduce the mass to lower values at the endpoint. It is not clear that this evolution occurs since, for instance, the nitrogen-rich WN stars do not have systematically larger masses than the carbon-rich WC stars, as one might expect in the simplest evolution schemes (Massey 1982). The best evidence for a change in mass with time is derived from the decreasing mass ratio in Wolf-Rayet/O star binary systems, in which case the presence of a companion may affect the evolution. The evolution to lower masses with time hypothesized by Moffat may thus arise only in binary, not single Wolf–Rayet stars. It remains possible that a late, very heavy mass loss phase will reduce even a $10\ M_\odot$ Wolf-Rayet star to 4–6 $M_\odot$ at the time of explosion. The question then arises whether this only occurs at the expense of drastically reducing the helium abundance in violation of the helium enrichment strongly suggested by the atmosphere calculations, at least for SN Ib events. This notion is supported by theoretical calculations which suggest that stars which are sufficiently massive ($\gtrsim 30\ M_\odot$) to eject their hydrogen envelope in a wind and spontaneously become Wolf–Rayet stars are apt to have ejected their helium layers in the wind as well before the explosion (Prantzos *et al* 1986). On the other hand, recent calculations by Langer (1989) suggest that the convective core retreats as a result of mass loss during the core helium burning stage. This may leave some helium behind in the atmosphere even if the total amount of mass loss is severe. A key question is then the quantity of helium. The atmosphere calculations show that atmospheres with 10% helium by mass give virtually identical spectra to pure oxygen atmospheres and that the result resembles a SN Ic. The deduced large helium abundances still suggest that the progenitors of SN Ib are less massive than $\sim 30\ M_\odot$, although SN Ic could arise in more massive progenitors than SN Ib on this basis. On the other hand, the light curves of SN Ic events tend to decline more rapidly than those of SN Ib events, suggesting that they have less, not more, mass to trap gamma rays. The rather large rates of SN Ib and Ic events, compared with SN II, also suggest that the progenitors are less massive than $\sim 30\ M_\odot$.

If SN Ib/c are helium rich, they must have lost their hydrogen envelopes, but if they arise from progenitors with main sequence mass $\lesssim 30\ M_\odot$, then they are unlikely to have lost their envelopes spontaneously in a wind. The only reasonable remaining conclusion is that they lost their envelopes to a binary companion. The strong suggestion seems to be that SN II (perhaps only SN II-P?) arise by the evolution of moderately massive ($\sim$10–25 $M_\odot$) single stars, and that SN Ib/c occur by essentially the same evolution in binary systems where the companion can catalyse the requisite envelope loss. Van den Bergh (1988) interprets the fact that SN Ib occur about a third as often as SN II to mean they come from a higher mass range. Alternatively, this fraction could be just the fraction of all stars with the proper binary parameters to convert a potential SN II to a SN Ib.

It is not clear how one should go about searching for evidence that all SN Ib/c occur in binaries, but it is clearly important to do so. One possibility is suggested by study of the H–R diagrams of the Galaxy, LMC, and SMC. Theory predicts that there should be a gap in the distribution to the blue of the main sequence caused by the rapid evolution before massive stars ($\sim 20\ M_\odot$) ignite a hydrogen burning shell. There is some evidence for this gap in observed H–R diagrams for these three galaxies, but the gap is not as depopulated as the standard theory would indicate. A possible means to fill it is to invoke binary evolution since the addition of helium-rich matter from a companion will move the red edge of the gap to the blue, thus filling it in (Tuchman and Wheeler 1990). The stars in this gap are observed to be helium rich (Kudritzki *et al* 1989). This may mean that all the stars in the gap are helium-enriched binary companions of systems in which the originally more massive star has already exploded as a SN Ib or Ic.

It is equally important to try to identify appropriate progenitor systems. Some Wolf–Rayet stars may form from lower mass progenitors in binaries. Alternatively, hot hydrogen-deficient stars have suggestive properties, although it is not clear that there are sufficient numbers of them in the Galaxy to account for the expected rate (Uomoto 1986). Candidate progenitor systems should, at least in some cases, have reasonably dense circumstellar nebulae to account for the observed radio emission from some SN Ib events.

Stars of different mass subjected to mass loss in a wind or to a binary companion will display different compositions at the time of explosion. An array of massive star models needs to be explored with atmosphere calculations to determine more precisely the sort of structure that will produce a theoretical spectrum in agreement with observations, not just at a single epoch, but as a function of time. Such calculations will constrain not only the mass of the core, but also the degree of mass loss the progenitor star must have undergone prior to explosion. Analysis of the supernebular phase will give an independent set of constraints on the structure, composition, and kinematics.

Care must be taken in future calculations to let the observed spectra and light curves be the guide. The evidence presented for the heterogeneity in terms of the composition of SN Ib and Ic may carry over into the mass, luminosity, and kinetic energy as well. Care should thus be exercised in assuming that all SN Ib or Ic follow precisely the same optical or IR colour evolution, or that the properties of the explosion are necessarily identical. The actual structure of SN Ib and Ic may not fit preconceived notions from either observations or theory. SN 1987A and its suggestions of significant mixing are an object lesson in point. Likewise, evolutionary calculations may provide more self-consistent initial models, but given uncertainties associated with assumptions of spherical symmetry, and a general lack of knowledge of the effects of time-dependent convection on the very late stages of stellar evolution, it will not be surprising if current theory does not give a precise fit to observations.

In contrast, SN Ib and SN Ic represent an immensely valuable new laboratory to study the late stages of massive star evolution by providing a source of direct observational constraints to guide theoretical studies. Type II supernovae are thought to be produced, in some instances at least, by core collapse, but the physically interesting regions are shielded from view by the enshrouding hydrogen envelope and, except for SN 1987A, only the most basic kinematic constraints have emerged. SN Ib in the early atmosphere and late supernebular phases may now give a way to determine the density, temperature, and composition structure of the cores of massive stars. This in turn will surely lead to important new insights into the physical processes in the star prior to and during the explosion and ultimately, perhaps, to an understanding of the explosion mechanism itself.

## Appendix. The supernova atmosphere program

The physical conditions in supernova atmospheres are quite unlike those found in conventional stellar atmospheres and winds and a different treatment of the atmosphere is necessary. The computer programs which have been developed by Harkness to model supernova atmospheres are based on the solution of the special relativistic co-moving radiative transfer equation appropriate for a spherically symmetric expanding medium. The expansion of a supernova is approximately homologous and it is important to retain all the terms in the transfer equation (describing aberration and advection in addition to Doppler terms), to correctly account for the effects of differential expansion at large optical depths. The use of a fully relativistic equation of transfer incurs no overheads with respect to a full order $(v/c)$ approximation. In any case, there are no precise upper limits on the observable expansion velocities so that relativistic effects may often be significant at lower optical depths or before maximum light when rapidly moving outer layers may dominate the spectrum. The co-moving approach offers considerable advantages for describing more complex physical processes and also removes the need to define frame dependent 'expansion opacities' (Karp *et al* 1977, Karp 1980), although a subsequent observer frame calculation must be performed to obtain the emergent flux distribution.

In the application to supernovae it is probably unnecessary to account for the hydrodynamical effects of the radiation field at times later than a couple of days after the explosion, so currently no attempt is made to solve the complete radiation-hydrodynamic problem. The main purpose of the programs is to study the conditions from shortly after the explosion to times at least 20–30 days after the explosion when the ejecta as a whole become optically thin.

Mihalas (1980) gives a full account of the co-moving radiative transfer equation and a method of solution derived from the earlier work of Mihalas *et al* (1976). The basic method of solution follows exactly the techniques outlined in these two papers. The co-moving transfer equation in spherical geometry is

$$\frac{\gamma}{c}(1+\beta\mu)\frac{\partial I}{\partial t}+\gamma(\mu+\beta)\frac{\partial I}{\partial r}+\gamma(1-\mu^2)\left[\frac{(1+\beta\mu)}{r}-\frac{\gamma^2}{c}(1+\beta\mu)\frac{\partial\beta}{\partial t}-\gamma^2(\mu+\beta)\frac{\partial\beta}{\partial r}\right]\frac{\partial I}{\partial\mu}$$
$$-\gamma\left[\beta\frac{(1-\mu^2)}{r}+\frac{\gamma^2}{c}\mu(1+\beta\mu)\frac{\partial\beta}{\partial t}+\gamma^2\mu(\mu+\beta)\frac{\partial\beta}{\partial r}\right]v\frac{\partial I}{\partial v}$$
$$+3\gamma\left[\frac{\beta(1-\mu^2)}{r}+\frac{\gamma^2}{c}\mu(1+\beta\mu)\frac{\partial\beta}{\partial t}+\gamma^2\mu(\mu+\beta)\frac{\partial\beta}{\partial r}\right]I=\eta-\chi I. \tag{1}$$

The time-dependent terms are neglected (i.e. the atmosphere is assumed to be quasi-steady), but can be included with no formal changes in the structure of the equations at any given time. The initial step in the solution is to recognize that the co-moving transfer equation is linear in the spatial derivatives. This implies that frequency independent characteristics can be defined along which a single spatial operator is a perfect differential, e.g.

$$\frac{\mathrm{d}I}{\mathrm{d}s}=\frac{\mathrm{d}r}{\mathrm{d}s}\frac{\partial I}{\partial r}+\frac{\mathrm{d}\mu}{\mathrm{d}s}\frac{\partial I}{\partial\mu}. \tag{2}$$

The appropriate characteristics are defined by the simultaneous solution of

$$\frac{\mathrm{d}r}{\mathrm{d}s}=\gamma(\mu+\beta) \tag{3a}$$

$$\frac{\mathrm{d}\mu}{\mathrm{d}s}=\gamma(1-\mu^2)\left[\frac{(1+\beta\mu)}{r}-\gamma^2(\mu+\beta)\frac{\partial\beta}{\partial r}\right] \tag{3b}$$

which reduce to parallel straight lines of constant impact parameter for $\gamma\to 1$, $\beta\to 0$.

Along these characteristics the transfer equation can be reduced to a two-dimensional form,

$$\frac{\partial i(s,\nu)}{\partial s}-a(s,\nu)\frac{\partial I(s,\nu)}{\partial\nu}=\eta(s,\nu)-\chi'(s,\nu)I(s,\nu) \tag{4}$$

where $s$ measures path length along a characteristic trajectory and

$$a(s,\nu)=\gamma\left[\frac{\beta(1-\mu^2)}{r}+\gamma^2\mu(\mu+\beta)\frac{\partial\beta}{\partial r}\right]\nu \tag{5}$$

$$\chi'(s,\nu)=\chi(s,\nu)+3\gamma\left[\frac{\beta(1-\mu^2)}{r}+\gamma^2\mu(\mu+\beta)\frac{\partial\beta}{\partial r}\right]. \tag{6}$$

The equations are integrated using a high-order Runge–Kutta method (Fehlberg 1968). The characteristics are chosen to have a tangent point on each spherical surface on a radial grid determined by the expected resolution requirements of a particular model. Additional rays which have a tangent point on spherical shells of radius less than the 'core' (or minimum model radius) are required to provide complete angular resolution and a central ray (at impact parameter zero) is also necessary. The core intersecting rays are generated with impact parameters chosen to provide sufficient resolution in $\mu$ to accurately represent the 'forward peaking' of the radiation field at large radii.

The transfer equation must now be solved along the characteristic trajectories running from surface to surface for rays which do not intersect the 'core' and surface to core and core to surface for those that do. In some cases, e.g. models of SN Ia events, the 'core' may be transparent. Solution of the transfer equation on the complete set of characteristics generates the total solution for $I(r, \mu, \nu)$, which can be integrated on the radial shells of the model grid to provide the moments of the radiation field

$$[J_\nu(r), H_\nu(r), K_\nu(r), N_\nu(r)] \equiv \tfrac{1}{2}\int_{-1}^{1} I(r, \mu, \nu)\mu^n \, d\mu \qquad (n=0, 1, 2, 3). \tag{7}$$

The transfer equation is backward-differenced in frequency and rewritten as

$$\frac{dI_K}{dt} + I_K = S_K^* \tag{8}$$

where

$$S_K^* = \frac{\eta_K + \alpha_{K-1/2} I_{K-1}}{\chi'_K + \alpha_{K-1/2}} \tag{9}$$

and

$$\alpha_{K-1/2} = a(s, \nu_K)/(\nu_{K-1} - \nu_K) \tag{10}$$

and defining

$$dt_K \equiv (\chi'_K + \alpha_{K-1/2})\, ds \tag{11}$$

on a frequency grid $\{\nu_K\}$ indexed from high to low frequency. For continuum transfer

$$\eta(s, \nu_K) = \chi_a(r, \nu_K) B(T(r), \nu_K) + \chi_s(r, \nu_K) J(r, \nu_K). \tag{12}$$

The integration along each characteristic is performed semi-analytically using a cubic spline representation of the integrand and a cubic Hermite spline method (Mihalas and Mihalas 1978) to evaluate the integration weights. To reduce computational costs, the steps in $dt$ are obtained analytically assuming that $\chi'_K + \alpha_{K-1/2}$ is piecewise linear in the log. (For grey models the defining equation for $t$ is integrated simultaneously with the equations defining the characteristics.) The discrete form of these equations can be organized to reduce to a set of linear equations which is solved for $J_\nu$. For the continuum solution along each characteristic one can write

$$\boldsymbol{I}_K = \boldsymbol{\Lambda}_K \boldsymbol{S}_K^* + \boldsymbol{E}_K \tag{14}$$

$$\boldsymbol{S}_K^* = \boldsymbol{\Xi}_K \boldsymbol{J}_K + \boldsymbol{\psi}_K + \boldsymbol{\Theta}_K \boldsymbol{I}_{K-1}. \tag{15}$$

From the definition of $J_K$ as a sum of $I_K$ over all angles, and hence all characteristics

$$J_{jk} = \sum_{i_+, i_-} I_{ijk} W_{ij}^{(0)} \tag{16}$$

where the $W_{ij}^{(m)}$ are appropriately chosen integration weights for equations (7), one can develop a final system of linear equations of the form

$$\boldsymbol{C}_K \boldsymbol{J}_K = \boldsymbol{D}_K \tag{17}$$

which is solved for $J_K$. Given $J_K$, the co-moving source function $S_K$ can be evaluated and the full solution for the co-moving specific intensity along each characteristic can be reconstructed. The full solution for the co-moving specific intensity is required in order to evaluate higher moments and to advance the overall process to the next frequency point. The frequency recursion is started by assuming that the specific intensity is zero at some sufficiently high frequency.

With a suitable choice of frequency grid, frequency integrals of the co-moving moments can be accumulated to determine the radiation energy density, pressure and radiative forces on the matter. These integrals are needed to couple the solution with the momentum and energy balance equations for the fluid.

The practical problems lie in the appropriate selection of the radius, core impact parameter and frequency grids. The radial grid should ideally be chosen to provide at least ten points per decade in optical depth. The initial grid is obtained by assuming that the major source of continuous opacity is electron scattering (a very good approximation in most cases appropriate to SN Ia models). In general the electron scattering grid gives an adequate zoning because the density profile is typically a very steep function of radius. Choice of an outer limiting radius at some small optical depth is not so critical, again because of the generally steep density profile.

Given an initial radial grid and the gamma-ray deposition function, the grey co-moving transfer equation can be solved to obtain a grey-body temperature distribution. This grid can be used to start the full non-grey calculation. At this point we have $\rho(r)$, $v(r)$ from the explosion model and an approximate $T(r)$ on some grid $\{r_j\}$. The explosion model also specifies the radial dependence of elemental abundances so an estimate can be made of the strengths of important lines and a further check made to ensure satisfactory radial resolution of major features. The radial dependence of line opacity and broadening parameters is required for a suitable determination of the frequency grid which must accommodate the likely range of relative opacities and feature widths (in frequency) with appropriate resolution. In general, the frequency grid requires many more points than a static atmosphere model although the co-moving formulation uses the same physics, because of the frequency derivative term in the transfer equation.

To compute the emergent flux from a model one integrates the observer frame source function on a set of parallel rays, normally taken to have tangent points on the same radial grid used in the co-moving solution. The observer frame extinction and emissivity are simply related to the co-moving values by

$$\chi(\mu_{\text{obs}}, \nu_{\text{obs}}) = (\nu_{\text{cm}}/\nu_{\text{obs}})\chi_{\text{cm}}(\nu_{\text{cm}}) \tag{18}$$

$$\eta(\mu_{\text{obs}}, \nu_{\text{obs}}) = (\nu_{\text{obs}}/\nu_{\text{cm}})^2 \eta_{\text{cm}}(\nu_{\text{cm}}) \tag{19}$$

where for each point along the ray the observed frequency, $\nu_{\text{obs}}$, is related to the frequency, $\nu_{\text{cm}}$, in the co-moving frame by $\nu_{\text{obs}} = \gamma(1+\beta\mu)\nu_{\text{cm}}$.

The radial grid of the co-moving solution is too coarse for zoning in the observer frame because of the very large variation in co-moving frequency sampled along a ray in the observer frame. Lines, for example, are confined to a small physical region of each ray and must be explicitly located and zoned on a fine mesh in order to resolve them; however, the co-moving solution naturally provides the most information at just those regions along the observer frame ray where lines are formed. Since the observer frame values of emissivity and extinction are *defined* at any point by the co-moving solution and the above relations, the integration is performed using a linear spline representation. A linear spline does not have problems with 'overshooting' which affect higher-order representations where there is a very rapid change with frequency over a line.

## References


Arnett W D 1969 *Astrophys. Space Sci.* **5** 280
—— 1980 *Astrophys. J. Lett.* **230** L37
—— 1982 *Astrophys. J.* **253** 785
Arnett W D, Bahcall J N, Kirshner R P and Woosley S E 1989 *Ann Rev. Astron. Astrophys.* **27** 629
Arnett W D, Branch D and Wheeler J C 1985 *Nature* **314** 337
Axelrod T S 1980a *PhD Thesis* University of California, Santa Cruz
—— 1980b *Type I Supernovae* ed J C Wheeler (Austin, TX: University of Texas) p 80
—— 1988 Springer Lecture Notes in Physics **305** 375
Baade W and Zwicky F 1934a *Proc. Natl. Acad. Sci. USA* **20** 259
—— 1934b *Proc. Natl. Acad. Sci. USA* **20** 254
—— 1934c *Phys. Rev.* **45** 138
—— 1934d *Phys. Rev.* **46** 76
Bandiera R 1987 *Astrophys. J.* **319** 885
Barbon R, Cappellaro E and Turatto M 1984 *Astron. Astrophys.* **135** 27
—— 1989 *Astron. Astrophys.* **226** 421
Barbon R, Ciatti F and Rosino L 1973 *Astron. Astrophys.* **29** 57
—— 1979 *Astron. Astrophys.* **72** 287
Barkat Z, Livne E, Swartz D and Wheeler J C 1990 *Supernovae* eds J C Wheeler, T Piran and S Weinberg (Singapore: World Scientific) p 133
Barratt R, Baus M and Hansen J P 1986 *Phys. Rev. Lett.* **56** 1063
Begelman M C and Sarazin C L 1986 *Astrophys. J. Lett.* **302** L59
Benetti S, Capellaro E and Turatto M 1990 *IAU Circular no.* 4967
Benz W, Bowers R I, Cameron A G W and Press W H 1990 *Astrophys. J.* **348** 647
Benz W, Cameron A G W and Bowers R I 1990 *Astrophys. J.* in press
Bertola F 1962 *Contributions of the Asiago Observatory* 128
—— 1964 *Annals d'Astrophysique* **27** 319
Bertola F, Mammano A and Perinotto M 1965 *Contributions of the Asiago Observatory* No. 174
Bionta R M *et al* 1987 *Phys. Rev. Lett.* **14** 1494
Blinnikov S I and Khokhlov A M 1986a *Sov. Astron. Lett.* **12** 131
—— 1986b *Sov. Astron.* **13** 364
Bragaglia A, Greggio L and Renzini A 1988 *ESO Mess.* **52** 35
Branch D 1981 *Astrophys. J.* **248** 1076
—— 1986 *Astrophys. J. Lett.* **300** L51
—— 1987 *Astrophys. J. Lett.* **316** L81
—— 1988a *Springer Lecture Notes in Physics* **305** 281
—— 1988b *The Extragalactic Distance Scale* ed S van den Bergh and C J Pritchett (Provo, UT: Brigham Young University Press) p 146
—— 1990 *Supernovae* ed A Petschek (Berlin: Springer) p 30
Branch D, Buta R, Falk S W, McCall M L, Sutherland P G P, Uomoto A K, Wheeler J C and Wills B J 1982 *Astrophys. J. Lett.* **252** L61
Branch D, Doggett J B, Nomoto K and Theilemann F-K 1985 *Astrophys. J.* **294** 619
Branch D, Drucker W and Jeffrey D J 1988 *Astrophys. J. Lett.* **330** L117
Branch D, Falk S W, McCall M L, Rybski P, Uomoto A K and Wills B J 1981 *Astrophys. J.* **244** 780
Branch D, Lacy C H, McCall M L, Sutherland P G, Uomoto A K, Wheeler J C and Wills B J 1983 *Astrophys. J.* **270** 123
Branch D and Nomoto K 1986 *Astron. Astrophys.* **164** L13
Branch D and Venkatakrishna K L 1986 *Astrophys. J. Lett.* **306** L21

Bruenn S W 1973 *Astrophys. J. Lett.* **183** L125
Buchler J-R, Wheeler J C and Barkat Z 1971 *Astrophys. J.*, **167** 465
Buta R J and Turner A 1983 *Pub. Astron. Soc. Pacific* **95** 72
Cadoneau R, Sandage A and Tammann G A 1985 *Springer Lecture Notes in Physics* **224** 151
Cahen S, Schaeffer R and Cassé M 1986 *Nucleosynthesis and its Implications on Nuclear and Particle Physics* ed J Audouze and N Mathieu (Dordrecht: Reidel) p 243
Caldwell C N and Oemler A Jr 1981 *Astrophys. J.* **86** 1424
Caldwell N and Phillips M M 1989 *Astrophys. J.* **338** 789
Canal R, Isern J and Labay J 1980 *Astrophys. J. Lett.* **241** L33
Canal R, Isern J and López R 1988 *Astrophys. J.* **330** L113
Cappellaro E and Turatto M 1988 *Astron. Astrophys.* **190** 10
Chevalier R A 1981a *Astrophys. J.* **246** 267
—— 1981b *Fund. Cosmic Phys.* **7** 1
—— 1982 *Astrophys. J. Lett.* **259** L85
—— 1984 *Astrophys. J. Lett.* **285** L63
Chevalier R A and Plait P C 1988 *Astrophys. J. Lett.* **301** L109
Ciardullo R, Ford H C, Neill J D, Jacoby G H and Shafter A W 1987 *Astrophys. J.* **318** 520
Clark D H and Stephenson F R 1977 *The Historical Supernovae* (Oxford: Pergamon)
Colgate S A, Petschek A G and Kreise J T 1980 *Astrophys. J.* **237** L81
Conti P S 1982 *Wolf–Rayet Stars: Observations, Physics, Evolution* ed C W H De Loore and A J Willis (Dordrecht: Reidel) p 3
Couch R G and Arnett W D 1975 *Astrophys. J.* **196** 791
de Vaucouleurs G 1979 *Astrophys. J.* **227** 729
—— 1988 *Bull. Am. Astron. Soc.* **20** 672
de Vaucouleurs G and Corwin H G Jr 1985 *Astrophys. J.* **295** 287
de Vaucouleurs G, de Vaucouleurs A and Corwin H G Jr 1976 *Second Reference Catalog of Bright Galaxies* (Austin, TX: Texas University Press)
de Vaucouleurs G and Pence W D 1978 *Astrophys. J.* **83** 1163
Dennefeld M 1982 *IAU Circular Nos. 3739, 3740*
Doggett J B and Branch D 1985 *Astrophys. J.* **90** 2303
Dopita M A private communication
—— 1988 *Space Sci. Rev.* **46** 225
Elias J H, Frogel J A, Hackwell J A and Persson S E 1981 *Astrophys. J. Lett.* **251** L13
Elias J H, Mathews K, Neugebauer G and Persson S E 1985 *Astrophys. J.* **296** 379
Ensman L M and Woosley S E 1988 *Astrophys. J.* **333** 754
Evans R O 1986 private communication
Fesen R A, Hamilton A J S and Saken J M 1989 *Astrophys. J. Lett.* **341** L55
Fesen R A, Wu C-C, Leventhal M and Hamilton A J S 1988 *Astrophys. J.* **327** 164
Filippenko A V 1988a *Astrophys. J.* **96** 1941
—— 1988b *Proc. Astron. Soc. Australia* **7** 540
—— 1990 *IAU Circular No. 4953*
Filippenko A V, Porter A C, Sargent W L W and Schneider D P 1986 *Astrophys. J.* **92** 1341
Filippenko A V and Sargent W L W 1985 *Nature* **316** 407
—— 1986 *Astrophys. J.* **91** 691
Foss D, Wade R A and Green R F 1990 *Astrophys. J.* in press
Fowler W A 1986 *Quart. J R Astron. Soc.* **28** 87
Fransson C 1988 *Springer Lecture Notes in Physics* **305** 383
Fransson C and Chevalier R A 1987 *Astrophys. J. Lett.* **332** L15
—— 1989 *Astrophys. J.* **343** 323
Frogel J A, Gregory B, Kawara K, Laney D, Phillips M M, Terndrup D, Vrba F and Whitford A E 1987 *Astrophys. J. Lett.* **315** L129
Fugimoto M Y and Taam R E 1982 *Astrophys. J.* **260** 249
Gamow G and Schoenberg M 1941 *Phys. Rev.* **59** 539
Gaskell C M 1990 preprint
Gaskell C M, Cappellaro E, Dinerstein H L, Garnett D R, Harkness R P and Wheeler J C 1986 *Astrophys. J. Lett.* **306** L77
Gaskell C M and Koratkar A P 1989 *IAU Circular No. 4759*
Graham J R 1986 *Mon. Not. R. Astron. Soc.* **220** p 27
—— 1987 *Astrophys. J.* **315** 588
—— 1988a *Astrophys. J. Lett.* **326** L51

—— 1988b *Astrophys. J. Lett.* **335** L53
Graham J R, Meikle W P S, Allan D A, Longmore AP J and Williams P M 1986 *Mon. Not. R. Astron. Soc.* **218** 93
Hachisu I, Eriguchi Y and Nomoto K 1986 *Astrophys. J.* **308** 161
Hachisu I, Kato M and Saio H 1989 *Astrophys. J. Lett.* **342** L19
Hamilton A J S and Fesen R A 1988 *Ap. J.* **327** 178
Hamilton A J S, Sarazin C L and Szymkoviak A E 1986 *Astrophys. J.* **300** 698
Hamilton A J S, Sarazin C L, Szymkoviak A E and Vartanian M H 1985 *Astrophys. J. Lett.* **297** L5
Harkness R P 1981 *D. Phil. Thesis* University of Oxford (unpublished).
—— 1983 *Mon. Not. R. Astron. Soc.* **204** 45
—— 1985 *Supernova as Distance Indications* ed N Bartel (Berlin: Springer) p 183
—— 1986 *Radiation Hydrodynamics in Stars and Compact Objects* ed D Mihalas and K-H A Winkler (Berlin: Springer) p 166
—— 1987 *Relativistic Astrophysics* ed M P Ulmer (Singapore: World Scientific) p 413
—— 1990 *Supernovae* ed S E Woosley (New York: Springer) in press
Harkness R P and Wheeler J C 1988 *Proc. Astron. Soc. Australia* **7** 431
—— 1990 Classification of Supernovae *Supernovae* ed A Petschek (Berlin: Springer) p 1
Harkness R P, Wheeler J C, Margon B, Downes R A, Kirshner R P, Uomoto A, Barker E S, Cochran A L, Dinerstein H L, Garnett D R and Levreault R M 1987 *Astrophys. J.* **317** 355
Hernanz M, Isern J, Canal R, Labay J and Mochkovitch R 1988 *Astrophys. J.* **324** 331
Hirata K *et al* 1987 *Phys. Rev. Lett.* **14** 1490
Holm A V, Wu C C and Caldwell J 1974 *Pub. Astron. Soc. Pacific* **86** 296
Hoyle F and Fowler W A 1960 *Astrophys. J.* **132** 565
Huang Y L 1987 *Pub. Astron. Soc. Pacific* **99** 461
Huang Y L and Thaddeus P 1986 *Astrophys. J.* **309** 804
Hummer D and Rybicki G 1971 *Mon. Not. R. Astron. Soc.* **152** 1
Iben I Jr. 1982 *Astrophys. J.* **253** 248
—— 1988 *Astrophys. J.* **324** 355
Iben I Jr., and Tutukov A V 1984 *Astrophys. J. Suppl.* **54** 335
—— 1985 *Astrophys. J. Suppl.* **58** 661
Ichimaru S, Iyetomi H and Tanaka S 1987 *Phys. Rep.* **149** 95
Itoh H, Masai K and Nomoto K 1988 *Astrophys. J.* **334** 279
Karp A H 1980 *Supernovae Spectra* eds R Meyerott, G H Gillespie (New York: American Institute of Physics) 125
Karp A H, Lasher G, Chan K L and Salpeter E E 1977 *Astrophys. J.* **214** 161
Khokhlov A M 1989 *Sov. Sci. Rev. E. Astrophys. Space Phys.* **8** 1
Kimeridze G N and Tsvetkov D Yu 1986 *Astrofizika* **25** 279 (transl. *Astrophysics* **25** 513)
Kirshner R P 1990 *Supernovae* ed A Petschek (Berlin: Springer) in press
Kirshner R P, Arp H C and Dunlap J R 1976 *Astrophys. J.* **207** 44
Kirshner R P and Oke J B 1975 *Astrophys. J.* **200** 574
Koyama K, Tsunemi H, Becker R H and Hughes J P 1987 *Pub. Astron. Soc. Japan* **39** 437
Kudritzki R P, Gabler A, Gabler R, Groth H G, Pauldrach A W A and Puls J 1989 The Physics of Luminous Blue Variables, *IAU Colloquium No. 113* eds K Davidson, A F J Moffat and H J G L M Lamers (Dordrecht: Kluwer) p 67
Langer N 1989 *Astron. Astrophys.* **220** 135
Linblad P O 1983 *IAU Circular No. 3897*
MacDonald J 1984 *Astrophys. J.* **283** 241
Maeder A and Lequeux J 1982 *Astron. Astrophys.* **114** 409
Massey P 1982 *Wolf-Rayet Stars: Observations, Physics, Evolution*, ed C W H De Loore and A J Willis (Dordrecht: Reidel) p 251
Mazurek T J 1973 *Astrophys. Space Sci.* **23** 365
—— 1980 *Type I Supernovae* ed J C Wheeler (Austin, TX: University of Texas) p 182
Mazurek T J Meier D L and Wheeler J C 1977 *Astrophys. J.* **215** 518
Mazurek T J and Wheeler J C 1980 *Fund. Cosmic Phys.* **5** 193
McLaughlin D B 1960 *Stellar Atmospheres* ed J L Greenstein (Chicago: University of Chicago Press) p 585
Meyerott R E 1980 *Astrophys. J.* **239** 257
Mihalas D 1980 *Astrophys. J.* **237** 574
Mihalas D, Auer L H and Mihalas B R 1978 *Astrophys. J.* **41** 513
Mihalas D, Kunasz P B and Hummer D G 1976 *Astrophys. J.* **206** 515
Miller D L and Branch D 1990 preprint

Miller G E and Scalo J M 1979 *Astrophys. J. Suppl.* **220** 1001
Miller W and Wheeler J C 1990 in preparation
Minkowski R 1939 *Astrophys. J.* **89** 156
—— 1940 *Pub. Astron. Soc. Pacific* **52** 206
—— 1941 *Pub. Astron. Soc. Pacific* **53** 224
Moffat A F J, Niemela V S and Marraco H G 1990 *Astrophys. J.* **348** 232
Moffat A F J, Vogt N, Paquin G, Lamontagne R and Barrera L 1986 *Astrophys. J.* **91** 1386
Müller E and Arnett W D 1982 *Astrophys. J. Lett.* **261** L109
Niemela V S, Ruiz M T and Phillips M M 1985 *Astrophys. J.* **289** 52
Nomoto K 1982a *Astrophys. J.* **253** 798
—— 1982b *Astrophys. J.* **257** 780
—— 1984 *Astrophys. J.* **277** 791
—— 1985 *Proc. 12th Texas Symp. on Relativistic Astrophysics*, ed M Livio and G Shaviv (New York: New York Academy of Sciences) p 294
Nomoto K and Iben I Jr., 1985 *Astrophys. J.* **297** 531
Nomoto K, Narai K and Sugimoto D 1979 *Pub. Astron. Soc. Japan* **31** 287
Nomoto K and Shigeyama T 1989 *Particle Astrophysics: Forefront Experimental Issues* ed E B Norman (Singapore: World Scientific) p 198
Nomoto K, Shigeyama T and Hashimoto M 1988a *Springer Lecture Notes in Physics* **305** 490
Nomoto K, Shigeyama T, Kumagai S and Hashimoto M 1988b *Proc. Astron. Soc. Australia* **7** 431
Nomoto K and Sugimoto D 1977 *Pub. Astron. Soc. Japan* **29** 765
Nomoto K, Sugimoto D and Neo S 1976 *Astrophys Space Sci.* **39** L37
Nomoto K, Thieleman F-K and Wheeler J C 1984a *Astrophys. J. Lett.* **279** L23
Nomoto K, Thielemann F-K and Yokoi K 1984b *Astrophys. J.* **286** 644
Oemler A and Tinsley B M 1979 *Astrophys. J.* **84** 985
Oliva E 1987 *Astrophys. J. Lett.* **321** L45
Oran E S, Young T R, Boris J P, Picone J M and Edwards D H 1982 *Proc. 19th Int. Symp. Combustion* (Pittsburgh, PA: The Combustion Institute), p 573
Ostriker J P 1982 *Supernovae: A Survey of Current Research*, ed M J Rees and R J Stoneham (Dordrecht: Reidel) p 49
Paczyński B E 1973 *Astrophys. J. Lett.* **15** 147
—— 1985 *Cataclysmic Variables and Low Mass X-ray Binaries*, ed D Q Lamb and J Patterson (Dordrecht: Reidel) p 1
Panagia N 1985 *Springer Lecture Notes in Physics* **224** 14
Panagia N and Laidler V 1989 *Supernova Shells and their Birth Events* ed W Kundt (Berlin: Springer), in press
Panagia N, Sramek R A and Weiler K W 1986 *Astrophys. J. Lett.* **285** L63
Panagia N *et al* 1986 preprint
Pearce E C, Colgate S A and Petschek A G 1988 *Astrophys. J. Lett.* **325** L33
Petschek A 1990 *Supernovae* (New York: Springer)
Phillips M M and 27 others 1987 *Pub. Astron. Soc.* **99** 592
Pinto P A 1988 *PhD Thesis* University of California, Santa Cruz
Pinto P A and Woosley S E 1988 *Astrophys. J.* **329** 820
Porter A C and Filippenko A V 1987 *Astrophys. J.* **93** 1372
Prantzos N, Doom C, Arnould M and de Loore C 1986 *Astrophys. J.* **304** 695
Pskovskii Y P 1977 *Sov. Astron.* **21** 675
Richter O-G and Rosa M 1984 *Astron. Astrophys.* **140** L1
Richtler T and Sadler E M 1983 *Astron. Astrophys.* **128** L3
Robinson E L and Shafter A W 1987 *Astrophys. J.* **322** 296
Rupen M P, van Gorkom J H, Knapp G R, Gunn J E and Schneider D P 1987 *Astrophys. J.* **94** 61
Saffer R A, Liebert J and Olszewski E 1988 *Astrophys. J.* **334** 947
Saio H and Nomoto K 1985 *Astron. Astrophys.* **150** L21
Scalo J M 1986 *Fund. Cosmic. Phys.* **11** 1
Schaeffer R, Cassé M and Cahen S 1987 *Astrophys. J. Lett.* **316** L31
Schlegel E M and Kirshner R P 1989 *Astrophys. J.* **98** 577
Schneider D P, Mould J R, Porter A C, Schmidt M, Bothun G D and Gunn J E 1988 *Pub. Astron. Soc. Pacific* **99** 1167
Schurmann S R 1983 *Astrophys. J.* **267** 779
Schweizer F and Middleditch J 1980 *Astrophys. J.* **241** 1039
Shara M M, Livio M, Moffat A F J and Orio M 1986 *Astrophys. J.* **311** 163
Sramek R A and Weiler K W 1990 *Supernovae* ed A Petschek (New York: Springer) p 76

Sramek R A, Panagia N and Weiler K W 1984 *Astrophys. J. Lett.* **285** L63
Stevenson D J 1980 *J. Physique Suppl.* **41** C2-53
Strom R G 1988 *Mon. Not. R. Astron. Soc.* **230** 331
Stryker L L, Gunn J E and Tinsley B M 1981 *Astrophys. J.* **249** 48
Sutherland P G and Wheeler J C 1984 *Astrophys. J.* **280** 282
Swartz D A, Harkness R P and Wheeler J C 1989 *Nature* **337** 439
Taam R E 1980 *Astrophys. J.* **242** 749
Tammann G A 1982 in *Supernovae: A Survey of Current Research* ed M J Rees and R J Stoneham (Dordrecht: Reidel) p 371
Thielemann F-K, Nomoto K and Yokoi K 1986 *Astron. Astrophys.* **158** 17
Tinsley B M 1980 *Type I Supernovae* ed J C Wheeler (Austin TX: University of Texas) p 196
Tornambé A and Matteucci F 1986 *Mon. Not. R. Astron. Soc.* **223** 69
Trimble V 1982 *Rev. Mod. Phys.* **54** 1183
—— 1983 *Rev. Mod. Phys.* **55** 511
Tsunemi H, Yamashita K, Masai K, Hayakawa S and Koyama K 1986 *Astrophys. J.* **306** 248
Tsvetkov D Yu 1985 *Sov. Astron.* **29** 211
—— 1986 *Sov. Astron. Lett.* **12** 328
—— 1987 *Sov. Astron. Lett.* **13** 376
Tuchman Y and Wheeler J C 1990 *Astrophys. J.* in press
Tully R B 1988 *Nearby Galaxies Catalog* (Cambridge: Cambridge University Press)
Tutukov A V and Yungleson L R 1979 *Acta. Astron.* **29** 665
Twarog B A and Wheeler J C 1982 *Astrophys. J.* **261** 636
—— 1986 *Astrophys. J.* **316** 153
Uomoto A 1986 *Astrophys. J. Lett.* **310** L35
Uomoto A and Kirshner R P 1985 *Astron. Astrophys.* **149** L7
—— 1986 *Astrophys. J.* **308** 685
van den Bergh S 1988 *Astrophys. J.* **327** 156
van den Bergh S, McClure R D and Evans R 1987 *Astrophys. J.* **323** 44
van den Bergh S and Younger P F 1985 *Astron. and Astrophys. Suppl.* **61** 365
Verdenet M 1984 *IAU Circular No. 3936*
Wagoner R F 1981 *Astrophys. J. Lett.* **250** L65
Weaver T A, Axelrod T S and Woosley S E 1980 *Type I Supernovae* ed J C Wheeler (Austin, TX: University of Texas) p 113
Webbink R F 1984 *Astrophys. J. Lett.* **277** 355
Wheeler J C 1981 *Rep. Prog. Phys.* **44** 85
—— 1990a *Supernovae* ed J C Wheeler, T Piran and S Weinberg (Singapore: World Scientific) in press
—— 1990b *Proc. McDonald Observatory 50th Anniversary Symposium, Frontiers of Stellar Evolution* ed D L Lambert (Provo: Astronomical Society of the Pacific), in press
—— 1990c *Nonlinear Astrophysical Fluid Dynamics* ed J-R Buchler (New York: New York Academy of Sciences) in press
Wheeler J C, Branch D and Falk S W J 1980 *Proc. Texas Workshop on Type I Supernovae*, ed J C Wheeler (Austin, TX: University of Texas) p 199
Wheeler J C and Harkness R P 1986 *Galaxy Distances and Deviations from Universal Expansion* ed B F Madore and R B Tully (Dordrecht: Reidel) p 45
Wheeler J C, Harkness R P, Barkat Z and Swartz D 1986 *Pub. Astron. Soc. Pacific* **98** 1018
Wheeler J C, Harkness R P, Barker E S, Cochran A L and Wills D 1987a *Astrophys. J. Lett.* **313** L69
Wheeler J C, Harkness R P and Cappellaro E 1987b in *Relativistic Astrophysics*, ed M P Ulmer (Singapore: World Scientific), p 402
Wheeler J C and Levreault R 1985 *Astrophys. J. Lett.* **294** L17
Wheeler J C, Li Z, Sutherland P G and Swartz D 1987c *Astrophys. J.* **316** 733
Wheeler J C, Piran T and Weinberg S 1990 *Supernovae* (Singapore: World Scientific)
Whitelock P 1983 *IAU Circular No. 3898*
Wood P R and Faulkner D J 1987 *Proc. Astron. Soc. Australia* **7** 75
Woosley S E 1988 *Astrophys. J.* **330** 218
—— 1990 *Supernovae* (New York: Springer) in press
—— 1990 *Supernovae* ed A Petschek (New York: Springer) p 182
Woosley S E, Axelrod T S and Weaver T A 1984 *Stellar Nucleosynthesis* ed C Chiosi and A Renzini (Dordrecht: Reidel) p 263
Woosley S E, Taam R E and Weaver T A 1986 *Astrophys. J.* **301** 601

Woosley S E and Weaver T A 1982 *Essays in Nuclear Astrophysics* ed C Barnes, D. Clayton and D. Schramm (Cambridge: Cambridge University Press) p 377

—— 1986a *Radiation Hydrodynamics in Stars and Compact Objects*, ed D Mihalas and K-H A Winkler (Berlin: Springer) p 91

—— 1986b *Ann. Rev. Astron. Astrophys.* **24** 205

Wu C-C, Leventhal M, Sarazin C L and Gull T R 1983 *Astrophys. J. Lett.* **269** L5

Xu Y, Sutherland P G, McCray R A and Ross R R 1988 *Astrophys. J.* **327** 197

Young T R and Branch D 1988 *Nature* **333** 305

—— 1989 *Astrophys. J. Lett.* **342** L79

Zwicky F 1965 *Stars and Stellar Systems* vol. VIII ed L H Aller and D B McLaughlin (Chicago, IL: University of Chicago Press) p 367

Rep. Prog. Phys. **52** (1989) 1421–1473. Printed in the UK

# The supernova 1987A in the Large Magellanic Cloud

Wolfgang Hillebrandt and Peter Höflich

Max-Planck-Institut für Physik und Astrophysik, Institut für Astrophysik, Karl-Schwarzschild-Strasse 1, 8046 Garching, FRG

**Abstract**

Two years after the discovery of supernova 1987A in the Large Magellanic Cloud the huge amount of new and exciting information is still being analysed by astronomers and astrophysicists all over the world. In this review we shall try to summarise the present situation concerning both the observational facts and their theoretical interpretation. In particular, we shall focus our attention on the information contained in the light curve and the spectra, and in the detection of neutrinos. It will be shown that many previously open questions have now been answered, but also that many new problems await a solution. These include the unexpected blue appearance of the progenitor star, its presupernova evolution, the explosion mechanism, apparent deviations from spherical symmetry and their origin, properties of matter at high densities and temperatures, etc. All these questions are far from being fully understood. Nevertheless, an attempt is made to give at least some hints on how solutions may be obtained.

This review was received in May 1989.

0034-4885/89/111421+53$10.00 

**Contents**

## 1. Introduction

The modern era of supernova research began about 100 years ago on 31 August 1885, when Hartwig discovered a 'nova' near the centre of the Andromeda galaxy (M31); this nova became invisible about 18 months later. In 1919 Lundmark estimated the distance to M31 to be about $7\times10^5$ light-years, and by that time it became obvious that Hartwig's 'nova' had been several thousand times brighter than a galactic nova. A similar event was observed in 1895 in NGC 5253 ('nova' Z Centauri), which appeared to be approximately five times brighter than the entire galaxy. Systematic searches for supernovae performed predominantly by Zwicky led to the discovery of 54 supernovae in the years up to 1956 and, due to improved observational techniques, 82 further supernovae in the years 1958 to 1963, all of course in external galaxies. Recent estimates obtained from historical supernovae in our own galaxy, radio observations of old remnants, and supernovae in external galaxies give an average rate of about 1 supernova per $70\pm30$ years in giant spirals like our own (Tammann 1982, van den Bergh *et al* 1987). For a typical supernova outburst the optical luminosity is about $10^{10}$ solar luminosities around maximum light and the 'duration' of the event is on average of the order of one year. Integrated energies range from $10^{48}$ to $10^{50}$ erg ($10^{41}$–$10^{43}$ J) in electromagnetic radiation, which is only of the order of a few per cent of the total kinetic energy of the ejected material. Typical expansion velocities are found to be of the order of $10^4$ km s$^{-1}$.

From the light curves and the spectra one has to distinguish at least two different types of supernovae. Those that show strong Balmer lines of hydrogen are called type II supernovae and those where hydrogen lines are absent are classified as type I. The standard interpretation is that the progenitors of type I events are low mass stars with no or very little hydrogen, presumably white dwarfs or helium stars. Type II supernovae, on the other hand, are believed to be exploding massive stars ($M\gtrsim 8M_\odot$; $M_\odot\approx 2\times 10^{33}$ g) with extended H envelopes. A strong argument in favour of this hypothesis is the fact that type I supernovae are found also in elliptical galaxies with old populations of stars whereas type II supernovae occur only in spiral arms. Moreover, mass estimates obtained from x-ray observations of young galactic supernova remnants are consistent with this general picture.

If this interpretation is correct then type II supernovae should ultimately get their energy from the increase in gravitational binding of a collapsing stellar core, and they should be related to the formation of compact objects such as neutron stars or, possibly, black holes. Moreover, if the progenitors are indeed quite massive, type II supernovae will eject large amounts of matter that has undergone nuclear processing and is enriched in heavy elements. This has led to the suggestion that supernovae are the main site of galactic nucleosynthesis. Finally, supernova shock fronts are thought to be the place where the galactic cosmic rays are accelerated, they may induce star formation in dense interstellar clouds, and supernovae may even be used as cosmic distance indicators. Given these far-reaching implications it is not surprising that many attempts have been made in the past to identify the mechanism that causes massive stars to explode, but so far no answer has been given that is accepted by all researchers working in this field. Some of the difficulties are caused by the fact that, whatever the right mechanism is, it has to be very inefficient, because the gravitational binding energy of a neutron star is much larger than the energy observed in the explosion. Moreover, for none of the historical Galactic supernovae or more recent extragalactic ones has the progenitor star been identified. One therefore had to rely mainly on theoretical models of stellar evolution rather than on good observational data.

When, on 23 February 1987, a supernova explosion was observed in the Large Magellanic Cloud (LMC), a neighbour of our own galaxy, at a distance of roughly 50 kpc only, this event caused a lot of excitement in the astronomical community, because not since Kepler's supernova in 1604 has a supernova been so close. There was some hope, therefore, that most of the open questions in supernova theory could now be answered. However, it soon became clear that this was not the case. Moreover, supernova 1987A raised new problems, which will be discussed in some detail in the following sections.

Spectra taken during the second night showed Balmer lines of hydrogen, indicating that the supernova was of type II. Its position coincided with that of a blue supergiant, Sanduleak $-69°202$. When the UV radiation from the supernova weakened a few weeks after the explosion it became obvious that this star had indeed disappeared. From its spectral class (B3) and its luminosity class (Ia) one could conclude that the main-sequence mass of the progenitor had been close to 20 solar masses. The first big surprise was that, in contrast to theoretical expectations, it was a blue rather than a red supergiant. Possible explanations will be discussed in § 2, together with further properties of the progenitor inferred from the light curve and the spectra. Moreover, interpretations of the late light curve and spectra, including X-ray and $\gamma$-ray emission, will be reviewed.

Section 3 is devoted to the explosion mechanism of massive stars. It had always been believed that the central cores of stars as massive as Sk $-69°202$ should collapse to neutron star densities. Most of the binding energy of a newly born neutron star was thought to be radiated away in the form of neutrinos during the formation process. Therefore, it did not come as a surprise that neutrinos were detected a few hours before the optical outburst because the theoretical models predicted that a few neutrinos should be detected by present neutrino detectors for a supernova at a distance of 50 kpc. The arguments leading to this conclusion are fairly simple. Suppose that in the centre of a collapsing star a neutron star of roughly $1.5 M_{\odot}$ is born. Initially this newly born neutron star will be very hot and, therefore, its gravitational energy will only be of the order a few times $10^{52}$ erg. Some fraction of this energy may lead to the observed supernova outburst and mass ejection. The major part, however, has to be carried away by weakly interacting particles such as neutrinos. Moreover, during cooling and contraction the gravitational binding energy will increase to a few times $10^{53}$ erg and again the gain in binding will be radiated away by neutrinos on timescales of a few seconds. Because typical neutrino energies will be around 10 MeV, a few events should be detectable in water Čerenkov detectors like the KAMIOKANDE experiment, and they have indeed been detected.

Although, therefore, the arguments in favour of a core collapse are very convincing, many questions remain open. In particular, we have not learnt anything about the explosion mechanism itself from the neutrino observations. These questions will also be addressed in § 3.

SN 1987A exploded in a region of the LMC where the stellar population is dominated by young objects. Such regions are known to contain a variety of interstellar clouds and dust. It was therefore expected that the interaction of the supernova radiation with circumstellar and interstellar material would give rise to observable effects, some of which have indeed been detected and will be discussed in § 4. In particular, we will concentrate on the interpretation of radio observations, the narrow UV and optical lines, and the recently observed light echoes. Some speculations related to the speckle image seen in spring 1987 will also be discussed.

SN 1987A has not only been the subject of hundreds of papers in the astronomical literature, it has also motivated a lot of activity in nuclear and particle physics. It has renewed interest in the equation of state of hot nuclear matter, attempts have been made to derive fundamental properties of neutrinos, and constraints on exotic particles have been given. All these questions will be addressed in § 5. Finally, in § 6, some future prospects will be discussed, including the detectability of a possible neutron star remnant of SN 1987A, and its impact on stellar evolution theory, nucleosynthesis and supernova models.

## 2. Properties of SN 1987A inferred from the light curve and the spectra

SN 1987A was discovered by Ian Shelton at Feb. 24.2 (UT), 1987 (*IAU Circular* 4316). It is the first supernova that has been monitored at all wavelengths from $\gamma$-rays up to the radio range starting very early after the explosion.

Photons emitted from SN 1987A were first detected on Feb. 23.443 on a photograph at $6.4^{\mathrm{m}}$ (magnitude) in the visual wavelength range, i.e. about 3.84 hours after the detection of neutrinos by the IMB and KAMIOKANDE experiments (Bionta *et al* 1987, Hirata *et al* 1987). Observations carried out by the amateur astronomer A Jones from New Zealand about 1.2 hours after the detection of the neutrinos showed no object brighter than about 7.0 to $7.5^{\mathrm{m}}$ in the field near the supernova. This sets an important limit on the visual brightness of SN 1987A when the shock reached the photosphere. Unfortunately, no good spectral information is available for the first day, during which the photospheric temperature decreased from about 500 000–1 000 000 K to about 12 000–14 000 K as was predicted by hydrodynamic models (e.g. Arnett 1987, Shigeyama *et al* 1988, Woosley *et al* 1987a, b). Observations made by Ryker (1987) at Feb. 23.52 allow for an estimate of $B-V$ to be in the range between $-0.2$ and $0.0^{\mathrm{m}}$ at a visual brightness of $5.3^{\mathrm{m}}$. The first more accurate information on the spectral slope was available due to *UBV* colour measurements on Feb. 24.454 (Cropper and Bailey 1987), and the first optical spectra were obtained by various groups at about 1.5 days after the neutrino detections (e.g. Catchpole *et al* 1987, Fosbury *et al* 1987, Tyson and Boeshaar 1987).

Since then, spectral observations in the optical have been performed continuously by groups at the South African Astronomical Observatory (SAAO), at the Anglo-Australian Telescope (AAO), at the European Southern Observatory (ESO), at the Cerro Tololo Inter-American Observatory (CTIO) and at the Bochumer Telescope in Chile (e.g. Catchpole *et al* 1987, Couch 1988, Danziger *et al* 1988a, b, Feast 1988, Hanuschik and Dachs 1987a, b, 1988, 1989, Hamuy *et al* 1988, Menzies *et al* 1987, Tyson and Boeshaar 1987, Whitelock *et al* 1988). Spectral observations in the IR wavelength range have been made by different groups beginning in March 1987 (e.g. Aitken *et al* 1988, Bouchet *et al* 1987, Dwek 1988, Larson *et al* 1987, McGregor *et al* 1988, Meikle *et al* 1987a, b, c, 1989). Radio observations are available from 24 February 1987 (e.g. Manchester 1987, Turtle *et al* 1987). UV observations have been performed with the International Ultraviolet Explorer (IUE) since 25 February 1987 (e.g.

Cassatella 1987, Fransson *et al* 1987, Kirshner *et al* 1987, Panagia *et al* 1987, Wamsteker *et al* 1987), x-rays and $\gamma$-rays since May 1987 (e.g. Cook *et al* 1988, Gehrels *et al* 1987, Dotani *et al* 1987, Sandie *et al* 1988, Sunyaev *et al* 1987)

The available data allow a continuous spectral analysis of the envelope and measurement for the first time of the bolometric light curve, at least for times after about one week. The early light curve, however, is not well determined because of the uncertain corrections of the intrinsic flux due to interstellar reddening, which most strongly affect the UV wavelength region. In addition, the progenitor star could be identified (see below), which for the first time makes it possible to link the analysis of the expanding envelope of SN 1987A to the structure and evolution of the progenitor in a consistent way.

### *2.1. The progenitor star*

Astrometric determinations of the position of SN 1987A and comparisons with pre-outburst observations allowed for the identification of the progenitor to be one of the components of the multiple system Sanduleak $-69°202$ (Sk $-69°202$) (RA = 5 h 35 m 49.992 s, decl. = $-69°\ 17'\ 58.08''$; Walborn *et al* 1987, West *et al* 1987). This position coincides well with the position of the main component, which was surrounded by two companions at relative distances of $1.6''$ and $2.9''$, corresponding to projected distances of about $1.2\times10^{18}$ cm and $2.1\times10^{18}$ cm, respectively.

Surprisingly, the main component of this system was a blue supergiant of spectral class B3 and luminosity class Ia (Rousseau *et al* 1978) at $V = 12.24^{\mathrm{m}}$, instead of a red supergiant which was thought to be the progenitor of a type II supernova. The observed colour indices of Johnson's filter system (1966) ($U - B = -0.65^{\mathrm{m}}$, $B - V = +0.04^{\mathrm{m}}$; Isserstedt 1975, West *et al* 1987) indicated an effective temperature between 14 000 and 17 000 K. No significant change in the brightness of Sk $-69°202$ had been found on 502 Harvard blue patrol plates taken between 1899 and 1953 and on 59 plates taken in 1971, 1978-9 and 1981-6. However, the data do suggest that Sk $-69°202$ may have been varying by $\pm0.3$–$0.5^{\mathrm{m}}$ over that interval (Hazen 1987). On the basis of the observed colours and brightness a total luminosity can be estimated to be of the order of $3$–$6\times10^{38}$ erg s$^{-1}$, assuming a distance of 49 kpc for the LMC (Graham and Nemec 1984, Walker 1987, Walker and Mack 1988). Spectral observations ($3700 \leqslant \lambda \leqslant 4840$ Å) using objective prism plates (resolution 460 Å mm$^{-1}$) were obtained by Wamsteker at the ESO Schmidt telescope in 1977. Additional spectrograms were taken by the ESO astrograph in 1972-3 in the wavelength range between 3850 and 4150 Å at a resolution of 110 Å mm$^{-1}$ (Rousseau *et al* 1978). No outstanding chemical anomaly was observed, although a blend seen at about 4026 Å, which may be attributed to He I or N II, is stronger than in comparison stars of about the same spectral class (Gonzales *et al* 1987, Walborn *et al* 1987).

IUE observations performed by Gilmozzi *et al* (1987) and Walborn *et al* (1987) definitely showed that the B3Ia star had disappeared and, consequently, that the B3Ia star had been the progenitor of SN 1987A. This result was confirmed later by Kirshner (1988). The core-mass-luminosity relation derived from stellar models (Arnett 1987, Hillebrandt *et al* 1987b, Nomoto *et al* 1987, Truran and Weiss 1987), together with the spectral type derived from the observed stellar spectrum and the *UBV* colours, allow the determination of its He-core mass, and one finds $5$–$7M_{\odot}$, indicating that the main-sequence mass of the progenitor was around $20M_{\odot}$, with an uncertainty of about $\pm4M_{\odot}$.

### 2.2. Presupernova evolution of massive stars and Sk −69°202

Different evolutionary scenarios have been constructed starting from the main sequence with masses of about 16–24$M_\odot$ consistent with the luminosity of Sk −69°202 (see above), all of which have a blue supergiant as their final result. The differences in the various models can be attributed to different assumptions concerning mass loss, convection theory and metallicity (see figure 1).

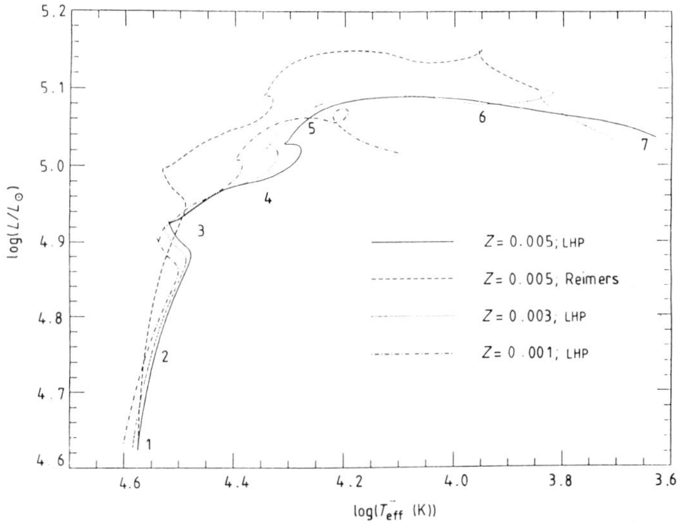


**Figure 1.** Evolution of a 20$M_\odot$ star assuming different metallicities and mass loss rates. LHP: mass loss rate above 6000 K as given by Lamers *et al* (1976), and as given by Reimers (1975) otherwise; Reimers: mass loss rate as given by Reimers (1975). (From Weiss 1989.)

(i) Brunish and Truran (1982) were the first to draw attention to the influence of the metallicity on the evolutionary tracks of supergiants in the Hertzsprung–Russell diagram (HRD). They have demonstrated that stars of 15 and 30$M_\odot$ and a very low metallicity never evolved to the red supergiant phase, but remained as blue supergiants, provided the mass loss rate was very small. In fact, the metallicity of the LMC is smaller than solar by a factor of two to three (Dufour *et al* 1982, Dufour 1984, Kudritzki *et al* 1987). Therefore, a number of different groups reinvestigated and extended the calculations for various metallicities and confirmed the earlier results (Arnett 1987, Hillebrandt *et al* 1987b, Saio *et al* 1988, Truran and Weiss 1987, 1988). In particular, it has been shown (see figure 1) that a somewhat lower metallicity than observed as a mean value in the LMC is necessary to explain that the progenitor was blue. Furthermore, spectral analyses of the supernova envelope (see below) give strong evidence for a metallicity lower than solar by a factor of three to four. All these models, however, have difficulties in explaining the existence of a high-density, slowly expanding circumstellar envelope, which strongly argues for an earlier, quite recent red giant phase (§§ 4.1.1–4.1.3). Moreover, most models evolved to the red supergiant stage if more realistic assumptions concerning mass loss were made and ended their lives there (Weiss 1989, El Eid *et al* 1987, Saio *et al* 1988).

(ii) The influence of strong mass loss on the evolution of massive stars has been investigated by several authors (e.g. Arnett 1980, Chiosi and Maeder 1986). It has been found that a star returns from the red supergiant phase to the blue as soon as it has lost most of its hydrogen-rich envelope. This mechanism has been investigated as a possible explanation for the blue appearance of Sk $-69°202$ by Maeder (1987). He calculated a $20M_{\odot}$ star which first evolved to the red supergiant phase and then returned to the blue supergiant stage once it had lost the main part of its hydrogen-rich envelope, i.e. less than $3M_{\odot}$ remained. However, we know from the spectral observations (§ 2.4.2) and the light curve models (§ 2.3.3) that certainly more than 7 (i.e. probably about 10) solar masses of a hydrogen-rich envelope have been retained.

(iii) Mixing of different stellar layers is another possible way to obtain blue supergiants because of its effect on the opacities and subsequently on the energy transport and on the density structure of the envelope. Saio *et al* (1988) have published models in which an *ad hoc* assumed mixing of helium into the hydrogen-rich layers during the red supergiant phase resulted in a return to the blue supergiant stage, i.e. it forced a blue–red–blue evolution. Their model is very encouraging, because a mixing of helium into the outer envelope of post-main-sequence stars is in good agreement with the high He/H ratios observed in some blue supergiants in the LMC (Kudritzki *et al* 1987). Furthermore, such a mixing would also be consistent with the high N/C and C/O ratios in the shell of Sk $-69°202$ (§ 4.1.2). On the other hand, it seems to be rather difficult to understand a substantial (nearly instantaneous) mixing of helium into the hydrogen-rich layers unless the star rotates near the break-up velocity (Weiss *et al* 1988), which is unlikely for the progenitor of SN 1987A. In this respect it would be extremely important if the discovery of a rapidly spinning neutron star in SN 1987A (Middleditch *et al* 1989, Kristian *et al* 1989) could be confirmed.

(iv) Woosley *et al* (1988) and Weiss (1989) have successfully generated blue progenitor models with a quarter of solar heavy element abundances and with the choice of the Ledoux criterion for convection instead of the Schwarzschild criterion that was used in most models discussed so far. Although their stellar evolution tracks are in good agreement with the observations of Sk $-69°202$ (§ 2.1), one has to note the sensitivity of the evolutionary calculations to the assumed helium abundance. In fact, a $15M_{\odot}$ star did not return to the blue if a slightly different helium abundance of the hydrogen-rich layer was assumed (Woosley *et al* 1988).

So, in conclusion, the evolution of Sk $-69°202$ has up to now not been understood. This is mainly due to the uncertainties in the theory of convection, mixing processes during stellar evolution (e.g. the influence of rotation on the stellar evolution) and the inconsistent treatment of the mass loss during different stages of the evolution (Weiss 1989, Woosley *et al* 1988). However, it now seems to be very likely that various physical effects are responsible for the blue appearance of this star, namely the low metallicity of the LMC, the low mass loss rate of the progenitor and also mixing processes.

### 2.3. *The light curve*

*2.3.1. Observations of the* UV*, optical and* IR *light curve.* As mentioned earlier, photons emitted from SN 1987A were first detected on Feb. 23.443 on a photograph at $V = 6.4^{m}$, i.e. about 3.84 hours after the neutrino detections by the KAMIOKANDE and IMB experiments (§ 2.2), indicating that SN 1987A brightened up from $12.24^{m}$ during this time. In addition, very early detections have been reported by Zoltowski (1987).

A compilation of the observations before Feb. 24.5, 1987 is given in table 1. Photographs obtained at about 1.5 to 2 hours after the neutrino detection show no object brighter than about $5^m$ at the position of Sk −69°202. Another limit on the visual brightness may be due to the non-detection by the experienced amateur astronomer A Jones from New Zealand, who observed the 30 Doradus complex with a small telescope about 1.2 hours after the neutrino events but did not see SN 1987A. On the other hand, he would have seen it if the visual magnitude were brighter than $7.5^m$ at that time (Wampler *et al* 1987). These non-detections set important limits on the visual luminosity of SN 1987A when the shock reached the photosphere (§ 2.3.3).

Beginning on Feb. 24.454, photometric measurements using different filter systems and spectral observations were available from the UV to the radio wavelength range (Bouchet *et al* 1987, Burki *et al* 1989, Catchpole *et al* 1987, 1988, 1989, Cristiani *et al* 1987, Danziger *et al* 1988a, b, 1989, Hamuy *et al* 1988, Helt *et al* 1987, Menzies *et al* 1987, Panagia *et al* 1987, Whitelock *et al* 1988; see figure 2).

The complete sample of observations allows the bolometric light curve (BLC) to be constructed for the first time. This is of major importance because in other cases the total flux has to be estimated from certain assumptions which may not be valid. The BLC (figure 3) was constructed on the bases of the flux distributions observed by the South African Astronomical Observatory (SAAO) group after Feb. 24.5, 1987 (Catchpole *et al* 1987, 1988, 1989, Menzies *et al* 1987, Whitelock *et al* 1988) and the Cerro Tololo Inter-American Observatory (CTIO) groups (Hamuy *et al* 1988, Suntzeff *et al* 1988). Uncertainties are due mainly to the fact that all filters were calibrated to standard stars, which differ significantly from the supernova spectra (Menzies *et al* 1987). This uncertainty can be overcome at least for the optical wavelength range if the visual flux is integrated by well calibrated spectra (Hanuschik and Dachs 1988, 1989). However, the size and frequency dependence of the interstellar reddening (§ 4.2) causes some problems (Menzies *et al* 1987), in particular during the very early stages ($t \leq 1$ week) when most of the flux is emitted at far ultraviolet wavelengths (Menzies *et al* 1987, Höflich 1988a). Discrepancies between different determinations can also be due to the methods used for the flux integration. It has been demonstrated that discrepancies of the order of 12% have to be expected if the same data are used but spline fits instead of black-body approximations are applied (Catchpole *et al* 1987). In general, however, the deviations between the various constructed light curves are rather small for later stages (Terndrup *et al* 1988).

**Table 1.** Fluxes (in magnitudes) of SN 1987A observed in the *B* and *V* filter system of Johnson (1966) after the neutrino detections by KAMIOKANDE and IMB (in $10^4$ s) before the first spectra were obtained.

| Time | $V$ | $U-B$ | $B-V$ | Reference |
|---|---|---|---|---|
| ≈0.64 | $\leq 7.5^m$ | — | — | Jones (1987) |
| 1.11 | $6.4 \pm 0.1^m$ | — | — | McNaught (1987) |
| 2.64 | $6.11 \pm 0.2^m$ | — | — | Zoltowski (1987) |
| 4.49 | $5.3^m$ | — | $-0.2$–$0.0^m$ | Ryker (1987) |
| 6.97 | $5.0 \pm 0.1^m$ | — | — | Garrison and Shelton (1987) |
| 8.61 | $4.77 \pm 0.1^m$ | — | — | Garrison and Shelton (1987) |
| 9.83 | $4.81^m$ | $-0.836^m$ | $0.085^m$ | Moreno and Walker (1987) |
| 12.82 | $4.63^m$ | $-0.79^m$ | $0.11^m$ | Winkler (1987) |

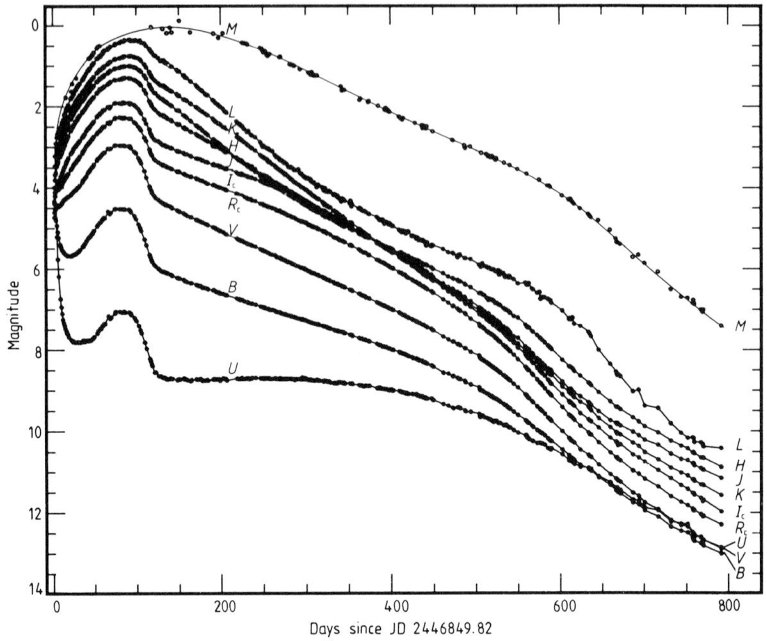


**Figure 2.** Light curves for SN 1987A in different filters of Johnson's (1966) system observed by Catchpole *et al* (1989). (From Catchpole *et al* 1989.)

*2.3.2. X-ray and γ-ray observations.* X-rays and γ-rays have been observed by the satellites Ginga and Kvant (Dotani *et al* 1987, Sunyaev *et al* 1987, Tanaka 1988, Trümper *et al* 1988), the Solar Maximum Mission satellite (SMM) (Matz *et al* 1988a, b) and balloon-borne detectors (Cook *et al* 1988, Gehrels *et al* 1988, Rester *et al* 1988, Sandie *et al* 1988).

In contrast to γ-rays and x-rays observed from other supernova remnants, the main components in SN 1987A must be attributed to radioactive decay instead of being produced by the interaction of the expanding matter with the interstellar medium or by a central pulsar.

The first direct evidence (besides the optical light curve) for radioactive decay is due to observations with SMM (Matz *et al* 1988a, b, c). At the beginning of August 1987, the γ-ray spectrometer on the SMM had detected significant net line fluxes at ≈847 keV in the background-subtracted spectrum of SN 1987A (Matz *et al* 1988a). There was also some evidence for flux in the 1238 keV line. These lines originate from the radiative de-excitation of $^{56}$Co. No positive detection was possible on 19 April 1987 (Sood *et al* 1988). These lines were predicted to be seen from supernovae as soon as the expanding envelope has become sufficiently optically thin (Bodansky *et al* 1968, Colgate and McKee 1969).

The first appearance of the γ-ray lines coincides well in time with the detections of hard x-rays by the Ginga and MIR satellites (Dotani *et al* 1987, Sunyaev *et al* 1987). Note that the Ginga team had observed the region of SN 1987A from day 2 after break-out onwards but had no positive detection before August 1987. This radiation

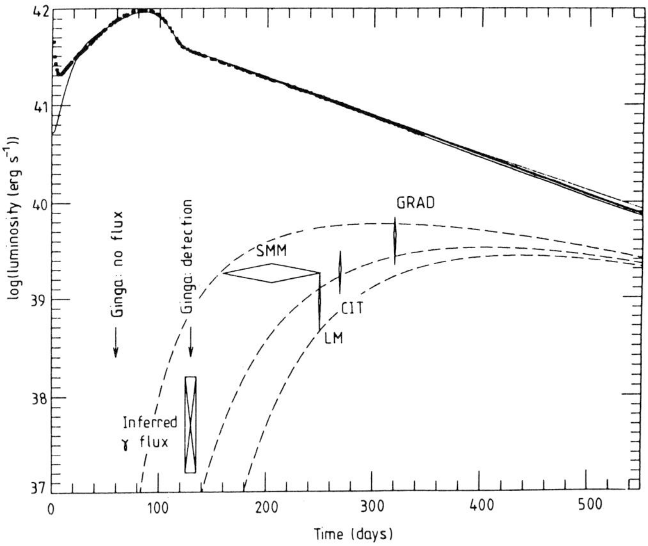


**Figure 3.** Bolometric light curve (BLC) of SN 1987A derived from the data of SAAO (broken curves, Catchpole *et al* (1987)) and $\gamma$-ray line light curves measured by the Ginga, SMM, LM, CIT and GRAD missions (see § 2.3.2). For comparison, the BLC (full curve) and the $\gamma$-ray line fluxes (broken curves) calculated for a $15M_\odot$ supernova model are shown. It was assumed that during the explosion $0.07M_\odot$ of $^{56}$Co are mixed up to a relative radius of 0.1, 0.2 and 0.4, respectively. (From Arnett and Fu 1989.)

can be well understood as a consequence of comptonisation of $\gamma$-rays inside the expanding envelope. However, this hard radiation occurred much earlier than had been expected from hydrodynamic models (Arnett 1987, 1988a, b, Woosley 1988a, b, c). On the other hand, the early emergence of x-rays and $\gamma$-rays can be modelled (§2.3.3) if (*ad hoc*) mixing of $^{56}$Co into the hydrogen-rich envelope is assumed (Ebisuzaki and Shibazaki 1988, Gebenev and Sunyaev 1988a, b, Itoh *et al* 1987a, b, Kumagai *et al* 1988, Pinto and Woosley 1988a, b, Sutherland *et al* 1988). Although the very early occurrence of x-rays and $\gamma$-rays can be explained by spherically symmetric models if mixing processes are assumed (see also § 2.3.3), these calculations run into problems as far as later observations are concerned. The nearly constant hard x-ray flux observed by the Ginga satellite up to day 570 is in conflict with the models which predicted a strong decrease by a large factor (Tanaka 1988). A possible solution may be that the absorption by newly synthesised heavy elements is effectively reduced owing to the formation of chemically inhomogeneous clumps (Kumagai *et al* 1988), as is also implied from the observed IR spectra (§ 2.4.2).

Soft x-rays have also been observed by Ginga. This component may be explained by thermal emission from the collision of the expanding envelope with circumstellar matter (Itoh *et al* 1987a, b, Masai *et al* 1987). In January 1988, the soft component flared up (Tanaka 1988). To model this increase the presence of a dense shell at about a distance of $3\times10^{16}$ cm was postulated (Masai *et al* 1988). Its existence may either be explained by fragmentation in the expanding stellar wind of the progenitor, or alternatively it may be attributed to pre-existing matter. Additional evidence for the existence of such a dense shell at about this distance may also come from the observations of the 'mystery' spot (§ 4.1.4).

*2.3.3. Bolometric light curve models.* The BLC of SN 1987A is quite unusual with respect to other type II supernovae. After a rapid decline during the first week, it increased very slowly until about day 84, then reached a broad maximum which was fainter by about a factor of 10 to 100 than a normal type II supernova at maximum. Finally, it showed a long linear decline in magnitude with mean half-life of about 77 days until day 300, and then a somewhat faster decay.

Despite these peculiarities the BLC (figure 3) can be well understood from the same general picture as other type II supernovae (Falk and Arnett 1977, Arnett 1980, 1982, Nomoto and Shigeyama 1988, Woosley 1988a, b, c). In particular, its evolution can be divided into five main parts:

(1) When the shock breaks through the photosphere, the outer layers are heated up. The optical depth of this region is small. Therefore, internal energy stored from the shock wave can be radiated away immediately. At this time, most of the radiation is emitted in the far UV, i.e. a UV light flash is to be expected with a duration of the order of a few minutes to one hour.

Although this UV radiation was not observed directly, Fransson and Lunquist (1989) have drawn attention to the fact that the ionisation stages seen in the circumstellar envelope constrain the colour temperature at shock break-out to values in the range 400 000–800 000 K (§ 4.1.2). This colour temperature is in agreement with the effective temperatures predicted from hydrodynamic models ($T = 200\,000$–$800\,000$ K; Arnett 1987, Shigeyama *et al* 1988, Woosley *et al* 1988).

The very early visual light curve can be reproduced by spherically symmetric hydrodynamic models if a total kinetic energy of the shock wave of the order of $1.5 \pm 0.5 \times 10^{51}$ erg is asumed and if the radiation field is approximated by a black body (Arnett 1988a, b, Shigeyama *et al* 1988, Utrobin 1988; Woosley *et al* 1988). However, it should be noted that the supernova atmosphere is strongly scattering-dominated so that the colour temperature may be significantly higher than the effective temperature would imply (Höflich *et al* 1986, Shigeyama *et al* 1988). Futhermore, using the density and temperature structure of model 14E1 of Nomoto *et al* (1989), detailed atmospheric calculations (Höflich 1989a, § 2.4.2) for the very early light curve have shown that the scattering optical depth during the very early stages is quickly changing due to the fast change of the densities at the photosphere, implying a strongly time-dependent change of the bolometric and visual luminosity of up to about two magnitudes relative to a black body.

In any case, even without a detailed analysis it can be seen that the results of the hydrodynamic models would not agree with the observations if the initial shock had started at the time when a few neutrinos were measured in the Mont Blanc detector (Aglietta *et al* 1987a) but they are consistent with the time the IMB and KAMIOKANDE neutrinos were detected (Arnett 1987, Shigeyama *et al* 1988, Woosley 1988a).

(2) Beginning on 25 February 1987, the UV flux was measured by the IUE satellite (Kirshner *et al* 1987, Panagia *et al* 1987). The UV flux of SN 1987A decreased during the first three days by about a factor of 1000 whereas the optical flux increased. All together, the bolometric luminosity showed only small changes in comparison to 'normal' type II supernovae. This second phase, lasting for about 25 days (Shigeyama *et al* 1988), can be characterised by a free, nearly homologous expansion and adiabatic cooling of the envelope. Energy losses due to photon diffusion are less important.

At about the time when the shock passed the photosphere of the progenitor roughly half of the total energy deposited by the shock in the envelope was stored in thermal and ionisation energy. However, because of the long diffusion timescales, most of the internal energy is converted into expansion energy. This means that only a small fraction (of the order of a few per cent) of the thermal energy is radiated away in the form of photons. The energy release in radiation (see above) depends strongly on the diffusion timescale. Therefore, the low bolometric luminosity of SN 1987A relative to 'normal' type II supernovae can be explained as a consequence of the more compact structure of the progenitor (Arnett 1987, Hillebrandt *et al* 1987b).

Adiabatic expansion then leads to a fast decrease of the effective temperature and an increase of the radius, which cause an increase of the visual luminosity. Because the energy released in the hydrogen-rich envelope is about constant, the resulting bolometric luminosity shows only small changes with time. During these first two stages the luminosity is governed mainly by the release of shock energy.

(3) During the following stages, the diffusion timescale decreases because of the decreasing density. Therefore, recombination and energy input from radioactive decay in the expanding envelope can contribute to the luminosity and it increases up to maximum light. Obviously, the diffusion timescale now governs the structure and consequently the peak luminosity of the light curve. It therefore depends strongly on the total mass of the envelope and also on the opacities, i.e. the chemical composition of the expanding envelope. In particular, the importance of a consistent treatment of recombination with respect to the diffusion of radiation should be noted (Falk and Arnett 1977, Nomoto *et al* 1989, Schaeffer *et al* 1987, Woosley 1988a, b, c).

Woosley (1988a) has demonstrated that the time at which the maximum luminosity is reached depends sensitively on the total mass of the hydrogen-rich envelope because of its influence on the diffusion timescale. He was able to determine the total mass of the hydrogen-rich envelope to be of the order of about ten solar masses, consistent with the spectral analyses (§ 4.2.2). These results were also confirmed by other groups (Arnett 1987, 1988a, b, Nomoto *et al* 1987, Utrobin 1988). However, surprisingly, the light curves constructed from simplified analytical models by Arnett (1980, 1982, 1987) and Fu (1988) show much better agreement with the observed light curve than the more sophisticated numerical calculations. This difference is mainly due to the fact that in the analytical models a depth dependence of the opacity as indicated by the stellar structure is replaced by a mean depth opacity. In fact, the discrepancies between the observations and the detailed numerical models vanish if (*ad hoc*) mixing of different layers of the progenitor star is introduced. Furthermore, mixing of layers of different chemical composition is also needed in order to explain the early occurrence of x-rays and $\gamma$-rays (§ 2.3.2) (Arnett 1988a, b, Arnett and Fu 1989, Ebisuzaki and Shibazaki 1988, Nomoto *et al* 1987, 1988a, b, 1989). We recall here that some mixing prior to the explosion is probably also needed to understand the evolution of the progenitor star (§ 2.2.2) and is certainly needed to explain the observed optical and IR spectra (§ 2.4.2). Possible processes will be discussed in § 2.5.

(4) In the next stage the evolution of the light curve is mainly governed by the release of radioactive decay energy, as long as the highly energetic $\gamma$-photons are mainly comptonised, and the decrease after maximum light can be understood as a consequence of the increase of the helium abundance and the exhaustion of recombination energy. Beginning from about day 110 the light curve was governed by the radioactive decay of $^{56}Co$, because its exponential decline (Catchpole *et al* 1987, 1988, 1989, Hamuy *et al* 1988, Whitelock *et al* 1988) with a mean half-life of 77 days is

identical to the decay time of $^{56}$Co (Martin 1987). Using the distance of 49 kpc to the LMC (Walker 1987, Walker and Mack 1988) a total initial mass of radioactive Co can be estimated, and one finds of the order of $0.07 M_{\odot}$, more or less independent of the details of the hydrodynamic models (Arnett 1988a, b, Nomoto and Shigeyama 1988, Shigeyama *et al* 1988, Woosley 1988a, b, Woosley *et al* 1988).

Although the light curve models based on the assumption of mixing of some core material into the hydrogen-rich layers give a very good representation of the bolometric light curve including the early occurence of high energetic photons, the observed $\gamma$-ray and x-ray luminosity was expected to decrease in the following months. However, the highly energetic radiation remained at nearly constant flux up to day 570 (§ 2.3.2). This discrepancy can be removed if in addition inhomogeneous mixing of radioactive Co is assumed (Shigeyama *et al* 1988). Evidence for this effect is also seen in the optical and IR spectra (§ 2.4.2).

(5) Beginning at about day 300, the $\gamma$-photons are no longer completely comptonised, i.e. the envelope became partially optically transparent. Therefore, the bolometric light curve (which does not take the high-energy radiation into account) decreased faster than expected from the $^{56}$Co decay.

No additional energy input has been confirmed up to now (September 1989). Most surprisingly, no evidence for a pulsar has been found so far in the light curve, i.e. the total luminosity of such a possible object could be limited to be less than a few times $10^{37}$ erg, corresponding to a luminosity significantly less than that of the Crab pulsar. However, besides a pulsar, additional energy sources are to be expected in the future, e.g. radioactive heating due to $^{57}$Co which was produced in smaller amounts ($2$–$3 \times 10^{-3} M_{\odot}$) than $^{56}$Co but has a much longer mean half-life (227 days) and therefore is expected to dominate the energy input from radioactive decay after about 3–4 years (Hashimoto *et al* 1989, Nomoto *et al* 1988b, 1989, Pinto and Woosley 1988b). Further deviations may also occur from the dissipation of kinetic energy once the expanding envelope starts to interact with accumulated interstellar matter.

### 2.4. *The observed spectra*

The observed spectra give direct information on the physical and chemical conditions and the geometry of the expanding envelope of the supernova at a given time. Deeper layers of the expanding envelope are observable at later times. Therefore, a detailed analysis of the observed spectra allows one to determine the velocity at the photosphere, the density structure and chemical composition as a function of depth, and the mass of the hydrogen-rich envelope. Furthermore, the chemical profiles and the lineshapes yield information on the mixing of different layers during stellar evolution or during the explosion, and may give some information on the geometry and the physical processes which govern the emitting region of the envelope. In addition, polarisation measurements give hints of deviations from spherical geometry.

The evolution of the observed spectra of SN 1987A can be divided into two main stages. (i) During the first few months the flux distribution was governed mainly by the continuum radiation modified by lines. (ii) At later times the spectra showed a 'nebular-like' phase, i.e. the flux was due mainly to emission lines in the optical as well as in the infrared wavelength region.

*2.4.1. Models for the atmospheric diagnostic.* In order to understand the early spectral behaviour of type II supernovae and SN 1987A in particular, detailed model atmospheres are needed. In principle, the same models as for 'standard' stellar atmospheres may be applied. However, there are a number of difficulties due to the physical properties which distinguish supernova envelopes substantially from stellar photospheres: (i) the density profile cannot be assumed to be given by hydrostatic equations; (ii) typical particle densities are much lower than in typical stellar atmospheres; (iii) very large velocities and velocity gradients have to be taken into account; and (iv) the radial extent may become important.

In order to address these problems a variety of models have been developed during the last few years (e.g. Branch *et al* 1981, Hempe 1981, Hershkowitz *et al* 1986, Shaviv *et al* 1984, Höflich *et al* 1986, Lucy 1987a, b, 1988). It has been shown that the effects due to sphericity are substantial even for the continuum-forming region, at least if the density profile is not characterised by very steep gradients (which are assumed by Hershkowitz *et al* 1986) and if hydrogen is partially ionised (Shaviv *et al* 1984, Höflich *et al* 1986). It has been demonstrated that the size of the continuum flux shows significant deviations of the order of 15–50% from black-body radiation if the colour temperature in the optical wavelength region is taken as the effective temperature (Hempe 1981, 1984, Shaviv *et al* 1984, Höflich *et al* 1986, Höflich 1989a).

In general, Balmer lines of hydrogen are the most prominent features in the optical wavelength region. It has turned out that these Balmer lines cannot be well reproduced by models which assume pure scattering for hydrogen (Branch *et al* 1981) or in which the interlocking between different energy levels is neglected (Lucy 1987a, b). However, the hydrogen lines can be well represented in a full non-LTE treatment (LTE: occupation numbers and ionisation equilibria are given by the local thermodynamic equilibrium), i.e. in calculations that take the collisional processes and the coupling between the energy levels and ionisation stages of different elements into account. This has been shown for type II supernovae (Hempe 1981, 1984) and in particular for SN 1987A (Höflich 1987, 1988a, b). Furthermore, the blue shift of the absorption minima of the lines cannot be attributed to a certain expansion velocity (Bildsten and Wang 1988, Hempe 1984, Höflich 1987b, Phillips 1988). Finally, the blanketing effects due to elements other than hydrogen are most important at UV, optical and IR wavelengths, at least during the phase when hydrogen is no longer completely ionised (Chilukuri and Wagoner 1988, Lucy 1987a, b, 1988, Höflich 1988a, b), although it is sufficient to treat the large number of important lines in a pure scattering approximation in order to include line blanketing for the flux distribution in the UV and optical wavelength range (Harkness and Wheeler 1988, Wheeler *et al* 1988). Models in which it is assumed that the occupation numbers and ionisation equilibria are given by the local thermodynamic equilibrium are very useful if attempts are made to identify lines and to give rough estimates of elemental abundances (Branch *et al* 1981, Williams 1987). However, detailed atomic models have to be used for quantitative analyses, at least if only a small number of strong lines are used (Höflich 1987, 1988a, Lucy 1987b).

The early spectra of SN 1987A have been interpreted by several groups, but the models were based on different approximations and assumptions. Therefore we will discuss them in some detail here in order to clarify the reliability of the various approaches.

Branch (1987a, b) has used power-law densities in order to allow for the identification of certain features in the spectra and to give some guidance to the temperature dependence of the ionisation stages that are to be expected. Spherical geometry, LTE population numbers and pure scattering approximations for lines were assumed, and the radiation field was assumed to be given by a diluted black body. Although these approximations do not seem to be very accurate for atmospheres of type II supernovae (see above), Branch could demonstrate the usefulness of this method for the identification of certain features in the spectra and for estimates of the ionisation stages at a certain time. A similar model was used for the identification of certain elements in the spectra by Williams (1987, 1988) and for the analysis of the optical and UV spectra by Harkness and Wheeler (1988) and Wheeler *et al* (1988).

Chilukuri and Wagoner (1988) have applied a similar model to interpret the continua, assuming plane parallel atmospheres. They allowed for deviations from LTE for hydrogen, but detailed balance was assumed for all line transitions. The density structures were described by a scale height and line scattering due to heavier elements was included. These models have mainly been used in order to determine the scale height and effective temperatures, and as an additional result, the distance to SN 1987A (§ 2.4.5).

Eastman and Kirshner (1988) have carried out an analysis of the very early spectra, based on a simple theory for the formation of hydrogen lines (Kirshner 1988), mainly to get some estimates on the lower limit of the mass of the hydrogen-rich envelope. Recently, these calculations have been repeated by means of more detailed non-LTE models for hydrogen (Eastman 1989).

In contrast to these models, Lucy (1987a, b, 1988) has used Monte Carlo techniques developed for stellar wind problems (Abbott and Lucy 1985). Spherical geometry and radiative equilibrium were assumed, and the density profiles were taken from hydrodynamic calculations of Arnett (see § 2.3.3) or were assumed to be power laws. Line formation occurred from coherent scattering in the fluid frame and was treated in the narrow-line limit (Sobolev approximation). This method allows for the treatment of a very large number (tens of thousands) of lines, including a consistent treatment of the velocity fields. However, his models did not take the coupling between different levels into account, and he did not calculate the continua consistently. Previously, these models have been applied widely and very successfully, mainly in order to calculate synthetic line spectra.

Spherically symmetric non-LTE models of scattering-dominated photospheres of type II supernovae have been applied by Höflich (1987, 1988a, b, 1989a, b). As a first-order approximation for the first three months density profiles were assumed to be given by the homologous expansion of the structure of a B3I star. A comparison between the density structures calculated from hydrodynamic calculations (Woosley 1988a, model 10H) and the ones used showed reasonable agreement except for the very first day and for the layers inside the hydrogen-rich region. This can be understood on the basis of analytic solutions for self-similar expansions (Sedov 1959). For stages later than three months, power-law density profiles were used instead. In this approach the expansion velocity $v_{ex}$ is given by a linear function of the distance from the centre of the star. The radiation transport equation is solved in a comoving frame scheme (Mihalas *et al* 1975, 1976) both for the lines and for the continua, including a finite linewidth. The statistical (i.e. NLTE) and the radiation transport equations are solved consistently for the continua and for the line transitions for several elements (H, He, C, N, O, Ne, Mg, ...), assuming up to three ionisation stages for each element. Coupling between different elements and ions as well as line blanketing due to a few thousand additional lines were also included. These models have been applied to the analysis of the spectra of the first months, mainly, in order to determine the density structure and the chemical composition of various layers of the star and to analyse observed line asymmetries, and finally to determine the effective temperature, the photospheric radius, the velocity fields, etc (figure 4) of SN 1987A.

For the late phases the use of the nebular theory (Osterbrock 1974) seems to be very attractive (Fransson 1987, 1988) because it can be assumed that all the lines and the continua are optically transparent, i.e. the radiation transport equation has not been solved but the emission line flux can be determined directly by adding all local components, if the ionisation stage and the corresponding source function are known. Therefore, the total mass of each element can be determined. One disadvantage of this method, however, is the fact that the depth-dependent information comes in only quite indirectly due to line broadening. Although in the nebular case an analysis of the spectra is simpler than for the earlier stages (see above) there are also some severe new problems. No (or only very indirect) information about the density and temperature structure can be derived from the models. Therefore, in general, these quantities are taken as being constant. In addition, assumptions concerning the ionisation equilibria and consequently the excitation and ionisation mechanisms enter and strongly affect the analysis. In this respect the use of IR lines is less model-dependent. The particle density in the line-forming regions of SN 1987A is rather high compared with 'classical' H II regions even at later stages, implying that the thin LTE assumption for the population numbers is relatively accurate for IR lines, even if the lines are optically thin. Using the IR lines has an additional advantage because line blanketing is less pronounced in this region (Meikle *et al* 1988). Unfortunately, accurate determinations of the total mass of several elements (e.g. C, O, Si) are not available at present because uncertainties in the ionisation and level excitation equilibria are still too large.

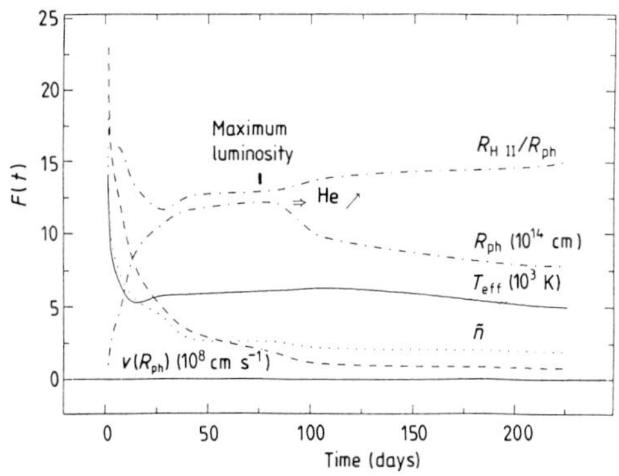


**Figure 4.** The radius $R_{ph}$ at which the optical depth is 1 for true absorption at 5000 Å, the effective temperature $T_{eff}$, the velocity field $v$ at $R_{ph}$ and the power-law index $\bar{n}$ of the corresponding density slope and distance $R_{H II}$ up to which hydrogen is mainly ionised are given as functions of time (in days since the explosion). The data have been derived from the atmospheric diagnostics of SN 1987A by Höflich (1988b).

*2.4.2. Interpretation of the observed spectra.* The early observed spectra of SN 1987A showed a number of peculiarities with respect to 'normal' supernovae of type II in the ultraviolet (Panagia 1984) as well as in the optical wavelength region (Branch *et al* 1981).

(i) During the first few days the energy distribution of the flux showed very rapid variations in both the optical and ultraviolet wavelength regions but it changed only slowly during the following few months. In particular, a UV flux deficit with respect to the optical flux was seen.

(ii) Already in the very early stages strong lines were observed in the spectra which showed a very rapid time evolution (e.g. in the spectra of SN 1979c hydrogen lines first occurred several weeks after the maximum of light (Branch *et al* 1981)).

(iii) The UV flux decreased by about a factor of 1000 during the first few days (Cassatella 1987, Kirshner *et al* 1987, Panagia *et al* 1987, Wamsteker *et al* 1987). After about two weeks all the radiation below 2000 Å originated from the companion stars of the B3Ia supergiant (Gilmozzi *et al* 1987). From this time on, an unusual UV flux deficit was observed relative to the flux at optical wavelengths. In the following weeks the UV flux remained nearly constant.

The relatively rapid evolution of the spectra also at optical wavelengths followed by small changes later on can be understood as a consequence of the rapid increase of the photospheric radius and of the strong coupling of the expansion of the photosphere to the expanding matter. Simultaneously the effective temperature decreased rapidly. One to two weeks after the explosion, the geometrical dilution of the material and the recombination of hydrogen outside a certain radius resulted in a much slower change of $R_{ph}$ and, consequently, in smaller changes with time of the spectral energy distribution, as long as the continuum was determined by electron scattering (figure 4, Höflich 1988a, b). The relatively small variations of the bolometric luminosity can be explained as being due to additional energy released when the temperature dropped below the recombination temperature of hydrogen (§ 3.3.3).

The observed UV flux deficit is mainly a result of line blanketing in the UV (Lucy 1987a, b, 1988). A metallicity of 1/2.75 times solar, as derived from H II regions in the LMC (Dufour 1984), is consistent with the early UV spectra for most of the elements others than H and He. However, the strong UV lines are saturated. Therefore, this inferred metallicity should be regarded as an upper limit only. Note that the UV line blanketing can only become strong if Thompson scattering does not dominate the line opacities in the line-forming region (Höflich 1987, 1988a, b). Therefore, the difference between the 'normal' spectral behaviour of type II supernovae and SN 1987A can be understood as being due to the relatively steep density gradient in the progenitor, and to the smaller luminosity, which resulted in a very early hydrogen recombination phase, i.e. as a consequence of the compact structure of the progenitor.

The evolution of the optical spectra (figures 5 and 6) has been monitored from about Feb. 24.5, 1987 until now by various groups at the AAO, CTIO, ESO and the Bochumer Telescope in Chile (Catchpole *et al* 1987, 1988, 1989, Couch 1988, Danziger *et al* 1988a, b, Feast 1988, Fosbury *et al* 1987, Hanuschik and Dachs 1987a, b, 1988, 1989, Menzies *et al* 1987, Phillips 1988, Tyson and Boeshaar 1987). The first observed spectra (Feb. 24.5, 1987) showed mainly broad, highly red-shifted absorption features with little emission, which can be attributed to H I and the He I 2p–3d transition of the triplet system. The wings of H$\alpha$ corresponded to line shifts of the order of about 25 000 km s$^{-1}$. Within the first week, strong P-Cygni features of hydrogen lines developed, and the He I line disappeared. However, new absorption features occurred which could be identified as being due to Na I, Ca II, Fe II, Sc II and Ba II (Williams 1988). In the following weeks, strong line blanketing due to a large number of weak lines developed. The spectral evolution up to about the end of June 1988 can be described by a decrease of the line shifts due to the decreasing expansion velocity at

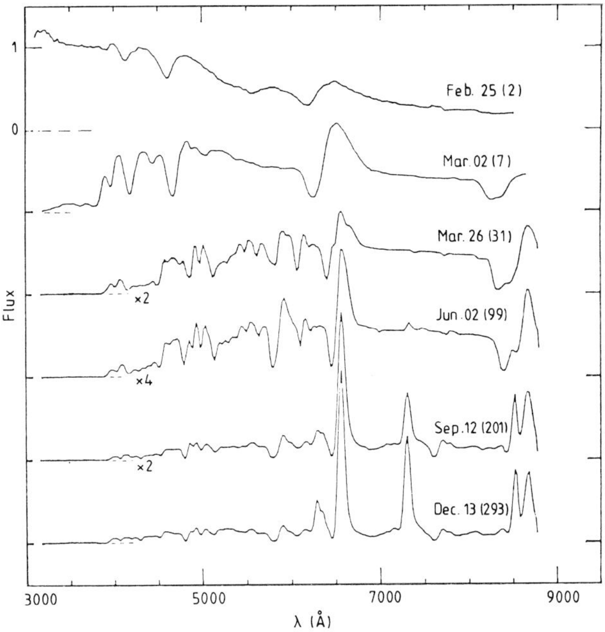


**Figure 5.** Time evolution of the spectra of SN 1987A between 3000 and 9000 Å (days 2–31: Hanuschik and Dachs (1988); day 99: Hanuschik and Dachs (1989); days 201 and 293: S Leimeswenger and R Patzel, personal communication).

the photosphere, an increase of the emission of H$\alpha$ relative to its absorption, and the development of a P-Cygni-like profile of the resonance line of Na I at about 5860 Å, starting near the end of March. However, the overall energy distribution showed only minor changes up to about June 1987. The following evolution was characterised by a strong decrease of the continuum flux relative to the emission features mainly of H$\alpha$, Na I and the forbidden [O I] line.

The optical spectra including the hydrogen lines that were observed during the first seven months can be reproduced by NLTE models (Höflich 1987, 1988a, b, figure 7). The early occurrence of strong hydrogen lines confirms a steep density gradient and the early recombination phase of hydrogen (see above). Therefore, these lines can be regarded as a direct consequence of the spectral type of the progenitor. In order to best reproduce the observed hydrogen lines, a statistical velocity field of the order of 5–20% of the expansion velocity has to be assumed. This should not be regarded as microturbulence, but may reflect the existence of deviations from sphericity on small scales (see below for the polarisation). One third of the solar abundance is needed for all heavy elements except for some s-process elements (table 2). For Na, Sc and Ba, which have been identified in the spectra, overabundance factors of about 5, 2 and 10 relative to their solar values have to be chosen, with a possible uncertainty of about a factor of two, in order to fit the spectra. However, there is no evidence for abundance gradients in the hydrogen-rich layers. The observed increase of the H$\alpha$ emission can be attributed to a decrease of the density slopes. The decrease of the continuum flux relative to the line emission after maximum light and the shift of

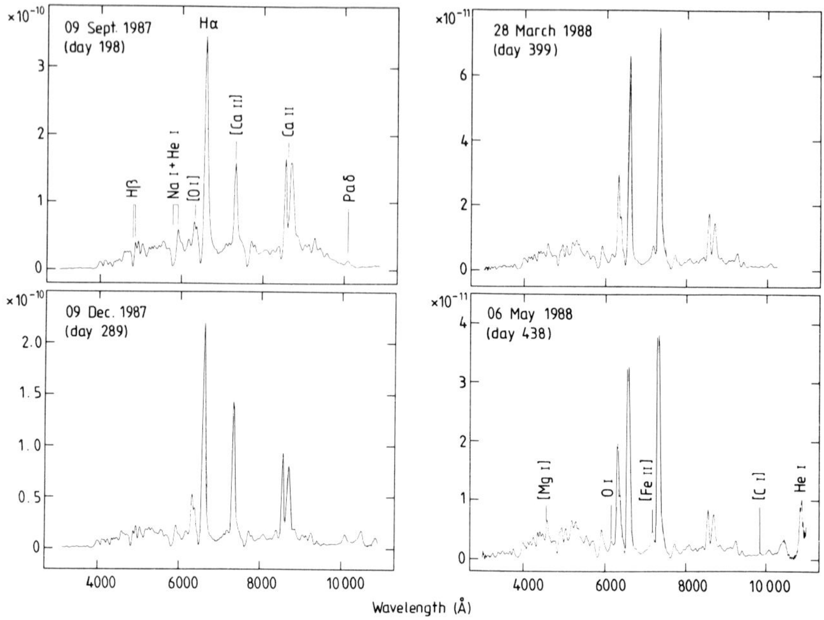


**Figure 6.** Late time spectra of SN 1987A obtained by the CTIO 1.5 m telescope at times approximately 200, 300, 400 and 450 days after the explosion. (From Terndrup 1988.)

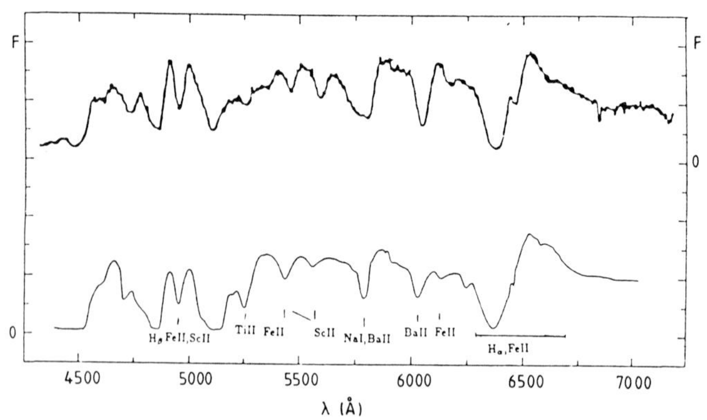


**Figure 7.** Spectrum of SN 1987A observed by Menzies *et al* (1987) on 23 May in comparison with the reddened, synthetic spectrum (lower spectrum, shifted by one unit; $E_{B-V} = 0.15^{\mathrm{m}}$) calculated by Höflich (1988b); see figure 4. In addition, identifications of some strong features are given. Note, however, that most of the lines are strongly blended with a large number of weak lines.

**Table 2.** Abundances relative to solar values at the photosphere derived from the optical spectra (see figures 4 and 7) on 23 March, 16 April, 14 May, 5 June and 1 October 1987, i.e. models I–V, respectively. For all elements other than those given in the table one third of solar abundances had to be assumed in order to fit the spectra.

| Element | Model I | Model II | Model III | Model IV | Model V |
|---|---|---|---|---|---|
| He | — | — | — | — | 5.0 |
| C | — | — | — | — | $\geqslant 25$ |
| O | — | — | — | — | $\geqslant 25$ |
| Na | 7.0 | 5.0 | 4.0 | 4.0 | 4.0 |
| Sc, (V, Cr, Sr) | 2.5 | 2.0 | 2.0 | 2.0 | 2.0 |
| Ti | 1.0 | 1.0 | 1.0 | 1.0 | 0.8 |
| Ba | 20 | 10 | 10 | 8.0 | 8.0 |

the energy distribution to higher temperatures were caused by a decrease of the photospheric radius, which resulted from the decrease of the luminosity and the increasing helium abundance. Starting about October 1987 helium-rich matter was seen at the photosphere (table 2). However, the presence of some amount of H at that time requires significant mixing of the inner few solar masses of the hydrogen-rich envelope with deeper layers during the explosion. This result is consistent with the light curve models and, in particular, with the observations of the hard radiation (§§ 4.3.2, 4.3.3).

Lucy (1987a, b, 1988) has performed a similar analysis and found satisfactory agreement between observed and calculated optical spectra for the first few weeks, assuming an abundance of 1/2.75 times solar values for the heavy elements. However, in his models he also needed a large overabundance of Ba, ranging from a factor of about 20 times solar at day 20 to about 5 some weeks later. The most serious deficiency in his model is the neglect of interlocking between different energy levels of hydrogen, resulting in a poor modelling of the Balmer lines of hydrogen and the sodium lines.

The transition from the hydrogen-rich to the helium-rich matter occurred around October 1987. By integrating the continuity equation of particle density and the photospheric velocities the total mass of the hydrogen-rich layers of the progenitor could be estimated to be of the order of 9–11$M_\odot$ (Höflich 1987, 1988a, b). However, this value was derived on the basis of one specific calculation and therefore a comparison between different models would be helpful. Unfortunately, no other detailed model is available up to now. A comparison can only be made with work by Eastman and Kirshner (1988), who used their model (see § 2.4.1) in order to determine the total mass of material with velocities larger than 6000 km s$^{-1}$ from their analysis of the early spectra and obtained a value of 1.8$M_\odot$. The corresponding mass of Höflich's (1987, 1988a) calculations was about 1.5$M_\odot$. Therefore, the total mass as derived from the optical spectra seems to be not too strongly affected by the different physical approximations.

Note that the value of the mass of the hydrogen-rich layers as determined from the spectral analysis is independent of the light curve models but gives essentially the same result (7–10$M_\odot$; Arnett 1988a, b, Shigeyama *et al* 1988, Woosley 1988a). Therefore, the value of 10$M_\odot$ should be regarded as quite safe, and, consequently, it can be assumed that the progenitor has undergone only moderate mass loss prior to the explosion.

Spectroscopic IR data have been obtained and published by most of the major southern observatories such as AAO (Bailey *et al* 1988, Meikle *et al* 1988), CTIO (Elias *et al* 1988), ESO (Danziger *et al* 1988a, b, Oliva *et al* 1988), MSSSO (McGregor 1988) and SAAO (Catchpole *et al* 1987, 1988, 1989, Menzies *et al* 1987, Whitelock *et al* 1988) and from the Kuiper Airborne Observatory (Larson *et al* 1987, 1988, Rank *et al* 1988a, b). The early spectra in the IR were characterised by a strong continuum on which hydrogen lines were superposed having a P-Cygni-like structure. Beginning in June 1987, more emission features began to develop gradually. Although hydrogen emission lines from the Paschen, Brackett and Pfund series still dominated the spectra, lines due to heavier species such as He I, C I, O I, Mg I, Na I, Si I, K I, Ca I-II, Fe II, Co II, Ni I and Sr II began to develop.

Starting in June 1987, carbon monoxide ($CO, CO^+$) (Danziger *et al* 1989, Feast 1988, Elias *et al* 1988) and SiO (Rank *et al* 1988a, b) appeared. The analysis of Spyromilio *et al* (1989) suggests that the CO was expanding at velocities of $\approx$500–2000 km $s^{-1}$, indicating that the CO has to be associated with the expanding supernova envelope and not with the circumstellar environment. Assuming a simplified model for the ejecta and the ionisation state (LTE), the mass seen in CO could be derived to be $10^{-5} M_\odot$ (Spyromilio *et al* 1988). Synthetic spectra of the CO features have been calculated from the same models which gave a good representation of the optical spectra (see the above discussion). The large overabundances of C and O by more than a factor of 25 relative to hydrogen obtained in that way give additional evidence for mixing processes (Sharp and Höflich 1988). However, the absolute value should be regarded as uncertain because there may be significant deviations from LTE for the formation of CO (Petuchowski *et al* 1988).

The temperatures derived from the IR lines and from the continua (Danziger *et al* 1988a, b, Elias *et al* 1988, Meikle *et al* 1989, Phillips 1988) are consistent with the values derived from the optical spectra. Here, most observers have used the work of Branch (1987a, b) as a guide to the ionisation balance that can be expected (§ 2.4.1). Analyses of the Co II lines using the nebulae theory (Osterbrock 1974) led to an initial mass of $^{56}$Co of about $0.07 M_\odot$, which is consistent with the mass estimated from the light curve. This means that in spring 1988 all Co was seen in the IR spectra.

From about July to December 1987, the spectra changed significantly from continua superposed by lines to emission line spectra in a 'nebular-like' phase in the optical and IR wavelength regions. From then on, the observed spectra (Catchpole *et al* 1989, Danziger *et al* 1989, Feast 1988, Whitelock *et al* 1988) resemble closely those observed from other supernovae at similar stages of their evolution (Filippenko 1989, Panagia 1984). Features due to H$\alpha$ and forbidden lines of the IR doublet [Ca II] (7291 and 7323 Å), the [O I] doublet (6300.23 and 6363.88 Å) and others became stronger with time (see figures 5 and 6), indicating a decreasing density.

It has been noted that in high-resolution spectra accumulated since October 1987 (Danziger *et al* 1989) the profiles of the [O I] doublet at about 6300 and 6363 Å contained significant structures which remain stable over several months (Dopita 1988). This clearly indicates the existence of some chemical inhomogeneities or geometrical clumpiness of the expanding inner regions of the envelope. Additional strong evidence for the fact that mixing processes must have occurred comes from observations by the Kuiper Airborne Observatory during two flights in November 1987 and spring 1988 (Erickson *et al* 1988, Rank *et al* 1988a, b, Witteborn *et al* 1989) where strong emission lines of Ni I-II, Ar II, Co II, Fe I-II, Co and SiO were detected. These strong lines have also been seen in several ground-based observations (see above). The full width at half maximum of the roughly symmetric line cores corresponds to about 3000 km s$^{-1}$, showing that these elements move with about the velocities observed for the photospheric region in May 1987 (figure 8). This confirms that at least some Co has been mixed with the inner layers of the hydrogen-rich region. However, at about May 1987 no enrichment of Co was seen in the optical spectra (Höflich 1988a, b), which means that the covering factor due to Co has been relatively small and that Co was mixed in the form of bubbles. Possible mechanisms will be discussed in § 2.5.

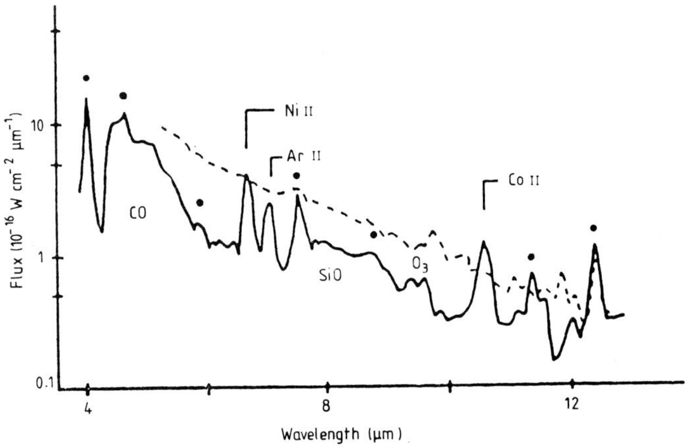


**Figure 8.** Spectrum of SN 1987A from 4000 to 12 500 nm on 12 November 1987 (full curve) and 21 April 1987 (broken curve) observed by Rank *et al* (1988b). (From Rank *et al* 1988b.)

*2.4.3. Line asymmetries.* The observed spectra do not only give information on the chemical composition and the density structure but also may give some hints on deviations from sphericity and/or on physical mechanisms which govern the physical situation at the photosphere. Asymmetries seen in several lines at optical and IR wavelengths provide important information on the physical conditions in the envelope and on its structure.

Between 14 March and about July 1987, intrinsic structures in the H$\alpha$ profile were observed (Catchpole *et al* 1987, Hanuschik and Dachs 1988, Hanuschik 1988, Menzies *et al* 1987, Phillips and Heathcote 1989) that are correlated with asymmetries of IR hydrogen lines (Larson *et al* 1987, 1988). Additional observations by Phillips and Heathcote (1989) in both the optical and IR wavelength regions have confirmed these data. These peculiarities were seen at about the same Doppler shift for all IR hydrogen lines. Therefore, the region of their origin seems to be nearly stationary in a reference frame moving with the matter.

Several attempts have been made to explain the H$\alpha$ features, including geometrical asymmetries (shells, clumps, etc (Larson *et al* 1987); jets (Rees 1987)), additional excitation due to $^{56}$Co (Lucy 1988) or perturbations in the non-LTE level populations by self-blanketing of hydrogen lines (Höflich 1988a). In the relativistic jet model proposed by Rees (1987) the H$\alpha$ emission is attributed to the thermalisation of fast moving particles or hard radiation originating near the region of the jet where the envelope is punctured by the outflowing plasma. Arguments against this interpretation are the gradual emergence of the features and the observed shift of the peak emission to negative velocities (Lucy 1987).

In contrast, Lucy (1988) has suggested that the hard radiation, presumably from the decay of radioactive cobalt, may penetrate into the atmospheric layers at these epochs, resulting in an extra source of ionisation near the photosphere. This model is strongly supported by the observations of the Co lines, which have been seen from the beginning of November 1987 with a velocity corresponding to the velocity of the matter at the photosphere in April/May 1987 (figure 4). However, it may be difficult to understand why the structures vanished after June 1987, although x-rays and $\gamma$-rays were detected only after August 1987 (see § 2.3.2). Furthermore, from an excitation just above the photosphere strong blue-shifted but less strong red-shifted features would be expected. In contrast, all IR lines show a much stronger red-shifted than blue-shifted feature.

An alternative explanation was given by Höflich (1988a), who attributes these features to effects of selective line blanketing on the level population of hydrogen. According to him, the main contribution is due to blanketing of the Pa$\alpha$ with the Br$\delta$ and Br$\varepsilon$ lines, resulting in a local perturbation of the third and fourth levels of hydrogen, i.e. of the upper level of H$\alpha$ and the lower levels of the Paschen and Bracket series. This effect depends on the Doppler shift and, therefore, explains why the observed features were almost stationary in the comoving frame of the expanding matter and why shifted structures of about the same Doppler shift were seen in all Bracket and Paschen lines and in H$\alpha$. However, the interaction between Bracket and Paschen lines is strongly influenced by the magnitude of the assumed statistical velocity field and, therefore, is not model-independent. Furthermore, a similar structure observed in the Na I line could not be explained consistently (Hanuschik 1988, Phillips and Heathcote 1989). So, in conclusion, a convincing explanation of these additional features has still not been found.

*2.4.4. The polarisation measurements.* There is also some evidence for the existence of significant deviations from sphericity. During the first few months several groups measured frequency-dependent linear polarisations of up to about 0.5% in the continua as well as in lines (Barrett 1987, Benvenuto *et al* 1987, Mendez *et al* 1988, Schwarz and Mundt 1987, Schwarz 1987). Jeffery (1987) was the first to derive from the polarisation data of the Balmer lines an axis ratio of the order of about 0.6–0.8 assuming a prolate ellipsoid. By assuming a prolate ellipsoid and by using time-dependent density profiles from Woosley *et al* (1987b), Mendez *et al* (1988) obtained slightly smaller values (0.8–0.92) from the continuum data. They were successful not only in determining the degree of polarisation at a certain time, but were able to reproduce in addition its time dependence over the first 50 days. In addition, they used the frequency dependence observed in the continuum polarisation. From these data they derived an ellipticity of the order of 0.8–0.92 seen at an inclination of about 15°. They used the analytical formulae given by Cassinelli *et al* (1987), i.e. in their analysis they included the depolarisation factor due to the extension of the central source, but applied the optically thin approximation according to Brown and McLean (1977).

Höflich *et al* (1989) have criticised this approach by arguing that the influence of optical depth effects may be important. By using a Monte Carlo technique, they determined the polarisation for pure Thomson scattering atmospheres. Density profiles implied from the atmospheric diagnostics were taken, but with an outer radius given by the recombination radius of hydrogen (figure 4). The light was assumed to be unpolarised at the thermalisation optical depth and this was taken as the inner boundary. A somewhat smaller axis ratio ($E = 0.94$) was needed for an oblate ellipticity in order to fit the size and time behaviour in the continuum. Note, however, that the same quantitative agreement can be achieved for a prolate ellipsoid seen pole-on ($E \approx 0.85$–$0.90$). Furthermore, there is no reason to assume rotational symmetry other than for the sake of simplicity. It can be concluded, therefore, that the polarisation data allow only for a rough estimate of global deviations from sphericity. They may be of the order of 5–20%.

We conclude by noting the interesting coincidence that the anisotropy derived from the polarisation data is in agreement with the size of the assumed statistical velocity (see above), indicating a common origin of both the polarisation and the statistical line broadening in small-scale deviations from sphericity.

*2.4.5. Supernovae as distance indicators.* We conclude this section with a discussion of SN 1987A as a good test case for using atmospheric models of supernovae as distance indicators. Because supernovae are the brightest single celestial objects and may at maximum reach the same luminosity as a whole galaxy, they can be used as standard candles (Kirshner and Kwan 1974) to determine the distances to galaxies by the Baade–Wesselink method (Baade 1926, Wesselink 1946) and they may provide a tool to calibrate other methods and, especially, to fix the constant $H_0$ of Hubble's law which gives the relation between the observable expansion velocity and the distances of galaxies. However, to apply the method the intrinsic fluxes of supernovae have to be known. These must be derived from atmospheric models. Since the distance to the LMC is fairly well known, SN 1987A can provide a unique opportunity to test the reliability of the absolute intrinsic fluxes derived from atmospheric models.

Several groups have estimated the distance $D$ to SN 1987A from their supernova model atmospheres. Chilukuri and Wagoner (1988) have obtained a distance of $43 \pm 4$ kpc mainly by using the UV flux. By using pure hydrogen atmospheres Höflich (1987) derived from the optical spectra of the first week $D$(SN 1987A) to be $46 \pm 4$ kpc. If the line blanketing effects due to heavier elements were taken into account, $D$(SN 1987A) was slightly modified to $48 \pm 4$ kpc on the basis of spectral fits for the first four months (Höflich 1988b, 1989a). The differences in these estimates may be explained as being due to extension effects, which were not taken into account by Chilukuri and Wagoner (1988). However, the good agreement with the distances derived from other methods ($D$(LMC) $\approx 49$ kpc (Walker 1987, Walker and Mack 1988)) indicates that type II supernovae may now be used as quite reliable distance indicators, at least if sophisticated atmospheric models are applied.

By means of a much simpler approach Branch (1988) has derived a distance of 55 kpc from the fluxes in the early spectra, assuming black-body radiation for the distribution and magnitude of the intensities. This result makes it tempting to use black-body fits in future. However, it should be noted that from detailed calculations an overestimate of about 20% is to be expected if the density gradient is large (i.e. extension effects are less important), and if the temperature is relatively low, but the error will become much larger for flatter density gradients and higher luminosities of other type II supernovae (Shaviv *et al* 1984, Höflich 1989a).

*2.5. Models for the mixing process*

Strong inhomogeneous mixing processes of different layers are evident from the bolometric light curve (§ 2.3.3), from the observation of x-rays and $\gamma$-rays (§ 2.3.2) and from the observed spectra (§ 2.4). However, in the models mixing had to be introduced *ad hoc* in order to be able to fit the observations. Therefore, several groups have searched for possible mechanisms. Nagasawa *et al* (1988) found Rayleigh–Taylor instabilities and fragmentation in supernova models based on polytropic density profiles and a three-dimensional smooth particle hydrodynamic code. Müller *et al* (1989a) have repeated these computations by means of both analytical methods (Chandrasekhar 1961) in one dimension and numerical simulations using various codes and methods in both two and three dimensions, but could not confirm the earlier results. Therefore, they concluded that the instabilities found by Nagasawa *et al* (1988) were caused by their numerical method rather than by fluid dynamic effects. Their conclusion has recently been confirmed by smooth particle hydrodynamic simulations of Benz and Thielemann (1989).

However, Arnett *et al* (1989) and Müller *et al* (1989b) found that Rayleigh–Taylor instabilities (figure 9) might occur out to the hydrogen-rich envelope if a realistic stellar structure was used. These instabilities led to small-scale structures in both the density and chemical composition. Although so far these calculations have only been

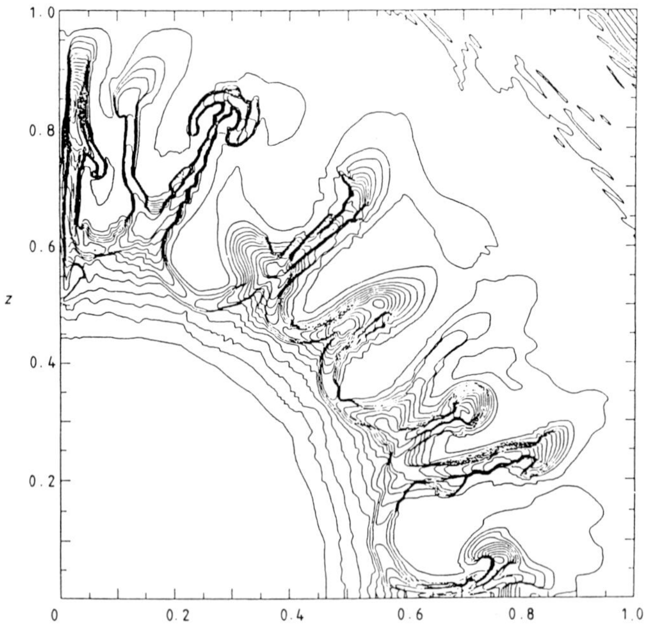


**Figure 9.** Density contours (5% spacing) showing Rayleigh–Taylor instabilities 9814 seconds after the explosion. An explosion energy of $2 \times 10^{51}$ erg and a B3 supergiant structure were assumed. The 'mushroom head' shapes that are characteristic of Rayleigh–Taylor instabilities are evident. (From Arnett *et al* 1989.)

performed under the assumption of rotational symmetry, it is very likely that the instability persists in less symmetric situations.

Another possible mechanism for mixing may be the expansion of hot bubbles of radioactive nickel (e.g. Arnett 1988a, b, Woosley 1988b, c). In this picture, energy release by radioactive decay of $^{56}$Ni during the early stages may cause instabilities and may result in mixing of different inner layers.

## 3. The explosion mechanism

Prior to SN 1987A, there was no direct evidence that type II supernovae will lead to the formation of neutron stars, although several supernova remnants are known that do contain a neutron star (Crab, Vela, RCW 103, etc). But in none of those cases do we know beyond doubt that the events were indeed of type II. Moreover, the remnant of an explosion of a rather massive star, Cas A, apparently does not contain a neutron star. The assumption, therefore, that massive stars ($M \gtrsim 8M_\odot$) are the progenitors of type II supernovae and neutron stars rests mainly on theoretical models of stellar evolution (for recent reviews see, e.g., Hillebrandt (1987a) and Blinnikov *et al* (1988)). On the other hand, a significant fraction of all Galactic supernovae may escape visual detection, as did Cas A some 350 years ago, and may be discovered only by their neutrino emission. It is interesting to note that, if SN 1987A had exploded near the centre of our own galaxy, it would probably have been seen as a neutrino source only. In the following subsections we shall briefly discuss the theoretical models that have been developed over the last 20 years.

### *3.1. Core collapse models of type II supernovae*

Several ways have been suggested which may, under favourable circumstances, convert a few per cent or less of the gravitational binding energy of a forming neutron star into outward motion of the stellar envelope, among which at present two scenarios seem to be the most promising ones. Firstly, it has been proposed that a hydrodynamic shock wave created by the rebounding neutron star may release enough energy in the stellar envelope (Arnett 1977, 1983, Baron *et al* 1985a, b, Bruenn 1987, Hillebrandt 1982, Hillebrandt *et al* 1984, Imshennik and Nadyozhin 1983). However, as will be discussed later, this mechanism requires very special initial conditions and/or special properties of the nuclear equation of state. It may, therefore, work only for a narrow range of stellar masses, and in particular not for stars as massive as Sk $-69°202$ (Hillebrandt 1987b). Secondly, it has been found that neutrinos leaking out of a newly born neutron star may heat up certain mass zones of the stellar mantle sufficiently to revive a shock that otherwise would not reach the envelope (Wilson 1985, Bethe and Wilson 1985, Wilson *et al* 1986). This effect, however, is extremely sensitive to the neutrino energies at very small 'optical depth', and therefore numerical models may not be very reliable. Moreover, the explosion energy found in these simulations may be too low to explain the very energetic outburst of SN 1987A. We shall come back to these questions later in § 3.2.

The outcome of computer simulations of stellar collapse will depend on the pre-collapse stellar models, the adopted microphysics input data and the numerical methods. Given the fact that the various computations use different initial models, different equations of state and/or different numerical schemes it is not surprising that the results also differ considerably. A simple estimate can clarify these difficulties. From the observations of SN 1987A we know that almost the entire binding energy of the newly born neutron star ($\gtrsim 10^{53}$ erg) was radiated away by neutrinos. The kinetic energy of the ejecta ($\simeq 10^{51}$ erg; see § 2) and the energy in electromagnetic radiation ($\lesssim 10^{49}$ erg; see § 2) were small corrections at the level of a few per cent or less. It is this small correction in the energetics of the outburst that numerical models try to determine.

There are, however, several features that are the same in all computations using the standard theory of weak interactions and non-rotating stellar models. During the collapse phase the entropy stays low ($S \lesssim 2k_B$/nucleon) and therefore all models collapse to nuclear matter density. Moreover, about that fraction of the stellar core corresponding to the Chandrasekhar mass ($M_{CH} \simeq 5.72 M_\odot y_e^2$; $y_e$ = electron concentration) collapses homologously, with $v = a(t)r$, where $a(t)$ is independent of $r$, in agreement with analytical considerations (Goldreich and Weber 1980). Consequently, a sonic point must exist and the matter outside the sonic point has supersonic velocities close to free-fall velocity. Since $M_{CH}$ is proportional to $y_e^2$ it will decrease during collapse. This decrease depends on the entropy of the initial model, on electron capture rates, the neutrino transport scheme and the equation of state, and has turned out to be different in most computations.

At core-bounce the central density of the star is in general only slightly higher than nuclear matter density ($\rho_c \gtrsim 3 \times 10^{14}$ g cm$^{-3}$) unless rather soft EOS are used (Baron *et al* 1987). The inner core, defined by the condition that its velocity is subsonic, is stopped on a sound-crossing time ($\lesssim 1$ ms) once nuclei in the centre of the star have dissolved into a homogeneous fluid of free nucleons and the EOS stiffens ($\gamma \simeq 2.5$–$3$). Because the outer material is still falling with supersonic velocity, a shock must form near the sonic point at $M_{CH}$ (or a radius of about 20 km). The unshocked inner core does not expand against the ram pressure of the supersonic outer layers and achieves a hydrostatic equilibrium soon after bounce. From energy conservation one can estimate that the energy that is put into the shock (Brown *et al* 1982) is typically

$$E_{\text{shock}} \simeq E_B^{\text{ic}} \simeq (4\text{–}8) \times 10^{51} \text{ erg} \tag{3.1}$$

where $E_B^{\text{ic}}$ is the binding energy of the unshocked inner core, in agreement with the results of numerical simulations. Here, the main source of uncertainties is the stiffness of the EOS near nuclear matter density but neutrino transport can also change these numbers considerably.

The outgoing shock wave is heavily damped by energy losses due to nuclear photodissociations, which cost about $8 \times 10^{18}$ erg g$^{-1}$, and neutrino losses once the shock has passed the neutrino sphere. Neglecting neutrino losses for a moment, we find from equation (3.1) that at most $0.5 M_\odot$ of heavy nuclei can be dissociated by the shock, and a necessary condition for successful propagation is

$$M_{\cdot\text{Fe}\cdot} - M^{\text{ic}} \lesssim 0.5 M_\odot \tag{3.2}$$

where $M_{\cdot\text{Fe}\cdot}$ is the mass of the original iron core. Numerical models predict $y_e \lesssim 0.38$ and thus $M^{\text{ic}} \lesssim 0.7 M_\odot$. It follows immediately that only stellar models with iron-core masses less than $1.2 M_\odot$ can lead to prompt explosions. Stellar evolution calculations (Hillebrandt *et al* 1984, Nomoto 1984, Woosley 1986) have shown that this condition limits the mass range of possible progenitor stars to values between 8 and $12 M_\odot$ (or at most $15 M_\odot$) on the main sequence, and the only successful collapse computations

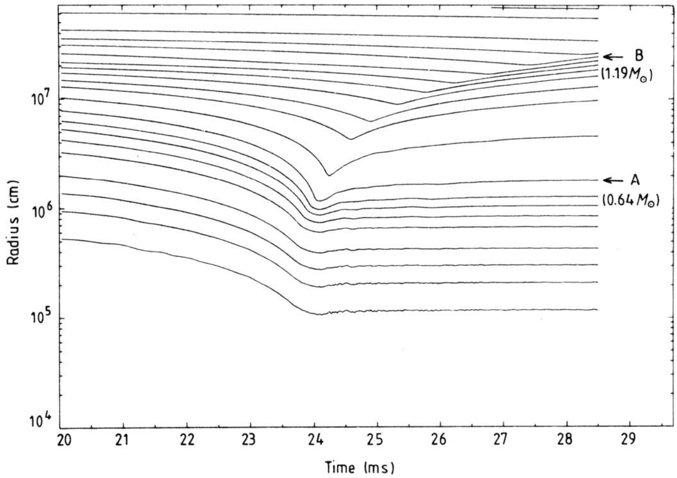


**Figure 10.** Radius against time for various mass zones of a collapsing and exploding stellar model of $9M_\odot$. A is the unshocked inner core and B is the first zone that reaches escape velocity. (From Hillebrandt *et al* 1984.)

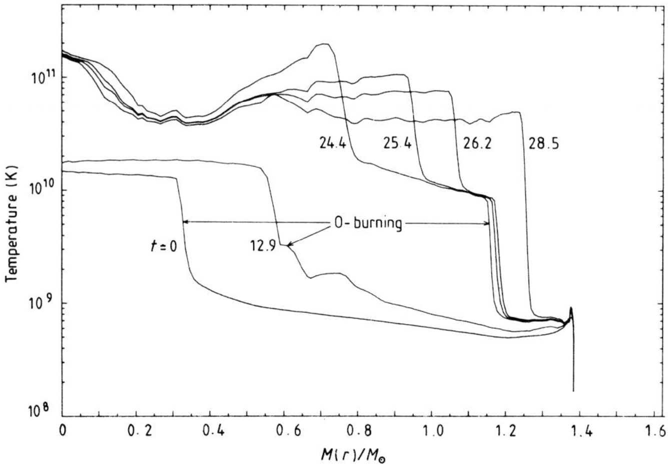


**Figure 11.** Snapshots of temperature against mass in units of solar masses for the stellar model shown in figure 10. The curves are labelled with the time in ms measured from the beginning of the computations. The drop in temperature in the oxygen-burning shell is also indicated. (From Hillebrandt *et al* 1984.)

have indeed been performed with stars in this mass range (Hillebrandt 1987a, Baron *et al* 1987).

Some results of such a computation are shown in figures 10 and 11. It should be noted, however, that in other simulations even the most favourable initial conditions do not lead to prompt explosions (Wilson 1988), probably because different equations of state and different neutrino transport schemes were used. Therefore we reach the conclusion that at present we cannot answer the question whether or not stars can explode by the core-bounce mechanism and we have to keep this unpleasant situation in mind when we try to interpret the neutrino events detected from SN 1987A.

Among researchers in the field, however, there is general agreement that stars with iron cores more massive than about $1.4M_\odot$ cannot explode by the core-bounce mechanism. Conventional models of stellar evolution therefore exclude main-sequence masses above about $15M_\odot$. Figures 12 and 13 show a typical example of a simulation where the shock stalled at about $1.2M_\odot$

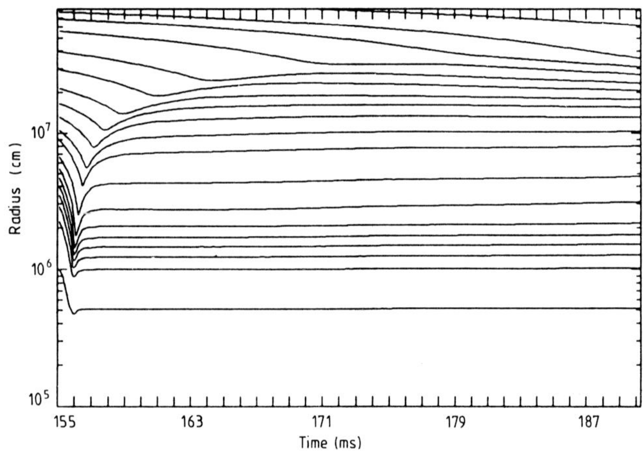


**Figure 12.** Same as figure 10, but for a stellar model of $20M_\odot$. It is obvious that this time the shock stalls at about 300 km. (From Hillebrandt 1987a.)

Because some supernova remnants (e.g., Cas A and Pup A) show large oxygen overabundances, indicating main-sequence masses of at least $20M_\odot$, alternative explosion mechanisms had been searched for even prior to SN 1987A. A possibility that has been discussed extensively is the so-called delayed explosion model (Wilson 1985, Bethe and Wilson 1985). The idea is that a few hundred milliseconds after core-bounce energy transport by neutrinos may revive a stalled shock. At that time the edge of the unshocked inner core ($\simeq 0.7M_\odot$) will be at a radius of about 20 km, and the density and temperature there are around $10^{12}$ g cm$^{-3}$ and 2–3 MeV, respectively. The shock has changed to an almost standing accretion shock at $1.4M_\odot$ and a radius of about 500 km, the temperature and density just behind the shock being $10^7$ g cm$^{-3}$ and 0.8 MeV, respectively. The shocked matter is irradiated by a neutrino flux orginating from the neutrino sphere at 30–70 km. Typical neutrino energies will be around 6–10 MeV. The energy gain per gram of irradiated matter at radius $R$ much

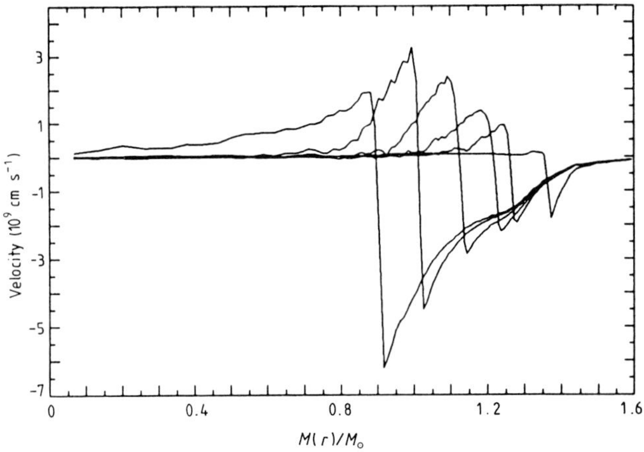


**Figure 13.** Snapshots of velocity against mass for the model of figure 12. The curves, from the left, correspond to times 156.2, 156.6, 157.6, 159.9, 163.2 and 202.6 ms, respectively. (From Hillebrandt 1987a.)

larger than $R_\nu$ of the neutrino sphere is given by

$$\dot{E}_{\text{gain}} = K_{\text{abs}} \frac{L_\nu}{4\pi R^2} \tag{3.3}$$

where $L_\nu$ is the neutrino luminosity and $K_{\text{abs}}$ is the absorption opacity. Each gram of matter will also lose energy by thermal emission of neutrinos, $\dot{E} \sim T^4 K_{\text{emiss}}$. Since the neutrino luminosity is proportional to $T^4 R_\nu^2$ and $K_{\text{abs}}$ and $K_{\text{emiss}}$ are proportional to $T_\nu^2$ and $T^2$, respectively, the net energy gain can be written as

$$\frac{dQ}{dt} = \frac{7}{16} acK_{\text{abs}} T_\nu^4 \left[\frac{1}{4}\left(\frac{R_\nu}{R}\right)^2 - \left(\frac{T}{T_\nu}\right)^6\right] \tag{3.4}$$

where the temperatures are in MeV, $a = 1.37 \times 19^{26}$ erg $c^{-1}$ cm$^{-3}$ MeV$^{-4}$ and $K_{\text{abs}}$ in CGS units is $6 \times 10^{20} \langle(\varepsilon_\nu / m_e c^2)^2\rangle$ cm$^2$ g$^{-1}$ (Lattimer and Burrows 1984). The maximum matter temperature that can be obtained from this heating mechanism can be estimated from the equilibrium condition $dQ/dt = 0$, and we find roughly 1 MeV from typical stellar parameters. However, it is clear that the quantity of key importance is the neutrino temperature at the neutrino sphere, which, due to uncertainties in the flux-limited diffusion schemes, is not very well determined. Small changes in $T_\nu$ can change the net heating considerably.

Wilson (1985) and Wilson *et al* (1986) have computed the hydrodynamic evolution of several stellar models for approximately 1 s after core-bounce and found explosions in all cases considered (see also figure 14). In most cases, however, the explosion energy ($\lesssim 4 \times 10^{50}$ erg) was too low to account for typical type II supernova light curves and in particular for the rather high kinetic energy seen in SN 1987A. Only for rather massive stars, $M \gtrsim 25 M_\odot$, did explosive oxygen burning add enough energy to the explosion to explain a typical outburst. This problem can be understood from a simple

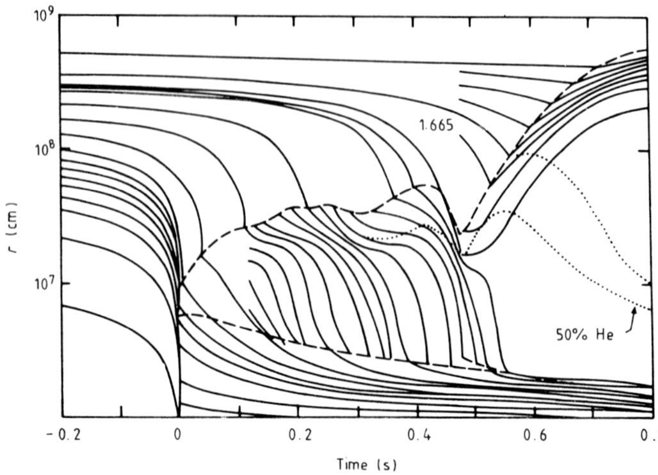


**Figure 14.** Radius against time for selected mass zones of a model of $25M_\odot$ (from Wilson 1985). Time is measured from core-bounce. The position of the shock front (upper broken curve) and that of the neutrino sphere are also shown. The stalled shock is revived by neutrino heating 0.5 s after bounce.

argument. Neutrino heating proceeds on a timescale much longer than the hydrodynamic timescale. Once, due to neutrino heating, a sufficient overpressure has been built up behind the accretion shock the heated zones will start to expand and further heating will be turned off. The thermal energy needed to build up such an overpressure will be of the order of the binding energy of the overlaying material, i.e., a few times $10^{50}$ erg. Consequently, one expects that the explosion energy is of the same order.

Recently Wilson (personal communication) repeated his computations with a slightly modified neutrino transport scheme and no longer found explosions. This, as well as results obtained by Hillebrandt (1985), indicates that the delayed explosion mechanism may not work at all. So we are left with the problem that also (and maybe in particular) for massive stars, $M \simeq 20M_\odot$, the explosion mechanism is not well understood. It may well be that we have to invent more complicated scenarios in order to be able to solve these problems. For example, the neglect of rotation, magnetic fields and/or nuclear energy generation may be an oversimplification (see, e.g., Müller and Hillebrandt 1981, Le Blanc and Wilson 1970, Symbalisty 1984, Bodenheimer and Woosley 1983), or some of our basic assumptions concerning the EOS (Baron *et al* 1987) or the theory of weak interactions are incorrect.

### *3.2. Neutrinos from supernova 1987A*

As was already mentioned in the introduction and in § 2, two neutrino pulses were detected prior to the optical outburst of SN 1987A, at February 23.12 (UT) (Aglietta *et al* 1987a) and at February 23.32 (UT) (Hirata *et al* 1987, Bionta *et al* 1987). There is also a report of neutrino events close to the second burst in the Baksan detector (Alexeyev *et al* 1987). Two photographs taken at about February 23.44 (UT)

(McNaught 1987) show the supernova at a visual magnitude of about $6.4^m$ only $2.8 \times 10^4$ s and $1.1 \times 10^4$ s after the neutrino events were discovered. This close correlation indicates that the neutrinos were indeed emitted from the exploding star.

The pulse seen in the Mont Blanc detector consisted of five events spread over $\Delta t \simeq 7$ s with measured positron energies between 5.8 and 7.8 MeV (see also table 3). From the second pulse, about 4.7 hours later, KAMIOKANDE detected 12 neutrinos spread over $\Delta t \simeq 13$ s, whereas IMB saw eight neutrinos with $\Delta t \simeq 6$ s. The recorded electron (positron) energies in the second burst were significantly higher and ranged from 6 to 35 MeV (KAMIOKANDE) to 20 to 40 MeV (IMB) (see table 3). It is interesting to note that the angular distribution of electrons (positrons) was not isotropic, as one would expect from the reaction $\bar{\nu}_e + p \rightarrow e^+ + n$. In particular, the high-energy events ($E_e \gtrsim 20$ MeV) seem to be strongly forward peaked away from the LMC. In the case of IMB this effect was not caused by the fact that about 25% of the photomultipliers were inoperative when the neutrinos were detected, as recent Monte Carlo simulations

**Table 3.** Properties of neutrino events detected from SN 1987A.

| Experiment | Event number | Time (UT) (23 February) | Electron energy (MeV) | Electron angle (deg) |
|---|---|---|---|---|
| Mont Blanc† | 1 | $2^h53^m36^s.79$ | 6.2 | — |
| | 2 | $40^s.65$ | 5.8 | — |
| | 3 | $41^s.01$ | 7.8 | — |
| | 4 | $42^s.70$ | 7.0 | — |
| | 5 | $43^s.80$ | 6.8 | — |
| KAMIOKANDE | 1 | $7^h35^m41^s.00$ (?) | $20.0 \pm 2.9$ | $18 \pm 18$ |
| | 2 | $41^s.11$ | $13.5 \pm 3.2$ | $40 \pm 27$ |
| | 3 | $41^s.30$ | $7.5 \pm 2.0$ | $108 \pm 32$ |
| | 4 | $41^s.32$ | $9.2 \pm 2.7$ | $70 \pm 30$ |
| | 5 | $41^s.51$ | $12.8 \pm 2.9$ | $135 \pm 23$ |
| | 6 | $41^s.69$ | $6.3 \pm 1.7$ | $68 \pm 77$ |
| | 7 | $42^s.54$ | $35.4 \pm 8.0$ | $32 \pm 16$ |
| | 8 | $42^s.73$ | $21.0 \pm 4.2$ | $30 \pm 18$ |
| | 9 | $42^s.92$ | $19.8 \pm 3.2$ | $38 \pm 22$ |
| | 10 | $50^s.22$ | $8.6 \pm 2.7$ | $122 \pm 30$ |
| | 11 | $51^s.43$ | $13.0 \pm 2.6$ | $49 \pm 26$ |
| | 12 | $53^s.44$ | $8.9 \pm 1.9$ | $91 \pm 39$ |
| IMB† | 1 | $7^h35^m38^s.37$ | $41 \pm 7$ | $80 \pm 10$ |
| | 2 | $41^s.79$ | $37 \pm 7$ | $44 \pm 15$ |
| | 3 | $42^s.02$ | $28 \pm 6$ | $56 \pm 20$ |
| | 4 | $42^s.52$ | $39 \pm 7$ | $65 \pm 20$ |
| | 5 | $42^s.94$ | $36 \pm 9$ | $33 \pm 15$ |
| | 6 | $44^s.06$ | $36 \pm 6$ | $52 \pm 10$ |
| | 7 | $46^s.38$ | $19 \pm 5$ | $42 \pm 20$ |
| | 8 | $46^s.96$ | $22 \pm 5$ | $104 \pm 20$ |
| Baksan | 1 | $7^h36^m11^s.82$ | $12 \pm 2.4$ | — |
| | 2 | $12^s.25$ | $18 \pm 3.6$ | — |
| | 3 | $12^s.25$ | $23.3 \pm 4.7$ | — |
| | 4 | $19^s.51$ | $17 \pm 3.4$ | — |
| | 5 | $20^s.92$ | $20.1 \pm 4.0$ | — |

† Mont Blanc data are taken from Aglietta *et al* (1987b); the revised IMB data were presented by T Haines at the Aspen Conference on SN 1987A (January 1988).

have shown (Svoboda *et al* 1987). Finally the Baksan group reported five events delayed by 30 s relative to the IMB time, with energies from 12 to 24 MeV.

The detections of neutrinos from a type II supernova are the first proof that the cores of massive stars do collapse to neutron star densities. However, certain aspects of the observed neutrino pulses are difficult to understand. Therefore, we will discuss the experimental results in some detail in the following subsections.

*3.2.1. Significance and consistency of the neutrino detections.* According to Hirata *et al* (1987), the rate at which events like the one observed in the KAMIOKANDE detector can arise from statistical fluctuations is of the order of one event every $7\times10^7$ years. There can be no doubt that the detected neutrinos were indeed emitted from SN 1987A. Although the statistical significance of the IMB detection is much less and gives a rate of one event of multiplicity 8 every three years from background noise, the close coincidence in time with the KAMIOKANDE detection is a strong argument in favour of the interpretation that these neutrinos also came from SN 1987A. Unfortunately, only a computer clock was used by the KAMIOKANDE group. The arrival time of the neutrinos in their detector, therefore, is uncertain by about $\pm1$ minute. It is interesting to note that the Baksan detection lies within this window.

With respect to the first neutrino events, the conclusions are less certain. Aglietta *et al* (1987a) state that their background rate of finding five positrons above threshold in a 10 s interval is one every 1.5 years. This translates into a probability of $3\times10^{-4}$ of finding five events in a 10 s interval in a 4 h window around the assumed time of the supernova explosion. Since at the time of the Mont Blanc detection no signal significantly above background was recorded by the (larger) KAMIOKANDE experiment, one cannot exclude the possibility that the first pulse was not real, but was due to random noise.

It is somewhat difficult to check the consistency of the various experiments from statistics alone, because they differ in their volume, their threshold energies and in the way positrons are detected. Generally speaking, the count rate of the KAMIOKANDE detector should be higher than that of all other detectors if the average neutrino energy is between 8 and 15 MeV. As can be seen from table 3 the average positron energy seen by the Mont Blanc detector was 6.7 MeV. Taking into account the sensitivity of the KAMIOKANDE detector at those energies ($\leqslant30\%$) one would conclude that KAMIOKANDE should have seen about 20 events from the first burst. The fact that this was not the case is commonly used as an argument to omit the Mont Blanc events. On the other hand, we are dealing with the statistics of small numbers, and the count rate of Mont Blanc may have been accidentally high.

The same arguments hold for the Baksan experiment, which has a fiducial mass of 200 tons only (as compared with the 2140 tons of KAMIOKANDE) and a threshold energy of about 10 MeV. They found one event close to the Mont Blanc time, but five events near the KAMIOKANDE/IMB time. Again, the latter result is at best only marginally consistent with the KAMIOKANDE detections because it requires a higher neutrino luminosity at the source.

IMB and KAMIOKANDE can be compared in a similar way. Taking into account the detection efficiency of IMB for events above 20 MeV and the fact that 25% of the photomultipliers did not work at the time of the neutrino burst, the total number of events above 20 MeV in the IMB detector would be $22\pm5$. Since the volume of KAMIOKANDE is about a factor of 2.3 smaller, one would expect $10\pm3$ events, which should be compared with the actual measurement of $4\pm2$ neutrinos.

So we reach the conclusion that although there are definitely inconsistencies among the various neutrino detections, the problem of the statistics of a small number of events does not allow us to rule out one or the other. The absence of a clear signal in the KAMIOKANDE detector during the first burst is certainly a strong argument against regarding this pulse to be real, but it is not a proof. It cannot be excluded from first principles that the neutrino energy in this burst was indeed low, in which case the higher detection efficiency of the Mont Blanc experiment at low energies could explain the observed number of events. However, there remains also the problem that the light curve models (§ 3.3.3) indicate that the explosion began around the time of the KAMIOKANDE and IMB detections.

*3.2.2. Neutrino energetics.* We will next assume as a working hypothesis that both neutrino pulses were real detections and will investigate the energetics of the events in order to check whether or not they are in conflict with astrophysical expectations. We only present rough estimates here and will refer to more detailed discussions in passing.

We will first discuss the implications from the Mont Blanc detection. As can be seen from table 3, two of the detected events had energies near threshold (around 5 MeV) and the first of them preceded the others by four seconds. We will therefore assume that the total number of events was $4 \pm 2$. Because of the much higher cross section of the $(\bar{\nu}_e, \mathrm{p})$ reaction compared with $(\nu_e, \mathrm{e})$ scattering we will assume that the incident reaction was $\bar{\nu}_e + \mathrm{p} \rightarrow \mathrm{e}^+ + \mathrm{n}$, for which the cross section is $9.5 \times 10^{-44}$ $(E_\nu/\mathrm{MeV})^2\ \mathrm{cm}^2$, where $E_\nu$ is the incident neutrino energy in MeV. Note that for this reaction the incident neutrino energy will on average be only 10–20% higher than the observed positron energy. By taking the number of free protons in the Mont Blanc experiment $(\simeq 8 \times 10^{30})$ we can estimate the integrated flux at the detector and obtain $(5 \pm 3) \times 10^{12}$ $(E_\nu/\mathrm{MeV})^{-2}\ \mathrm{cm}^{-2}$. For the distance to the LMC ($\simeq$50 kpc) this corresponds to a total number of $\bar{\nu}_e$ emitted from the source of about $(1.2 \pm 0.7) \times 10^{60}$ $(E_\nu/\mathrm{MeV})^{-2}$. The total energy is then $E_{\mathrm{tot}} \simeq (1.9 \pm 1.1) \times 10^{54}$ erg $(E_\nu/\mathrm{MeV})^{-1}$. Assuming an average neutrino energy of 6 MeV in order to be consistent with the (non-) detection of KAMIOKANDE we end up with a total energy emitted in the form of $\bar{\nu}_e$ of about $(3 \pm 2) \times 10^{53}$ erg. This number is in fact a lower limit because one probably has to go to even lower mean energies in order to satisfy the conditions set by the KAMIOKANDE experiment. If about the same energy was emitted in $\nu_e$, $\nu_{\mu,\tau}$ and $\bar{\nu}_{\mu,\tau}$, the total energy in the burst increases by a factor of six and thus would be in conflict with the maximum binding energy of a neutron star ($\lesssim 6 \times 10^{53}$ erg) (Burrows 1987). This problem can only be overcome if it is assumed that either the neutrino emission was beamed towards us or the luminosity of $\bar{\nu}_e$ was higher than that of all other kinds. Neither possibility, however, is completely excluded.

An analysis similar to the one presented here has been performed by Sato and Suzuki (1987) and many others (e.g., Bahcall *et al* 1987, Burrows and Lattimer 1987, Arnett and Rosner 1987, De Rujula 1987) for the KAMIOKANDE and IMB data, assuming again that all neutrinos detected were $\bar{\nu}$. It was found that the KAMIOKANDE data can be fitted by a thermal neutrino spectrum corresponding to a temperature of $2.8 \pm 0.3$ MeV, whereas the IMB data require a significantly higher temperature of $4.6 \pm 0.7$ MeV, again indicating that the data are only marginally consistent. The total energy in three neutrino flavours is then $(2.9 \pm 0.6) \times 10^{53}$ erg for KAMIOKANDE and $(1.5^{+1.2}_{-0.6}) \times 10^{53}$ erg for IMB. If some of the events are considered to be noise, this energy estimate is reduced by a factor of about two. If, on the other hand, some of the events are assumed to be caused by ($\nu$, e) scattering, the energy in the burst would increase by about a factor of two. In any case, the data are consistent with the assumption that a neutron star of about $1.5M_\odot$ was born in SN 1987A and has radiated away a significant fraction of its binding energy in thermal neutrinos during the first few seconds of its life. It is also apparent, however, that this conclusion is still uncertain, because in the case of Mont Blanc we have to rely on the poor statistics of a few events and in the case of KAMIOKANDE and IMB we do not know for sure which (if any) of the detections were due to ($\nu$, e) scattering.

One may ask, moreover, whether or not the assumption of a thermal spectrum with zero chemical potential describes the neutrino energy distribution sufficiently well. In order to answer this question Janka (1987) and Janka and Hillebrandt (1989a) have performed Monte Carlo simulations of the neutrino transport in type II supernova explosions. Based on one particular model (Hillebrandt 1985) they could show that, in fact, the results of the Monte Carlo simulations are not well fitted by Fermi–Dirac distributions, even if one allows for a non-vanishing neutrino chemical potential. In particular they have demonstrated that the assumption of a thermal spectrum overestimates the number of neutrinos both in the high-energy tail of the distribution function and at low energies. Much better fits could be obtained if an additional parameter were introduced to generate the first three moments of the neutrino energy distribution (see figure 15), indicating the importance of geometrical effects. This result is not surprising because the thickness of the shell where neutrinos decouple from the matter is not small compared with the radius of the newly born neutron star, at least for the first few seconds after core-bounce.

Based on these results Janka and Hillebrandt (1989b) have reanalysed the KAMIOKANDE and IMB data. They performed maximum likelihood tests, varying the neutrino chemical potential, their temperature and the total energy emitted in the form of electron antineutrinos from SN 1987A (see figures 16). As can be seen from the figures, the KAMIOKANDE (K2) and IMB data are only weakly consistent, which is in agreement with earlier investigations. In particular, the IMB data indicate a significantly higher temperature and lower total energy. This may be due to the possibly non-thermal nature of the neutrino pulse or may indicate that the IMB detector, because of its higher threshold energy, has predominantly detected neutrinos from the very early cooling phase, whereas KAMIOKANDE has seen neutrinos from a different (and larger) time interval. Unfortunately, the detection time of KAMIOKANDE is uncertain by about 1 or 2 minutes and therefore the latter hypothesis cannot be proven. In any case, the analysis of Janka and Hillebrandt (1989b) suggests that the neutrinos must have been emitted with higher average energies and that the total energy flow through the detectors might have been somewhat lower than estimates based on the assumption of black-body emission.

### *3.3. Non-standard scenarios*

Given the apparent difficulties of the standard core collapse models one may search for alternative explosion mechanisms. Rotation, for example, may alter the general picture outlined in § 3.1 considerably. A non-rotating stellar core will collapse until the adiabatic index $\gamma$ exceeds the critical value of $\frac{4}{3}$ around nuclear matter density. For a rigidly rotating stellar core the critical adiabatic index is given by

$$\gamma_c = \frac{4}{3} - \frac{2}{9}\frac{I\omega^2}{|E_{\rm pot}|} < \frac{4}{3} \qquad (3.5)$$

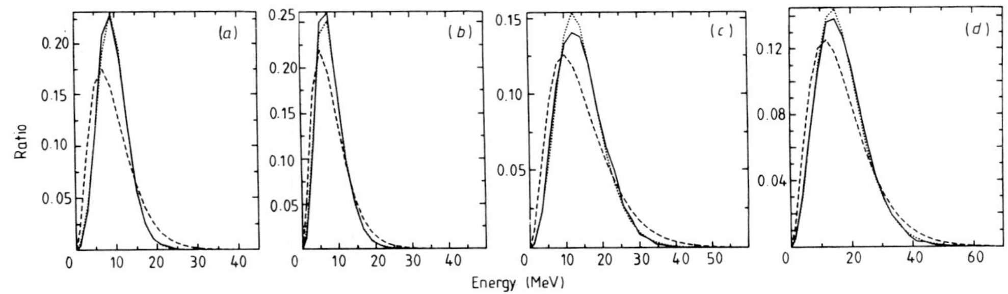


**Figure 15.** Thermal fits of neutrino distribution functions obtained from Monte Carlo transport simulations. The full curves give the computed spectra. The broken curves are Fermi–Dirac distributions which give locally the same number density and energy density. These latter spectra are close to black-body distributions. The dotted curves are fits to the first three moments of the calculated energy distribution. (*a*) Electron neutrino spectrum 12 ms after core-bounce; (*b*) electron neutrino spectrum 0.315 s after core-bounce; (*c*) electron antineutrino spectrum at 0.315 s; (*d*) muon neutrino spectrum at 0.315 s. (From Janka and Hillebrandt 1989a.)

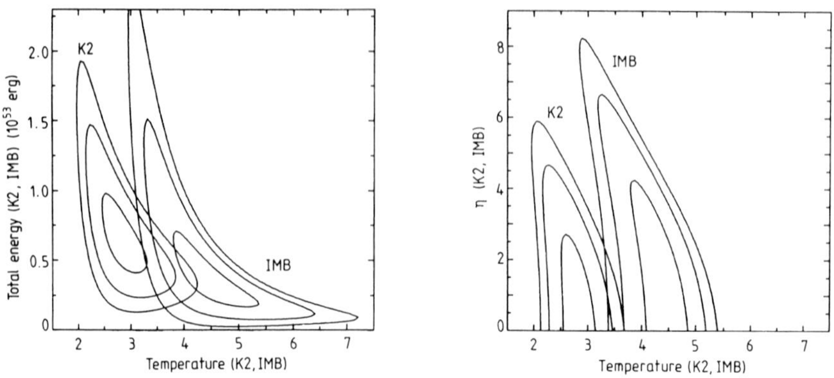


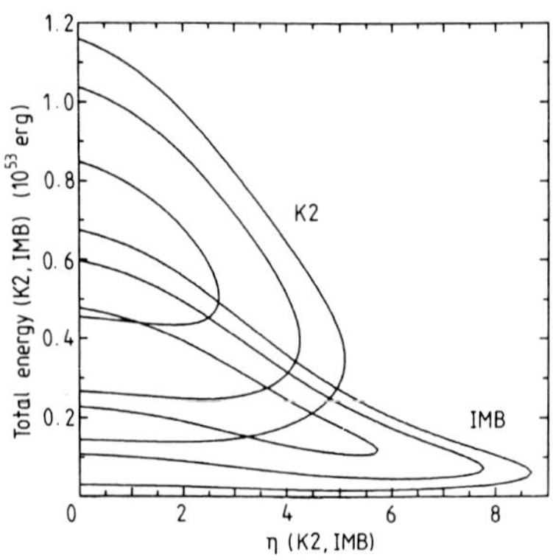


**Figure 16.** Maximum likelihood regions in the space of chemical potential, temperature and total energy for neutrinos observed by the KAMIOKANDE (K2) and IMB detectors. The cut planes are located at the best fit values for the degeneracy parameter $\eta$ (=chemical potential/$k_B T_\nu$), total energy emitted and neutrino temperature in MeV for both detections. The contours mark 68%, 95% and 99% likelihood volumes. It is apparent that the data are consistent with zero chemical potential but do not exclude finite values. Moreover, the two detections are only weakly consistent. (From Janka and Hillebrandt 1989.)

where $I$ is the moment of inertia, $\omega$ is the angular velocity and $E_{\text{pot}}$ is the gravitational energy. So a rotating core may not collapse to nuclear matter density provided the equation of state is sufficiently stiff and/or the angular velocity is sufficiently high. Moreover, non-spherical velocity fields will lead to non-spherical shocks, which may introduce large-scale circulations and thus a more efficient neutrino transport. Finally, rapidly rotating cores will be unstable to triaxial deformations or even fragmentation. Collapsing rotating cores were first modelled by Le Blanc and Wilson (1970), Müller *et al* (1980), Müller and Hillebrandt (1981) and Tholine (1984). More recently Symbalisty (1984) and Mönchmeyer and Müller (1989) have repeated those computations using more realistic initial stellar models and equations of state. So far, however, the conclusion has always been that no prompt explosion triggered by rotation alone has been found. On the contrary, it seems that certain initial angular momentum distributions inside a collapsing stellar core rather weaken the chance of getting a prompt explosion (Mönchmeyer and Müller 1989).

Recently, Middleditch *et al* (1989) (see also Kristian *et al* (1989) for more details) have reported the detection of submillisecond optical pulsations from SN 1987A which, if real, definitely require the presence of a neutron star. Moreover, the observed frequency of the pulsations of approximately 2 kHz, if interpreted as being due to rotation, would require that the pulsar be spinning very rapidly, close to break-up. Implications from these exciting observations for equations of state of neutron stars will be discussed in § 5. For our present discussion it is important to note that, so far, core collapse computations have been performed only for rather slowly rotating stellar models. One may therefore speculate that the results obtained do not apply to SN 1987A but have to be extended in order to include non-axisymmetric effects. We will come back to this question later in § 6. The observations of Middleditch *et al* (1989) together with the low bolometric luminosity of SN 1987A ($\lesssim$ a few times $10^{38}$ erg $s^{-1}$) in December 1988 can be used to set upper limits on the strength of the magnetic field of the pulsar and one obtains $B \lesssim 5 \times 10^9$ G. Therefore, it is obvious that magnetic fields were dynamically unimportant in SN 1987A and the explosion was certainly not driven by magnetic effects, and in particular the supernova models proposed by Bisnovatyi-Kogan (1971, 1980), Meier *et al* (1976) and Müller and Hillebrandt (1979) cannot be applied to this supernova.

Fast rotation will also affect the neutrino signal emitted from a newly born neutron star. Janka and Mönchmeyer (1988) have demonstrated that an observer will detect more neutrinos if he sees the neutron star pole-on because he sees a larger surface area. Consequently, he will overestimate the total neutrino energy in that case. Janka and Mönchmeyer (1988) therefore conclude that, if a rapidly spinning neutron star was born in SN 1987A, we should have seen it almost pole-on. Otherwise the energy detected by the KAMIOKANDE experiment would be in conflict with the maximum binding energies allowed by reasonable equations of state for neutron star matter (see § 2.2.2).

We conclude this section by noting that fast rotation will also change the dynamics of the mass zones outside the original iron core considerably and may, in fact, lead to a thermonuclear explosion of the oxygen shell as was shown by Bodenheimer and Woosley (1983). Their explosion mechanism may work if, supported by centrifugal forces, the oxygen shell moves inwards with subsonic velocities and contains enough mass ($M_O \gtrsim 1.5$–$2 M_\odot$). Both conditions may have been fulfilled for Sk −69°202. The prediction from this model is that less than approximately $0.5 M_\odot$ of oxygen should be observed in the supernova ejecta, but unfortunately the abundance determinations (Danziger *et al* 1989) are still not sufficiently accurate to prove or disprove this scenario. Explosive oxygen burning may, nonetheless, explain the observed early fragmentation of the supernova shell and the mixing of radioactive $^{56}$Co with material of high velocity (see, e.g., § 2.4.2).

## 4. The interaction of SN 1987A with its surroundings

### *4.1. Observations of the circumstellar matter*

The stellar surroundings are highly influenced by the progenitor star. Therefore, an analysis of the circumstellar matter provides important information on the evolution of the progenitor star Sk −69°202.

*4.1.1. The radio observations.* On 25 February 1987 radio emission was detected firstly with Australian telescopes by Turtle *et al* (1987). It showed a peak flux of about 100 mJy at about 1 GHz three days after the initial event, and then declined on a similar timescale. No radio emission could be measured after about one month. In addition, very-long-baseline interferometry (VLBI) observations (Jauncey *et al* 1987) at day 5.2 allowed the radio sphere to be measured; it was found to be larger than 1.2 milliarcsec, corresponding to $8.3 \times 10^{14}$ cm, indicating an expansion velocity of the radio-emitting region of greater than 18 500 km $^{-1}$. The occurrence of radio emission at early stages and its maximum intensity is in contrast to previously observed type II supernovae (Weiler *et al* 1986). Usually, the maximum radio luminosity is reached several months to years after the maximum of the optical light curve, and its peak luminosity is larger by a factor of 100 to 1000 and varies on timescales of months to years. However, the radio emission from SN 1987A can be attributed to the same mechanisms operating also in other supernovae (Chevalier 1982). The emission and the expansion velocity of the emitting gas can be understood as being due to synchrotron radiation from a thin shell near the edge of the expanding medium (Chevalier 1987, Manchester 1987, Storey and Manchester 1987). The differences between SN 1987A and other type II supernovae, therefore, are caused by the absence of a high-density envelope near the star which would be optically thick at radio wavelengths. Note that lower densities are to be expected as a consequence of a blue supergiant progenitor with a wind of high velocity and low density instead of a red supergiant with a dense wind of low velocity.

*4.1.2. The narrow* UV *lines.* As discussed in § 2.2, there is strong evidence that Sk −69°202 has gone through a red supergiant phase during its presupernova evolution. Observations with IUE from 24 May 1987 onwards show several narrow emission features ($\nu \leq 30$ km s$^{-1}$) from multiply ionised C, N and O (Cassatella 1987, Fransson 1988, Kirshner *et al* 1987, Wamsteker *et al* 1987). These lines show a line strength increasing with time up to April 1988 (Fransson *et al* 1988). A spectroscopic analysis based on the nebular theory indicates particle densities of about $3 \times 10^4$ cm$^{-3}$, which can be attributed to the shock region where the blue and red supergiant winds interact. In addition, a CNO mass fraction of 0.3 and a strong nitrogen enhancement ($N/C \approx 5$; $N/O \approx 2$ relative to solar values) are found, indicating that we see CNO-processed matter in the dense wind.

The high ionisation stages of the shell can be attributed to the strong extreme ultraviolet (EUV) burst (Lunquist and Fransson 1988, Fransson and Lunquist 1989) which, according to hydrodynamic calculations, is to be expected (§ 2.3.3) when the shock breaks through the photosphere. Therefore, the peak radiation temperature (4–8 $\times 10^5$ K) derived from the observed ionisation stages and the time evolution and flux in the UV lines may impose some constraints on hydrodynamic models.

*4.1.3. The narrow optical lines.* Systems of narrow forbidden lines in high-resolution spectra ($\lambda/\delta\lambda = 2.4 \times 10^4$; Wampler 1988, Wampler and Richichi 1989) of O III have been detected in the optical and near-IR wavelength ranges. A velocity spread of 15 km s$^{-1}$ (FWHM) was measured. The derived densities and velocities agree well with those implied from the UV lines. Therefore, it seems very likely that these optical lines are formed in the same region. The extent of the emitting region can be determined to be about 2 arcsec, corresponding to a radius of about $8 \times 10^{17}$ cm. Because the shock between the blue and red supergiant wind is travelling at a velocity of about 50 km s$^{-1}$ (Fransson *et al* 1988, Chevalier 1988), one can estimate the time that has passed since the red supergiant phase and one finds 5000–7000 years.

*4.1.4. The mystery spot.* A more uncertain observation of the circumstellar surroundings has been the so-called 'mystery spot'. Speckle interferometric observations by AAO and CTIO (Karovska *et al* 1987, Marcher *et al* 1987, Meikle *et al* 1987a; Nisenson *et al* 1987) of SN 1987A showed for three months beginning five weeks after the explosion the presence of a second object ('mystery spot') with about 10% of the supernova's luminosity at that time and a distance of about 60 milliarcsec, corresponding to about $5 \times 10^{16}$ cm. The change in position of the secondary speckle image implied a projected velocity of about 0.3–0.5*c*. Hillebrandt *et al* (1987a) have suggested as a possible explanation that this second object may reflect the interaction of the early UV burst with a nearby dense cloud, and predicted that a strong soft x-ray burst should be seen about one year after the explosion. In fact, such a burst was indeed detected in January 1988, but its x-ray luminosity was an order of magnitude fainter than had been predicted. These observations indicate the presence of some pre-existing dense matter (§ 2.3.2) and the overestimate of the x-ray brightness may be due to the (unrealistic) assumption of a spherical cloud by Hillebrandt *et al* (1987a). Therefore, although the correlation may appear to be quite convincing, uncertainties inherent in applications of the speckle technique should be kept in mind. It may well be that the mystery spot has been an artefact rather than a real object.

### *4.2. The interstellar matter towards SN 1987A*

SN 1987A allows us to investigate the properties of the interstellar matter in the direction of the Large Magellanic Cloud. Firstly, a detailed analysis of this matter is of great importance for the determination of the size and frequency distribution of the flux from SN 1987A. In this context, a reinvestigation of the interstellar polarisation is essential for the use of the continuum and line polarisation measurements of SN 1987A as quantitative indicators of deviations from sphericity (§ 2.4). Secondly, because of its brightness, SN 1987A is a very useful 'background candle' and allows us to determine the density and velocity distributions, the ionisation stages and the chemical structure of the interstellar medium towards the LMC. Its light probes gas in both the disc and halo of our own Galaxy and in the LMC.

The interstellar reddening has been determined by different groups. Danziger *et al* (1988b) have derived a colour index $E_{B-V}$ of about $0.2 \pm 0.06^{m}$ by measuring the ratios of emission line fluxes in H$\alpha$ and H$\beta$ of H II regions near the position of Sk −69°202. This value is in good agreement with the one given by Panagia *et al* (1987) on the basis of the intrinsic reddening of B supergiants near the position of SN 1987A. However, by the same method but by using the measured colours of Sk −69°202 (Isserstedt 1975), Woosley (1988a) derived a colour index $E_{B-V}$ of $0.13^{m}$. On the basis of the interstellar line system at 5790 Å, Wampler derived a mean value of $E_{B-V} = 0.15^{m}{}^{+3}_{-6}$ (Wampler *et al* 1987, Wampler 1988). Spectral analyses of SN 1987A (§ 4.2) favour values of $E_{B-V}$ of the order of about 0.13 and $0.15^{m}$ (Lucy 1987a, b, Höflich 1988b). However, the better agreement between calculated and observed spectra should only be regarded as a hint in favour of these values because of the uncertainties in the spectral diagnostics. So, in conclusion, the total reddening $E_{B-V}$ may be somewhere in the range from 0.09 to $0.26^{m}$. Note that this uncertainty will induce only minor errors in the correction of the bolometric light curve of SN 1987A at later stages, but may imply major errors (up to a factor of two) in the observed light curve during the first few days when most of the luminosity was emitted in the UV.

The interstellar polarisation of the visual light ($\lambda_{max} = 5400$ Å) in the direction of Sk −69°202 was determined to be about 0.41% (Schmidt 1976) and 0.42% on the basis of a number of bright stars in the LMC. A similar value (about 0.5%) was derived for the nearby star Sk −69°203 (Mendez *et al* 1988). According to studies of the LMC region, the interstellar polarisations of both the Galaxy and the LMC behave normally

(Clayton *et al* 1983, Schmidt 1976). Therefore, the frequency relation of Serkowski *et al* (1975) can be used.

At early times SN 1987A was sufficiently bright to study in some detail the interstellar medium (ISM) in the UV wavelength region (Blades *et al* 1988, De Boer *et al* 1987, Grewing *et al* 1987). A rich variety of different ionisation stages and elements could be identified (C I, II, IV; O I; Mg I, II; Al II–III; Si II, IV; S II; Cl I; Cr II; Mn II; Fe II; Ni II; Zn II). Although the interstellar absorption line spectrum towards SN 1987A appears similar to those towards other LMC stars (Fitzpatrick and Savage 1983), lines of very highly ionised species have been identified which, because of their high velocities, can be attributed most likely to the LMC (Blades *et al* 1988, Dupree *et al* 1987). At about 6375 Å an absorption line has been detected which has to be attributed to a forbidden feature of Fe X distributed between 205 and 370 km s$^{-1}$ (D'Odorico *et al* 1987), indicating a blob of a very hot gas ($T \approx 10^6$ K) in the ISM. These lines may either be due to highly ionised gas within the 30 Doradus complex or, alternatively, they may be due to ionisation caused by the EUV flash at the shock break-out of SN 1987A.

In addition, optical observations with high resolution have been performed (Andreani *et al* 1987, Andreani and Vidal-Madjar 1987, D'Odorico *et al* 1987, Magain 1987, Vladilo and Molaro 1987). At least 27 different components of optical lines of Ca II, Na I, K I and Ca I could be identified covering the velocity range 0–300 km s$^{-1}$ (Vidal-Madjar *et al* 1987, Magain 1987) and indicating the existence of well separated interstellar clouds.

From the beginning of February 1988 onwards, double arcs were seen around SN 1987A at angular distances of about 32 and 47 arcsec, which about one month later showed an expansion of about 3 arcsec/month for the outer ring (Crotts 1988, Rosa *et al* 1988, Suntzeff *et al* 1988, West 1988, see figure 17). Previously, similar arcs have been seen around Nova Persei (Ritchey 1901, 1902). The geometry and theory of such light echoes were discussed a long time ago (Couderc 1939), and they were also expected for SN 1987A (Schaefer 1987). They can be explained as light echoes on dust particles from the time of maximum luminosity of SN 1987A. Spectra of the arcs show about the same spectra as seen for SN 1987A at maximum light and therefore

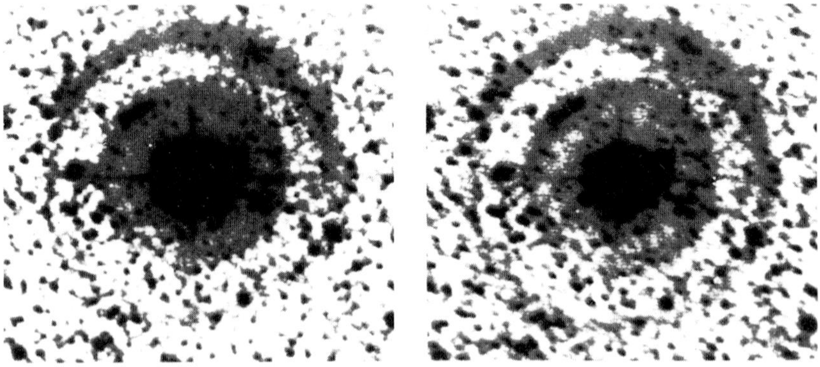

**Figure 17.** CCD images of the double light echo around SN 1987A, discovered by the ESO 3.6m telescope on 13 February 1988 (left) and 16 March 1988, by M Rosa and Chr Gouiffes (West 1988).

confirm this interpretation. Distances of about 115 and 305 pc in front of SN 1987A could be derived for the two clouds, confirming the cloudy structure of the interstellar matter near SN 1987A.

## 5. Implications from SN 1987A for nuclear and particle physics

The observations of SN 1987A can be used, and have been used, to place new constraints on important physical quantities such as the nuclear equation of state, the properties of neutrinos and the existence of exotic particles. In the following subsections we will briefly review the conclusions that can be drawn.

### *5.1. The equation of state of nuclear matter*

As has been discussed in § 3, soft equations of state at densities beyond nuclear saturation increase the chance of prompt explosions. In particular Baron *et al* (1985a, b) have argued that for successful shock propagation the nuclear incompressibility has to be significantly below the values obtained from fits to the giant dipole resonance of lead and other nuclei (Sharma *et al* 1989) and from empirically obtained effective nucleon–nucleon interactions. The rather high explosion energy of SN 1987A, indicating a prompt rather than a delayed explosion, together with the presumably large iron-core mass of the progenitor seem to be in favour of their arguments. On the other hand, since it is extremely uncertain by which mechanism massive stars explode, it seem to be dangerous to draw such far-reaching conclusions.

Certainly, if it should be confirmed that SN 1987A left behind a neutron star spinning at a rate of 2000 revolutions per second, this would rule out most stiff equations of state for neutron star matter such as the tensor interaction model or the mean-field approach of Pandharipande and Smith (1975), even if effects of rotation on the neutron star's structure are incorporated (Friedmann *et al* 1986). The main reason for this result is fairly simple. Soft equations of state tend to give neutron star models of smaller radii than stiff ones. Consequently such models can rotate faster before their surface layers reach Keplerian velocity. They have the disadvantage, however, that the maximum mass they predict for a stable slowly rotating neutron star may be in conflict with neutron star masses observed in binary systems (see, e.g., Shapiro and Teukolsky 1983 and references therein). Both observations will therefore leave a very narrow range of 'allowed' equations of state. In any case, the data published by Kristian *et al* (1989) indicate that the pulsar born in SN 1987A is close to the critical mass of a neutron star and may even be beyond the critical mass of a non-rotating object. One may speculate, therefore, that it will collapse further to a black hole after it has lost a significant fraction of its angular momentum by gravitational and/or electromagnetic radiation.

### *5.2. Fundamental properties of neutrinos*

The neutrino signal from SN 1987A has been used to place new constraints on fundamental properties of neutrinos, including rest masses, lifetimes, magnetic moments, mixing angles, etc.

In order to obtain an estimate of the electron neutrino rest mass we can proceed as follows. If neutrinos are emitted from a source at a distance $d$ with certain energy distribution on timescales short compared with the spread of the measured signal, a rest mass can be inferred from the relation (Hillebrandt *et al* 1987c)

$$m_0 = 4.28\left(\frac{E_\nu^{\min}}{7\ \mathrm{MeV}}\right)\left(\frac{d}{52\ \mathrm{kpc}}\right)^{-1/2}\left(\Delta t\frac{\alpha^2}{\alpha^2-1}\right)^{1/2} \tag{5.1}$$

where $\alpha = E_\nu^{\max}/E_\nu^{\min}$ is the observed energy band width of neutrinos. If we do not use a model for the intrinsic time spread of the signal, then (5.1) gives an upper limit for the rest mass. From the IMB and KAMIOKANDE data (table 3) one obtains, allowing for possible errors in the energy determination, $m_0 \leqslant 33\ \mathrm{eV}/c^2$ and $15\ \mathrm{eV}/c^2$, respectively, but the data are also compatible with a zero rest mass (Arnett and Rosner 1987, Bahcall and Glashow 1987, Kolb *et al* 1987), in particular if we allow for the possibility that the last three KAMIOKANDE events as well as the last two IMB events, which were well separated from the main pulse, were caused by background noise. In any case, the supernova data are fully consistent with tritium decay experiments (Fritschi *et al* 1986). Therefore, only if it should turn out that SN 1987A left behind a black hole rather than a neutron star can we hope to get model-independent information on finite neutrino rest masses.

The fact that we have seen neutrinos from the supernova leads to a lifetime of the $\nu_e$ of

$$\tau \gtrsim 140\ \mathrm{h}\ (m_0c^2/\mathrm{eV}) \tag{5.2}$$

in the comoving frame of the neutrinos, but better limits exist for solar neutrinos (Raffelt 1985). The absence of a $\gamma$-ray burst associated with SN 1987A (Vestrand *et al* 1987) can, however, be used to place limits on decay modes $\nu_H \to \nu_L + \gamma$ of heavy neutrinos into light ones, which are essentially independent of the particular supernova model and apply to any massive weakly interacting particle emitted with a rate comparable to that of the $\bar{\nu}_e$. For temperatures below 5 MeV von Feilitzsch and Oberauer (1987) obtain a lifetime larger than $10^{14}$ s $(m_Hc^2/\mathrm{eV})$, which would also apply to heavy $\mu$- and $\tau$-neutrinos, provided $m_H - m_L \leqslant 2m_e$, for which $\nu_H \to \nu_L + \gamma$ is the only possible decay mode. For larger masses the limits are less stringent (Takahara and Sato 1987) because other decay modes become possible.

Most papers in which an attempt is made to place limits on other neutrino properties such as mixing angles (Nötzold 1987, Kuo and Pantaleone 1987) or magnetic moments (Nötzold 1988) suffer from the fact that those estimates are based on the standard core collapse scenario and, therefore, have to be taken with some care. For example, if the first two of the KAMIOKANDE events are attributed to $\nu_e$ from the deleptonisation burst, the absence of the Mikheyev–Smirnov–Wolfenstein effect (Mikheyev and Smirnov 1985, Wolfenstein 1978) places very strong constraints on neutrino masses and mixing angles (Nötzold 1987). However, this assumption may be in conflict with the standard scenario, as was discussed in § 3.2.2.

### *5.3. Constraints on exotic particles*

The fact that the total energy emitted in the neutrino burst from SN 1987A is in good agreement with the expected binding energy of a roughly $1.5M_\odot$ neutron star can be used to set new and interesting constraints on the properties of other weakly interacting light exotic particles (see, e.g., Raffelt 1988 for a review).

Generally speaking, the argument goes as follows (Ellis and Olive 1987). Let us suppose that those particles are light enough ($m \leqslant 10$ MeV) to be produced in a thermal bath in the hot supernova core. If their mean free path is smaller than a typical dimension of the proto neutron star but larger than the mean free path of the neutrinos they will be emitted with approximately a black-body distribution from a radius smaller than the neutrino sphere. Consequently, because the temperature is a rapidly decreasing function of radius, their luminosity will be higher than the neutrino luminosity, and the total energy carried away by those particles will increase with decreasing interaction strength. Because the total energy available is fixed this effect should weaken the neutrino signal and shorten its timescale. If, on the other hand, their mean free path is larger than the neutron star's radius they can escape freely and their luminosity will decrease with their interaction strength, but again the energy available for the neutrinos will be reduced.

These general arguments have been used by various groups to set limits on the coupling of the axion to nucleons and on its mass (Raffelt and Seckel 1988, Turner 1988, Mayle *et al* 1988, Burrows *et al* 1988). Of course, these estimates are not model-independent and depend on both the supernova model and the parameters of the axion model adopted. The upper limits obtained for the dimensionless Yukawa coupling constant to the nucleon range from a few times $10^{-12}$ to several times $10^{-11}$. The upper bounds on the axion mass vary between about $10^{-4}$ and $10^{-3}$ eV. This is an interesting range because axions can be of importance in cosmology as candidates for the dark matter, provided their mass is of the order of $10^{-5}$ eV.

Methods similar to the one outlined here have also been applied to constrain the properties of other hypothetical particles such as right-handed neutrinos (Barbieri and Mohapatra 1988, Raffelt and Seckel 1988) or majorons (Choi *et al* 1988, Fuller *et al* 1988, Grifols *et al* 1988). For example, the laboratory upper bound on the coupling of the right-handed neutrinos has been improved by about three orders of magnitude, and the dimensionless coupling constant of the majoron to electron neutrinos (in the model of Gelmini and Roncadelli (1981)) has been constrained to values of less than $3\times10^{-4}$, which is consistent with laboratory upper bounds obtained from double $\beta$-decay (Doi *et al* 1985).

## 6. Future prospects

Astronomers will continue to monitor SN 1987A at all wavelength bands for decades to come. Optical and infrared observations will reveal more information on the element abundances in the ejecta, on the formation of dust and grains and, for the first time, on the dynamic evolution of a very young supernova remnant. In some 10–20 years the supernova shock will begin to interact with the circumstellar shell and it can then be studied as a bright x-ray and radio source. SN 1987A will allow an accurate determination of the distance to the LMC and thus will help to calibrate the cosmic distance scale.

The most important question, however, is: did the supernova really leave behind an extremely rapidly spinning, weakly magnetised neutron star? Such a possibility had not been anticipated by theoretical models and, therefore, estimates of the time at which the pulsar should become visible may have been off by an order of magnitude. A confirmation of the observations of Middleditch *et al* (1989) would have far-reaching consequences for theories of stellar evolution and supernova explosions, for nuclear equations of state and, possibly, also for the interpretation of the observational data presented in this review, including the interpretation of the neutrino signals.

SN 1987A is generally being considered as a very unusual event in many respects, such as the blue nature of its progenitor and its low visual peak luminosity. But this may be due to selection effects favouring the discovery of bright explosions. For example, if SN 1987A had exploded near the position of Cas A it would have been detected only as a neutrino source and, possibly, an infrared source. If the standard interpretation is correct and sub-luminous type II supernovae are likely if the progenitor stars have lower-than-solar metallicities, one would have to revise present-day estimates of supernova rates. Certainly, future automated supernova search programs will help to answer this question.

With respect to theoretical models, SN 1987A has raised more new questions than it has answered. The apparent early fragmentation of the supernova envelope is not yet understood. If it was caused by a Rayleigh–Taylor instability in the expanding shell, as is suggested by recent numerical simulations, this would impose strong constraints on progenitor models. Moreover, even if we ignore the possibility of a rapidly rotating progenitor star, the observed (and unexpected) strong mixing calls for two- or even three-dimensional supernova models.

SN 1987A has also raised new interest in stellar evolution models. The presumably rather complicated history of Sk $-69°202$ cannot be understood in terms of standard stellar evolution scenarios, the main uncertainties being the correct treatment of convection, mixing and mass loss. Further complications may arise if rotation has to be included.

With regard to nucleosynthesis, detailed computations of synthetic spectra will, for the first time, allow us to determine rather accurately the elemental abundances ejected from a star which may be typical for the enrichment of galaxies with heavy elements, and to compare them with theoretical predictions.

Finally, the fact that the collapse of the core of a massive star did lead to a type II supernova explosion still lacks a reliable explanation. Although numerical simulations were successful in predicting the right neutrino luminosity and spectra, they have failed so far in explaining the very energetic explosion. Whether this failure is due to our incomplete knowledge of the nuclear equation of state and stellar evolution theory, or whether some important ingredients are still missing in our present supernova models is certainly an open question. We may hope, however, that SN 1987A will stimulate new activities which finally will lead to convincing answers.

## References


Abbott D C and Lucy L B 1985 *Astrophys. J.* **288** 679

Aglietta M *et al* 1987a *Europhys. Lett.* **3** 1315

—— 1987b *SN 1987A* ed. I J Danziger (Garching: ESO) p 207

Aitken D K, Smith C H, James S D, Roche P F, Hyland A R and McGregor P J 1988 *Mon. Not. R. Astron. Soc.* **231** 7p

Alexeyev E N, Alexeyeva L N, Krivosheina I V and Volchenkov V I 1987 *SN 1987A* ed. I J Danziger (Garching ESO) p 237

Andreani P, Ferlet R and Vidal-Madjar A 1987 *Nature* **326** 770

Andreani P and Vidal-Madjar A 1987 *SN 1987A* ed. I J Danziger (Garching: ESO) p 517

Arnett W D 1977 *Astrophys. J.* **218** 815

—— 1980 *Astrophys. J.* **237** 541

—— 1982 *Astrophys. J.* **253** 541

—— 1983 *Astrophys. J.* **263** L55

—— 1987 *Astrophys. J.* **319** 136

—— 1988a *Astrophys. J.* **331** 377
—— 1988b *Proc. 4th George Mason Astrophysics Workshop: SN 1987A in the LMC* ed. M Kafatos and A G Michalitsianos (Cambridge: Cambridge University Press) p 301
Arnett W D and Fu A 1989 *Astrophys. J.* **340** 396
Arnett W D and Rosner J L 1987 *Phys. Rev. Lett.* **58** 1906
Arnett W D, Fryxell B A and Müller E 1989 *Astrophys. J. Lett.* **341** L63
Baade W 1926 *Astron. Nachrichten* **228** 359
Bahcall J N, Dar A and Piran T 1987 *Nature* **326** 135
Bahcall J N and Glashow S L 1987 *Nature* **326** 476
Bailey J, Barton J R, Conray P, Hiller D J, Hyland A R, Jones T J Shortridge K and Whittard D 1988 *Pub. Astron. Soc. Pac.* **100** 1178
Barbieri R and Mohapatra R N 1988 *Report UMPP 88-205*
Baron E, Bethe H A, Brown G E, Cooperstein J and Kahana S 1987 *Phys. Rev. Lett.* **59** 736
Baron E, Cooperstein J and Kahana S 1985a *Phys. Rev. Lett.* **55** 126
—— 1985b *Nucl. Phys.* A **440** 744
Barrett P 1987 *SN 1987A* ed. I J Danziger (Garching: ESO) p 173
Benvenuto O, Clocciatiie A, Feinstein A, Garcia B, Luna H, Mendez M and Morrell N 1987 *IAU Circular* 4358
Benz W and Thielemann F-K 1989 in preparation
Bethe H A and Wilson J R 1985 *Astrophys. J.* **295** 11
Bildsten L and Wang J C L 1988 *Proc. 4th George Mason Astrophysics Workshop: SN 1987A in the LMC* ed. M Kafatos and A G Michalitsianos (Cambridge: Cambridge University Press) p 116
Bionta R M *et al* 1987 *Phys. Rev. Lett.* **58** 1494
Bisnovatyi-Kogan G S 1971 *Sov. Astron.* **14** 652
—— 1980 *Ann. NY Acad. Sci.* **336** 389
Blades J C, Wheatley J M, Panagia N, Grewing M, Pettini M and Wamsteker W 1988 *Astrophys. J.* **334** 308
Blinnikov S I, Chugai N N, Golenetskii S V, Lozinskaya T A and Mazets E P 1988 *Astrophys. Space Phys. Rev.* **6** 197
Bodansky D, Clayton D D and Fowler W A 1968 *Astrophys. J. Suppl.* **16** 299
Bodenheimer P and Woosley S E 1983 *Astrophys. J.* **269** 281
Bouchet P, Stanga R, Moneti A, Le Bertre G Th, Manfoid J, Silvestro G and Slezak G 1987 *SN 1987A* ed. I J Danziger (Garching: ESO) p 79
Branch D 1987a *Astrophys. J.* **320** L23
—— 1987b *Astrophys. J.* **320** L121
—— 1988 *Proc. Symp. on Extragalactic Distance Scales* ed. S van den Bergh at press
Branch D, Falk S, McCall M, Rybski P, Uomoto A K and Will B J 1981 *Astrophys. J* **244** 780
Brown G E, Bethe H A and Baym G 1982 *Nucl. Phys.* A **375** 481
Brown J C and McLean I S 1977 *Astron. Astrophys.* **57** 141
Bruenn S W 1987 *Phys. Rev. Lett.* **59** 938
Brunish W M and Truran J W 1982 *Astrophys. J.* **256** 247
Burki G, Cramer N, Burnet M, Rufener F, Pernier B and Richard C 1989 *Astron. Astrophys.* at press
Burrows A 1987 *Astrophys. J.* **318** L57
Burrows A and Lattimer J M 1987 *Astrophys. J.* **318** L63
Burrows A, Turner M S and Brinkmann R P 1989 *Phys. Rev.* D **39** 1020
Cassatella A 1987 *SN 1987A* ed. I J Danziger (Garching: ESO) p 101
Cassinelli J P, Nordsiek K H and Murison M A 1987 *Astrophys. J.* **317** 290
Catchpole R M *et al* 1987 *Mon. Not. R. Astron. Soc.* **229** 15p
—— 1988 *Mon. Not. R. Astron. Soc.* **231** 75p
—— 1989 *Mon. Not. R. Astron. Soc.* **237** 55p
Chandrasekhar S 1961 *Hydrodynamic and Hydromagnetic Stability* (Oxford: Clarendon)
Chevalier R A 1982 *Astrophys. J.* **259** 302
—— 1987 *SN 1987A* ed. I J Danziger (Garching: ESO) p 481
—— 1988 *Nature* **332** 514
Chilukuri M and Wagoner R V 1988 *IAU Coll. 108: Atmospheric Diagnostic of Stellar Evolution* ed. K Nomoto (Berlin: Springer) p 295
Chiosi C and Maeder A 1986 *Ann. Rev. Astron. Astrophys.* **24** 329
Choi K, Kim C W, Kim J and Lam W P 1988 *Phys. Rev.* D **37** 3225
Clayton G C, Martin P G and Thompson I 1983 *Astrophys. J.* **265** 194
Colgate S A and McKee C 1969 *Astrophys. J.* **157** 623

Cook W R, Palmer D, Prince T, Schindler C, Starr C and Stone E 1988 *IAU Circular* 4527
Couch W J 1988 *Proc. 4th George Mason Astrophysics Workshop: SN 1987A in the LMC* ed. M Kafatos and A G Michalitsianos (Cambridge: Cambridge University Press) p 60
Couderc P 1939 *Ann. d'Astrophys.* **2** 271
Cristiani S *et al* 1987 *SN 1987A* ed. I J Danziger (Garching: ESO) p 65
Cropper M and Bailey J A 1987 *IAU Circular* 4319
Crotts A 1988 *IAU Circular* 4561
Danziger I J, Bouchet P, Fosbury R A E, Gouiffes C, Lucy L B, Moorwood A F M, Oliva E and Rufener F 1988a *Proc. 4th George Mason Astrophysics Workshop: SN 1987A in the LMC* ed. M Kafatos and A G Michalitsianos (Cambridge: Cambridge University Press) p 37
Danziger I J, Bouchet P, Gouiffes C and Rufener F 1988b *SN 1987A: One Year Later* ed. M Greco (Paris: Editions Frontières) p 3
—— 1989 *Big Bang, Active Galactic Nuclei and Supernovae* ed. S Hayakawa and K Sato (New York: Universal Academic) p 429
De Boer K S, Grewing M, Richter M, Wamsteker T, Gry C and Panagia N 1987 *Astron. Astrophys.* **177** L37
De Rujula A 1987 *Phys. Lett.* **193B** 525
D'Odorico S, Molaro P, Pettini M, Stathakis R and Vladilo G 1987 *SN 1987A* ed. I J Danziger (Garching: ESO) p 525
Doi M, Kotani T and Takasugi E 1985 *Prog. Theor. Phys. Suppl.* **83** 1
Dopita M A 1988 Personal communication
Dotani T *et al* 1987 *Nature* **330** 230
Dufour R J 1984 *IAU Symp. 108: Structure and Evolution of the Magellanic Clouds* ed. S van den Bergh and K de Boer (Dordrecht: Reidel) p 353
Dufour R J, Shields G A and Talbot R J 1982 *Astrophys. J.* **252** 461
Dupree A K, Kirschner R P, Nassiopoulos G E, Raymond J C and Sonneborn G 1987 *Astrophys. J.* **320** 597
Dwek E 1988 *Proc. 4th George Mason Astrophysics Workshop: SN 1987A in the LMC* ed. M Kafatos and A G Michalitsianos (Cambridge: Cambridge University Press) p 240
Eastman R G 1989 *Particle Astrophysics, Berkeley* ed. E B Norman (Singapore: World Scientific) p 779
Eastman R and Kirshner R P 1988 *BAAS Vancouver Late Papers* at press
Ebisuzaki T and Shibazaki N 1988 *Astrophys. J.* **328** 699
El Eid M, Baraffe I and Weiss A 1987 *SN 1987A* ed. I J Danziger (Garching: ESO) p 283
Elias J H, Gregory B, Phillips M M, Williams R E, Graham J R, Meikle W P S, Schwartz R D and Wilking B 1988 *Astrophys. J.* **331** L9
Ellis J and Olive K 1987 *Phys. Lett.* **193B** 525
Erickson E F, Haas M R, Colgan S W J, Lord S D, Burton M G, Wolf J, Hollebach D J and Werner M 1988 *Astrophys. J.* **330** L39
Falk S W and Arnett W D 1977 *Astrophys. J. Suppl.* **33** 515
Feast M W 1988 *Proc. 4th George Mason Astrophysics Workshop: SN 1987A in the LMC* ed. M Kafatos and A G Michalitsianos (Cambridge: Cambridge University Press) p 51
Filippenko A V 1989 *Particle Astrophysics, Berkeley* ed. E B Norman (Singapore: World Scientific) p 177
Fitzpatrick E L and Savage B D 1983 *Astrophys. J.* 93
Fosbury R A E, Danziger I J, Lucy L B, Gouiffes C and Critiani S 1987 *SN 1987A* ed. I J Danziger (Garching: ESO) p 139
Fransson C 1987 *SN 1987A* ed. I J Danziger (Garching: ESO) p 467
—— 1988 *Proc. Astron. Soc. Aust.* 7 520
Fransson C, Cassatella A, Gilmozzi R, Kirshner R P and Panagia N 1988 *Astrophys. J.* submitted for publication
Fransson C, Grewing M, Cassatella A, Panagia N and Wamsteker W 1987 *Astron. Astrophys.* **177** L33
Fransson C and Lunquist P 1989 *Astron. Astrophys.* at press
Friedmann J L, Ipser J R and Parker L 1986 *Astrophys. J.* **304** 115
Fritschi M *et al* 1986 *Phys. Lett.* **173** 485
Fuller G, Mayle R and Wilson J R 1988 *Astrophys. J.* B **332** 826
Fu A 1988 *Proc. Astron. Soc. Aust.* **7** 505
Garrison R and Shelton I 1987 *IAU Circular* 4330
Gebenev S A and Sunyaev R A 1988a *Sov. Astron. Lett.* **13** 945
—— 1988b *Sov. Astron. Lett.* **14** 675
Gehrels N, MacCallum C J and Leventhal M 1987 *Astrophys. J.* **317** L73
Gehrels N *et al* 1988 *IAU Circular* 4593
Gelmini G and Roncadelli M 1981 *Phys. Lett.* **99B** 411

Gilmozzi R, Cassatella A, Clavel J, Fransson C, Gonzalez R, Gry C, Panagia N, Talavera A and Wamsteker W 1987 *Nature* **328** 318

Goldreich P and Weber S V 1980 *Astrophys. J.* **238** 991

Gonzales R, Wamsteker W, Gilmozzi R, Walborn N and Lauberts A 1987 *SN 1987A* ed. I J Danziger (Garching: ESO) p 33

Graham J A and Nemec J M 1984 *IAU Symp. 108: Structure and Evolution of the Magellanic Clouds* ed. S van den Bergh and K de Boer (Dordrecht: Reidel) p 37

Grewing M, Blades J C, Panagia N, Pettini M and Wamsteker W 1987 *SN 1987A* ed. I J Danziger (Garching: ESO) p 511

Grifols J A, Masso E and Peris S 1988 *Report UAB-FT-190/88*

Hamuy M I, Suntzeff N B, Gonzalez R and Martin G 1988 *Astron. J.* **95** 63

Hanuschik R W 1988 *Proc. Astron. Soc. Aust.* **7** 446

Hanuschik R W and Dachs J 1987a *SN 1987A* ed. I J Danziger (Garching: ESO) p 153

—— 1987b *Astron. Astrophys.* **182** L29

—— 1988 *Astron. Astrophys.* **205** 135

—— 1989 *Astron. Astrophys.* at press

Harkness R P and Wheeler J C 1988 *Proc. Astron. Soc. Aust.* **7** 431

Hashimoto M, Nomoto K and Shigeyama T 1989 *Astrophys. J.* at press

Hazen M L 1987 *IAU Circular* 4367

Helt B E, Franco G A P and Florentin Nielsen R 1987 *SN 1987A* ed. I J Danziger (Garching: ESO) p 89

Hempe K 1981 *Astron. Astrophys.* **98** 19

—— 1984 *Mitt. d. Astron. Ges.* **60** 107

Hershkowitz R, Lindner E and Wagoner B 1986 *Astron. Astrophys.* **301** 220

Hillebrandt W 1982 *Supernovae: A Survey of Current Research* ed. M J Rees and R J Stoneham (NATO ASI Series C, vol. 90) (Dordrecht: Reidel) p 123

—— 1985 *Proc. 3rd Workshop on Nuclear Astrophysics* vol. 199, ed. W Hillebrandt *et al* (Garching: MPA) p 59

—— 1987a *High Energy Phenomena around Collapsed Stars* ed. F Pacini (NATO ASI Series C, vol. 195) (Dordrecht: Reidel) p 73

—— 1987b *SN 1987A* ed. I J Danziger (Garching: ESO) p 301

Hillebrandt W, Höflich P, Kafka P, Müller E, Schmidt H U, Truran J W and Wampler J 1987c *Astron. Astrophys.* **180** L20

Hillebrandt W, Höflich P, Schmidt H U and Truran J W 1987a *Astron. Astrophys.* **186** L10

Hillebrandt W, Höflich P, Truran J W and Weiss A 1987b *Nature* **327** 597

Hillebrandt W, Nomoto K and Wolff R G 1984 *Astron. Astrophys.* **133** 175

Hirata K *et al* 1987 *Phys. Rev. Lett.* **58** 1490

Höflich P 1987 *Proc. 4th Workshop on Nuclear Astrophysics, Lecture Notes in Physics* vol. 287, ed. W Hillebrandt *et al* (Berlin: Springer) p 307

—— 1988a *Proc. Astron. Soc. Aust.* **7** 434

—— 1988b *IAU Coll. 108: Atmospheric Diagnostic of Stellar Evolution* ed. K Nomoto (Berlin: Springer) p 388

—— 1989a *Particle Astrophysics Workshop, Berkeley* ed. C Pennypacker at press

—— 1989b *Proc. 5th Workshop on Nuclear Astrophysics* ed. W Hillebrandt and E Müller at press

Höflich P, Sharp Ch M and Zorec J 1989 *Particle Astrophysics, Berkeley* ed. E B Norman (Singapore: World Scientific) p 186

Höflich P, Wehrse R and Shaviv G 1986 *Astron. Astrophys.* **163** 105

Imshennik V S and Nadyozhin D K 1983 *Astrophys. Space Phys. Rev.* **2** 75

Isserstedt J 1975 *Astron. Astrophys. Suppl.* **19** 259

Itoh H, Hayakawa S, Masai K and Nomoto K 1987a *Publ. Astron. Soc. Japan* **39** 529

—— 1987b *Nature* **330** 233

Janka H-T 1987 *Proc. 4th Workshop on Nuclear Astrophysics, Lecture Notes in Physics* vol. 287, ed. W Hillebrandt *et al* (Berlin: Springer) p 319

Janka H-T and Hillebrandt W 1989a *Astron. Astrophys. Suppl.* **78** 375

—— 1989b *Astron. Astrophys.* at press

Janka H-T and Mönchmeyer R 1988 *Astron. Astrophys.* **209** L5

Jauncey D L *et al* 1987 *Nature* **334** 412

Jeffery D J 1987 *Nature* **329** 419

Johnson H L 1966 *Ann. Rev. Astron. Astrophys.* **4** 197

Jones A 1987 Personal communication to J Wampler

Karovska M, Nisenson P, Noyes R and Papaliolos C 1987 *IAU Circular* 4604

Kirshner R P 1988 *Proc. 4th George Mason Astrophysics Workshop: SN 1987A in the LMC* ed. M Kafatos and A G Michalitsianos (Cambridge: Cambridge University Press) p 87

Kirshner R and Kwan J 1974 *Astrophys. J.* **193** 27

Kirshner R P, Sonneborn G, Crenshaw D M and Nassiopoloulos G E 1987 *Astrophys. J.* **320** 602

Kolb E W, Stebbins A J and Turner M S 1987 *Phys. Rev.* D **35** 3598

Kristian J A *et al* 1989 *Nature* **338** 234

Kudritzki R P, Groth H G, Butler K, Husfeld D, Becker S, Eber F and Fitzpatrick E 1987 *SN 1987A* ed. I J Danziger (Garching: ESO) p 39

Kumagai S, Itoh M, Shigeyama T, Nomoto K and Nishimura J 1988 *Proc. 4th George Mason Astrophysics Workshop: SN 1987A in the LMC* ed. M Kafatos and A G Michalitsianos (Cambridge: Cambridge University Press) p 413

Kuo T K and Pantaleone J 1987 *Phys. Lett.* **198B** 406

Lamers H J G L M, van Heuvel E P J and Petterson J A 1976 *Astron. Astrophys.* **49** 327

Larson H P, Drapatz S, Mumma M J and Weaver H A 1987 *SN 1987A* ed. I J Danziger (Garching: ESO) p 147

Larson H P, Drapatz S, Mumma S and Weaver H A 1988 *Proc. 4th George Mason Astrophysics Workshop: SN 1987A in the LMC* ed. M Kafatos and A G Michalitsianos (Cambridge: Cambridge University Press) p 74

Lattimer J M and Burrows A 1984 *Problems of Collapse and Numerical Relativity* ed. D Bancel and M Signore (NATO ASI Series C, vol. 134) (Dordrecht: Reidel) p 147

Le Blanc J M and Wilson J R 1970 *Astrophys. J.* **161** 541

Lucy L B 1987a *Astron. Astrophys.* **182** L31

—— 1987b *SN 1987A* ed. I J Danziger (Garching: ESO) p 417

—— 1988 *Proc. 4th George Mason Astrophysics Workshop: SN 1987A in the LMC* ed. M Kafatos and A G Michalitsianos (Cambridge: Cambridge University Press) p 323

Lunquist P and Fransson C 1988 *Astron. Astrophys.* **192** 221

Maeder A 1987 *SN 1987A* ed. I J Danziger (Garching: ESO) p 251

Magain P 1987 *Nature* **329** 606

Manchester R N 1987 *Astron. Astrophys.* **171** 205

Marcher S J, Meikle W P S and Morgan B L 1987 *IAU Circular* 4391

Martin M J 1987 *Nuclear Data Sheets* **51** 67

Masai K, Hayakawa S, Itoh H, Nomoto K and Shigeyama T 1988 *IAU Coll. 108: Atmospheric Diagnostic of Stellar Evolution* ed. K Nomoto (Berlin: Springer) p 450

Masai K, Kumagai S, Shigeyama T and Nomoto K 1987 *Nature* **330** 235

Matz S M, Share G H, Leising M D and Chupp E L 1988b *IAU Circular* 4568

—— 1988c *IAU Circular* 4618

Matz S M, Share G H, Leising M D, Chupp E L, Vestrand W T, Purcell W R, Strickman M S and Reppin C 1988a *Nature* **331** 416

Mayle R, Wilson J R, Ellis J, Olive K, Schramm D N and Steigman G 1988 *Phys. Lett.* **203B** 188

McGregor P 1988 *Proc. Astron. Soc. Aust.* **7** 450

McGregor P J, Hyland A R and Hiller D J 1988 *Astrophys. J* **324** 1071

McNaught R H 1987 *IAU Circular* 4389

Mendez M, Clocchiatti A, Benvenuto G, Feinstein C and Marraco U G 1988 *Astrophys. J.* **334** 295

Meier D L, Epstein R I, Arnett W D and Schramm D N 1976 *Astrophys. J.* **204** 869

Meikle W P S, Allen D A, Spyromilio J and Varani G F 1989 *Mon. Not. R. Astron. Soc.* **238** 193

Meikle W P S, Marcher S J and Morgan B L 1987a *IAU Circular* 4391

—— 1987b *Nature* **329** 608

—— 1987c *SN 1987A* ed. I J Danziger (Garching: ESO) p 197

Menzies J W *et al* 1987 *Mon. Not. R. Astron. Soc.* **227** 39p

Middleditch J, Pennypacker C, Morris D E, Muller R A, Perlmutter S and Sasseen T 1989 *IAU Circular* 4735

Mihalas D, Kunasz R B and Hummer D G 1975 *Astrophys. J.* **202** 4

—— 1976 *Astrophys. J.* **206** 5

Mikheyev S P and Smirnov A Yu 1985 *Yad. Fiz.* **42** 1441

Mönchmeyer R and Müller E 1989 *Timing Neutron Stars* ed. H Ögelmann and E van den Heuvel (NATO ASI Series C, vol. 262) (Deventer: Klumer) p 549

Moreno B and Walker S 1987 *IAU Circular* 4316

Müller E and Hillebrandt W 1979 *Astron. Astrophys.* **80** 147

—— 1981 *Astron. Astrophys.* **103** 358

Müller E, Hillebrandt W, Orio M, Höflich P, Mönchmeyer R and Fryxell B A 1989a *Astron. Astrophys.* **220** 167

Müller E, Hillebrandt W, Orio M, Höflich P, Mönchmeyer R, Fryxell B A and Arnett D 1989b *Proc. 5th Workshop on Nuclear Astrophysics* ed. W Hillebrandt and E Müller (Garching: MPA) p 103
Müller E, Rozyczka M and Hillebrandt W 1980 *Astron. Astrophys.* **81** 288
Nagasawa M, Nakamura T and Miyama S 1988 *Publ. Astron. Soc. Japan* **40** 691
Nisenson P, Karovska M, Noyes R and Papaliolios C 1987 *Bull. Am. Astron. Soc.* **19** 950
Nomoto K 1984 *Astrophys. J.* **277** 791
Nomoto K, Hashimoto M, Shigeyama T, Kumagai S, Yamaoka H and Saio H 1989 *Big Bang, Active Galactic Nuclei and Supernovae* ed. S Hayakawa and K Sato (New York: Universal Academic) p 495
Nomoto K and Shigeyama T 1988 *Proc. 4th George Mason Astrophysics Workshop: SN 1987A in the LMC* ed. M Kafatos and A G Michalitsianos (Cambridge: Cambridge University Press) p 273
Nomoto K, Shigeyama T, Kumagai S and Hashimoto S 1988b *Proc. Astron. Soc. Aust.* **7** 490
Nomoto K, Shigeyama T and Hashimoto M 1987 *SN 1987A* ed. I J Danziger (Garching: ESO) p 325
—— 1988a *IAU Symp. 108: Atmospheric Diagnostic of Stellar Evolution* ed. K Nomoto (Berlin: Springer) p 319
Nötzold D 1987 *Phys. Lett.* **196B** 315
—— 1988 *Phys. Rev.* D **38** 1658
Oliva E, Moorwood F M and Danziger I J 1988 *ESO Messenger* **50** 18
Osterbrock D E 1974 *Astrophysics of Gaseous Nebulae* (San Francisco: Freeman)
Panagia N 1984 *Supernovae as Distance Indicators, Lecture Notes in Physics* vol 224 (Berlin: Springer) p 14
Panagia N, Gilmozzi R, Clavel J, Barylak M, Gonzales Riesta R, Lloyd C, Snas Fernandez de Cordoba L and Wamsteker W 1987 *Astron. Astrophys.* **177** L25
Pandharipande V R and Smith R A 1975 *Nucl. Phys.* A **175** 225
Petuchowski R *et al* 1988 in preparation
Phillips M M 1988 *Proc. 4th George Mason Astrophysics Workshop: SN 1987A in the LMC* ed. M Kafatos and A G Michalitsianos (Cambridge: Cambridge University Press) p 16
Phillips M M and Heathcote S R 1989 *Mon. Not. R. Astron. Soc.* at press
Pinto P and Woosley S E 1988a *Astrophys. J.* **329** 820
—— 1988b *Nature* **333** 534
Raffelt G 1985 *Phys. Rev.* D **31** 3002
—— 1988 *Proc. XXIV int. Conf. on High Energy Physics* ed. R Kotthaus and J Kühn (Berlin: Springer) at press
Raffelt G and Seckel D 1988 *Phys. Rev. Lett.* **60** 1793
Rank D M, Gregman J, Witteborn F C, Cohen M, Lynch D K and Russell R W 1988a *Astrophys. J.* **325** L1
Rank D M, Pinto P A, Woosley S E, Bregman J D, Witteborn F C, Axelrod T S and Cohen M 1988b *Nature* **331** 505
Rees M J 1987 *Nature* **328** 207
Reimers D 1975 *Mem. Soc. R. Liège Ser.* **8** 369
Rester A C, Eichhorn G, Coldwell R L, Trombka J I, Starr R and Lasche G P 1988 *IAU Circular* 4535
Ritchey G W 1901 *Astrophys. J.* **14** 167
—— 1902 *Astrophys. J.* **15** 129
Rosa M, Gouiffes C and Ruiz M-T, 1988 *IAU Circular* 4564
Rousseau J, Martin N, Prevout L, Reheirot E, Robin A and Brunet J P 1978 *Astron. Astrophys. Suppl.* **31** 243
Ryker E 1987 Personal communication to J Wampler
Saio H, Kato M and Nomoto K 1988 *Astrophys. J.* **331** 388
Sandie W G, Nakoano G H, Chase L F, Fishman G J, Meegan C A, Wilson R B and Paciesas W 1988 *Proc. 4th George Mason Astrophysics Workshop: SN 1987A in the LMC* ed. M Kafatos and A G Michalitsianos (Cambridge: Cambridge University Press) p 366
Sato K and Suzuki H 1987 *Phys. Rev. Lett.* **58** 2722
Schaefer B E 1987 *Astrophys. J.* **323** L47
Schaefer R, Casse M and Cahen S 1987 *Astrophys. J.* **316** L31
Schmidt Th 1976 *Astron. Astrophys. Suppl.* **24** 357
Schwarz H E 1987 *SN 1987A* ed. I J Danziger (Garching: ESO) p 167
Schwarz H E and Mundt R 1987 *Astron. Astrophys.* **177** L4
Sedov L I 1959 *Similarity and Dimensional Methods in Mechanics* (New York: Academic) p 260
Serkowski K, Mathewson D S and Ford V L 1975 *Astrophys. J.* **196** 261
Shapiro S L and Teukolsky S A 1983 *Black Holes, White Dwarfs, and Neutron Stars* (New York: Wiley)
Sharma M M, Stocker W, Gleissl P and Brack M 1989 *Preprint*
Sharp Ch and Höflich P 1989 *Highlights of Astronomy* vol. 8, p 207
Shaviv G, Wehrse R and Wagoner R V 1984 *Astrophys. J.* **289** 198
Shigeyama T and Ebisuzaki T 1988 *Astrophys. J.* **327** L9

Shigeyama T, Nomoto K and Hashimoto M 1988 *Astron. Astrophys.* **196** 141
Sood R K *et al* 1988 *Proc. 4th George Mason Astrophysics Workshop: SN 1987A in the LMC* ed. M Kafatos and A G Michalitsianos (Cambridge: Cambridge University Press) p 425
Spyromilio J, Meikle W P S, Learner R C M and Allen D A 1989 *Nature* at press
Storey M C and Manchester R N 1987 *Nature* **329** 421
Suntzeff N B, Heathocote S, Weller W G, Caldwell N, Huchra J P, Olowin R P and Chambers K C 1988 *Nature* **334** 135
Sunyaev R *et al* 1987 *Nature* **330** 327
Sutherland P, Xu Y, McCray R and Ross R P 1988 *IAU Symp. 108: Atmospheric Diagnostic of Stellar Evolution* ed. K Nomoto (Berlin: Springer) p 394
Svoboda R *et al* 1987 *SN 1987A* ed. I J Danziger (Garching: ESO) p 315
Symbalisty E 1984 *Astrophys. J.* **285** 729
Takahara M and Sato K 1987 *Mod. Phys. Lett.* A **2** 293
Tammann G A 1982 *Supernovae: A Survey of Current Research* ed. M J Rees and R J Stoneham (NATO ASI Series C, vol. 90) (Dordrecht: Reidel) p 371
Tanaka Y 1988 *IAU Coll. 108: Atmospheric Diagnostic of Stellar Evolution* ed. K Nomoto (Berlin: Springer) p 399
Terndrup D M, Elias J H, Gregory B, Heathcote S R, Phillips M M and Williams N B 1988 *Proc. Astron. Soc. Aust.* **7** 412
Tholine J E 1984 *Astrophys. J.* **285** 721
Trümper J *et al* 1988 *Proc. 4th George Mason Astrophysics Workshop: SN 1987A in the LMC* ed. M Kafatos and A G Michalitsianos (Cambridge: Cambridge University Press) p 355
Truran J W and Weiss A 1987 *SN 1987A* ed. I J Danziger (Garching: ESO) p 271
—— 1988 *Proc. 4th George Mason Astrophysics Workshop: SN 1987A in the LMC*, ed. M Kafatos and A G Michalitsianos (Cambridge: Cambridge University Press) p 322
Turner M S 1988 *Phys. Rev. Lett.* **60** 1797
Turtle A J, Campbell-Wilson D, Bunton J D, Jauncey D L, Kesteven M J, Manchester R N, Norris R P, Storey M C and Reynolds J E 1987 *Nature* **327** 38
Tyson J A and Bocshaar P G 1987 *Publ. Astron. Soc. Pac.* **99** 905
Utrobin V P 1988 *Atominform* p 20
van den Bergh S, McClure R D and Evans R 1987 *Astrophys. J.* **323** 44
Vestrand W T, Gosh A and Chupp E 1987 *IAU Circulars* 4338, 4340, 4365
Vidal-Madjar A, Anderani P, Cristiani S, Ferlet R, Lanz T and Vladilo G 1987 *Astron. Astrophys.* **177** L17
Vladilo G and Molaro P 1987 *SN 1987A* ed. I J Danziger (Garching: ESO) p 539
von Feilitzsch F and Oberauer L 1987 *Phys. Lett.* **200B** 580
Walborn N R, Lasker B M, Laidler V G and Chu Y-H 1987 *Astrophys. J.* **321** L41
Walker A R 1987 *Mon. Not. R. Astron. Soc.* **225** 627
Walker A R and Mack P 1988 *Astron. J.* **96** 872
Wampler E J 1988 *SN 1987A: One Year Later* ed. M Greco (Paris: Editions Frontières) p 17
Wampler E J and Richichi A 1989 *Astron. Astrophys.* **217** 31
Wampler E J, Truran J W, Lucy L B, Höflich P and Hillebrandt W 1987 *Astron. Astrophys.* **182** L51
Wamsteker W, Panagia N, Barylak M and Cassatella A 1987 *Astron. Astrophys.* **177** L21
Weiler K W, Sramek R A, Panagia N, van der Hulst J M and Salvati M 1986 *Astrophys. J.* **301** 790
Weiss A 1987 *Astron. Astrophys.* **185** 165
—— 1989 *Astrophys. J.* **339** 365
Weiss A, Hillebrandt W and Truran J W 1988 *Astron. Astrophys.* **197** L11
Wesselink A J 1946 *Bull. Astron. Inst. Neth.* **368** 91
West R M 1988 *ESO Messenger* **52** 13
West R M, Lauberts A, Jorgensen H E and Schuster H-E 1987 *Astron. Astrophys.* **177** L1
Wheeler J C, Harkness R P and Barkat Z 1988 *IAU Symp. 108: Atmospheric Diagnostic of Stellar Evolution* ed. K Nomoto (Berlin: Springer) p 305
Whitelock P A *et al* 1988 *Mon. Not. R. Astron. Soc.* **234** 5p
Williams R E 1987 *Astrophys. J.* **320** L117
Williams R E 1988 *Proc. 4th George Mason Astrophysics Workshop: SN 1987A in the LMC* ed. M Kafatos and A G Michalitsianos (Cambridge: Cambridge University Press) p 355
Wilson J R 1985 *Numerical Astrophysics* ed. J Centrella *et al* (Boston: Jones and Bartlett) p 422
—— 1988 *Talk presented at the Aspen Conference on SN 1987A*
Wilson J R, Mayle R W, Woosley S E and Weaver T A 1986 *Ann. NY Acad. Sci.* **479** 267
Winkler H 1987 *IAU Circular* 4320

Witteborn F, Bregman J D, Wooden D, Pinto P A, Rank D M, Woosley S E and Cohen M 1989 *Astrophys. J. Lett.* **338** L9

Wolfenstein L 1978 *Phys. Rev.* D **17** 2369

Woosley S E 1986 *Nucleosynthesis and Chemical Evolution, Saas Fee Lecture Notes* ed. B Hauck *et al* (Geneva Observatory) p 1

—— 1988a *Astrophys. J.* **330** 218

—— 1988b *Proc. 4th George Mason Astrophysics Workshop: SN 1987A in the LMC*, ed. M Kafatos and A G Michalitsianos (Cambridge: Cambridge University Press) p 289

—— 1988c *IAU Symp. 108: Atmospheric Diagnostic of Stellar Evolution* ed. K Nomoto (Berlin: Springer) p 361

Woosley S E, Pinto P A and Ensman L 1987a *Astrophys. J.* **324** 664

—— 1987b *Astrophys. J.* **324** 466

Woosley S E, Pinto P A and Weaver T A 1988 *Proc. Astron. Soc. Aust.* **7** 355

Zoltowski F 1987 *IAU Circular* 4389

Rep. Prog. Phys. (1992) 377–430. Printed in the UK

# Neutrino astronomy

Y Totsuka

Institute for Cosmic Ray Research, University of Tokyo, 3-2-1 Midoricho, Tanashi, 188 Tokyo, Japan

**Abstract**

Observations of extraterrestrial neutrinos are reviewed. Solar neutrinos and supernovae neutrinos have been successfully detected and provided valuable information for astronomy and particle physics. Observation of the 1.9 K cosmic background neutrinos would be most challenging. It probably requires finite neutrino masses and ultrahigh-energy neutrinos of cosmological and perhaps of grand unified theory (GUT) related origin. Ultrahigh-energy neutrinos have not yet been observed, but they will provide a clue about the origin of cosmic rays, and probably exciting information about cosmology and GUT.

This review was received in February 1991.

0034-4885/92/030377+54$18.00 

## Contents

## 1. Introduction

In February 1987 neutrinos from the supernova SN1987A in the Large Magellanic Clouds were successfully observed at the two large water-type Cerenkov detectors called Kamiokande [1] and IMB [2] (4500 t and 8000 t, respectively). An underground scintillation detector, the Neutrino Telescope in Baksan (200 t) [3], has also observed several burst-like events, which were presumably produced by neutrinos from SN1987A. Another scintillation detector, LSD in Mt Blanc (90 t), also saw several events probably correlated with SN1987A, but about five hours earlier than the above three observations. What happened in the LSD detector at that time is still a matter of debate.

This, already historic, event tells us the two facts inherent to neutrino astronomy. (i) neutrinos actually play a significant role in stellar death and probably in many other phenomena in astronomy and cosmology; (ii) neutrinos are quite elusive. To detect neutrinos requires a large detector with an extremely low background and high sensitivities. For example, Kamiokande observed 11 interacting events while the detector had $10^{16}$ incident neutrinos, namely the detection efficiency of $10^{-15}$, which should be compared with the detection efficiency of about 0.5 for optical observations.

Despite the great difficulties, the successful observation of a supernova by means of neutrinos is now considered to have opened a complete new field of neutrino astronomy. The great distance to SN1987A, 160 000 light years (lyr), proved that neutrino astronomy is capable of studying our whole Galaxy.

The Sun is the nearest star to us, and therefore the best object with which to study every detail of stellar physics. In particular, the energy source of the Sun is of great interest not only for stellar theory but also for human lives on the Earth. Direct experimental study of the energy generation is, however, impossible with conventional astronomical means, because the source lies deep in the core of the Sun. If the Sun indeed generates energy from the thermonuclear-fusion process as stellar theory claims, an enormous number of neutrinos must be emitted. The basic fusion process is to create a $^4$He nucleus from four protons, which must also produce two positrons in order to conserve electric charge. The weak interaction then demands the simultaneous production of two electron neutrinos ($\nu_e$) to conserve fermionic and leptonic number. The fusion processes occurring in the Sun are actually more complicated, basically consisting of several two-body reactions, electron captures and $\beta$-decays [5].

In the 1960s an experiment in the Homestake gold mine began to observe solar neutrinos ($E_{\nu_e} > 0.81$ MeV). This experiment, led by R Davis Jr [6–8], has continued to operate up to the present time. Solar neutrinos were successfully observed and the experiment produced two surprising results: (i) the neutrino interaction rate in the detector is only $27 \pm 4\%$ of the theoretical prediction; (ii) there is an apparent anticorrelation of the solar-neutrino yield with the sunspot number, that is the solar activity.

The Homestake experiment must be called the pioneering one of neutrino astronomy. The great difficulty of neutrino observation prohibited other experiments to follow and to confirm Davis's important results until 1990. The Kamiokande group [9, 10], which was the first to observe the SN1987A neutrinos, has finally seen solar neutrinos with $E_{\nu_e} > 7.5$ MeV. Their results are: (i) the neutrino yield is $46 \pm 5 \pm 6\%$ of the theoretical prediction, and (ii) no anticorrelation of the neutrino yield with the sunspot number (<36% at 90% CL). Thus, Kamiokande confirmed the deficit in the solar-neutrino flux, but failed to confirm the anticorrelation with the sunspot number.

If the two results are considered simultaneously apart from the controversial anticorrelation, one is led to the important conclusion, namely the most probable solution to the solar-neutrino problem is the occurrence of the resonant matter oscillations (the Mikheyev–Smirnov–Wolfenstein mechanism) [11–18] in the solar interior, which is a direct consequence of finite neutrino masses and of a mixing between at least two neutrino species.

The solar neutrino experiment tells us another aspect of neutrino astronomy, that is the observation of astronomical neutrinos leads to an investigation of the neutrino properties, especially their masses, mixings, magnetic moments, etc. Neutrinos are the least well understood particles among the quarks and leptons, and a full understanding of them will lead to a new development of elementary-particle physics, notably a way to the long postulated Grand Unified Theories. Thus the importance in neutrino astronomy is to clarify the mutual influences between astronomy and particle physics.

We showed two examples of neutrino astronomy, and in fact these two are the only ones that have achieved successful observations of astronomical neutrinos. The reason is obvious. Since neutrinos are neutral particles, they must interact with detector material and produce detectable signals. Then the extremely small interaction cross sections allow us only to observe either a nearby star, namely the Sun, or an object that emits an enormous number of neutrinos, namely a type-II supernova. Even if a supernova emits $3\times10^{53}$ erg of energy in the form of neutrinos, corresponding to $10^3$ times the total solar luminosity integrated over 4.5 billion years, it is not possible to extend the observation to the nearby galaxy, M31 in the Andromeda constellation, 2.1 million lyr away from us. Thus the range of neutrino astronomy is quite limited.

Despite these difficulties it is of the utmost importance to extend the reach of neutrino astronomy further, because it is the only way to look directly at the deep interior of the two objects, the Sun and type-II supernovae.

Are there other observable neutrino sources in the Universe? Let us first consider low-energy neutrinos ($E_\nu < 100$ MeV). We find two promising diffuse sources of cosmological origin. One is the cosmic background neutrinos (CBN) [19] which decoupled and were left behind from the expansion of the Universe at about 1 s after the Big Bang. They obey the Fermi–Dirac distribution with the temperature $T_\nu = 1.9$ K and the number density $n_\nu = 100\ \mathrm{cm}^{-3}$ per species at the present epoch. Note that there are three neutrino species with light masses ($m_\nu < 1$ MeV). The direct observation of CBN provides evidence for the state of the Universe when it was only 1 s old. It is a true cosmological observation and its effect will be enormous.

Can one detect the CBN? The answer is almost certainly no. Since the neutrino cross section with matter decreases in proportion to $E_\nu p_\nu$, the mean free path of neutrinos greatly exceeds the size of the Universe (the Hubble radius is $c/H_0 = 3h_0^{-1}$ Gpc $\sim 10^{10}$ lyr), and there is no way to see the CBN interactions in a detector of finite size, unless the neutrinos had an exceptionally large magnetic moment ($\mu_\nu \sim 10^{-10}\ \mu_{\mathrm{B}}$) [20] as envisaged from the solar-neutrino problem. The large value of $\mu_\nu$ would induce electromagnetic interactions between neutrinos and electrons much larger than weak interactions, and would cause the emission of detectable radiowave signals. However, recent observations of solar neutrinos at Kamiokande [10] and SAGE [21] render this large value of $\mu_\nu$ somewhat unlikely.

There might be a remote chance of observing the CBN, if the following two assumptions are met [22]: (1) the tau neutrino ($\nu_\tau$) has a finite mass sufficiently large to produce a closed Universe ($m_{\nu_\tau} \sim 10$ eV) and (2) there exist ultrahigh-energy (UHE) tau neutrinos perhaps of cosmological origin. Then at the energy of UHE $\nu_\tau$ satisfying $E_{\nu_\tau} = m_{Z^0}^2/2m_{\nu_\tau}$, $m_{Z^0}$ being the $Z^0$ mass, a strong resonance occurs, which results in a dip in the energy spectrum of UHE $\nu_\tau$. Although one must first clarify these two assumptions, observation of such a dip in the energy spectrum would be dramatic evidence for the CBN.

The second diffuse neutrinos are relic neutrinos produced by past type-II supernovae which have been exploding since the epoch of galaxy formation [23–26]. Their energies are much higher than the CBN, being about 8 MeV, so that we could call them high-energy CBN (HECBN). Their flux could be as large as 100 cm$^{-2}$ s$^{-1}$. Knowing the characteristics of neutrinos emitted from stellar collapse, observation of the HECBN will clarify the past supernova rate and the amount of red shift, which may identify the time of galaxy formation and may provide a rough estimate of the expansion rate of the Universe or the density parameter $\Omega_0$. If the stellar activity was higher at the epoch of the galaxy formation, the mean energy of the HECBN would become even lower due to larger red shift. Thus the HECBN, if observed, will shed a light on the unknown and mysterious epoch of the galaxy formation.

Their observation, however, confronts the rather formidable background of antielectron neutrinos ($\bar{\nu}_e$) from nuclear reactors ($E_{\bar{\nu}_e} < 8$ MeV) [27] and of solar neutrinos ($E_{\nu_e} < 15$ MeV). Atmospheric neutrinos become dominant for $E_\nu > 50$ MeV. Indeed the energy range between 15 and 50 MeV is a very quiet window for diffuse neutrinos, and there is a good chance to see the HECBN in future experiments.

Figure 1 shows the energy spectra of various astronomical and terrestrial neutrinos with energies of 0.1 MeV to $10^3$ GeV [28]. The CBN is not shown here because of its extreme low energies.

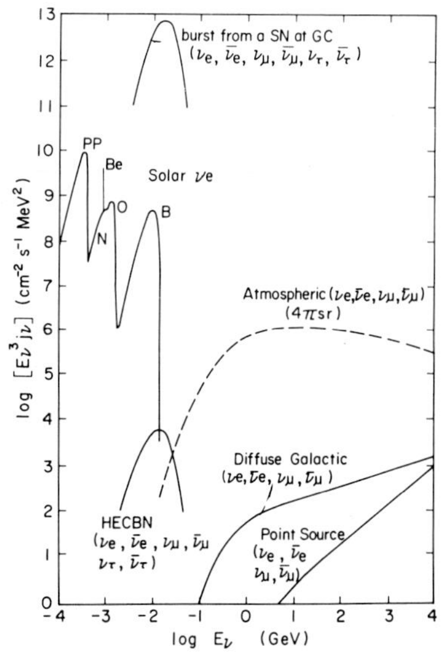


**Figure 1.** Overview of possible astronomical neutrinos for energies from 0.1 MeV to $10^3$ GeV (from [28]).

Let us now look at the higher energy regions ($E_\nu > 100$ MeV). Atmospheric neutrinos are produced in the atmosphere by the primary cosmic rays, namely

$$p + O, N \rightarrow \pi^{\pm} + X \rightarrow \bar{\nu}_\mu(\nu_\mu) + \mu^{\pm} + X$$
$$\rightarrow \nu_\mu + \bar{\nu}_\mu + \bar{\nu}_e(\nu_e) + e^{\pm} + X.$$

The flux is of the order of 1 neutrino $cm^{-2}\,s^{-1}$ for $E_\nu > 1$ GeV, and has an energy dependence of about $E_\nu^{-3}$ at high energies ($E_\nu \geqslant 10$ GeV) [29–34]. Figure 2 shows a typical calculated flux of the atmospheric neutrinos [29]. The atmospheric neutrinos are unfortunately too bright to look beyond the atmosphere at energies between 0.1 and $10^3$ GeV.

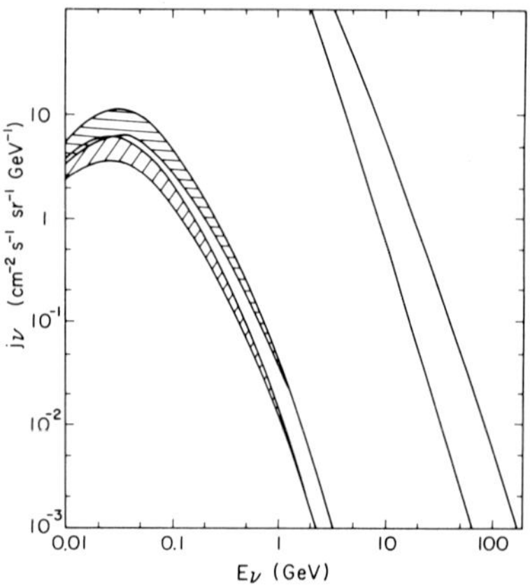


**Figure 2.** Energy spectra of vertical atmospheric neutrinos. The upper curve is the flux of $\nu_\mu + \bar{\nu}_\mu$ and the lower of $\nu_e + \bar{\nu}_e$. The top of each band is for the solar minimum and the bottom for the solar maximum (from [29]).

The energy region between $10^3$ and $10^6$ GeV is again open for extraterrestrial neutrinos. It is now generally agreed upon that ultrahigh-energy cosmic rays are accelerated in an expanding gas of a supernova or a supernova remnant. The acceleration mechanism is considered to be the old Fermi acceleration which predicts $E^{-2}$ for the energy spectrum of cosmic rays [35, 36]. Deconfinement of these cosmic rays in our Galaxy steepens the spectrum, resulting in the observed shape of $E^{-2.7}$. There is another possible acceleration site, namely near a rapidly rotating neutron star or a black hole with strong magnetic fields ($\approx 10^{12}$ G) [37]. Accelerated particles interact with the surrounding gases of a companion star and produce UHE $\gamma$ and UHE $\nu$ as decay products of $\pi^0$ and $\pi^{\pm}$, respectively [38, 39, 40]. Thus one expects point sources of UHE $\gamma$ and UHE $\nu$, which are probably manifest as associated with bright x-ray and radio sources such as pulsars. Their energies are expected to follow the $E^{-2}$ spectrum. Their discoveries will thus provide the best evidence for the origin of UHE cosmic rays. Figure 3 [38] is the expected spectrum of UHE neutrinos from the x-ray binary Cyg X-3 assuming the total cosmic-ray luminosity of $10^{39}$ $erg^{-1}s$, which is close to the upper limit derived from observation of UHE $\gamma$-rays. Figure 1 also compares the Cyg X-3 flux with that of atmospheric neutrinos. In order to have a reasonable signal-to-noise

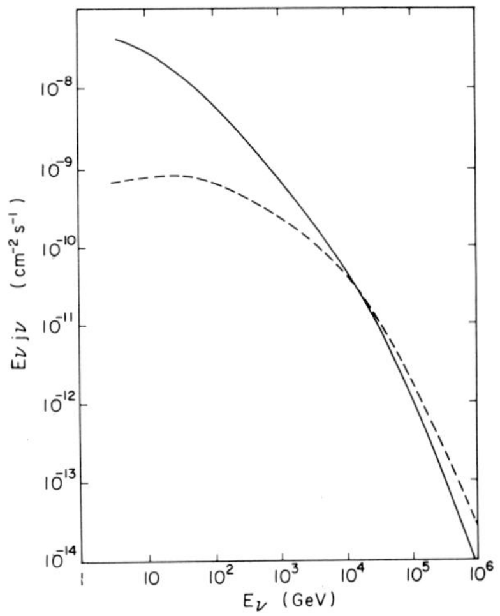


**Figure 3.** Neutrino spectra at the Earth from a $10^8$ GeV monoenergetic beam (broken line) and a power-law spectrum with slope two (full line). The distance to Cyg X-3 and the power are assumed to be 10 kpc and $10^{39}$ erg $s^{-1}$, respectively (from [38]).

ratio the atmospheric neutrinos must be reduced by at least a factor of $10^4$ at $10^3$ GeV, which could be achieved by looking at Cyg X-3 with an angular resolution of one degree.

The origin of cosmic rays is interesting, but it may not invoke new physics to explain the energy spectrum, composition and energetics up to $10^6$ GeV. There actually exist cosmic rays impinging upon the Earth with even higher energies ($\leqslant 10^{11}$ GeV) [41]. As is well known, these UHE cosmic rays probably have an extragalactic origin and consist of nucleons. They have a threshold of about $5 \times 10^9$ GeV to produce pions by interactions with the 2.7 K cosmic background radiation (CBR) [42, 43]. Therefore, above this energy the Universe becomes opaque for them with the corresponding mean-free path of a mere 8 Mpc, which is even shorter than the distance to the Virgo cluster (~20 Mpc). Of course it is quite unnatural that UHE nucleons exist only in the Local Cluster. If UHE nucleons are indeed uniformly distributed throughout the whole Universe, UHE neutrinos, which are decay products of $\pi^\pm$ photoproduced by interactions with 2.7 K CBR, would be as copious as the parent nucleons, because the Universe is essentially transparent for UHE neutrinos and thus many more UHE neutrinos might exist that were produced from a possible phase of active cosmic ray production in the past [44–47]. Figure 4 [45] illustrates a calculation that takes into account a bright phase at various red shifts $\bar{z}$, compared with the assumed nucleon spectrum ($\simeq$ observed cosmic ray spectrum). Thus the direct observation of UHE neutrinos of $E_\nu \geqslant 10^8$ GeV is quite important and one will obtain valuable information about the epoch of galaxy formation. It is interesting that both UHE neutrinos and HECBN could be used as a probe for the early Universe despite the vast difference, some 10 orders of magnitude, in their energies.

It is now speculated that conventional acceleration mechanisms like shock acceleration are not efficient enough for nucleons to acquire energies above $10^{11}$ GeV, even in the most probable sites like active galactic nuclei or hot spots in the radio jets of active radio galaxies [48–51]. Note that the energy scale for the grand unified theories (GUT) of about $10^{15}$ GeV is only four orders of magnitude larger than the supposed

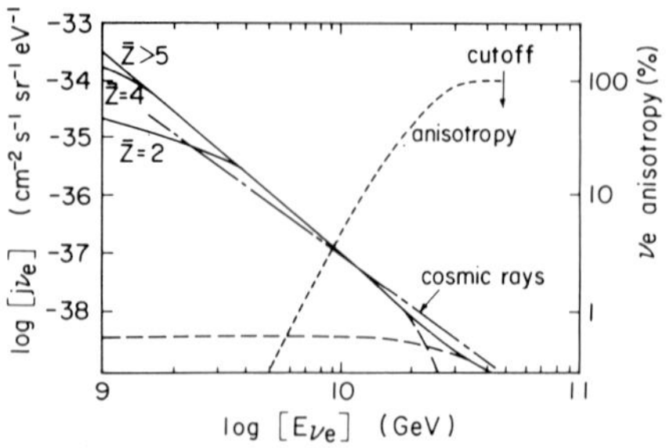


**Figure 4.** The predicted and fully normalized electron-neutrino spectrum for the diffuse + semilocal model fit to the spectrum corresponding to an injection index of 2.0. Note the occurrence of a neutrino anisotropy above $7 \times 10^{18}$ eV, which is a much smaller effect than the nucleon anisotropy which begins at about $10^{18}$ eV because of the relatively large diffuse contribution here. The lower broken curve is the Virgo component, and the UHE cut-off of the diffuse component is shown as the upper broken curve. The chain curve indicates the observed cosmic-ray spectrum (from [45]).

acceleration limit. If particles still exist above $10^{12}$ GeV, they will certainly be of cosmological origin and must be related to GUT, namely possible decays of false vacua (monopoles, cosmic strings, etc.), which had an energy density of the GUT scale and were left behind as defects during the vacuum phase transition. Indeed it has been speculated [52] that the superconducting cosmic strings with fermionic charge carriers would eject, in their last stages, high-mass particles which can subsequently decay to produce UHE neutrinos. The energies of UHE neutrinos would be as high as $10^{15}$ GeV as shown in figure 5, see [52]. Needless to say, observation of these GUT-related neutrinos would be the most spectacular event possible in neutrino astronomy. As we mentioned before, provided that the mass of $\nu_\tau$ is about $1 \sim 10$ eV, the Universe becomes opaque for UHE neutrinos at $E_\nu = m_Z^2/2m_\nu \simeq 10^{12} \sim 10^{13}$ GeV because of the resonant interaction at the $Z^0$ pole with $\nu_\tau$ in the 1.9 K CBN [22]. The neutrino spectrum has a large dip at this energy, which, however, has not been taken into account in figure 5. Such a distortion of the spectrum is also unambiguous evidence for the existence of CBN.

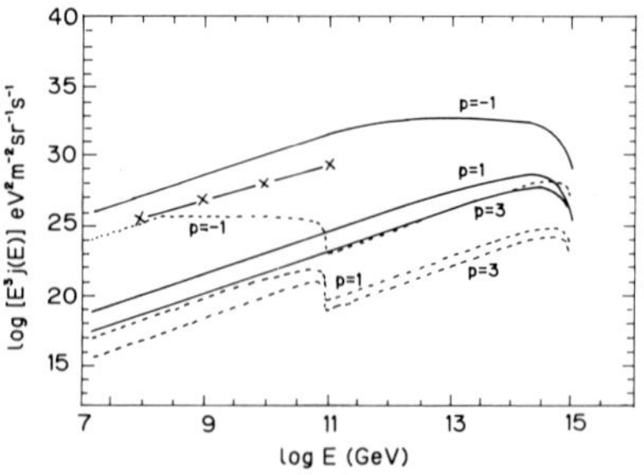


**Figure 5.** The UHE nucleon spectrum integrated over red shift for $M_F = 10^{15}$ GeV. The parameter $p$ specifies the evolution of the cosmic magnetic field as $B(z) = (1+z)^{-p+3/2}B_0$, $B_0 \approx 10^{-9}$ G (from [52]). The broken curve is the upper limit obtained by the Utah Fly's Eye group. The dotted lines are the corresponding spectra for nucleons with the same parameter $p$ (see section 5.3 and [129]).

So far we have outlined various astronomical sources of neutrinos from $1.6\times10^{-4}$ to $10^{24}$ eV and from real to exotic. Each source plays an important role in astronomy and cosmology, and observation of these astronomical neutrinos will provide exciting new results. In the following sections we review the results already achieved and the prospects for observations of these neutrinos. Section 2 describes supernova neutrinos and SN1987A with future expectations. Section 3 is devoted to recent progress in the study of solar neutrinos, which is perhaps the most exciting topic at present. Section 4 deals with a brief description of the recent progress in atmospheric neutrinos. Section 5 is devoted to high-energy and ultrahigh-energy neutrinos and the feasibility of their detection.

## 2. Supernova neutrinos and SN1987A

### 2.1. Stellar collapse

It has been well established that massive stars with $M \geqslant 8M_\odot$($M_\odot$ denotes the mass of the Sun) evolve to an onion skin type of structure with a central iron core of about the Chandrasekhar mass $5.76\,Y_e^2 M_\odot$, where $Y_e$ is the number of electrons per nucleon. The iron core radius is calculated to be about 1000 km. Once the core mass exceeds the Chandrasekhar mass of about $1.3M_\odot$, it becomes unstable and eventually collapses gravitationally. The collapse is halted due to repulsive nuclear forces when the density becomes so high that the nucleons almost react with each other. The core oscillates and shock waves are produced deep in the core ($\simeq 0.7M_\odot$), which propagate through the dense nuclear matter. This shock initiates the production of a burst of electron neutrinos $\nu_e$ (of duration about 10 ms) from the neutronization process $e^- + p \to \nu_e + n$. These $\nu_e$ escape freely. The core density becomes even higher and finally opaque even for neutrinos. Neutrinos are then thermalized in the core and can only escape from the surface of a sphere called the neutrino sphere. The initial radius of the neutrino sphere is calculated to be about 100 km. The remaining outer shell of the iron core accretes on this hot protoneutron star and its mass reaches 1.4–$1.6M_\odot$. As soon as the neutrino sphere forms, the heat is quickly carried away by the emission of thermal neutrinos through the processes $\gamma+\gamma\to\nu_i+\bar{\nu}_i$ and $e^+ + e^- \to \nu_i + \bar{\nu}_i$, where $\nu_i$ denotes $\nu_e$, $\nu_\mu$ and $\nu_\tau$. During the course of cooling, the radius of the protoneutron star ($R$) shrinks and finally settles down to about 10 km [53, 24, 54–56]. Figure 6 shows how each shell of a fixed mass collapses as a function of time [55]. It also shows the size of the neutrino sphere. The figure was made from a simulation of a massive star with main-sequence mass of 15 $M_\odot$. The accretion stops about 0.5 s after the collapse and the outer shell with $M > 1.7$ $M_\odot$ is blown out. Figure 7 shows the resultant neutrino luminosities calculated for the initial stellar masses of 12, 25, 50 and 100 $M_\odot$ [56]. Only the case of $25M_\odot$ shows the luminosities for $\nu_e(L_\nu)$, $\bar{\nu}_e(L_{\bar{\nu}})$ and one of $\nu_\mu$, $\bar{\nu}_\mu$, $\nu_\tau$ and $\bar{\nu}_\tau(L_\mu)$. The other cases show only $L_\nu$. The initial sharp peak corresponds to the neutronization burst, and it is remarkable that its luminosity does not depend sensitively on the initial mass. This arises from the fact that the initial collapse occurs at the core mass equal to the Chandrasekhar mass, which is almost independent of the progenitor mass. The total energies carried by the neutronization burst and $\bar{\nu}_e$ are about $3\times10^{51}$ and $5\times10^{52}$ erg, respectively, depending slightly on the progenitor mass [57]. Calculations also show that approximately equal energy is carried off by each species of neutrinos, that is $L_{\nu_e} = L_{\bar{\nu}_e} = L_{\nu_\mu} = L_{\bar{\nu}_\mu} = L_{\nu_\tau} = L_{\bar{\nu}_\tau}$ [55, 56]. Therefore the total energy

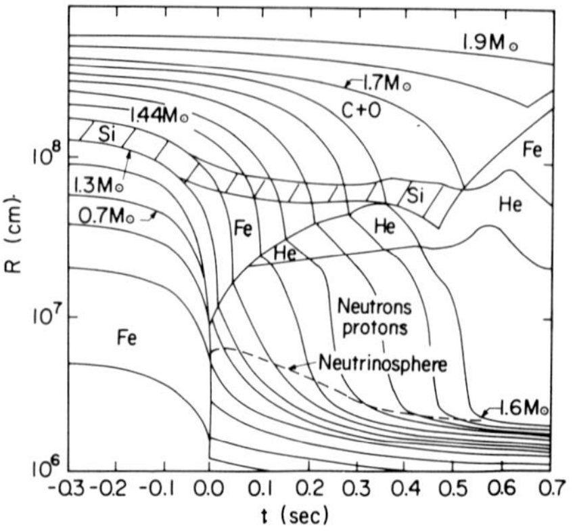


**Figure 6.** Diagram of collapse of equimass shells for a $15M_\odot$ progenitor. The radius of the neutrino sphere is also shown (from [55]).

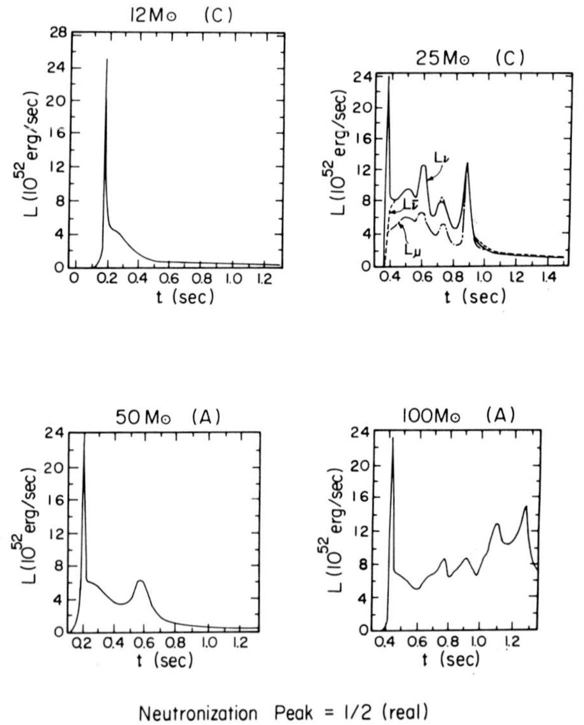


**Figure 7.** Electron neutrino luminosity ($L_\nu$) for specified progenitor masses. (A) and (C) denote different collapse models. The case for $25M_\odot$ also shows the $\bar{\nu}_e$ luminosity ($L_{\bar{\nu}}$) and the $\nu_\mu$ luminosity ($L_\mu$). The neutronization peak is scaled by a factor of a half (from [56]).

carried by all kinds of neutrinos is about $6\,L_{\bar{\nu}_e} \simeq 3\times 10^{53}$ erg, which of course corresponds to the gravitational energy $(\frac{3}{5})GM^2/R$. The mean energies of neutrinos were also calculated; approximately $E_{\nu_e} = 10$ MeV, $E_{\bar{\nu}_e} = 15$ MeV, and $E_{\nu_\mu} = E_{\bar{\nu}_\mu} = E_{\nu_\tau} = E_{\bar{\nu}_\tau} =$ 20 MeV [55, 56]. The energy distribution obeys, to a first approximation, the Fermi-Dirac distribution [55, 56, 58].

One must not forget the possibility of black holes forming. Normally it is considered that a star with a mass larger than 30 $M_\odot$ collapses all the way to the black hole. Since there is no clear observational evidence for the existence of black holes, it is important to be ready to observe the type of neutrino burst characteristic of black-hole formation. The neutrino burst from black-hole formation may have the following properties: (i) the existence of a neutronization burst, (ii) a short duration of the cooling phase, (iii) the abrupt disappearance of the burst, and (iv) lower mean energies of neutrinos owing to a strong gravitational red shift [24]. Direct observation of these properties will provide convincing evidence for black-hole formation.

### 2.2. *SN1987A*

The four underground experiments reported the observation of the neutrino burst correlated in time to optical sighting of SN1987A. Table 1 shows some essential parameters and results from the four experiments, Kamiokande [1, 59], IMB [2, 60] LSD in Mt Blanc [3] and the Neutrino Telescope at Baksan [4].

**Table 1.** Data for SN1987A reported by the four experiments Kamiokande [1], IMB [2], Mt Blanc [3] and Baksan [4].

| Laboratory | Depth (mwe) | Mass (t) | Material | Observation time (UT) | Number of events | $E_{th}$ (MeV)† | Background rate |
|---|---|---|---|---|---|---|---|
| Kamiokande | 2700 | 2140 | $H_2O$ | 7:35:35 ($\pm 1$ min) | 11 | 7.5 | $1.2\times 10^{-8}\,a^{-1}$ |
| IMB | 1570 | 6800 | $H_2O$ | 7:35:41 | 8 | 35 | ‡$5\times 10^{-3}\,a^{-1}$ |
| Mt Blanc | 4000 | 90 | $(CH_2)_n$ | 2:52:37 | 5 | 5 | $0.7\,a^{-1}$ |
| Baksan | 850 | 200 | $(CH_2)_n$ | 7:36:12 ($^{+2}_{-54}$ s) | 5 | 10 | $0.7\,d^{-1}$ |

†Threshold energy at 50% efficiency.
‡The rate contains edge clipping muons. Non-muon background would be similar with Kamiokande.

Since neutrinos are neutral particles they are detected indirectly by observing charged particles scattered by their interactions with detector materials. At the low energies concerned here, scattering off electrons ($\nu e\to \nu e$) and inverse-$\beta$ reactions ($\bar{\nu}(\nu)N\to e^{\pm}N$) are employed. Their cross sections are [61, 62]:

$$\sigma(\bar{\nu}_e p\to e^+ n) = 9.77\times 10^{-42}(E_\nu/10\ \mathrm{MeV})^2 \tag{1}$$

$$\sigma(\bar{\nu}_e e\to \bar{\nu}_e e) = 0.388\times 10^{-43}(E_\nu/10\ \mathrm{MeV}) \tag{2}$$

$$\sigma(\nu_e n\to e^- p) = 2.2\times 10^{-42}(E_\nu/10\ \mathrm{MeV})^2 \tag{3}$$

$$\sigma(\nu_e e\to \nu_e e) = 0.933\times 10^{-43}(E_\nu/10\ \mathrm{MeV}) \tag{4}$$

$$\sigma(\nu_\mu e\to \nu_\mu e) = 0.159\times 10^{-43}(E_\nu/10\ \mathrm{MeV}) \tag{5}$$

$$\sigma(\bar{\nu}_\mu e\to \bar{\nu}_\mu e) = 0.130\times 10^{-43}(E_\nu/10\ \mathrm{MeV}) \tag{6}$$

$$\sigma(\nu_e {}^{16}O\to e^-\,{}^{16}F) = 1.1\times 10^{-44}[E(\mathrm{MeV}) - 13]^2 \tag{7}$$

where the unit is cm$^2$, and obviously $\sigma(\nu_\tau e \rightarrow \nu_\tau e) = \sigma(\nu_\mu e \rightarrow \nu_\mu e)$ and $\sigma(\bar{\nu}_\tau e \rightarrow \bar{\nu}_\tau e) = \sigma(\bar{\nu}_\mu e \rightarrow \bar{\nu}_\mu e)$. The four experiments do not have free neutrons in the detector, and the inverse-$\beta$ reaction in $^{16}$O is not important for the low-energy neutrinos concerned here. Taking into account the relative numbers of protons and electrons, the ratio of event rates, $N(\bar{\nu}_e p)/N(\nu_e e) = 20:1(25:1)$ for water (liquid scintillator) at $E_\nu = 10$ MeV. Thus, $\bar{\nu}_e$ signals dominate over those of other neutrino species. However, scattering with electrons has an advantage, that is the strong directionality of the scattered electrons enables us to locate the position of the supernova.

Figure 8 shows the detection efficiencies of electrons for the four experiments. Their useful target masses are 2140, 6800, 90 and 200 t for Kamiokande, IMB, Mt Blanc and Baksan, respectively. Based on their low backgrounds, normally the results from Kamiokande and IMB are used for physics analyses. The observed time of the Mt Blanc experiment was about five hours earlier than the other three. The Mt Blanc and Baksan groups claim that their signals had a significant time correlation to those observed in the gravitational wave antennae of Rome and Maryland [63, 64]. This issue has not been resolved. It is clear that one must await the next supernova to draw any firm conclusions.

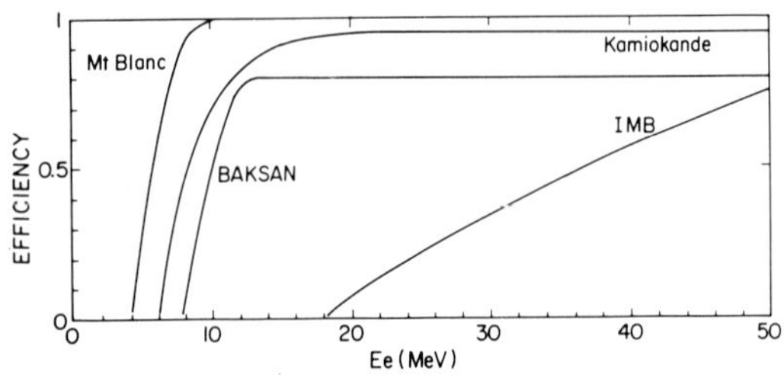


**Figure 8.** Detection efficiencies of the Mt Blanc [3], Kamiokande [1], Baksan [4] and IMB [2] detectors.

The numbers of events observed by Kamiokande and IMB are only eleven and eight, respectively. Despite the great volume of their detectors, the background rates that might fake the neutrino burst are extremely small. Figure 9 shows the frequency distributions of multiple events in a 10 s period, in which the number of photomultipliers activated was used as an energy indicator for Kamiokande [59, 2]. They nicely follow the Poisson distribution, and the extrapolated rate at the observed multiplicity is indeed negligibly small. The time sequence and energy of the bursts are shown in figure 10. Note that the relative timing of the two sets of data is unknown by one minute and they should not simply be folded in the figure, and note also that the IMB results has a much higher energy threshold. The angles with respect to SN1987A and the energies of the observed electrons are shown in figure 11. It was sometimes argued that events with higher energies tend to be emitted forward and they might have been produced by yet unknown new particles [60]. However, the number of data is too small to draw a definite conclusion, and the isotropic distribution is still allowed.

Taking into account the detection efficiency, the reaction cross section and the distance to SN1987A (50 kpc = 160 000 lyr), one can deduce the total neutrino luminosity ($L_{\bar{\nu}_e}$, $L_t = 6\,L_{\bar{\nu}_e}$) and the average temperature ($T_{\bar{\nu}_e}$) of the neutrino sphere of SN1987A. One can go a step further to include additional parameters like the time

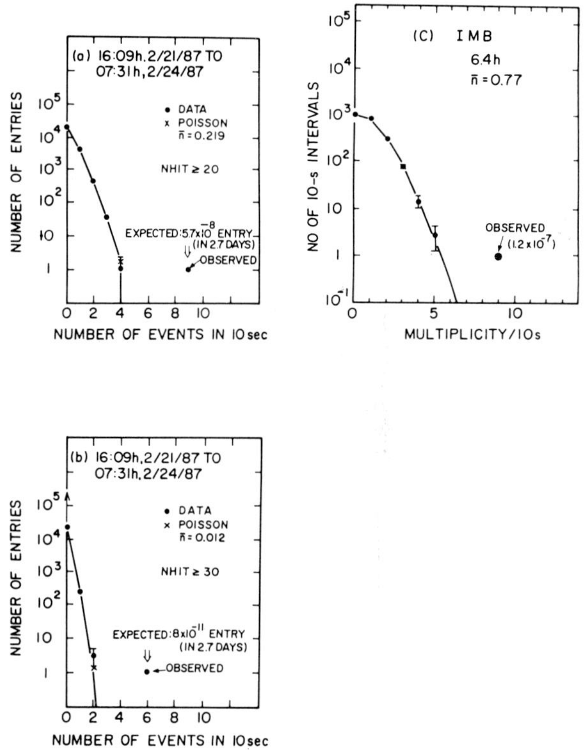


**Figure 9.** (a) Poisson distribution of the number of events with $N_{\rm hit} \geqslant 20 (E_{\rm e} \geqslant 7.5\,{\rm MeV})$ per 10 seconds in a 2.7 day period (Kamiokande data [59]). (b) Same for $N_{\rm hit} \geqslant 30(E_{\rm e} \geqslant 11.5\,{\rm MeV})$. (c) Same for the IMB data [2]. The point denoted as 'observed' corresponds to the neutrino burst.

dependence of $T_{\bar{\nu}_{\rm e}}$, such as

$$T_{\bar{\nu}_{\rm e}} = T^0_{\bar{\nu}_{\rm e}} \exp(-t/4\tau).$$

The radius of the neutrino sphere can be calculated from the Stefan–Boltzmann equation. However, one must remember that only 19 events were observed, and this is certainly too few with which to study the energy distribution. There were in fact a large number of papers that tried to extract these important parameters, and they gave more or less similar results.

In general one can assume the neutrino spectrum at the Earth [65] to be given by

$$\frac{{\rm d}N_\nu}{{\rm d}E_\nu {\rm d}t} = 4\pi R^2 \frac{g\pi c}{(hc)^3} \frac{1}{\exp[(E_\nu - \mu)/T] + 1} E_\nu^2 \frac{1}{4\pi D^2} \tag{8}$$

where $g = 1$ is the helicity weight of the neutrino species, $R$ the radius of the neutrino sphere, $T$ the temperature, $\mu$ the chemical potential and $D$ the distance to the supernova. $R$, $T$ and $\mu$ in general depend on time $t$. If one is only interested in the total luminosity $L_{\bar{\nu}_{\rm e}}$ and the average energy $E_{\bar{\nu}_{\rm e}}$, these do not depend much on small or moderate $\mu$; thus it is normally assumed that $\mu = 0$. However, in case that one wants to derive $R$

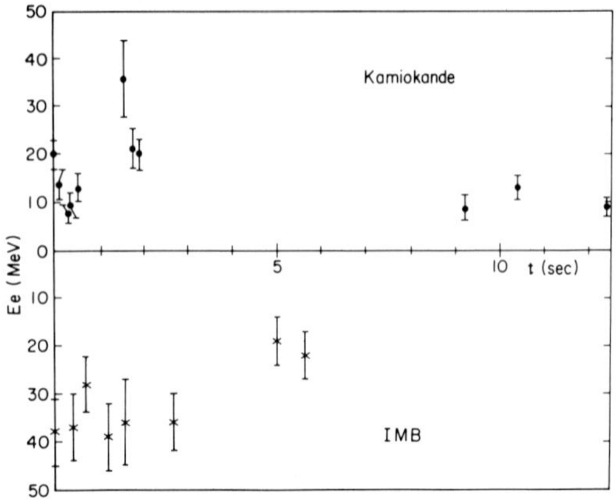


**Figure 10.** Time sequence and energy of the bursts observed by Kamiokande [1] and IMB [2].

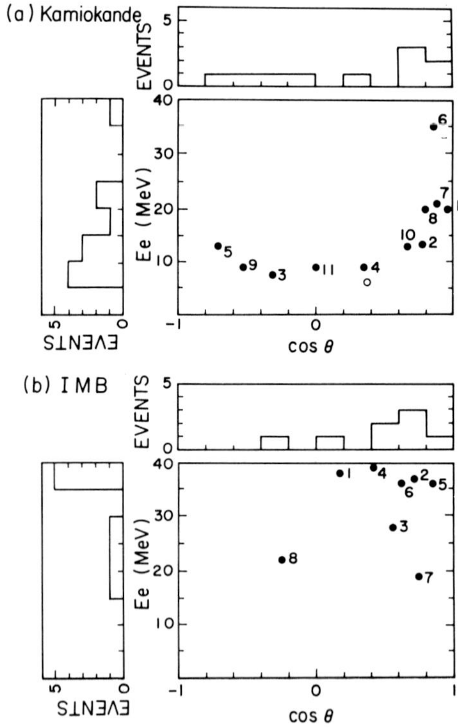


**Figure 11.** Angular and energy distributions of the observed electrons. (a) Kamiokande [59] and (b) IMB [2]. The point denoted by an open circle in (a) is below the analysis threshold and should not be taken seriously. The number attached to each event is a sequential number in chronological order.

from $L_{\bar{\nu}_e}$ and $E_{\bar{\nu}_e}$, even a small value of $\mu$ does influence the result, leading, for example, to a radius smaller by a factor of 0.63 (0.27) for $\mu = T(3T)$ [66]. Note that $\mu$ is calculated to be about 1-2.5 depending on $t$ [58].

Let us now show some examples of analysis, in which all the observed events are assumed to be $\bar{\nu}_e$.

*2.2.1. Time-independent T.* Two analyses give, with $\mu = 0$ and $D = 50$ kpc,

$$(T_{\bar{\nu}_e}, L_{\bar{\nu}_e}) = (3.7 \pm 0.4 \text{ MeV}, 4.8^{+2.1}_{-1.8} \times 10^{52} \text{ erg}) \quad \text{(ref. [66])}$$

$$= (4.0 \pm 0.5 \text{ MeV}, 3.4^{+1.4}_{-1.0} \times 10^{52} \text{ erg}) \quad \text{(ref. [58])}$$

for Kamiokande and IMB combined. If the Kamiokande and IMB data are fitted separately for the first analysis, the result is $(2.8 \pm 0.4 \text{ MeV}, 6.3^{+4.0}_{-3.1} \times 10^{52} \text{ erg})$ and $(4.2 \pm 1.0 \text{ MeV}, 4.5^{+12.0}_{-3.4} \times 10^{52} \text{ erg})$, respectively [66]. The small errors in the combined fit are simply the result of a poor overlap of the two values, although they agree within statistical error. Therefore the errors in the combined result must be taken cautiously. The difference in the above analyses is due to the different methods adopted for the analyses. By multiplying $L_{\bar{\nu}_e}$ by six, the total energy carried by neutrinos is

$$L_\nu = (2.5 \pm 1.2) \times 10^{53} \text{ erg}$$

which agrees very well with the expected value of $3 \times 10^{53}$ erg. The $\bar{\nu}_e$ temperature is also quite consistent with expectations.

*2.2.2. Time-dependent T.* Assume that $R$ is independent of time but that $T = T_0 \exp(-t/4\tau)$. The factor of four is introduced to give $L_\nu = L_0 \exp(-t/\tau)$, since $L_\nu$ is proportional to $T^4$.

The result is, for the combined Kamiokande and IMB data, [65],

$$T_0 = 4.5^{+1.1}_{-0.9} \text{ MeV}$$

$$\tau = 4.2^{+2.5}_{-1.5} \text{ s}$$

$$R = 23^{+22}_{-10} \text{ km}$$

$$L_{\bar{\nu}_e} = 5^{+8}_{-3} \times 10^{52} \text{ erg.}$$

The large error in $L_{\bar{\nu}_e}$ arises from the fact that after fitting $T_0$, $\tau$ and $R$, $L_{\bar{\nu}_e}$ is calculated by

$$L_{\bar{\nu}_e} = \int E_\nu \frac{\mathrm{d}N_\nu}{\mathrm{d}E_\nu \, \mathrm{d}t} 4\pi D^2 \, \mathrm{d}E_\nu \, \mathrm{d}t$$

$$= 0.58 \times 10^{52} \left(\frac{R}{10 \text{ km}}\right)^2 \left(\frac{\tau}{4 \text{ s}}\right) \left(\frac{T_0}{4 \text{ MeV}}\right)^4 \left(\frac{D}{50 \text{ kpc}}\right)^2 \text{ erg.} \quad (9)$$

As mentioned above, the value of $R$ could be as small as 10 km depending on the unknown value of $\mu$. Note that if the error estimate is taken seriously, the time-independent temperature $\tau = 0$ is not favoured from the data.

Since the chemical potential $\mu$ was not obtained and neither was a reliable value of $R$, it is not possible to choose the best equation of state (EOS) that governs the density and radius of the neutron star. Once the EOS is determined, the mass of the neutron star $(M)$ can be calculated.

### 2.3. Particle aspects of SN1987A

As discussed above, the data fit quite well the conventional theory on stellar collapse and cooling of a protoneutron star. Thus there is no room for new phenomena possibly related to particle physics, and the only limits are obtained on the $\bar{\nu}_e$ mass, the axion mass and the magnetic moment of $\bar{\nu}_e$, etc. After five years observation of SN1987A, only the limits on the axion mass survive competition with other estimates.

The axion is a pseudo-Nambu–Goldstone boson appearing in the symmetry-breakdown of the U(1) Peccei–Quinn symmetry [67–69], which was introduced to conserve product of charge conjugation and parity (CP) in the strong interactions. The mass of the axion is determined by the PQ symmetry-breaking scale $f_a$:

$$m_a = (0.62\ \text{eV})[10^7\ \text{GeV}/(f_a/N)]$$

where $N$ is the colour anomaly of the PQ symmetry or simply the number of quark doublets, $N=3$ [68, 69]. The axion couples to quarks, therefore nucleons (hadronic axion) and possibly charged leptons (the Dine–Fischler–Srednicki axion) [70] with a coupling strength of

$$g_{aN} \simeq 0.5\ m_N/(f_a/N) \approx 7.6\times10^{-8}(m_a/\text{eV})$$

with which the interaction Lagrangian is

$$L_{int} = (g_{an}/2m_N)(\bar{n}\gamma_\mu\gamma_5 n)\partial^\mu a + (g_{ap}/2m_N)(\bar{p}\gamma_\mu\gamma_5 p)\partial^\mu a \qquad (10)$$

where $n$, $p$ and $a$ are the wavefunctions of the neutron, proton and axion, respectively. In addition $g_{aN} \simeq g_{an} \simeq g_{ap}$ and $m_N$ is the nucleon mass. Axions are produced via $N-N-a$ bremsstrahlung, and if $g_{aN}$ is large enough to produce copious numbers of axions but small enough for them to stream freely out of the protoneutron star, the protoneutron star cools too rapidly to be compatible with the observations [69]. As $g_{aN}$ becomes large, axions are trapped and thermalized inside the protoneutron star just like neutrinos, and they are emitted thermally from the axiosphere. Even in this case $g_{aN}$ is constrained in such a way that the cooling time should not be too short [71]. Figure 12 shows the calculated duration of burst required for the expected number of neutrino events to achieve 90% of its asymptotic value [71]. Thus axion masses in the interval $10^{-3}$–3 eV are excluded. Combining this with other estimates, the allowed mass range (or $f_a$ range) is further squeezed, especially if the density parameter of the Universe $\Omega_0$ is of the order of 0.1 as observations suggest. Figure 13 [72] shows the present situation on the excluded region of $f_a$.

Note that a similar argument can be applied to any new particles that couple very weakly to nucleons or electrons.

### 2.4. Future prospects for supernova observations

A single observation of a supernova is by no means sufficient to solve every detail about stellar collapse. The next supernova will almost certainly occur in our galaxy, thus the distance to it will be less (~10 kpc) than that to SN1987A (50 kpc). One, however, may not be as lucky as in the case of SN1987A, namely it will most probably not be possible to see it optically because of thick clouds of dust in the galactic centre and arms. Therefore one must have other means, with which to estimate the distance to the supernova, such as combined infrared, radio and x-ray observations. It is still essential to have the direction to the supernova as accurately as possible. We discuss the number of observed events expected in the next generation of detectors for

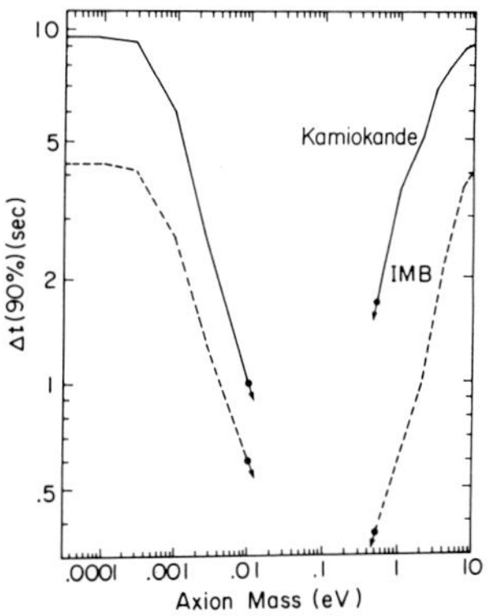


**Figure 12.** Expected neutrino burst duration $\Delta t_{90\%}$ in the Kamiokande and IMB detectors for axion-cooled nascent neutron-star models as a function of axion mass. The quantity $\Delta t_{90\%}$ is the time required for the expected number of neutrino events to achieve 90% of its asymptotic value. Based upon the expected duration of the neutrino burst, axion masses in the interval ($10^{-3}$–3 eV) are excluded by the Kamiokande and IMB data. Note that the burst duration in both detectors is reduced by about the same factor (from [71]).

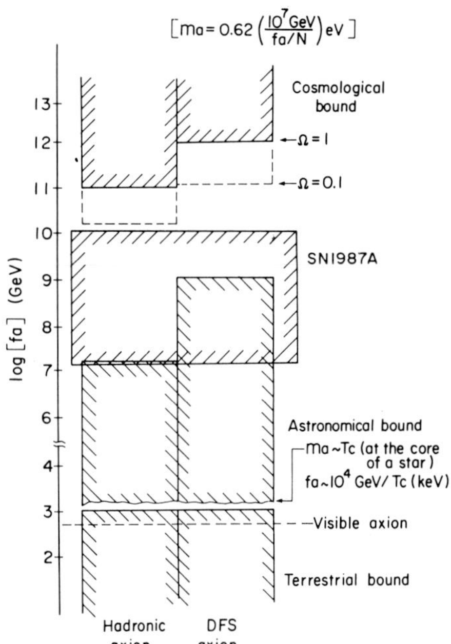


**Figure 13.** Excluded region (shaded) from various constraints for hadronic and DFS axions [72]. The smaller $\Omega$ extends the excluded region to smaller values of $f_a$, and hadronic axions are almost ruled out if the current estimate of $\Omega = 0.1$ is taken.

low-energy neutrinos, Super-Kamiokande [73, 74] and SNO, both of which have recently been funded [62, 75].

*2.4.1. Expected event rates from a supernova in our Galaxy.* Table 2 shows the event rates expected in the Kamiokande, Super-Kamiokande and SNO detectors from a supernova at the distance of 10 kpc. Some essential parameters of these detectors are shown in table 3 for reference. Compare the tabulated event rates with mere 19 $\bar{\nu}_e$ events observed in the Kamiokande and IMB detectors (2140+6800 t).

Several comments are in order.

(i) Signals of the neutronization burst are observable, about 13 events in Super-Kamiokande, unless $\nu_e$ oscillate to $\nu_\tau$ in the crust of the progenitor star. The total luminosity $L_n$ of the neutronization burst can be determined with an accuracy of $1/\sqrt{13}=30\%$ for Super-Kamiokande. As mentioned above, $L_n$ does not depend sensitively on the properties of the progenitor star, such as the mass and chemical composition and takes a value of about $3\times10^{51}$ erg [56, 57]. This leads to an

**Table 2.** Expected neutrino events from a supernova at 10 kpc away in the three detectors, Kamiokande [1], Super-Kamiokande [73, 74], and SNO [62, 75].

| | | Supernova neutrinos ($d$ = 10 kpc) | | | | |
|---|---|---|---|---|---|---|
| Detector | Mode | n | $\nu_e$ | $\bar{\nu}_e$ | $\nu_x$ | Remarks |
| Kamiokande | $\nu e\to\nu e$<br>$\bar{\nu}_e p\to e^+ n$ | 0.9 | 9 | 3<br>270 | 6 | Calculated based on the SN1987A data. |
| Super-Kamiokande | $\nu e\to\nu e$<br>$\bar{\nu}_e p\to e^+ n$ | 13 | 130 | 44<br>4000 | 80 | $\nu_e$ can be separated from $\bar{\nu}_e$ by the angular correlation to SN. $\nu_e$ can be distinguished from $\nu_x$ from energy spectrum. |
| SNO | $\nu_e d\to e^- pp$<br>$\nu e\to\nu e$<br>$\nu d\to\nu pn$<br>$\bar{\nu}_e d\to e^+ nn$<br>$\bar{\nu}_e p\to e^+ n$ | 2.3<br>1.1<br>0.9 | 23<br>11<br>9 | <br>4<br>21<br>37<br>210 | <br>7<br>110 | $\nu_e$ and $\bar{\nu}_e$ are distinguished by measuring $\nu_e$d and $\bar{\nu}_e$p, but are difficult to separate from $\nu_x$d. $\nu_x$ can be measured by $\nu_x$d, but its energy cannot be measured. |

*Notes*

n; neutronization neutrinos, which appear in the first 0.01 s of a SN explosion.

$\nu_x$; $\nu_x$ represents $\nu_\mu$, $\nu_\tau$ and their antiparticles.

Distance to a supernova is assumed to be 10 kpc (30000 light years).

Neutrons from $\nu_x d\to\nu_x pn$ are captured by Cl emitting 8 MeV $\gamma$-rays. These $\gamma$-rays cannot be well separated from electrons produced by $\nu_e d\to e^- pp$.

The ratio between n, $\nu_e$ and $\bar{\nu}_e$ is from M15C+M25C of K Sato and H Suzuki [57].

The expected $\nu_x$ signals in Kamiokande and Super-Kamiokande are calculated using $(1/6.5)\times2\times4\times\nu_e\times(1/2)$, where (1/6.5) is $\sigma(\nu_x)/\sigma(\nu_e)$, 2 is $E(\nu_x)/E(\nu_e)$, 4 comes from the fact that $\nu_x$ represents four types of neutrinos, and the final factor of 1/2 is the ratio of the total $\nu_x$ energy to the total $\nu_e$ energy.

The expected signals in SNO are scaled from those of Super-Kamiokande, assuming that the mean energies of $\nu_e$, $\bar{\nu}_e$ and $\nu_x$ are 10, 15 and 20 MeV, respectively.

**Table 3.** Some parameters of the three detectors, Kamiokande [1], Super-Kamiokande [73, 74] and SNO [62, 75].

| Detector | Type | Mode | $T_{th}$ (MeV) | Ability to measure | |
|---|---|---|---|---|---|
| | | | | Energy | Direction |
| Kamiokande | $H_2O$ Cerenkov | $\nu e \to \nu e$ | 7 | Yes | Yes |
| | SN, 2140 t | | | | |
| | $\nu_{atm}$, 1036 t | | | | |
| | $\nu_{Sun}$, 680 t | $\bar{\nu}_e p \to e^+ n$ | | Yes | No |
| Super-Kamiokande | $H_2O$ Cerenkov | $\nu e \to \nu e$ | $\leqslant 5$ | Yes | Yes |
| | SN, 32000 t | | | | |
| | Others, 22000 t | $\bar{\nu}_e p \to e^+ n$ | | Yes | No |
| SNO | $D_2O$ Cerenkov | $\nu_e d \to e^- pp$ | 6 | Yes | No |
| | $D_2O$, 1000 t | $\nu e \to \nu e$ | 6 | Yes | Yes |
| | $H_2O$, 1700 t | $\nu d \to \nu pn$ | 2.225 | No | No |
| | | $\bar{\nu}_e d \to e^+ nn$ | $\sim 5$ | Yes | No |
| | | $\bar{\nu}_e p \to e^+ n$ | $\sim 5$ | Yes | No |

*Notes*
$T_{th}$, threshold energy in kinetic energy;
SN, fiducial volume for the detection of supernova neutrinos;
$\nu_{atm}$, fiducial volume for the detection of atmospheric neutrinos;
$\nu_{Sun}$, fiducial volume for the detection of solar neutrinos.
$\nu$ represents $\nu_e$, $\nu_\mu$ and $\nu_\tau$. In case of $\nu e \to \nu e$, the cross section of $\nu_e$ is larger than those of $\nu_\mu$ and $\nu_\tau$.

approximate estimate for the distance to the supernova $D$,

$$D = (L_n / 4\pi\varepsilon_n)^{1/2} \tag{11}$$

where $\varepsilon_n$ is the total energy flux (erg cm$^{-2}$) observed in a detector. $D$ is determined with an accuracy of about 20%.

(ii) The number of electron scattering events ($\nu e \to \nu e$) is about 300 in the Super-Kamiokande detector. This is sufficiently large to determine the direction of the supernova with an accuracy of $27°/\sqrt{300} \approx 2°$. Thus one does not need help from optical observations to identify the supernova. Instead, with a fast on-line analysis of the scattering data one is able to inform optical observatories about the occurrence and position of the supernova before it becomes bright.

(iii) The number of $\nu_\tau + \bar{\nu}_\tau$ events is about 40 and 60 for Super-Kamiokande and SNO, respectively. Their arrival time distribution severely constrains the $\nu_\tau$ mass with a sensitivity of about 50 eV (for $D = 10$ kpc), which is an improvement of six orders of magnitude on the current $m_{\nu_\tau}$ limit. However, it is still larger than or comparable with the mass that is needed to close the Universe, $100 h_0^2 \Omega_0$ eV, where $h_0$ is the Hubble constant in units of 100 km s$^{-1}$ Mpc$^{-1}$ and $\Omega_0$ is the density parameter of the Universe.

(iv) If $\nu_\mu$ and $\nu_\tau$ have finite and non-negligible masses, as are needed to solve the solar-neutrino problem (the MSW mechanism, see section 3.5), the same MSW [76, 77] mechanism must also take place in the stellar matter of the progenitor. As is well known the MSW mechanism is specified by $\Delta m^2 = m_{\nu_H}^2 - m_{\nu_L}^2$ (H and L denote heavy and light, respectively), $\sin^2 2\theta$ ($\theta$ is the mixing angle between $\nu_H$ and $\nu_L$) and $n_e(r)$ (the electron number density as a function of radius $r$ in a star). If $\Delta m^2$ and $\sin^2 2\theta$ lie in the triangular region shown in figure 14 [77] assuming that $n_e(r)$ follows a typical

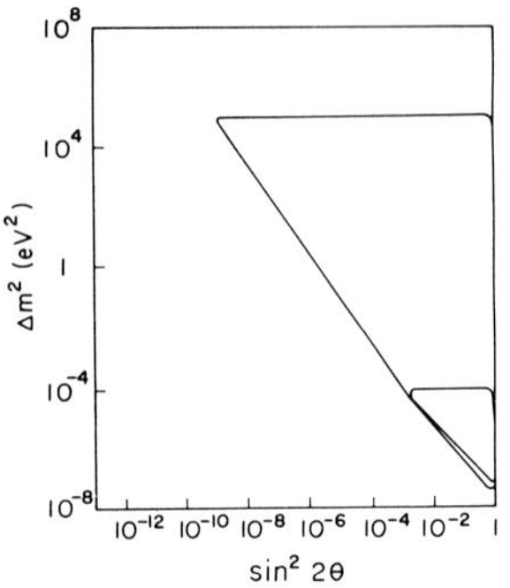


**Figure 14.** The MSW triangle for supernova, the region where the resonant flavour conversion is effective, is schematically illustrated as a large triangular contour. The neutrino sphere density is tentatively taken as $10^{11}$ g cm$^{-3}$. Also shown (small triangle) is the MSW triangle for the solar neutrinos (from [77]).

calculation, $\nu_e$ is fully converted to $\nu_\tau$, and $\nu_\mu$ and $\nu_\tau$ are converted to $\nu_e$ and $\nu_\mu$, respectively, for most cases (adiabatic transition). For the non-adiabatic transition $\nu_\mu$ and $\nu_\tau$ are converted to some admixture of $\nu_e$ and $\nu_\mu$, depending on the adiabaticity condition. The consequence of this is rather serious. The neutronization events disappear, because all $\nu_e$ are converted to $\nu_\tau$ and the cross section for $\nu_\tau e \rightarrow \nu_\tau e$ is one sixth of that for $\nu_e e \rightarrow \nu_e e$. For thermal neutrinos from the cooling phase, all $\nu_e$ are converted to $\nu_\tau$, and $\nu_\mu$ and $\nu_\tau$ are converted to $\nu_\mu$ and $\nu_e$ or to some admixture of them. This conversion influences the number of observed scattering events as well as their energy spectrum. Table 4 shows the calculated numbers of scattering events with or without the MSW mechanism [77]. The energy spectrum of recoil electrons becomes harder owing to interchange between $\nu_e$ and $\nu_\tau$ (note that the temperature for thermal

**Table 4.** Expected numbers of events in the proposed Super-Kamiokande detector from the neutrino burst from a future supernova at a distance of 10 kpc. The table includes the case with $T_{\nu_e} = T_{\bar{\nu}_e} = 3$ and 4 MeV, each with and without the effect of neutrino mixing. The event numbers are calculated using two different detection efficiencies with an effective threshold at electron energies of about 7.5 and 5 MeV, and the latter results are presented in parentheses (from [77]).

| $T_{\nu_e}$ (MeV) | Neutrino mixing | Absorption events $\bar{\nu}_e p \rightarrow ne^+$ | $\nu e$ and $\bar{\nu} e$ scattering events $\nu_e$[a] | $\bar{\nu}_e$[b] | $\nu_\mu + \nu_\tau$[c] | $\bar{\nu}_\mu + \bar{\nu}_\tau$[d] | Total scattering events a+b+c+d | $^{37}Cl$ |
|---|---|---|---|---|---|---|---|---|
| | No | | 85 (133) | 17 (33) | 49 (60) | 39 (49) | 190 (275) | 3 |
| 3 | | 4024 (4496) | | | | | | |
| | Yes | | 160 (192) | 17 (33) | 37 (51) | 39 (49) | 253 (325) | 9 |
| | No | | 118 (160) | 27 (44) | 57 (66) | 46 (55) | 248 (325) | 5 |
| 4 | | 6167 (6463) | | | | | | |
| | Yes | | 183 (209) | 27 (44) | 46 (58) | 46 (55) | 302 (366) | 13 |

$\nu_\tau$ is higher approximately twice that of thermal $\nu_e$ because of the difference in cross sections).

The effect of the MSW mechanism is welcome or unwelcome, depending upon what one is interested in, either welcome (neutrino properties) or unwelcome (mechanism of stellar collapse).

(v) The number of $\bar{\nu}_e$ events is large enough to identify the formation of black holes, in which case the normal burst-like nature of neutrino signals appears in the first instance (0.1 s to a few seconds) followed by a sudden disappearance of the neutrino signals [24]. This clearly indicates that neutrinos are swallowed in the event horizon. One will also see a reddening of the neutrino energies owing to a strong gravitational redshift, although detailed calculations still have to be worked out.

(vi) The large number of $\bar{\nu}_e$ events enables us to perform a detailed analysis based on equation (8), namely to determine $T_{\bar{\nu}_e}(t)$, $\mu_{\bar{\nu}_e}(t)$ and $R_{\bar{\nu}_e}(t)$ as a function of time $t$. They lead to an accurate determination of $L_{\bar{\nu}_e}$. There is a hope that these measured quantities will uniquely specify the best EOS of the nuclear matter, which is of fundamental importance for nuclear physics as well as the explosion mechanism of supernovae.

An example is shown in figure 15, where the $M-R$ relation is shown for a neutron star, a neutron star with quark matter in the core (denoted as a hybrid) and a star consisting of self-bound strange matter [78].

(vii) The detectors at Super-Kamiokande and SNO are still too small to see supernovae beyond a distance of a few 100 kpc.

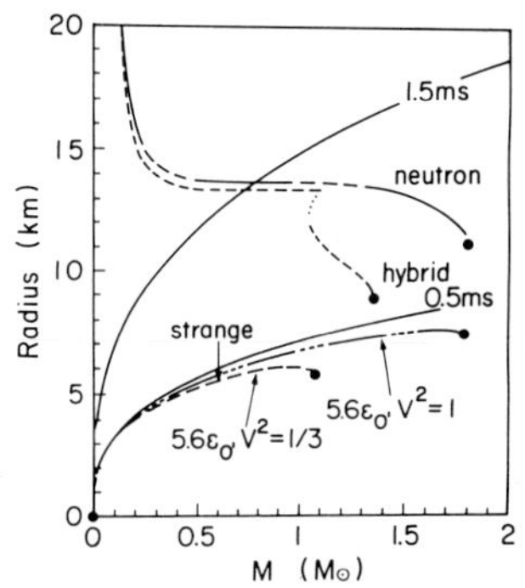


**Figure 15.** Generic relations for compact stars bound only by gravity (neutron and hybrid) and stars made of self-bound matter (strange). Stars below the solid lines can rotate at or faster than the indicated period. The equation of state for strange matter parametrized as $\varepsilon = p/v^2 + \varepsilon_b$, and star families labelled by the value of $\varepsilon_b$ and $v^2$. $\varepsilon_0$ is the nuclear density ($2.5\times10^{14}$ g cm$^{-3}$) (from [78]).

*2.4.2. Supernova rate.* When will the next supernova occur? This is a central question, and many authors have estimated the supernova rate in our Galaxy [79–89]. We are interested in type-II supernovae. Let us list some of these estimates.

(i) 0.034 $h_0^2$ a$^{-1}$ from supernova counting in galaxies beyond our galaxy [83].

(ii) $0.022 \pm 0.013\ \mathrm{a}^{-1}$ from counting of all bright galactic radio supernovae remnants with $100° < l < 260°$ ($l$ is the galactic longitude) and the number of supernovae observed optically in this longitude range during the last two millennia. [84].

(iii) $0.008 \sim 0.033\ \mathrm{a}^{-1}$ from the pulsar formation rate in our Galaxy [85].

(iv) $0.11^{+0.05}_{-0.02}\ \mathrm{a}^{-1}$ from the number of massive stars ($>5\ M_\odot$) divided by their lifetimes. Namely the rate of stellar death resulting in gravitational collapse $D(M_{\min})$ is

$$D(M_{\min}) = N \int_{M_{\min}}^{\infty} \mathrm{d}M \frac{q[R(M)]\Phi(M)}{T(M)} \tag{12}$$

where $N$ is the total number of stars in the disc ($\sim 7 \times 10^{10}$), $q(R)$ is the fraction of stars in the galactic disc that lie within a distance $R$ of the Sun, $\Phi(M)\,\mathrm{d}M$ is the observed probability that a Population-I star has a mass between $M$ and $M + \mathrm{d}M$, and $T(M)$ is the evolutionary lifetime of stars of mass $M$. The above rate was obtained based on the Bahcall and Soneira model for our Galaxy [79].

(v) $0.010^{+0.015}_{-0.006}\ \mathrm{a}^{-1}$ based on the same reasoning as (iv) but using only information about main-sequence stars and different luminosity functions [80].

(vi) $0.1^{+0.1}_{-0.05}\ \mathrm{a}^{-1}$ (a factor of two uncertainty) from the oxygen abundance in our Galaxy, the oxygen production rate per type-II supernova and the age of our Galaxy. Specifically, the metal abundance is 1%, the mass of the disc $10^{11} M_\odot$, the age $10^{10}$ a, and the oxygen product per supernova $1^{+1}_{-0.5} M_\odot$. These numbers yield a type-II supernova rate of $10^{-2} \times 10^{11}/(1 \times 10^{10}) = 0.1\ \mathrm{a}^{-1}$ [82].

(vii) $>0.08\ \mathrm{a}^{-1}$ based on a new counting method for historical supernovae [81].

(viii) $>0.03\ \mathrm{a}^{-1}$ possibly $\sim 0.1\ \mathrm{a}^{-1}$ from systematic counting of radio supernova remnants [86].

(ix) $\sim 0.1\ \mathrm{a}^{-1}$ from the observation of intense iron line emission (6.7 keV) from the galactic ridge as well as from the galactic centre [87–89].

Obviously each estimate suffers from a large degree of uncertainty. However, observation of SN1987A seems to make people optimistic about the estimation of the supernova rate. Perhaps it may not be overly optimistic to consider the rate to be as large as $0.1\ \mathrm{a}^{-1}$ for type-II supernovae. This is indeed very encouraging and there is a good chance of observing the next supernova around the turn of the century.

*2.4.3. High-energy cosmic background neutrinos (HECBN).* Let us show several arguments to estimate the flux of HECBN [23–25].

*2.4.3.1.* Assume a constant supernova rate in a comoving volume element, that is the supernova activity has been constant throughout time. Then the energy density of the antielectron-neutrino ($\bar{\nu}_e$) is

$$\rho_{\bar{\nu}_e} = n_G^0 R_{SN}^0 L_{\bar{\nu}_e}^0 \int_0^{t_0} \frac{R(t)}{R_0} \mathrm{d}t \tag{13}$$

where $n_G^0$, $R_{SN}^0$ and $L_{\bar{\nu}_e}^0$ are the density of galaxies in the Universe, the supernova rate per galaxy and the total $\bar{\nu}_e$ luminosity per supernova, respectively, all of which correspond to the values in the present epoch. The integral is to calculate the red shift of $\bar{\nu}_e$ energies, and $t_0$, $R(t)$ and $R_0$ are the age of the Universe, the scale factor at time $t$ and the present scale factor, respectively.

Assuming a matter-dominated Universe and denoting $z$ as the red shift factor,

$$z + 1 = R_0/R(t) \qquad \mathrm{d}z = H_0(1+z)^2\sqrt{1+\Omega_0 z}\,\mathrm{d}t \tag{14}$$

then

$$\rho_{\bar{\nu}_e} = n_G^0 R_{SN}^0 L_{\bar{\nu}_e}^0 \int_0^{z_0} \frac{1}{H_0(1+z)^3\sqrt{1+\Omega_0 z}} \, dz \tag{15}$$

where the present density fraction $\Omega_0$ is

$$\Omega_0 = \rho/\rho_c \qquad \rho_c = 3H_0^2/8\pi G \tag{16}$$

and $\rho_c$, $H_0$ are the critical density and the Hubble constant at the present age ($H_0 = 50 - 100$ km s$^{-1}$ Mpc$^{-1}$). Taking $\Omega_0 = 1$ or 0, letting $z_0 \to \infty$, and that noting $H_0 = 2/(3t_0)$ ($\Omega_0 = 1$) or $H_0 = 1/t_0$ ($\Omega_0 = 0$),

$$\rho_{\bar{\nu}_e}/(n_G^0 R_{SN}^0 L_{\bar{\nu}_e}^0 t_0) = \tfrac{3}{5} \qquad \Omega_0 = 1 \tag{17}$$

$$\rho_{\bar{\nu}_e}/(n_G^0 R_{SN}^0 L_{\bar{\nu}_e}^0 t_0) = \tfrac{1}{2} \qquad \Omega_0 = 0 \tag{18}$$

The average $\bar{\nu}_e$ energy is red shifted by the same factor, that is

$$E_{\bar{\nu}_e} = \tfrac{3}{5} E_{\bar{\nu}_e}^0 \qquad \Omega_0 = 1 \tag{19}$$

$$= \tfrac{1}{2} E_{\bar{\nu}_0}^0 \qquad \Omega_0 = 0 \tag{20}$$

where $E_{\bar{\nu}_e}$ and $E_{\bar{\nu}_e}^0$ are the average $\bar{\nu}_e$ energies at the present epoch and at the supernova explosion, respectively. The value in the real Universe is probably in between. The $\bar{\nu}_e$ flux is then,

$$F_{\bar{\nu}_e} = \rho_{\bar{\nu}_e} c / E_{\bar{\nu}_e} = n_G^0 R_{SN}^0 t_0 \frac{L_{\bar{\nu}_e}^0}{E_{\bar{\nu}_e}^0} c \tag{21}$$

which does not apparently depend on the evolution of the Universe. Note that

$$\frac{dF_{\bar{\nu}_e}}{dE_{\bar{\nu}_e}} = E_{\bar{\nu}_e}^2 \int dz \, n_G(t) R_{SN}(t) \frac{1+z}{H_0\sqrt{1+\Omega_0 z}} \times \int_{\text{Burst}} f((1+z)E_{\bar{\nu}_e}, \tau) 4\pi R(\tau)^2 \, d\tau, \tag{22}$$

where $f(E, t)E^2$ is the energy density spectrum of $\bar{\nu}_e$ at stellar collapse (see equation (8)),

$$f(E, t) = \frac{g\pi c^2}{(hc)^3} \left[ \exp\left( \frac{1}{T(t)} [E - \mu(t)] \right) + 1 \right]^{-1}. \tag{23}$$

Normalizing $n_G^0 = 0.02$ Mpc$^{-3}$, $R_{SN}^0 = 0.1$ a$^{-1}$, $t_0 = 15 \times 10^9$ a, $L_{\bar{\nu}_e}^0 = 3 \times 10^{53}/6$ erg and $E_{\bar{\nu}_e}^0 = 13$ MeV, one ends up with

$$F_{\bar{\nu}_e} = 75 \left( \frac{R_{SN}^0}{0.1} \right) \left( \frac{n_G^0}{0.02} \right) \left( \frac{t_0}{15} \right) \text{cm}^{-2}\,\text{s}^{-1} \tag{24}$$

and

$$E_{\bar{\nu}_e} = 8 \text{ MeV} \qquad \Omega_0 = 1 \tag{25}$$

$$E_{\bar{\nu}_e} = 6 \text{ MeV} \qquad \Omega_0 = 0 \tag{26}$$

for $E_{\bar{\nu}_e}^0 = 13$ MeV, which was obtained from SN1987A. An assumption was made here that the contribution of elliptical galaxies is negligible in deriving $n_G^0$ (type-II supernovae do not occur in ellipticals).

*2.4.3.2.* Reference [24] gives a good account of the material in this subsection. Assume that all baryons in the Universe had participated in star formation (which may not be a good assumption). Then the total number density of supernovae integrated over time $N_{SN}$ is

$$N_{SN} = \frac{\rho_c \Omega_b f(>8M_\odot)}{\bar{M}(>8M_\odot)} \tag{27}$$

where $f$ is the fraction of stars with mass larger than $8M_\odot$ in main-sequence stars, $\bar{M}$ is the average stellar mass for stars with mass larger than $8M_\odot$, $\Omega_b$ is the baryonic density parameter and $\rho_c\Omega_b$ corresponds to the baryonic density in the Universe. $\rho_c = 3H_0^2/8\pi G = 1.88 \times 10^{-29} h_0^2\ \mathrm{g\ cm^{-3}}$ and $H_0 = 100h_0\ \mathrm{km\ s^{-1}\ Mpc^{-1}}$. $N_{SN}$ is actually equal to $n_G^0 R_{SN}^0 t_0$.

Taking $f = 0.01$ and $\bar{M} = 15M_\odot$,

$$N_{SN} = 6.30 \times 10^{-65} \Omega_b h_0^2 (\mathrm{cm^{-3}}) \tag{28}$$

Thus,

$$F_{\bar{\nu}_e} = N_{SN} \frac{L^0_{\bar{\nu}_e}}{E^0_{\bar{\nu}_e}} c = 45 \left( \frac{\Omega_b h_0^2}{0.01} \right) \mathrm{cm^{-2}\ s^{-1}}. \tag{29}$$

Note that the latest estimate reads [90]

$$0.0097 < \Omega_b h_0^2 < 0.016. \tag{30}$$

*2.4.3.3.* Reference [24] gives a good account of the material in this subsection. A similar argument is applied for the metallicity in the Universe. The metallic elements mostly consist of oxygen, $^{16}O$. Then, the total number density of supernovae is

$$N_{SN} = Z\rho_c \Omega_b f(>8M_\odot) / W_{SN}(Z) \tag{31}$$

where $Z$ is the metal abundance and $W_{SN}(Z)$ is the metal product per type-II supernova. Putting $Z = 0.01$ and $W_{SN}(Z) = 1\ M_\odot$ as used in section 2.4.2(vi),

$$F_{\bar{\nu}_e} = 68(\Omega_b h_0^2/0.01)\ \mathrm{cm^{-2}\ s^{-1}}. \tag{32}$$

It is remarkable that the three derivations give consistent flux values. However, the real uncertainties may well amount to a factor of two. Thus $F_{\bar{\nu}_e}$ could be taken as

$$F_{\bar{\nu}_e} = 65^{+70}_{-40}\ \mathrm{cm^{-2}\ s^{-1}} \tag{33}$$

as a first guess.

Can one detect the HECBN? The answer is yes but difficult. The researchers at Kamiokande have searched for electron signals in an energy region between 15 and 50 MeV [91, 92]. This region is the most quiet region; below 15 MeV there is a steep-rising background resulting from external $\gamma$-rays and $\beta$-rays from unstable nuclei produced by cosmic-ray muons (spallation products), while above 50 MeV atmospheric neutrinos begin to dominate. Figure 16 represents a visible-energy spectrum after the elimination of various backgrounds [92]. Note that near 15 MeV the dominant background is now the solar neutrinos. The large background indicates that the experimental site at Kamioka (2700 metre water equivalent) is too shallow to reduce the number of spallation products sufficiently.

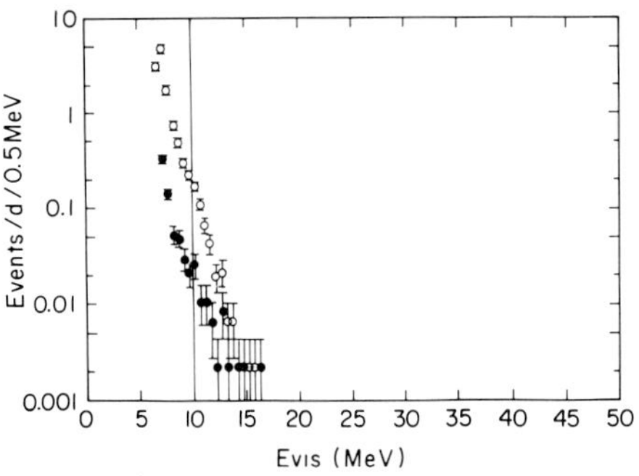


**Figure 16.** Differential energy spectrum of the remaining events in the Kamiokande detector for 594 days. The open circles represent the events of $\cos\theta_{Sun} < 0.88$ and the full circles represent those of $\cos\theta_{Sun} \geqslant 0.88$ (from [92]).

Applying equation (22) and equation (23) with $\mu = 0$, time-independent $T$ and taking into account all the detector parameters, one finds the upper limit on $F_{\bar{\nu}_e}$

$$F_{\bar{\nu}_e} < 780\ \mathrm{cm}^{-2}\,\mathrm{s}^{-1} \qquad 90\%\ \mathrm{CL} \tag{34}$$

for $T = 4$ MeV [92]. The result is based on the data sample corresponding to 1044 days and the background subtraction was made in a part of the data. Figure 17 shows the upper limit of $F_{\bar{\nu}_e}$ for various temperatures.

The present upper limit of $F_{\bar{\nu}_e}$ is larger by an order of magnitude than the expected flux, equation (33). In order to improve the experimental sensitivity one obviously needs a larger detector volume and less background. The ultimate background is, as mentioned earlier, solar neutrinos. Thus the search is limited only above 15 MeV, unless a positive identification of $\bar{\nu}_e$ from other neutrinos is made, which is possible by detecting neutrons from the reaction $\bar{\nu}_e + p \to e^+ + n$, for example, by the following scheme; dissolve NaCl in a detector, produce $\gamma$-rays with the reaction Cl(n, $\gamma$)Cl, and detect the $\gamma$-rays in coincidence (with a time gate of about 100 $\mu$s) with the $e^+$. The overall detection efficiency can reach 30–50%. Then one can come down to an energy of about 8 MeV for the search. Below this energy $\bar{\nu}_e$ produced by nearby nuclear reactors dominate [27]. There is a good chance that we shall be able to reach a flux level as low as 20 $\mathrm{cm}^{-2}\,\mathrm{s}^{-1}$ with the next generation detectors, such as Super-Kamiokande.

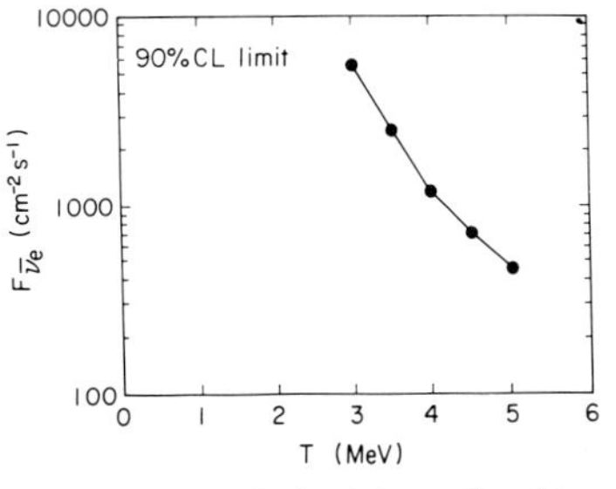


**Figure 17.** Upper limit of the total antielectron neutrino flux at 90% CL as a function of temperature $T$. $\Omega_0$ is set to unity (from [92]).

An active research effort is being made to observe diffuse cosmic x-ray background (CBX), [93, 94], which is presumably a relic from an active phase of galaxy formation. HECBN has, in a sense, a similar and complementary significance to CBX. Observation of both relics will provide profound knowledge about a dark and mysterious epoch of the early Universe.

## 3. Solar neutrinos

### 3.1. Neutrino generation

It is very well known that stars shine by nuclear fusion processes taking place deep within their stellar cores. In fact the standard stellar theory, with elaborated numerical calculations, explains observed stellar properties like the Hertzsprung–Russell diagram [95–97]. The underlying equations are, unless instabilities like convection take place, rather simple. Denoting the pressure, density, temperature and the interior mass at a distance $r$ from the centre of a star as $P(r)$, $\rho(r)$, $T(r)$ and $M(r)$, respectively, the balance between the gravitational and pressure forces implies that

$$\frac{\mathrm{d}P(r)}{\mathrm{d}r} = -\frac{GM(r)\rho(r)}{r^2}. \tag{35}$$

$M(r)$ and $\rho(r)$ are related by

$$\mathrm{d}M(r)/\mathrm{d}r = 4\pi r^2 \rho(r). \tag{36}$$

The net energy flux at $r$, $L(r)$, the energy per unit time that passes through a sphere of radius $r$ is, assuming transport by radiation

$$L(r) = -4\pi r^2 \frac{ac}{3} \frac{1}{\kappa\rho(r)} \frac{\mathrm{d}T(r)^4}{\mathrm{d}r} \tag{37}$$

where $\kappa$ is the opacity, which depends sensitively on the metal abundance, and $a$ and $c$ are the Stefan–Boltzmann constant and the velocity of light, respectively. The pressure $P(r)$ is given by

$$P(r) = \frac{a}{3} T^4 + \frac{1}{\mu} \frac{k\rho(r)T(r)}{m_{\mathrm{H}}} (1+D) \tag{38}$$

where $k$ is the Boltzmann constant, $\mu$ the molecular weight the inverse of which is the number of particles per nucleon relevant to the equation of state (2 for hydrogen(e and p), 3/4 for helium and 1/2 for fully ionized heavier elements), $m_{\mathrm{H}}$ the mass of the hydrogen atom, and $D$ a correction factor for the degeneracy of the electrons and for Debye–Hückel modifications to the equation of state [5].

Finally the energy generation is formulated,

$$\frac{\mathrm{d}L(r)}{\mathrm{d}r} = 4\pi r^2 \rho(r) \left( \varepsilon - T\frac{\mathrm{d}S}{\mathrm{d}t} \right) \tag{39}$$

where $S$ is the stellar entropy. The second term is actually negligible compared with the nuclear energy $\varepsilon$. The observed luminosity is then

$$L_\odot = \int_0^{R_\odot} \frac{\mathrm{d}L(r)}{\mathrm{d}r} \mathrm{d}r. \tag{40}$$

$\varepsilon$, the energy produced per unit time per unit mass is, [5],

$$\varepsilon = \sum_{ij} (Q - \langle q_\nu \rangle)_{ij} n_i(r) n_j(r) \langle \sigma v \rangle_{ij} \frac{1}{(1+\delta_{ij})}. \tag{41}$$

Here $Q$ is the $Q$-value of the reaction between nuclei $i$ and $j$, $\langle q_\nu \rangle$ is the average energy carried off by solar neutrinos, $n_i$ and $n_j$ are the number densities of nuclei of types $i$ and $j$, $\sigma$ is the total reaction cross section, $v$ is the relative velocity between particles $i$ and $j$. $\delta_{ij}$ is the Kronecker delta. The average $\langle \sigma v \rangle$ is carried out with the Maxwell-Boltzmann distribution. The cross section $\sigma$ has a very steep component from the Coulomb barrier. By singling out its effect, $\sigma$ can be written as

$$\sigma(E) = \frac{S(E)}{E} \exp(-2\pi\eta)$$

$$\eta = Z_i Z_j (e^2/\hbar v). \tag{42}$$

$S(E)$ depends only on nuclear effects and is a smooth function of $E$. $S(0) \equiv S_0$ is called the cross section factor. $\langle \sigma v \rangle$ can now be explicitly written as

$$\langle \sigma v \rangle = \left( \frac{8}{\pi \mu (kT)^3} \right)^{1/2} f_0 \int_0^\infty \mathrm{d}E\, S(E) \exp(-2\pi\eta - E/kT) \tag{43}$$

where $\mu$ is the reduced mass of the particles $i$ and $j$, and the quantity $f_0$ takes into account the screening effect. An order of estimate shows that the central temperature is about $10^7$ K or $kT \simeq 1$ keV. Taking into account the steep exponential term in equation (43), the effective temperature for the nuclear reactions is about 10 keV.

We must now know all the reactions relevant at this energy range. Table 5 shows the parameters for a series of reactions called the proton–proton chain [5].

There is another process that contributes to energy production in the Sun, namely the carbon-nitrogen-oxygen (CNO) cycle, which, however, plays a much smaller role than the p-p chain. Tables 6 shows the relevant parameters [5]. The values of $S_0$ are

**Table 5.** Reaction parameters for the proton–proton chain. The uncertainties for the cross section factors are indicated in parentheses; they correspond to errors of $3\sigma$ and have been calculated (from [5]).

| Reaction | $Q$ (MeV) | $\langle q_{\nu_e} \rangle$ (MeV) | $S_0$ (keV b) | $\mathrm{d}S/\mathrm{d}E$ (b) |
|---|---|---|---|---|
| $^1H(p, e^+ \nu_e)^2H$ | 1.442 | 0.265 | $4.07\ (1 \pm 0.05) \times 10^{-22}$ | $4.52 \times 10^{-24}$ |
| $^1H(p, e^-, \nu_e)^2H$ | 1.442 | 1.442 | † | |
| $^2H(p, \gamma)^3He$ | 5.494 | | $2.5 \times 10^{-4}$ | $7.9 \times 10^{-6}$ |
| $^3He(^3He, 2p)^4He$ | 12.860 | | $5.15(1 \pm 0.17) \times 10^{0.3}$ | $-0.9$ |
| $^3He(^4He, \gamma)^7Be$ | 1.586 | | $0.54(1 \pm 0.06)$ | $-3.1 \times 10^{-4}$ |
| $^7Be(e^-, \nu_e)^7Li$ | 0.862 | 0.862 | ‡ | |
| | 0.384 | 0.384 | | |
| $^7Li(p, \alpha)^4He$ | 17.347 | | $52\ (1 \pm 0.5)$ | 0 |
| $^7Be(p, \gamma)^8B$ | 0.137 | | $0.0243\ (1 \pm 0.22)$ | $-3 \times 10^{-5}$ |
| $^8B(e^+ \nu_e)^8Be^*$ | 17.980 | 6.710 | | |
| $^8Be^*(\alpha)^4He$ | | | | |
| $^3He(p, e^+ \nu_e)^4He$ | 19.795 | 9.625 | $8 \times 10^{-20}$ | |

† $R_{pep} \simeq 1.102 \times 10^{-4} (\rho/\mu_e) T_6^{-1/2} (1 + 0.02 T_6) R_{pp}$, where $\mu$ is the mean molecular weight per electron, $T_6 = T/10^6$ K, and $R_{ij} = n_i n_j \langle \sigma v \rangle_{ij} / (1 + \delta_{ij})$(cf. equation(41)), and $\rho$ is in units of g cm$^{-3}$.
‡ $R_{^7Bee^-} \simeq 5.54 \times 10^{-9} (\rho/\mu_e) T_6^{-1/2} [1 + 0.004(T_6 - 16)]$ s$^{-1}$.

**Table 6.** Reaction parameters for the CNO cycle. The uncertainties for the cross section factors are indicated in parentheses; they correspond to errors of $1\sigma$ (from [5]).

| Reaction | $Q$ (MeV) | $\langle q \rangle$ (MeV) | $S(0)$ (MeV b) | $S'(0)$ (b) | $S''(0)$ (barns/MeV) |
|---|---|---|---|---|---|
| $^{12}C(p, \gamma)^{13}N$ | 1.943 | | $1.45\,(1 \pm 0.15) \times 10^{-3}$ | $2.45 \times 10^{-3}$ | $6.80 \times 10^{-2}$ |
| $^{13}N(e^+\nu_e)^{13}C$ | 2.221 | 0.7067 | | | |
| $^{13}C(p, \gamma)^{14}N$ | 7.551 | | $5.50\,(1 \pm 0.15) \times 10^{-3}$ | $1.34 \times 10^{-2}$ | $9.87 \times 10^{-2}$ |
| $^{14}N(p, \gamma)^{15}O$ | 7.297 | | $3.32\,(1 \pm 0.12) \times 10^{-3}$ | $-5.91 \times 10^{-3}$ | $9.06 \times 10^{-3}$ |
| $^{15}O(e^+\nu_e)^{15}N$ | 2.754 | 0.9965 | | | |
| $^{15}N(p, \gamma)^{16}O$ | 12.128 | | $6.4\,(1 \pm 0.09) \times 10^{-2}$ | $3 \times 10^{-2}$ | 4.0 |
| $^{15}N(p, \alpha)^{12}C$ | 4.966 | | $78.0\,(1 \pm 0.17)$ | 351 | $1.11 \times 10^{4}$ |
| $^{16}O(p, \gamma)^{17}F$ | 0.600 | | $9.4\,(1 \pm 0.16) \times 10^{-3}$ | $-2.3 \times 10^{-2}$ | $6.0 \times 10^{-2}$ |
| $^{17}F(e^+\nu_e)^{17}O$ | 2.762 | 0.9994 | | | |

in fact much smaller than those in the p–p chain, simply because of the stronger Coulomb barrier.

In order to solve the stellar equations one further needs detailed information about the opacity $\kappa$ and the chemical composition of the Sun. Knowing, in addition, $L_\odot$, $M_\odot$ and $R_\odot$ one then carries out numerical calculations and evaluates all the parameters in the interior of the Sun, among which are the fluxes of the various neutrinos. The neutrino energies are determined only by the interaction properties, that is either monochromatic from two-body reactions or continuous from $\beta$-decay. Tables 7 and 8 [5] summarize the reactions for the p–p chain and the CNO cycle, respectively, together with end-point energies of the neutrinos. The termination percentage for the p–p chain is also shown.

Figure 18 shows the resultant fluxes of solar neutrinos from various sources as a function of neutrino energy [5]. The temperature dependences of various neutrinos

**Table 7.** The pp chain in the Sun. The average number of p-p neutrinos produced per termination in the Sun is 1.85 (from [5]).

| Reaction | Termination (%) | $\nu$ energy (MeV) |
|---|---|---|
| $p+p \rightarrow {}^2H+e+\nu_e$ | 100 | $\leqslant 0.420$ |
| or | | |
| $p+e^-+p \rightarrow {}^2H+\nu_e$ | 0.4 | 1.442 |
| ${}^2H+p \rightarrow {}^3He+\gamma$ | 100 | |
| ${}^3He+{}^3He \rightarrow \alpha+2p$ | 85 | |
| or | | |
| ${}^3He+4He \rightarrow {}^7Be+\gamma$ | 15 | |
| ${}^7Be+e^- \rightarrow {}^7Li+\nu_e$ | 15 | (90%) 0.861<br>(10%) 0.383 |
| ${}^7Li+p \rightarrow 2\alpha$ | 15 | |
| or | | |
| ${}^7Be+p \rightarrow {}^8B+\gamma$ | 0.02 | |
| ${}^8B \rightarrow {}^8Be^*+e^++\nu_e$ | 0.02 | $<15$ |
| ${}^8Be^* \rightarrow 2\alpha$ | 0.02 | |
| or | | |
| ${}^3He+p \rightarrow {}^4He+e^++\nu_e$ | 0.00002 | $\leqslant 18.77$ |

**Table 8.** The CNO cycle: the main reactions are shown for the carbon-nitrogen-oxygen cycle in the Sun (from [5])

| Reaction | $\nu$ energy (MeV) |
|---|---|
| $^{12}C+p \rightarrow ^{13}N+\gamma$ | |
| $^{13}N \rightarrow ^{13}C+e^{+}+\nu_e$ | $\leq 1.199$ |
| $^{13}C+p \rightarrow ^{14}N+\gamma$ | |
| $^{14}N+p \rightarrow ^{15}O+\gamma$ | |
| $^{15}O \rightarrow ^{15}N+e^{+}+\nu_e$ | $\leq 1.732$ |
| $^{15}N+p \rightarrow ^{12}C+\alpha$ | |

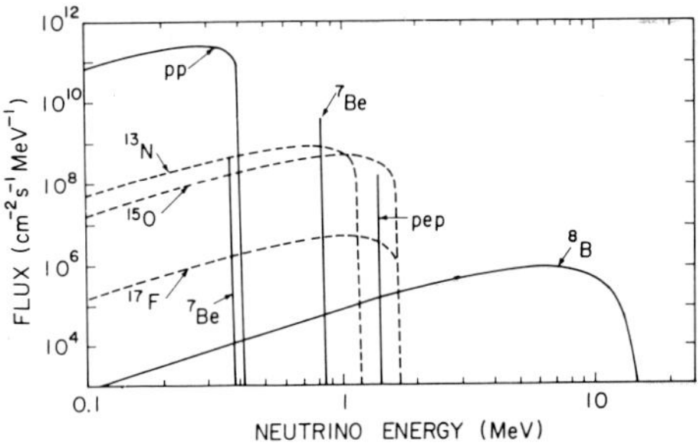


**Figure 18.** Solar-neutrino spectrum. The neutrino fluxes from continuum sources are given in units of $cm^{-2} s^{-1} MeV^{-1}$ at the Earth. The line fluxes (pep and $^{7}Be$) are given in units of $cm^{-2} s^{-1}$ (from [5]).

were also calculated and they are roughly $T^{18}$, $T^{8}$ and $T^{-1.2}$ for $^{8}B$, $^{7}B$ and pp neutrinos, respectively [5]. The dependence for pp neutrinos is negative because of the constraint that $L_{\odot}$ is fixed to the observed value and that the termination percentage of $^{3}He+^{4}He$ becomes large as the temperature rises, resulting in a smaller number of the reactions $pp \rightarrow De^{+}\nu$. It is interesting that the measurement of $^{8}B$ neutrinos determines the central temperature quite accurately. For example, the forthcoming facility at Super-Kamiokande will observe about 30 $^{8}B$ neutrino events per day. This event rate is sufficient to monitor the central temperature variation of 1% every day.

The calculation of the solar-neutrino fluxes does not include any new physics. The basic equations are simple and solvable. This calculation is therefore called the standard solar model (SSM). Therefore the calculated fluxes for pp and pep neutrinos are expected to be reliable. The $^{8}B$ neutrino flux, on the other hand, suffers a large uncertainty from an inaccurately known cross section factor $S_0(^{7}Be(p, \gamma)^{8}B)$ and to the influence of the less well-known heavy elements on the opacity $\kappa$. Table 9 shows the calculated solar-neutrino fluxes and their estimated $3\sigma$ uncertainties [5]. Note, however, that the shapes of their energy spectra are accurately known, because they are either monoenergetic or $\beta$-decay spectra.

These predicted fluxes now require experimental verification. At present there are only two experiments that have presented results. The third experiment recently presented a preliminary but very important result, and the fourth experiment is about to start observations. Their results are summarized below.

**Table 9.** Calculated solar-neutrino fluxes. The errors correspond to $3\sigma$ (from [5]).

| Source | Flux ($10^{10}\,\mathrm{cm}^{-2}\,\mathrm{s}^{-1}$) |
|---|---|
| pp | $6.0(1\pm0.02)$ |
| pep | $0.014(1\pm0.05)$ |
| hep | $8\times10^{-7}$ |
| $^{7}$Be | $0.47(1\pm0.15)$ |
| $^{8}$B | $5.8\times10^{-4}(1\pm0.37)$ |
| $^{13}$N | $0.06(1\pm0.50)$ |
| $^{15}$O | $0.05(1\pm0.58)$ |
| $^{17}$F | $5.2\times10^{-4}(1\pm0.46)$ |

### 3.2. *The Homestake experiment*

R Davis Jr and co-workers have been doing pioneering and ingenious work 1500 m underground (equivalent to 4100 m of water) in the Homestake gold mine for over twenty years [6–8]. The basic idea is to use the reaction

$$\nu_e + {}^{37}\mathrm{Cl} \rightarrow \mathrm{e}^- + {}^{37}\mathrm{Ar} \tag{44}$$

and to count the number of produced $^{37}$Ar atoms. The threshold energy of this reaction is $E_{\mathrm{th}} = 0.814$ MeV. Since its cross section depends on $(E_\nu - E_{\mathrm{th}})^2$, the dominant contribution comes from $^{8}$B neutrinos (see figure 18).

The expected yield is

$$Y = N_a \sum_i \int_{0.814} F_i(E)\sigma(E)\,\mathrm{d}E \tag{45}$$

where $N_a$ is the number of the target atoms, $F_i$ is the neutrino flux of the $i$th source, and $\sigma$ is the cross section of equation (44). The magnitude of $\sigma$ averaged over the $^{8}$B neutrino spectrum is [98]

$$\langle\sigma\rangle_{^8\mathrm{B}} = (1.35\pm0.1)\times10^{-42}\,\mathrm{cm}^2. \tag{46}$$

$Y/N_a$ is what experiment determines and is normally expressed in units of SNU (solar neutrino unit, $10^{-36}\,\mathrm{s}^{-1}\,\mathrm{atom}^{-1}$). With known $\sigma(E)$, the calculated capture rates $Y/N_a$ for various neutrino sources are shown in table 10, see [5]. The dominant contribution

**Table 10.** Calculated capture rates for a $^{37}$Cl detector. The uncertainty corresponds to estimated $3\sigma$ error (from [5]).

| Neutrino source | Capture rate (SNU) |
|---|---|
| pp | 0 |
| pep | 0.2 |
| hep | 0.03 |
| $^{7}$Be | 1.1 |
| $^{8}$B | 6.1 |
| $^{13}$N | 0.1 |
| $^{15}$O | 0.3 |
| $^{17}$F | 0.003 |
| Total | $7.9\pm2.6$ |

comes from $^8B$ neutrinos as expected and the rest is mainly from $^7B$ neutrinos ($^8B/^7B \approx 1/6$). The total capture rate is

$$Y/N_a = 7.9 \pm 2.6 \text{ SNU} \tag{47}$$

the error being with an uncertainty of $3\sigma$.

The Homestake detector consists of 615 t of liquid $C_2Cl_4$ in which $2.16 \times 10^{30}$ $^{37}Cl$ atoms are contained. Thus the expected event rate is $1.5 \pm 0.5\ d^{-1}$. The $^{37}Ar$ produced are then extracted by bubbling $^4He$ gas through the liquid and are fed into a small proportional counter. The reaction $e^- + {}^{37}Ar \rightarrow \nu_e + {}^{37}Cl$ simply proceeds as the inverse process to equation (44) with a half life of 35 d. The 2.82 keV Auger electrons that are produced by this electron capture are counted.

Figure 19 summarizes all the runs taken from 1970 to 1990, and also shows the theoretical prediction and the measured average value of the Ar production rate, $0.4 \pm 0.06\ d^{-1}$. The experimental result for the capture rate is

$$Y/N_a = 2.1 \pm 0.3 \text{ SNU} \tag{48}$$

Sometimes a value of $2.3 \pm 0.3$ SNU is also cited. Taking the ratio equation (48)/equation (47) the data/SSM prediction is

$$R_H = 0.27 \pm 0.04 \tag{49}$$

where the theoretical uncertainty is not considered.

This large discrepancy between theory and experiment is called the solar-neutrino problem and has been a central issue both in astronomy and in particle physics.

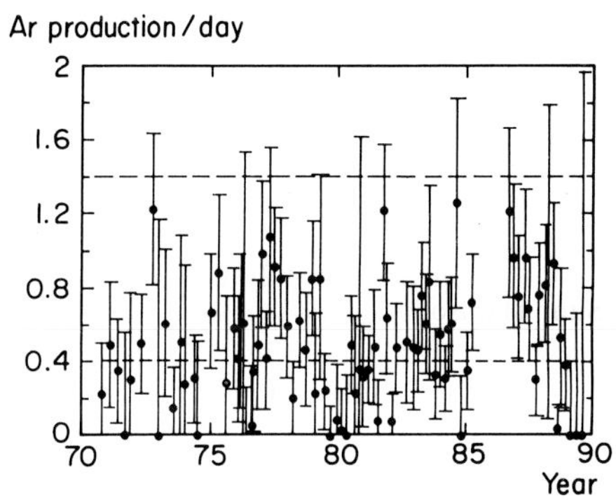


**Figure 19.** Results about Ar production rate ($d^{-1}$) from the Homestake experiment [6–8]. The data extend from 1970 to 1990. The observed average rate (dotted line) is shown together with the standard-solar-model (SSM) prediction (broken line).

Figure 20 shows the variation of the sunspot numbers over the same period. It has been argued that the Ar production rate anticorrelates with the sunspot number, notably high yields in 1977–1978 and 1987–1988 and low yields in 1980 and 1989, which roughly correspond to the solar minima in 1976–1977 and 1986–1987 and the solar maxima in 1980 and 1989–1990, respectively. Since it is now believed that the time variation of the solar activity is caused by the magnetic fields in the convection zone, that is the outer layer $r > 0.7R_\odot$, it is quite unlikely that the solar activity influences the energy generation or the neutrino output at the centre of the Sun. Therefore, if the anticorrelation is indeed correct, it must arise from some new interaction between magnetic field

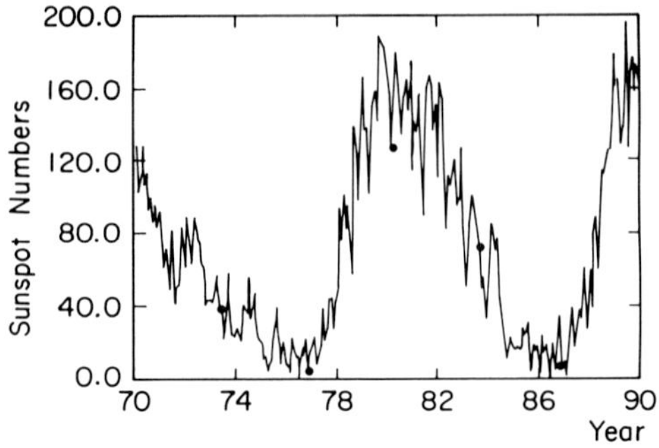


**Figure 20.** Monthly averaged sunspot numbers from 1970 to 1990, which should be compared with the results shown in figure 19.

and neutrino, that is a large neutrino magnetic moment [99]. However, the anticorrelation was not confirmed by the Kamiokande experiment for the period of 1987–1990, see [10]. More data from several experiments are clearly needed.

### *3.3. The Kamiokande experiment*

The Kamiokande experiment [9, 10, 100] has already been described in section 2.2, which had first observed the neutrino burst from SN1987A. Continuous work has been devoted to lower the minimum detectable energy of the Kamiokande detector since 1987, and the detector has been successfully operated to obtain useful data at the visible energy larger than 7.5 MeV. The energy threshold limits the observation only for $^8$B neutrinos. The experiment employs the reaction $\nu_e + e \rightarrow \nu_e + e$ (cf. equation (4)) and measures the position, direction and energy of the recoil electrons. Since the recoil electron has a sharp forward peak (though smeared somewhat by multiple Coulomb scattering in the detector), one can take advantage of the directional correlation between the recoil electron and the Sun. The isotropic background can thus be reduced by a factor of 10–20. Figure 21 shows the angular distributions of observed electrons (the forward direction is at $\theta_{Sun} = 0$) for two energy thresholds, $E_e \geqslant 7.5$ MeV and $E_e \geqslant 9.3$ MeV for the periods of June 1988–April 1990 (590 d) and of January 1987–April 1990 (1040 d), respectively [10]. Strong forward peaks are clearly visible above the background for both distributions. The solid histograms are the expected angular distributions from the calculation of Bahcall and Ulrich, see Bahcall [5]. The observed peaks are substantially less than the theoretical predictions, confirming the existence of the solar neutrino problem.

Figure 22 shows the energy spectrum of observed electrons [10]. Its shape is consistent with $^8$B $\beta$-decay but the absolute yield is less than the expected one. By applying a maximum-likelihood method to fit the individual data points, one obtains the ratio of observed/expected events,

$$R_K = \frac{\text{data}}{\text{SSM}} = 0.46 \pm 0.05 \pm 0.06 \qquad (50)$$

where the SSM adopts the central flux value calculated by Bahcall and Ulrich, the first and second errors correspond to statistical and systematic ones, respectively, and the

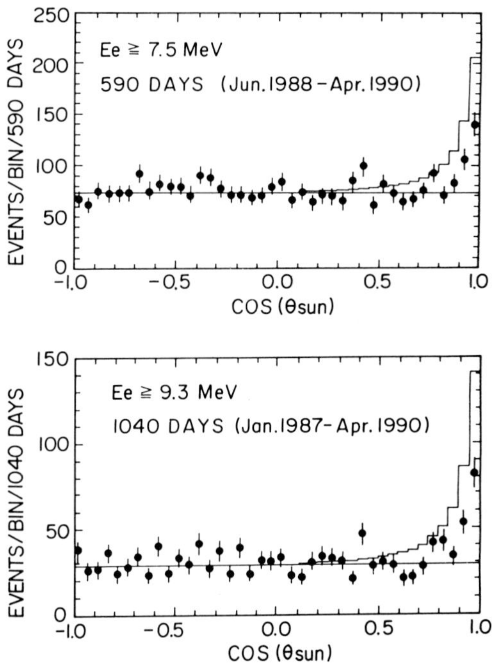


**Figure 21.** Angular distribution of the observed events in the Kamiokande detector. $\theta_{Sun}=0$ corresponds to the forward direction from the Sun. The peaks near $\cos\,\theta_{Sun}=1$ are the events caused by the solar neutrinos (from [10]).

energy threshold is $E_\nu \geqslant 7.5$ MeV. Figure 23 [10] shows the five data points, each corresponding to 200 d, for the period of January 1987–April 1990, in which the earliest two data points were obtained for $E_\nu \geqslant 9.3$ MeV, and the other three correspond to $E_\nu \geqslant 7.5$ MeV. Note that the ordinate is $R_K$, which does not depend on the threshold energy unless the spectral shape changes between $7.5 < E_\nu < 9.3$ MeV. The result shows no apparent time variation of the solar-neutrino flux, while the Sunspot number changed very much in the same period as shown in figure 24. The (anti) correlation of the Kamiokande data to the sunspot number is less than 36% at 90% CL. This limit does not seem to support the strong anticorrelation observed by the Homestake experiment (see figure 19). Continuous observation for another few years at Kamiokande will set a much more severe limit.

Finally figure 25 [100] shows the short-term variation of the Kamiokande data. No time variations are seen in seasonal, biannual and day–night subdivisions. Note that the seasonal and biannual variations are expected to arise from interactions between solar magnetic field and neutrino magnetic moment, [99] while the day–night effect would be induced by neutrino oscillations in the Earth (see below).

### *3.4. The SAGE experiment*

SAGE is an abbreviation of Soviet–American gallium experiment, which is at present being carried out with 30 t of metallic Ga, with another 30 t to come in the near future.

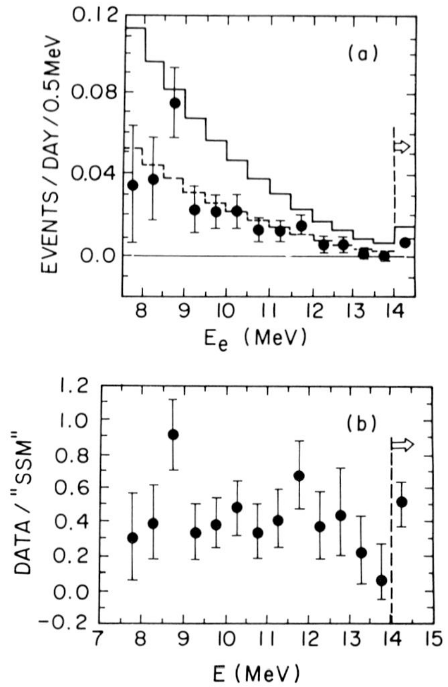


**Figure 22.** (a) Energy spectrum of recoil electrons obtained by the Kamiokande experiment. Electrons are scattered by solar neutrinos from the reaction $\nu_e + e \rightarrow \nu_e + e$. The standard-solar-model (SSM) prediction and 0.46 of the SSM prediction are shown by the full and broken histograms, respectively [10]. (b) The ratio of data/SSM is shown.

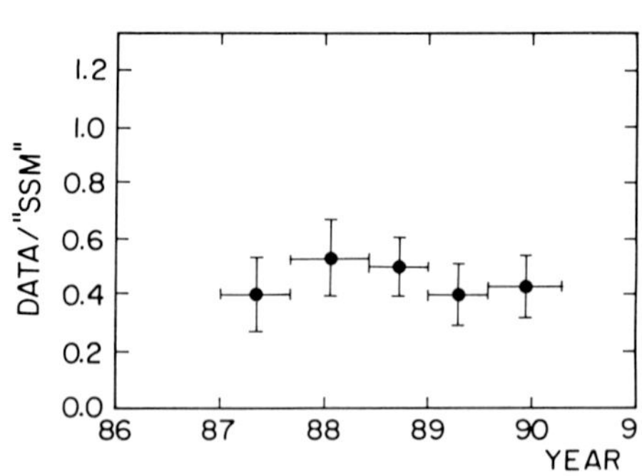


**Figure 23.** The ratio $R_K$ = data/SSM for Kamiokande's five data points, each of 200 day observations [10].

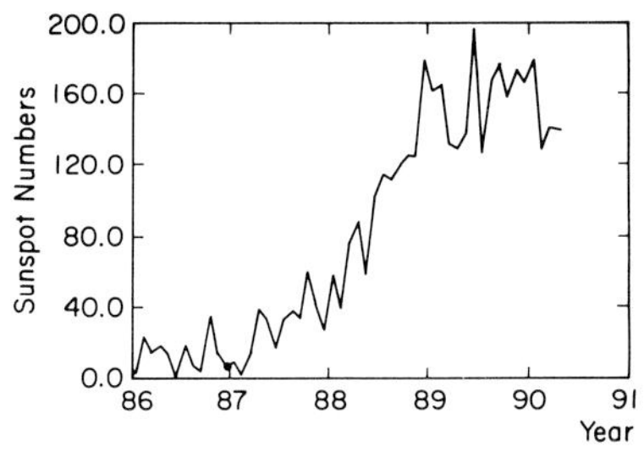


**Figure 24.** Monthly averaged sunspot number in the same period as shown in figure 23 [10].

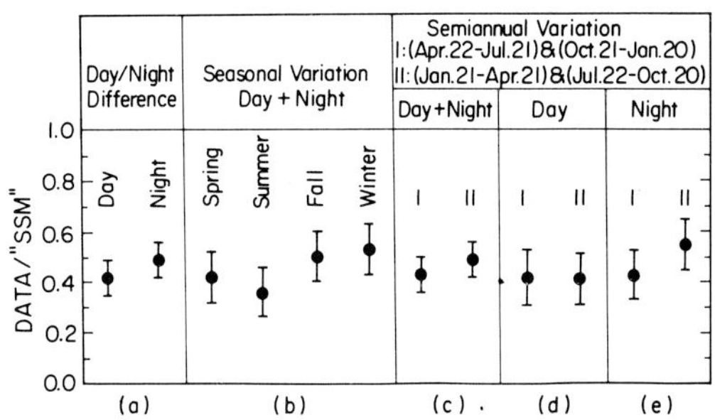


**Figure 25.** $R_K$ = data/SSM of the Kamiokande data for subdivisions of (a) day and night, (b) four seasons, (c) semiannual (period I, 4/22-7/21 and 10/21-1/20; period II, 1/21-4/21 and 7/22/10/20), (d) semiannual and daytime and (e) semiannual and night time. In period I the Kamiokande detector looks at the solar centre through the equator, while in period II off the equator by 4-7°. This division is made to search for a possible effect of magnetic fields $B_{\odot}(B_{\odot} \simeq 0$ near the equator).

The detector is located at the Baksan Neutrino Observatory, Russia. The experiment counts the number of $^{71}$Ge produced by the reaction

$$\nu_e + {}^{71}\text{Ga} \rightarrow e^- + {}^{71}\text{Ge}. \tag{51}$$

$^{71}$Ge goes back to $^{71}$Ga via electron capture with a half life of 11.43 d. The inverse reaction is accompanied by an emission of Auger electrons and x-rays. The unique feature of this reaction consists in the very low threshold energy, $E_{\text{th}} = 0.233$ MeV, which is sufficiently low to detect pp neutrinos ($E_\nu < 0.42$ MeV, see figure 18). The pp neutrino flux can be calculated much more accurately than the $^8$B neutrino flux (table 9), so that the experiment is able to test the SSM most reliably. If the experiment really detects a discrepancy from theory, it unambiguously indicates that neutrinos must be blamed for the solar-neutrino problem.

The cross section of equation (51) unfortunately has a rather large uncertainty, and it induces about a 15% uncertainty ($3\sigma$) to the calculated capture rate [5],

$$Y/N_a = 132^{+20}_{-17}\ \text{SNU}. \tag{52}$$

The breakdown of the various contributions is shown in table 11 [5]. The pp neutrinos dominate over the other sources.

**Table 11.** Capture rates predicted by the standard model for a $^{71}$Ga detector (from [5]).

| Neutrino source | Capture rate (SNU) |
|---|---|
| pp | 70.8 |
| pep | 3.0 |
| hep | 0.06 |
| $^{7}$Be | 34.3 |
| $^{8}$B | 14.0 |
| $^{13}$N | 3.8 |
| $^{15}$O | 6.1 |
| $^{17}$F | 0.06 |
| Total | 132 |

The experimental procedure is more complicated than the Cl experiment, because to extract Ge is not as simple as for Ar. 30 t of metallic Ga after an exposure for about a month are melted and mixed with HCl to form $GeCl_4$. $GeCl_4$ is subsequently converted to $GeH_4$ which is fed into a small proportional counter. The number of $^{71}$Ge atoms is counted by detecting x-rays and the Auger electrons from electron capture.

The SAGE experiment started in January 1990 and results have been presented [21]. The characteristic 11.4 d decay curve of $^{71}$Ge was not seen, and the best fit to the number of $^{71}$Ge for five months of operation yielded zero, which should be compared with the value of 18 expected from equation (52). The $^{71}$Ge counting is still dominated by the ambient background, and the upper limit of the capture rate is

$$Y/N_a < 55 \text{ SNU } (68\% \text{ CL}) \tag{53}$$

$$Y/N_a < 79 \text{ SNU } (90\% \text{ CL}). \tag{54}$$

Thus the experiment is about to confirm the deficit even in pp neutrinos, which, if confirmed, will have an enormous implication for neutrino physics.

A similar Ga experiment called GALLEX is about to start [101]. The two independent experiments will provide reliable results in the near future.

### 3.5. *Solution to the solar-neutrino problem*

We have presented two experimental results that clearly show the deficit of the solar-neutrino flux. They are recapitulated by using the ratio of data/SSM,

$$R_H = 0.27 \pm 0.04 \qquad E_\nu > 0.81 \text{ MeV} \tag{55}$$

$$R_K = 0.46 \pm 0.05 \pm 0.06 \qquad E_\nu > 7.5 \text{ MeV}. \tag{56}$$

The result from Baksan is

$$R_B < 0.42 \qquad 68\% \text{ CL}, \ E_\nu > 0.23 \text{ MeV} \tag{57}$$

$$R_B < 0.60 \qquad 90\% \text{ CL}, \ E_\nu > 0.23 \text{ MeV}. \tag{58}$$

Further results from Baksan are eagerly awaited.

How can one reconcile all these results? Is there something wrong in the rate of energy production of the Sun? Let us discuss this point first [106]. The Kamiokande experiment observes exclusively $^{8}$B neutrinos. The Homestake experiment also observes

$^8$B neutrinos as predominating, but it observes in addition $^7$Be and pep neutrinos, which amount to about one sixth of the $^8$B neutrinos. Since the $^7$Be and pep neutrino flux is calculated more reliably than the $^8$B one, their contribution can be subtracted from the data. Then the $^8$B neutrinos deduced from the Homestake experiment is (table 10),

$$R_{\mathrm{H}}(^8\mathrm{B}) \leqslant \frac{2.1-1.1-0.2}{6.1} = 0.13 \pm 0.05 \tag{59}$$

which should now be compared with Kamiokande value for $R_{\mathrm{K}}$ given by (56). It is obvious that these two results are not compatible with each other. Since we are discussing the same neutrino source (but with different energy ranges), any modification of the solar interior cannot explain both the Homestake and Kamiokande results simultaneously.

Another argument against non-standard solar models is that the $^8$B neutrino flux is much more sensitive to the central temperature ($\propto T^{18}$) than the $^7$Be one ($\propto T^8$). The temperature dependences of $R_{\mathrm{H}}$ and $R_{\mathrm{K}}$ are

$$\frac{\Delta R_{\mathrm{H}}}{R_{\mathrm{H}}} \leqslant \frac{6.1}{7.9}\,\frac{18\Delta T}{T} + \frac{1.8}{7.9}\,\frac{8\Delta T}{T} = \frac{15.7\Delta T}{T} \tag{60}$$

$$\frac{\Delta R_{\mathrm{K}}}{R_{\mathrm{K}}} = \frac{18\Delta T}{T} \tag{61}$$

and obviously

$$|\Delta R_{\mathrm{K}}/R_{\mathrm{K}}| > |\Delta R_{\mathrm{H}}/R_{\mathrm{H}}|.$$

In fact, in order to explain the Homestake result $\Delta R_{\mathrm{H}}/R_{\mathrm{H}} = -0.73 \pm 0.04$, $\Delta T/T$ is required to be $-0.046 \pm 0.003$, which then predicts that $\Delta R_{\mathrm{K}}/R_{\mathrm{K}} = -0.83 \pm 0.05$, namely $R_{\mathrm{K}} = 0.17 \pm 0.05$, in disagreement with observation ($0.46 \pm 0.05 \pm 0.06$).

Therefore, we can conclude that non-standard solar models are not a viable solution to the solar-neutrino problem. We are somewhat relieved by the conclusion that the Sun seems to be working as theory predicts.

We are then left to consider seriously some peculiar neutrino properties. Namely we are forced to consider particle physics beyond the standard model. The standard model of particle physics assumes that all neutrino masses are zero. However, to reduce the solar-neutrino flux inevitably requires finite neutrino masses. Thus the study of solar neutrinos is now treated as a particle physics rather than an astronomy problem, and the Sun is treated as a neutrino generator. This, like supernovae, again indicates a close interplay between astronomy and particle physics.

There are at present three possibilities of solving the solar-neutrino problem, namely neutrino oscillations in solar material with a high but exponentially decreasing density, [11–18] neutrino oscillations in the vacuum of space between the Sun and the Earth (so-called just-so oscillations) [102–104] and a peculiar decay of neutrinos [105].

A naive neutrino decay ($\nu_e \rightarrow$ something) is not allowed, because there is the firm constraint of $\gamma\tau \geqslant 10^5$ a derived from the observation of the $\bar{\nu}_e$ burst from SN1987A, where $\tau$ and $\gamma$ are the lifetime and Lorentz factor of $\bar{\nu}_e$. On the other hand, the solar-neutrino problem imposes $\gamma\tau \leqslant 500$ s ($10^{-5}$ a).

The most promising solution to the solar-neutrino problem the neutrino oscillations. Since the neutrino oscillation is a purely particle physics problem, let us describe it very briefly [107, 12].

Consider two species of neutrinos $\nu_e$ and $\nu_\mu$ which are not eigenfunctions of the mass matrix. Write the mass eigenfunctions as $\nu_1$ and $\nu_2$, that is

$$M\begin{pmatrix}\nu_1\\ \nu_2\end{pmatrix}=\begin{pmatrix}m_1 & 0\\ 0 & m_2\end{pmatrix}\begin{pmatrix}\nu_1\\ \nu_2\end{pmatrix} \tag{62}$$

and

$$\begin{pmatrix}\nu_e\\ \nu_\mu\end{pmatrix}=\begin{pmatrix}\cos\theta & \sin\theta\\ -\sin\theta & \cos\theta\end{pmatrix}\begin{pmatrix}\nu_1\\ \nu_2\end{pmatrix} \tag{63}$$

where $\theta$ is the mixing angle.

Consider low energies. Then

$$H=(p_\nu^2+M^2)^{1/2}+V=p_\nu+\frac{1}{2p}M^2+V. \tag{64}$$

The Schrödinger equation for the wavefunctions $\nu_e$ and $\nu_\mu$ are

$$\mathrm{i}\frac{\mathrm{d}}{\mathrm{d}t}\begin{pmatrix}\nu_e\\ \nu_\mu\end{pmatrix}=2\pi\begin{pmatrix}-\dfrac{\cos 2\theta}{L_v}+\dfrac{1}{L_e} & \dfrac{\sin 2\theta}{2L_v}\\ \dfrac{\sin 2\theta}{2L_v} & 0\end{pmatrix}\begin{pmatrix}\nu_e\\ \nu_\mu\end{pmatrix} \tag{65}$$

where constant terms are subtracted and $V_{\mu e}=0$. $L_v$ and $L_e$ are

$$L_v=\frac{4\pi E_\nu}{\Delta m^2}\left(\frac{\hbar}{c^3}\right) \tag{66}$$

$$L_e=\frac{1}{V_{ee}-V_{\mu\mu}}=\frac{\sqrt{2}\pi}{G_F n_e}(\hbar c). \tag{67}$$

Here we have considered the effective potential acting on $\nu_e$ owing to an extra W boson exchange term in neutral-current $\nu_e$e interactions [12]. $G_F$ and $n_e$ are the Fermi coupling constant and the electron number density, respectively. $L_v$ and $L_e$ can be expressed more explicitly as

$$L_v=2.48\frac{E_\nu(\mathrm{MeV})}{\Delta m^2(\mathrm{eV}^2)}\,m, \tag{68}$$

$$L_e=\frac{\sqrt{2}\pi}{G_F}\frac{\mu_e m_H}{\rho}\simeq\frac{3.3\times 10^7}{\rho(\mathrm{g\ cm^{-3}})}\,m, \tag{69}$$

where $\Delta m^2=m_{\nu_2}^2-m_{\nu_1}^2$, $\rho$ is the usual density and $\mu_e$ the mean molecular weight per electron, that is $\mu_e^{-1}$ is the number of electrons divided by the number of nucleons in the matter.

One can easily solve equation (65) when neutrinos traverse vacuum ($n_e=0$). One obtains the probability of neutrino conversion after traversal of a distance $L$,

$$P_{e\mu}=P_{\mu e}=|\langle\nu_\mu(t)|\nu_e\rangle|^2=\sin^2(2\theta)\sin^2(\pi L/L_v). \tag{70}$$

Suppose that $L=1$ AU (an astronomical unit is the distance from the Sun to the Earth), $L_v\simeq L/2$ and $\theta\simeq\pi/4$. Then $P_{\mu e}\simeq 1$, namely the original $\nu_e$ produced at the solar core are fully converted to $\nu_\mu$. Moreover, since $L_v$ is proportional to $E_\nu$, $P_{\mu e}$ also depends on $E_\nu$.

The neutrino capture reaction of equation (44) does not take place for $\nu_\mu$, and in addition the cross section for $\nu_\mu e\to\nu_\mu e$ is much less than that for $\nu_e e\to\nu_e e$ (cf equation (5)). Thus if the above case is indeed occurring, by choosing a suitable set of parameters

$\theta$ and $\Delta m^2$, one can simultaneously explain the Homestake and Kamiokande results. This is the solution given by the just-so oscillations [103, 104]. A detailed analysis gives tiny allowed regions in the $\Delta m^2 - \sin^2 2\theta$ space as shown in figure 26, [104], namely several small regions with $\Delta m^2 \simeq (0.5\text{–}1) \times 10^{-10}\ \mathrm{eV}^2$ and $\sin^2 2\theta > 0.8$.

Next let us consider neutrino oscillations in the solar material. Putting the calculated value of $n_e(r)$ into equation (67), one can solve equation (65) numerically for various values of $\Delta m^2$ and $\sin^2 2\theta$. Figure 27 is an example, where the survival probability $P_{ee} = 1 - P_{\mu e}$ is shown as a function of $E_\nu/\Delta m^2$ for $\sin^2 2\theta = 0.01$. Surprisingly there is a region in which the $\nu_e$ are almost completely converted to $\nu_\mu$ (more precisely to $\nu_2$), and also there is a wide region of partial conversion. This resonant conversion of $\nu_e$ to $\nu_\mu$ is what Smirnov and Mikheyev had found by elaborating the oscillations in matter developed by Wolfenstein, and it is widely known as the MSW mechanism [11, 12].

Another important aspect of the MSW oscillations is that the $\nu_2$ that have left the Sun traverse the Earth in the night and the MSW mechanism regenerates the $\nu_e$ during the traversal. Since such a regeneration does not take place in the day there is a region in the $\Delta m^2 - \sin^2 2\theta$ space that exhibits different event rates between day and night. This so-called day–night effect was studied at Kamiokande and was not found within the rather large statistical errors [100].

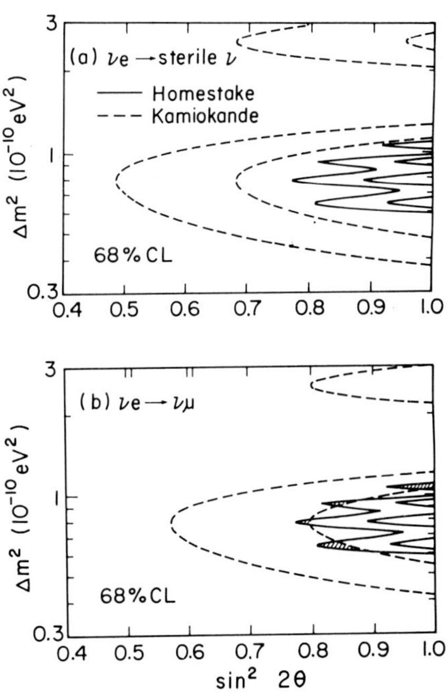


**Figure 26.** One-standard-deviation-allowed regions for the Homestake (full curves) and Kamiokande (broken curves) experiments in the space of the oscillation parameters $\sin^2 2\theta$ and $\Delta m^2$, for (a) $\nu_e \to \nu_x (x \neq \mu, \tau)$, and (b) $\nu_e \to \nu_\mu$. The overlap regions in (b) are denoted by crosshatching (from [104]). Note that the observed energy spectrum (figure 22) was not taken into account to obtain the region for Kamiokande.

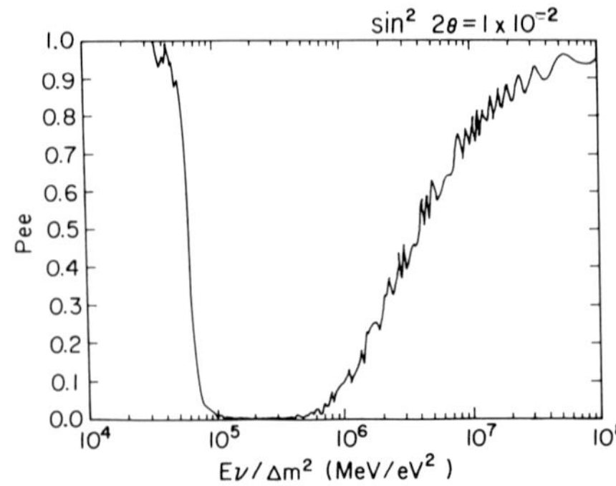


**Figure 27.** The survival probability $P_{ee} = |\langle \nu_e(t)|\nu_e\rangle|^2$ with $\sin^2 2\theta = 10^{-2}$ for solar neutrinos observed at the Earth.

The actual procedure with which to analyse the data is to calculate the event rates for a number of $\Delta m^2 - \sin^2\theta$ sets and to find an allowed region for each experiment, Homestake and Kamiokande. If there is a region allowed for both the experiments, the MSW mechanism gives a decent solution to the solar-neutrino problem. Figure 28 shows the result. What is shown here is the allowed regions with 90% confidence level for Homestake and Kamiokande. Also shown is the excluded region from the non-observation of the day–night effect by Kamiokande. Note that the electron energy distribution observed at Kamiokande (figure 22) was fully taken into account to derive the above results. There are two separated regions that satisfy the Homestake and Kamiokande results;

$$\Delta m^2 = 10^{-6\pm1}\,\mathrm{eV}^2 \qquad \sin^2 2\theta = 4\times10^{-2\pm0.2}(10^{-6}/\Delta m^2) \tag{71}$$

and

$$\Delta m^2 = 10^{-5.5\pm0.5}\,\mathrm{eV}^2 \qquad 10^{-7.5\pm0.5}\,\mathrm{eV}^2 \qquad \sin^2 2\theta = 0.6\pm0.2 \tag{72}$$

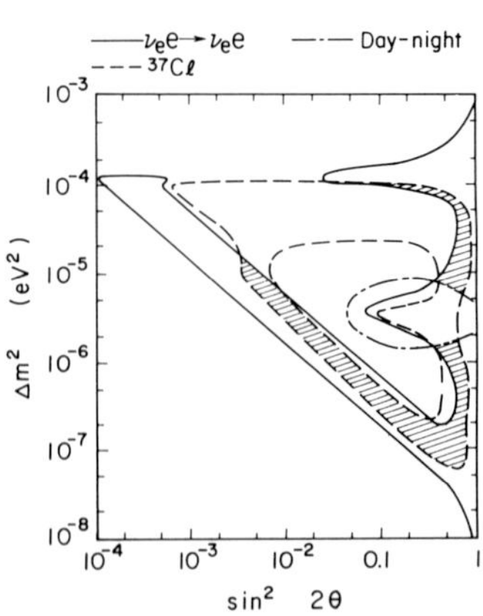


**Figure 28.** The MSW-allowed regions for the Homestake (broken line) and Kamiokande (full-line) experiments. The shaded region is allowed for both Homestake and Kamiokande. The region enclosed by the chain curve is excluded by non-observation of the day–night effect in Kamiokande.

where the allowed regions correspond to the 90% CL and a narrow strip near $\Delta m^2 = 10^{-4}$ eV$^2$ and $\sin^2 2\theta = 0.1$–$0.4$ was neglected.

One important point of caution is that the SSM calculation by Bahcall and Ulrich [5] was used to obtain figure 28, see [108]. However, since the theoretical uncertainty is almost completely correlated between the Homestake and Kamiokande experiments, it should not simply be added to the experimental errors. Instead one shifts the theoretical calculations within the allowed range and carries out the same analysis. Fortunately the shifts within the quoted $3\sigma$ errors do not wipe out the allowed regions and equations (71) and (72) are still viable solutions.

In summary the hypothesis of neutrino oscillations beautifully solves the long-standing solar-neutrino problem. The solution results in a very important implication for particle physics, that is finite neutrino masses. If indeed established, it is the first indication beyond the standard model of particle physics and could open a new window to the grand-unified theories.

There is also another possibility for neutrino decay. As mentioned above, a naive decay scenario does not work. One must elaborate the neutrino decay to evade the SN1987A constraint [105]. Consider the decay scheme.

$$\nu_2 \rightarrow \bar{\nu}_1 \phi \tag{73}$$

where $\phi$ is a scalar (pseudoscalar) boson. On the way from SN1987A to the Earth, a $\bar{\nu}_2$ part of $\bar{\nu}_e$ (proportional to $\sin\theta$) decays. However, at the same time, $\bar{\nu}_e$ is replenished by the decay of $\nu_e$, $\nu_\mu$ and $\nu_\tau$ according to equation (73). One can balance the total $\bar{\nu}_e$ on the Earth by selecting $\theta < 60$ deg [105].

Apply this decay scenario to the solar neutrinos. The survival probability can be written as [105]

$$P_{ee} = \cos^4\theta + \sin^4\theta \exp[-500\text{ s}/\tau(E)] \tag{74}$$

where 500 s is the flight time of a neutrino from the Sun to the Earth, and $\tau(E) = \gamma\tau$. The following values of $\theta$ and $\tau$ satisfy the observational constraints for the Homestake and Kamiokande detectors [103, 109]

$$\theta > 38\text{ deg}, \qquad \tau(10\text{ MeV}) < 1500\text{ s} \qquad (90\%\text{ CL}). \tag{75}$$

Thus the mixing angle is severely constrained to the range

$$38 < \theta < 60\text{ deg}. \tag{76}$$

The MSW mechanism is not irrelevant here, because the mass $m_2$ is constrained as $0.01 \leqslant m_2 \leqslant 0.1$ eV in order to have a sufficiently short lifetime, which is given by

$$\tau = 16\pi / g^2_{\bar{\nu}_1\nu_2\phi} m_2$$

where $g^2_{\bar{\nu}_1\nu_2\phi}$ is a coupling constant for the $\bar{\nu}_1 - \nu_2 - \phi$ vertex.

We have not discussed the anticorrelation of the Homestake data with the sunspot number. The issue is still controversial, and we must wait for the next few years for additional data.

### 3.6. *Future prospects*

The solar-neutrino problem is now a central issue of particle physics, and more data are eagerly awaited. SAGE in Baksan [21] and GALLEX in Gran Sasso [101] will soon provide important data. The neutrino oscillation hypothesis has definite predictions for these Ga experiments. Table 12 shows the expected capture rate predicted from the allowed regions shown in figure 26 [104]. Ga experiments should give a value of the deficit as large as the result at Kamiokande $R_{Ga} \simeq 0.5$. Figure 29 shows the capture rate in units of SNU expected from the MSW solution. Comparing figure 28 and 29, Ga experiments could still see the capture rate of 5–100 SNU. The SAGE upper limit (55 SNU at 68% CL) [21] already severely constrains the allowed region of Homestake and Kamiokande, favouring equation (71), that is only the non-adiabatic region.

The year 1996 will be important for solar-neutrino experiments. SNO, already funded in 1990, [62, 75] and Super-Kamiokande, funded in 1991, [73, 74] will commence full operation in 1996. Both experiments will observe $^8$B neutrinos with threshold energies of $E_{th} \simeq 6$ and 5 MeV, respectively. The relevant reactions are

$$\nu_e d \rightarrow e^- pp \tag{77}$$

for SNO and

$$\nu_e e \rightarrow \nu_e e \tag{78}$$

for Super-Kamiokande.

Both experiments have advantages and disadvantages.

**Table 12.** Predictions of $R$ = data/SSM for the best just-so $\nu_e \rightarrow \nu_\mu$ oscillation solutions (from [104]).

| $\Delta m^2$ ($10^{-10}$ eV$^2$) | $\sin^2 2\theta$ | $R_{Ga}$ |
|---|---|---|
| 0.50–0.52 | 1.00 | 0.37–0.42 |
| 0.63–0.66 | 0.83 | 0.45–0.48 |
| 0.77–0.80 | 0.79 | 0.48–0.50 |
| 0.91–0.93 | 0.83 | 0.47–0.51 |
| 1.04–1.08 | 0.95 | 0.40–0.47 |
| 1.06–1.09 | 1.00 | 0.37–0.41 |

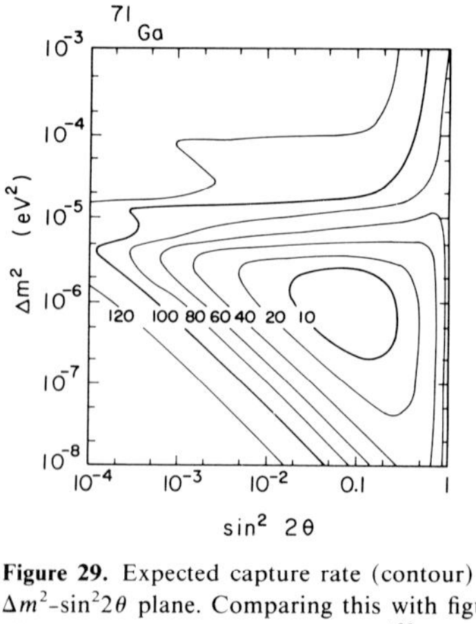


**Figure 29.** Expected capture rate (contour) for $^{71}$Ga experiment in units of SNU on the $\Delta m^2$-$\sin^2 2\theta$ plane. Comparing this with figure 28, the capture rate of 5–100 SNU is still allowed. The SSM prediction is $132^{+20}_{-17}$ SNU.

(i) Recoil electrons from equation (77) directly provide $\nu_e$ energies, while electron energies from equation (78) are severely smeared out owing to motion of the center-of-mass system.

(ii) Electrons from equation (78) have a sharp forward peak while equation (77) has only a mild enhancement in the backward direction,

$$\frac{1}{\sigma}\frac{d\sigma}{d\Omega} = 1 - \tfrac{1}{3}\cos\theta_{Sun}. \tag{79}$$

As mentioned above the forward angular peak improves the signal-to-noise ratio by more than an order of magnitude. Thus the background condition of SNO is more severe by this magnitude than that of Super-Kamiokande. This fact reflects the threshold energies.

(iii) Equation (77) has a larger cross section than equation (78) (cf equations (3) and (4)). The detector volume of SNO can thus be smaller by the cross-section factor than that of Super-Kamiokande. Some parameters of the two detectors were summarized in table 3.

The expected event rates are about 8000 and 2000 $a^{-1}$ for Super-Kamiokande and SNO, respectively. A large number of accumulated events enables one to measure the shape of the electron energy distribution accurately. Since the $^8B$ decay is accurately known, a distortion of the spectral shape, independent of the absolute flux value, presents clear evidence for new properties of neutrinos.

Figure 30 shows the ratio $R_{SN}$ = data/SSM for Super-Kamiokande as a function of electron energy for $(\Delta m^2, \sin^2 2\theta) = (3\times10^{-6}\,eV^2, 0.01)$, $(4\times10^{-7}\,eV^2, 0.1)$, $(1\times10^{-7}\,eV^2, 0.5)$ in the non-adiabatic region and $(5\times10^{-5}\,eV^2, 0.7)$ in the quasivacuum region (see figure 28), denoted as (1), (2), (3) and (4), respectively. The error bars attached represent statistical errors expected from operation of Super-Kamiokande for three years. One immediately sees a deviation from expectation for (1), (2) and (3) (deviation from the horizontal line). Moreover, (2) and (3) can be separated. (4) is indistinguishable from the expected shape, but the parameter set $(\Delta m^2, \sin^2 2\theta)$ of (4)

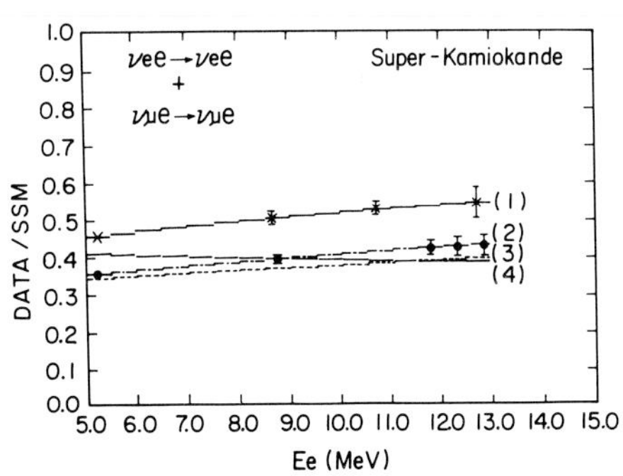


**Figure 30.** The expected $R_{SK}$ = data/SSM for the Super-Kamiokande detector. The parameters $(\Delta m^2(eV^2), \sin^2 2\theta)$ are (1) $(3\times10^{-6}, 0.01)$, (2) $(4\times10^{-7}, 0.1)$, (3) $(1\times10^{-7}, 0.5)$ and (4) $(5\times10^{-5}, 0.7)$. Typical error bars are shown for 0.5 MeV bins and for 3-year operation of the Super-Kamiokande detector. Note that $R_{SK}$ contains the contribution from $\nu_\mu + e \to \nu_\mu + e$, the cross section of which is about a sixth of $\nu_e + e \to \nu_e + e$. The energy resolution of the detector was taken into account.

induces a strong day–night effect and can be easily distinguished. (1) and (2) cannot be separated by the energy spectrum alone, but again the parameter set of (1) exhibits a day–night effect, which is small but still definitely observable in the Super-Kamiokande detector. Therefore, combining the electron energy spectra both in the day time and in the night time, Super-Kamiokande will be able to identify the MSW oscillations uniquely. SNO will observe a larger spectral distortion as shown in figure 31 and can carry out a similar analysis.

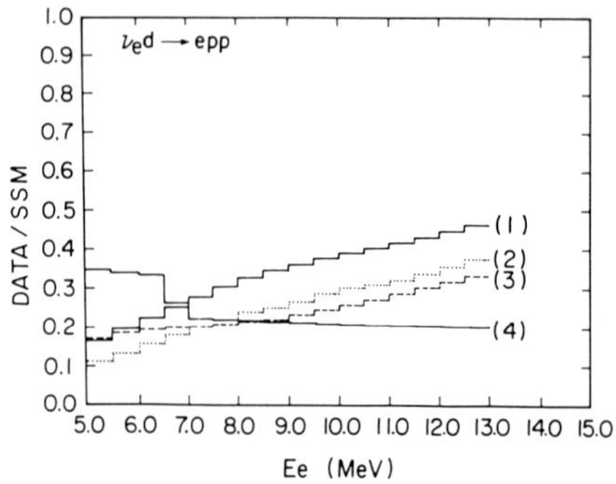


**Figure 31.** $R_{SNO}$ = data/SSM for the SNO detector. The parameters are the same as those in figure 30. The energy represents $\nu_e$ energies and the detector resolution was not taken into account. Only $\nu_e$ are considered because the relevant reaction is $\nu_e d \rightarrow e^- pp$.

The just-so oscillations also distort the spectral shape, but the distortion alone does not completely identify the just-so oscillations from the MSW solutions. The eccentricity of the orbit of the Earth induces an sinusoidal time variation (max. 7%) of the event rate (seasonal effect). The just-so oscillations affect the magnitude of this seasonal variation by a similar factor. Thus combining this with the spectral shape, Super-Kamiokande can distinguish the just-so oscillations from the MSW ones.

We believe that this type of precise measurement of electron energy spectra, together with day–night and seasonal variations, will completely solve the neutrino problem.

SNO and a new project called BOREX in Gran Sasso [110, 111] try to measure $\nu_\mu$ components by detecting the neutral-current reactions,

$$\nu_\mu d \rightarrow \nu_\mu pn \qquad nCl \rightarrow Cl\gamma(\text{observed}) \tag{80}$$

for SNO and

$$\nu_\mu B \rightarrow \nu_\mu B^* \qquad B^* \rightarrow B\gamma(\text{observed}) \tag{81}$$

for BOREX. Note that both experiments do not measure the $\nu_\mu$ energy but count only the number of events above a certain energy threshold. Moreover, the experiments will encounter enormous backgrounds and the feasibility is still unclear. Although observation of neutral-current events undoubtedly shows the existence of neutrino oscillations, the accurate measurement of spectral shape plus time variation seems to be sufficient to solve the solar-neutrino problem.

How long does it take to solve the solar-neutrino problem? Probably ten years. Is this then the end of the solar-neutrino astronomy? The answer is absolutely no and in fact it is the true beginning of solar-neutrino astronomy. With a precise knowledge of the neutrino properties one can now measure the true flux of solar neutrinos that confronts the solar model. Even more important is a long-term monitor of the solar-neutrino flux. Remember that Super-Kamiokande is able to measure a 1% variation in the central temperature of the Sun every day. Continuous observation of the solar-neutrino flux, perhaps for centuries, will test the possible time variation in the energy output of the Sun, which obviously is of the utmost importance for the very existence of life on the Earth.

## 4. Atmospheric neutrinos

Atmospheric neutrinos are terrestrial products and are a big obstacle to looking beyond the atmosphere. They have been observed in large underground detectors and have been analysed in order to see elusive proton decay signals. Observed events were mostly at energies below a few GeV, and separated events for $\nu_e + \bar{\nu}_e$ and $\nu_\mu + \bar{\nu}_\mu$ are available. The total number of events are about 1000 for the NUSEX [112], Frejus [113], IMB [114, 115] and Kamiokande [116, 117] experiments all combined.

Kamiokande found a deficit of $\nu_\mu + \bar{\nu}_\mu$ compared with the theoretical expectation [116, 117]. Note that the ratio $(\nu_\mu + \bar{\nu}_\mu)/(\nu_e + \bar{\nu}_e)$ is insensitive to the primary cosmic-ray spectrum and can be calculated with an accuracy better than 5% [117]. Kamiokande separated electron-like events from muon-like ones by looking at Cherenkov rings in detail. They also counted the number of $\mu$-decays. The result is summarized in the following way.

$$R_{\mu e} = \frac{(\mu - \text{like}/\text{e} - \text{like})_{\text{data}}}{(\mu - \text{like}/\text{e} - \text{like})_{\text{MC}}} = 0.61^{+0.09}_{-0.08} \pm 0.05 \tag{82}$$

and

$$R_{\text{decay}} = \frac{(\mu - \text{decay}/\text{no } \mu - \text{decay})_{\text{data}}}{(\mu - \text{decay}/\text{no } \mu - \text{decay})_{\text{MC}}} = 0.64^{+0.01}_{-0.09}, \tag{83}$$

based on 3.43 kt a of exposure and for $p < 1.2\ \text{GeV c}^{-1}$.

The parentheses in equation (82) are the ratio of events identified as muon-like $(\mu + \pi)$ to those as electron-like $(\text{e} + \gamma)$. The subscripts 'data' and 'MC' are the ratios for the observed data and for the expected ratios estimated by the Monte Carlo method, respectively. The parentheses in equation (83) are the ratio of events with observed $\mu$-decay signals to those without $\mu$-decay signals. These two ratios $R_{\mu e}$ and $R_{\text{decay}}$ naturally have a similar meaning.

The report from Kamiokande prompted the Frejus group to analyse their data in detail. Their recently published result gives a value of [113]

$$R_{\mu e} = 1.06^{+0.19}_{-0.16} \pm 0.08 \tag{84}$$

which is consistent with theory and is about $2\sigma$ away from the Kamiokande results. The NUSEX group also published the result [112],

$$R_{\mu e} = 0.99^{+0.35}_{-0.25} \tag{85}$$

again consistent with theory, but their statistics are too low for testing the Kamiokande result.

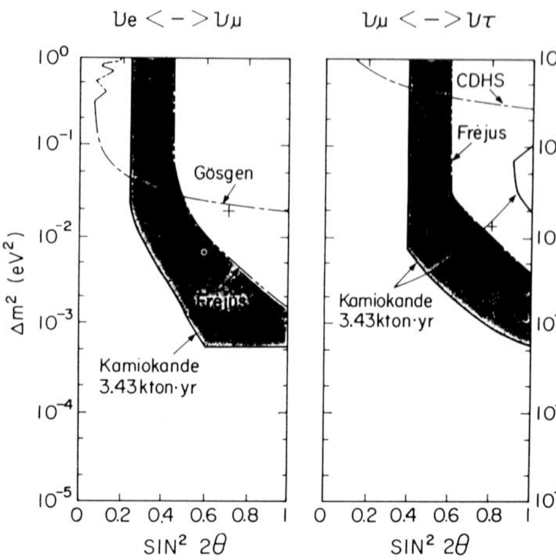


**Figure 32.** The allowed region from the observed atmospheric $\nu_\mu + \bar{\nu}_\mu$ and $\nu_e + \bar{\nu}_e$ in the Kamiokande detector, together with those regions excluded by other experiments. The shaded region is still allowed [74]. In comparison with figure 28, the oscillations $\nu_\mu \to \nu_\tau$ are favoured.

Very recently the IMB group reported their result based on 3.4 kt $a^{-1}$ of exposure. They followed the procedure developed by Kamiokande. They found,

$$R_{\mu e} = 0.67 \pm 0.09 \pm 0.08 \tag{86}$$

and

$$R_{\text{decay}} = 0.77 \pm 0.10 \tag{87}$$

which are quite consistent with the Kamiokande results, although the significance level is only about $2\sigma$.

The situation is controversial and more data are clearly needed. More importantly, the procedure for separating electrons from muons must be evaluated experimentally with accelerator beams. The Kamiokande group is planning: (i) to test the muon identification by using cosmic-ray muons that impinge and stop in the detector, and (ii) to evaluate the detector characteristics by accelerator $e/\mu$ beams.

The $\nu_\mu + \bar{\nu}_\mu$ deficit, if indeed true, can only be explained by neutrino oscillations. Figure 32 shows, in the $\Delta m^2$-$\sin^2 2\theta$ space, the region allowed by Kamiokande and the region excluded by Frejus. The shaded region is still allowed. If this is combined with the result on the solar-neutrino deficit, the $\nu_\mu \to \nu_e$ oscillations are ruled out for the explanation of the $\nu_\mu + \bar{\nu}_\mu$ deficit (see figures 28 and 32). Thus one is led to the $\nu_\mu \to \nu_\tau$ oscillations with $\Delta m^2 = 5 \times 10^{-4} \sim 0.5\ \text{eV}^2$ and $\sin^2 2\theta = 0.4 - 1$. This is still speculative and much more study is mandatory.

Atmospheric neutrinos have been considered to be a nuisance to extraterrestrial neutrinos, but they are now emerging as a clue to new physics.

## 5. Ultrahigh-energy (UHE) neutrinos

As discussed in section 1, the main purpose of UHE neutrino astronomy is divided into two; (i) to study the origin of UHE cosmic rays ($E \simeq 10^4 \sim 10^7$ GeV and point sources), and (ii) to search for GUT-related UHE neutrinos ($E > 10^9$ GeV and isotropic). This field is still at its infancy, and there is no existing detectors adequate to observe UHE neutrinos. Their estimate fluxes have a large uncertainty, and are extremely small. The interaction cross section of neutrinos rises in proportion to $E_\nu$ and this fact somewhat eases the formidable difficulties in the construction of large-scale detectors. However, the rise of the cross section is eventually weakened owing to the diminishing role of the W-boson mass in the propagator term. This occurs at about $10^4$ GeV, and above this energy the detector size must grow more rapidly as the flux decreases. Moreover, above $10^6$ GeV the cross section is large enough that the mean-free path of neutrinos in the Earth becomes shorter than the diameter of the Earth. Thus the Earth becomes opaque and the detection is only possible for down-going UHE neutrinos, thus significantly reducing the useful solid angle of the detector.

Despite these difficulties the physics underlying UHE neutrinos is so important that every effort must be made to realize the observation of UHE neutrinos.

### *5.1. Interaction cross sections*

The cross section of the reaction $\nu_\mu + N \rightarrow \mu^- +$anything has been calculated by taking into account the QCD corrections for the quark distribution in a nucleon [118, 119]. At ultrahigh energies cross sections for $\nu_e N \rightarrow e^- X$ and $\nu_\tau N \rightarrow \tau^- X$ have the same value as for $\nu_\mu N \rightarrow \mu^- X$. The cross section is expressed in terms of the scaling variables $x = Q^2/2M\nu$ and $y = \nu/E_\nu$ as

$$\frac{d^2\sigma}{dx dy} = \frac{2G_F^2 M E_\nu}{\pi}\left(\frac{M_W^2}{Q^2 + M_W^2}\right)^2 [xq(x, Q^2) + x(1-y)^2 \bar{q}(x, Q^2)] \qquad (88)$$

where $Q^2 = -(p_\nu - p_\mu)^2$, $\nu = E_\nu - E_\mu$, and $M$ and $M_W$ are the nucleon and intermediate-boson masses, respectively. The target $N \equiv (n+p)/2$ is an isoscalar nucleon, which is a good approximation for many materials. The quark distribution functions $q(x, Q^2)$ and $\bar{q}(x, Q^2)$ have been calculated in detail from the observed structure functions and their QCD evolution (scale breaking, i.e. $q$, $\bar{q}$ depend on $Q^2$) [120]. The cross section for the reaction $\bar{\nu}_\mu + N \rightarrow \mu^+ +$anything is obtained by simply interchanging $q(x, Q^2)$ and $\bar{q}(x, Q^2)$ in equation (88). The functions $q(x, Q^2)$ and $\bar{q}(x, Q^2)$ are the sums of contributions from various quarks,

$$q(x, Q^2) = d_v(x, Q^2) + d_s(x, Q^2) + s_s(x, Q^2) + b_s(x, Q^2) \qquad (89)$$

$$\bar{q}(x, Q^2) = u_s(x, Q^2) + c_s(x, Q^2) + t_s(x, Q^2) \qquad (90)$$

where the subscripts v and s label valence and sea contributions, and u, d, c, s, t and b denote quark flavours. Here the equality $\bar{u}_s(x, Q^2) = u_s(x, Q^2)$ etc has been used. At ultrahigh energies the contribution from sea quarks dominates over that from valence quarks, and $q(x, Q^2) \simeq \bar{q}(x, Q^2)$. Thus $\sigma(\nu N) \simeq \sigma(\bar{\nu} N)$ at ultrahigh energies. Figure 33 [118] shows the calculated cross sections from 10 to $10^{10}$ GeV. They are seen to deviate from a simple $E_\nu$ dependence and to have a less steep dependence on $E_\nu$. Figure 34 [119] shows the mean-free path of neutrinos in the Earth. The Earth becomes opaque above $10^6$ GeV as mentioned above. The cross section is about $10^{-33}$ cm$^2$ at $E_\nu = 10^6$ GeV and $3.5 \times 10^{-32}$ cm at $E_\nu = 10^{11}$ GeV. In addition,

$$\sigma(E_\nu) = 10^{-33}\left(\frac{E_\nu}{10^6 \text{ GeV}}\right)^{0.3} \text{cm}^2 \qquad (91)$$

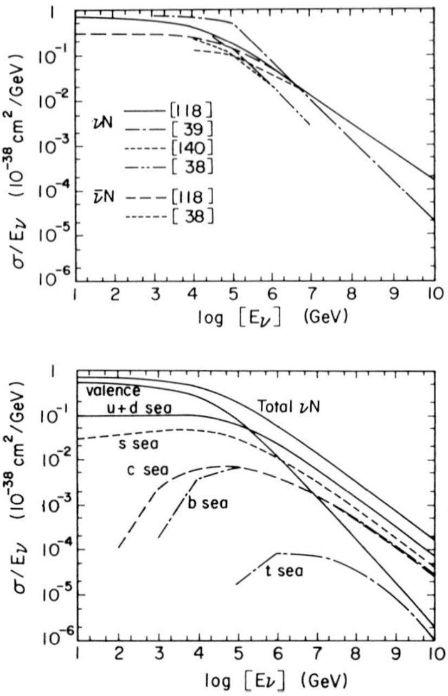


**Figure 33.** (a) The cross sections for $\nu_\mu N \rightarrow \mu^- X$ and $\bar{\nu}_\mu N \rightarrow \mu^+ X$, compared with other calculations. (b) The contributions of various quarks in the cross section of $\nu_\mu N \rightarrow \mu^- X$ (from [118]).

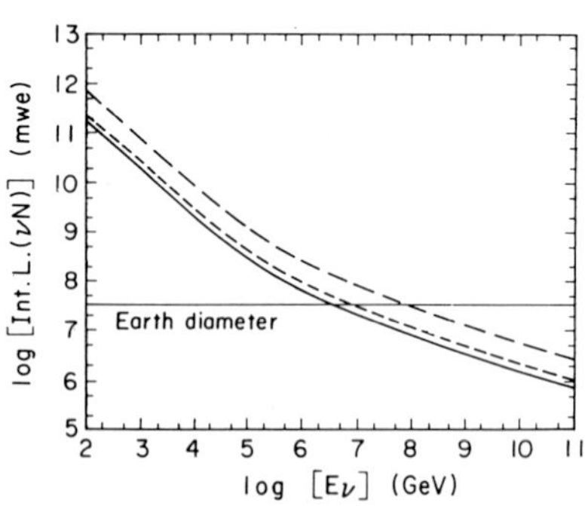


**Figure 34.** Interaction lengths in the Earth for neutrinos. Dotted line, charged-current interaction length; broken line, neutral-current interaction length; full line, total interaction length. The heavy horizontal line indicates a length of one Earth diameter (from [119]).

in the energy range above $10^5$ GeV. Neutrinos with $10^{11}$ GeV have a probability of $2\times10^{-5}$ (vertical incidence) with which they interact with air nuclei in the atmosphere.

### *5.2. Detection of up-going* UHE *muons*

Consider up-going $\nu_\mu$. They interact in the rock and produce muons, which can be detected with large-area underground detectors. The event rate in a detector of the cross-sectional area $A$ is formulated in the following way [39, 119]. The probability that a neutrino of energy $E_\nu$(TeV) interacts in the rock outside the detector and produces a muon that passes through the detector is

$$P_\mu(E_\nu)=\int_{E_{\min}}^{E_\nu} n\sigma_\nu(E_\nu)L(E_\mu)f(E_\mu)\,\mathrm{d}E_\mu \tag{92}$$

where $f(E_\mu)\,\mathrm{d}E_\mu$ is the probability that the muon produced has an energy between $E_\mu$ and $E_\mu+\mathrm{d}E_\mu$, $L(E_\mu)$ is the range of the muon of energy $E_\mu$, and $n$ is the number density of target nucleons. $E_{\min}$ is the minimum $E_\mu$ required to observe a muon in the detector. The event rate in the detector is

$$\Gamma_\mu=A\int P_\mu(E_\nu)j_\nu(E_\nu)S(E_\nu)\,\mathrm{d}E_\nu \tag{93}$$

where $j_\nu(E_\nu)$ is the neutrino flux (cm$^{-2}$ s$^{-1}$ TeV$^{-1}$) from a point source, and $S(E_\nu)$ is the shadowing factor or the opaqueness of the Earth,

$$S(E_\nu)=\exp[-l\sigma_\nu(E_\nu)N_\mathrm{A}] \tag{94}$$

where $l$ is the distance travelled through the Earth, and $N_\mathrm{A}$ is Avogadro's number.

Assuming the power law spectrum predicted for Cyg X-3 (see figure 3), and neglecting the shadowing effect $S=1$,

$$j_\nu(E_\nu)=4\times10^{-10}\left(\frac{E_\nu}{\mathrm{TeV}}\right)^{-2.1}\ \mathrm{cm}^{-2}\,\mathrm{s}^{-1}\,\mathrm{TeV}^{-1} \tag{95}$$

the event rate is [39]

$$\Gamma_\mu\simeq10^{-7}\left(\frac{A}{10^3\ \mathrm{m}^2}\right)\mathrm{s}^{-1}=3\left(\frac{A}{10^3\ \mathrm{m}^2}\right)\mathrm{a}^{-1} \tag{96}$$

which is detectable, for example, in the MACRO detector [121, 122].

So far IMB ($A=400$ m$^2$) [123], Kamiokande ($A=150$ m$^2$) [124] and several other underground detectors [125] have searched for up-going muons and have found no point sources above an isotropic background of atmospheric neutrinos. These detectors are too small considering the expected flux of equation (96).

There are several proposals to construct detectors with useful areas as large as $10^5$ m$^2$, such as DUMAND [126], LENA [127] and GRANDE [128]. They have $E_{\min}>10$ GeV, with which the background of low-energy atmospheric neutrinos can be significantly reduced. However, one must be careful about the limit inherent to this method of detection, namely it is in general not possible to measure muon energies $E_\mu$. The information that one can obtain is the direction of a point source, the integrated flux above $E_{\min}$ and the possible time variation of the flux. In this respect it is essential to have a UHE $\gamma$-detection system at the same time and to measure the energy spectrum of UHE $\gamma$ together with UHE $\nu_\mu$.

### 5.3. Detection of UHE neutrinos with $E_\nu > 10^6$ GeV

As mentioned above, the Earth is opaque at these energies and up-going muons are not expected. Down-coming neutrinos must be detected either by observing air showers initiated by UHE electrons from $\nu_e N \to eX$, or by observing a large energy deposit in an underground detector, again from $\nu_e N \to eX$. Electrons essentially carry most of the $\nu_e$ energies at UHE, thus the energy spectrum of incoming neutrinos is directly measurable. Since point sources are not expected at $E_\nu > 10^6$ GeV, neutrinos are of a cosmological origin and arrive isotropically. Therefore it is not necessary to deploy a number of detectors to cover the entire sky.

The Utah Fly's Eye group has searched for deeply-penetrating particles in the cosmic-ray flux for $E > 10^8$ GeV [129]. The strategy is to search for extensive air showers (EAS) with a zenith angle of more than 80°. These EAS must have traversed more than 3000 g cm$^{-2}$ of atmosphere before interacting. Since the interaction length of protons in the atmosphere is about 45 g cm$^{-2}$, one expects that these EAS are only generated by weakly interacting particles. The experiment failed to find penetrating EAS, and the resultant upper limits on the quantity

$$I(E_{\min}) = \int_{E_{\min}} \sigma_\nu j_\nu(E_\nu)\, dE_\nu \tag{97}$$

were $<10^{-45}$, $<3.8\times10^{-46}$, $<10^{-46}$ and $<3.8\times10^{-47}$ (s$^{-1}$ sr$^{-1}$) for $E_{\min} = 10^8, 10^9, 10^{10}$ and $10^{11}$ GeV, respectively [130]. Assuming $j_\nu = j_0 E_\nu^{-3}$, and using equation (91), one obtains the flux limit at each energy, which is shown as a broken line in figure 5. The limit is already sufficiently strong to rule out a part of the parameter region for superconducting cosmic strings advocated by Hill *et al* [52]. Note that the parameter $p$ was introduced to specify the evolution of the cosmic magnetic field as

$$B(z) = (1+z)^{-p+3/2} B_0 \tag{98}$$

where $B_0 \approx 10^{-9}$ G and $z$ is the red-shift factor. $p = -\frac{1}{2}$ corresponds to a primordial magnetic field energy density that scales as radiation density and $p > -\frac{1}{2}$ would correspond to the fields generated by galactic dynamos in recent epochs [52].

The method adopted by the Utah Fly's Eye group is potentially interesting and even larger scale detectors are certainly desirable.

Let us next consider direct observation of UHE $\nu_e$ in a huge underwater detector. The event rate is

$$\Gamma = \int_{E_{\min}} dE_\nu\, d\Omega\, j_\nu(E_\nu)\sigma_\nu(E_\nu)S(E_\nu, \Omega)\, N_A V \tag{99}$$

where $\Omega$ is the solid angle, and $V$ is the detector volume. $S = 1$ for down-going $\nu_e$. Assuming

$$j_\nu = j_0 \left(\frac{E_\nu}{10^6\ \text{GeV}}\right)^{-\gamma} \text{cm}^{-2}\,\text{s}^{-1}\,\text{sr}^{-1}\,\text{GeV}^{-1} \tag{100}$$

and using equation (91),

$$\Gamma = \frac{3.8\times10^{12}}{\gamma - 1.3}\left(\frac{E_{\min}}{10^6\ \text{GeV}}\right)^{-\gamma+1.3} j_0 \left(\frac{V}{\text{km}^3}\right) \quad (\text{s}^{-1}). \tag{101}$$

The detection limit could be $\Gamma \simeq 1\ \text{a}^{-1}$. Then assuming $\gamma = 3$ as above, and $E_{\min} = 10^8$ GeV,

$$j_0 V = 3.6\times10^{-17}\ \text{cm}^{-2}\,\text{s}^{-1}\,\text{sr}^{-1}\,\text{GeV}^{-1}\,(\text{km})^3. \tag{102}$$

Thus the limit $E_\nu^3 j_\nu$ at $10^8$ GeV is $3.6\times10^{19}(V/\mathrm{km}^3)^{-1}$ eV$^2$ m$^{-2}$ sr$^{-1}$ s$^{-1}$, which improves the Fly's Eye limit by a factor of $10^6$ (see figure 5). Note that a volume of 1 km$^3$ could be realized by the extended DUMAND detector, and even 100 km$^3$ might be possible with an acoustic detection technique.

It is stressed again that the real existence of neutrinos with $E_\nu > 10^{11}$ GeV will give us the possibility to detect 1.9 K cosmic background neutrinos (CBN) [22].

## 6. Conclusions

We have reviewed the achievements and future prospects of neutrino astronomy. Two kinds of extraterrestrial neutrinos, supernova neutrinos and solar neutrinos, have been successfully observed and their measured fluxes, as well as their energy spectra, shed a new light on astronomy and particle physics. However, the event rates currently observed by existing detectors are too low to draw definite conclusions on many aspects. In this respect it is welcome that the two large underground detectors, SNO and Super-Kamiokande have been funded and are expected to commence full operations in 1996. SNO has 1000 t of $D_2O$ as a useful volume and Super-Kamiokande has 22000 t of $H_2O$. They are complementary and will provide independent results about supernova neutrinos as well as solar neutrinos.

Ultrahigh-energy neutrinos with $E_\nu > 10^4$ GeV are awaiting future investigations. They will provide valuable information about the origin of cosmic rays and more importantly about the direct observation of GUT-related phenomena that are totally inaccessible by other means.

We have not discussed high-energy neutrinos possibly coming from the Sun and the Earth [130–135]. If dark matter consists of yet-unknown weakly interacting massive particles (WIMP), they are gravitationally attracted and eventually trapped by the Sun. If the number density of the WIMP becomes large, they start to annihilate each other and to emit neutrinos as an end product.

Monopoles, if they exist, are also trapped by the Sun, and induce proton decay via the Rubakov effect [136, 137] producing $\nu_e$ with energies of about 35 MeV from the decay chain [138, 139],

$$\mathrm{p}\rightarrow\pi^+\mathrm{X}\rightarrow\mu^+\nu_\mu\mathrm{X}\rightarrow\mathrm{e}^+\nu_e\bar{\nu}_\mu\nu_\mu\mathrm{X}. \quad (103)$$

These neutrinos have been searched for, so far with negative results. The next generation of underground detectors will continue the search.

Neutrino astronomy has just begun. The energy of interest ranges from $10^{-4}$ to $10^{24}$ eV. Many ingenious techniques must be developed to cover this whole energy range. It is almost certain that these efforts will be rewarded by completely surprising observations.

## Acknowledgments

The author thanks his colleagues in the Institute for Cosmic Ray Research, University of Tokyo for useful discussions. He especially thanks Jiro Arafune, Motohiko Nagano, Yozo Ohashi and Takaaki Kajita for many comments. He is also grateful to Yoshiyuki

Fukuda and Keiko S Hirata who spent some time calculating the MSW-modified event rates by solar neutrinos and the HECBN yield.

## References


[1] Hirata K *et al* 1987 *Phys. Rev. Lett.* **58** 1490
[2] Bionta R M *et al* 1987 *Phys. Rev. Lett.* **58** 1494
[3] Aglietta M *et al* 1987 *Europhys. Lett.* **3** 1315
[4] Alexeyev E N, Alexeyeva L N, Krivosheina I V and Volchenco V I 1988 *Phys. Lett.* **205B** 209
[5] Bahcall J N 1989 *Neutrino Astronomy* (Cambridge: Cambridge University Press)
[6] Davis R Jr 1987 *Proc. 7th Workshop on Grand Unification/ICOBAN'86* ed J Arafune (Singapore: World Scientific) p 237
[7] Davis R Jr 1989 *Proc. 13th Conf. on Neutrino Physics and Astrophysics, 'NEUTRINO'88'* ed J Schneps *et al* (Singapore: World Scientific) p 518
[8] Lande K *Proc. 14th Int. Conf. on Neutrino Physics and Astrophysics, 'NEUTRINO'90'* ed K Winter, to be published
[9] Hirata K S *et al* 1989 *Phys. Rev. Lett.* **63** 16
[10] Hirata K S *et al* 1990 *Phys. Rev. Lett.* **65** 1297
[11] Mikheyev S P and Smirnov A 1986 *Nuovo Cimimento* **C 9** 17
[12] Wolfenstein L 1978 *Phys. Rev.* D **17** 2369
[13] Hirata K S *et al* 1990 *Phys. Rev. Lett.* **65** 1301
[14] Rosen S P and Gelb J 1986 *Phys. Rev.* D **34** 969
[15] Parke S 1986 *Phys. Rev. Lett.* **57** 1275
[16] Baltz A and Weneser J 1988 *Phys. Rev.* D **37** 3364
[17] Bahcall J, Rosen S P and Gelb J 1987 *Phys. Rev.* D **35** 2976
[18] Kuo T K and Pantaleone J 1989 *Rev. Mod. Phys.* **61** 937
[19] Steigman G 1979 *Ann. Rev. Nucl. Part. Sci.* **29** 313
[20] Loab A and Starkman G D 1990 *IASSNS-AST 90/10*
[21] Abazov A I *et al* 1991 *Phys. Rev. Lett.* **67** 3332
[22] Weiler T 1982 *Phys. Rev. Lett.* **49** 234; 1984 *Astrophys. J.* **285** 495
[23] Krauss L M, Glashow S L and Schramm D N 1984 *Nature* **310** 191
[24] Woosley S E, Wilson J R and Mayle R 1986 *Astrophys. J.* **302** 19
[25] Schramm D N, Mayle R and Wilson J R 1985 *Enrico Fermi Institute* preprint no 85-47
[26] Dar A 1985 *Phys. Rev. Lett.* **55** 1422
[27] Lagage P O 1985 *Nature* **316** 420
[28] Lagage P O, 1985 *Service d'Astrophysique, CEN, Saclay* preprint
[29] Gaisser T K, Stanev T and Barr G 1988 *Phys. Rev.* D **38** 85
[30] Barr S, Gaisser T K, Lipari P and Tilav S 1988 *Phys. Lett.* **214B** 147
[31] Barr G, Gaisser T K and Stanev T 1989 *Phys. Rev.* D **39** 3532
[32] Lee H and Bludman S A 1988 *Phys. Rev.* D **37** 122
[33] Bugaev E V, Domogatsky G V and Naumov V A 1986 *Proc. Japan-U.S. Sem. 'Cosmic Ray Muon and Neutrino Physics/Astrophysics Using Deep Underground/Underwater Detectors'* ed Y Ohashi and V Z Peterson (Tokyo: Institute for Cosmic Ray Research, University of Tokyo)
[34] Lee H and Koh Y 1989 *Chungnam National University* preprint CNUPHY-1989-T1
[35] Blandford R and Eichler D 1987 *Phys. Rep.* **154** 1
[36] Blandford R and Ostriker J P 1980 *Astrophys. J.* **237** 793
[37] Eichler D and Vestrand W T 1986 *Nature* **319** 380
[38] Gaisser T K and Stanev T 1985 *Phys. Rev. Lett.* **54** 2265
[39] Kolb E W, Turner M S and Walker T P 1985 *Phys. Rev.* D **32** 1145; Errata 1986 *Phys. Rev.* D **33** 859
[40] Dar A 1985 *Phys. Lett.* **159B** 205
[41] Loh E 1990 *ICRR Int. Symp. on Astrophysical Aspects of Highest-Energy Cosmic Rays* (Kofu, Japan, 1990)
[42] Greisen K 1966 *Phys. Rev. Lett.* **16** 748
[43] Zatsepin G T and Kuzmin V A 1966 *Zh. Eksp. Teor. Fiz.* **4** 114
[44] Hill C T and Schramm D N 1983 *Phys. Lett.* **131B** 247
[45] Hill C T and Schramm D N 1985 *Phys. Rev.* D **31** 564
[46] Berezinsky V S and Grigoreva S I 1988 *Astron. Astrophys.* **199** 1

[47] Yoshida S and Teshima M 1990 *Institute for Cosmic Ray Research ICRR Report* 221-90-14
[48] Lovelace R V E 1976 *Nature* **262** 649
[49] Hillas A M 1984 *Ann. Rev. Astron. Astrophys.* **22** 425
[50] Biermann P L and Strittmatter P A 1987 *Astrophys. J.* **322** 643
[51] Takahara F 1990 *Prog. Theor. Phys.* **83** 1071
[52] Hill C T, Schramm D N and Walker T P 1987 *Phys. Rev.* D **36** 1007
[53] Colgate A and White R H 1966 *Astrophys. J.* **143** 626
[54] Burrows A and Lattimer J M 1986 *Astrophys. J.* **307** 178
[55] Mayle R 1985 *PhD Thesis* University of California, Berkeley (UCRL-53713)
[56] Mayle R, Wilson J R and Schramm D N 1987 *Astrophys. J.* **318** 288
[57] Sato K and Suzuki H 1987 *Phys. Rev. Lett.* **58** 2722
[58] Suzuki H 1989 *PhD Thesis* University of Tokyo
[59] Hirata K S *et al* 1988 *Phys. Rev.* D **38** 448
[60] Bratton C B *et al* 1988 *Phys. Rev.* D **37** 3361
[61] Arafune J and Fukugita M 1987 *Phys. Rev. Lett.* **59** 367
[62] Ewan G T *et al Department of Physics, Queen's University, Canada Report*, SNO-85-1
[63] Alexeyev E N *et al* 1989 *JETP Lett.* **49** 480
[64] Amaldi E *et al* 1989 *Proc. 14th Texas Symp. on Relativistic Astrophysics* p 561 (*Ann. N.Y. Acad. Sci.* 1989 **571**)
[65] Loredo T J and Lamb D Q 1989 *Proc. 14th Texas Symp. on Relativistic Astrophysics* p 601 (*Ann. N.Y. Acad. Sci.* 1989 **571**)
[66] Lattimer J M and Yahil A 1989 *Astrophys. J.* **340** 426
[67] Peccei R D and Quinn H R 1977 *Phys. Rev. Lett.* **38** 1440
Wilczek F 1987 *Phys. Rev. Lett.* **40** 279
Weinberg S 1987 *Phys. Rev. Lett.* **40** 223
[68] Kaplan D 1985 *Nucl. Phys.* B **260** 215
[69] Burrows A, Turner M and Brinkmann R P 1989 *Phys. Rev.* D **39** 1020
[70] Dine M, Fischler W and Srednicki M 1981 *Phys. Lett.* **104B** 199
[71] Burrows A, Ressell M T and Turner M S 1990 *Phys. Rev.* D **42** 3297
[72] Yoshimura M private communication
[73] Totsuka Y 1987 *Proc. Seventh Workshop on Grand Unification/ICOBAN'86* ed J. Arafune, (Singapore: World Scientific) p 118
[74] Totsuka Y 1990 *Proc. Int. Conf. Underground Physics Experiments*, (Tokyo: Institute for Cosmic Ray Research) p 129
[75] Beier E W and Sinclair D (eds) 1989 Scientific and technical description of the Mark II SNO detector *SNO-89-15*
[76] Arafune J, Fukugita M, Yanagida T and Yoshimura M 1987 *Phys. Lett.* **194B** 477
[77] Minakata H and Nunokawa H 1990 *Phys. Rev.* D **41** 2976
[78] Glendenning N K 1990 *Lawrence Berkeley Lab. preprint LBL-29101*
[79] Bahcall J N and Piran T 1983 *Astrophys. J.* **267** L77
[80] Ratnatunga K U and van den Bergh S 1989 *Astrophys. J.* **343** 713
[81] Li-Zhi Fang and Huang J 1990 *Fermilab-Pub-90/24-A*
[82] Arnett W D, Schramm D N and Truran J W 1989 *Astrophys. J.* **339** L25
[83] van den Bergh 1988 *Comment. Astrophys.* **12** 131
[84] van den Bergh 1988 *Astrophys. J.* **327** 156
[85] Lyne A G, Manchester R N and Taylor J H 1985 *Mon. Not. R. Astron. Soc.* **213** 613
[86] Sofue Y 1988 *Proc. Second Workshop on Elementary-Particle Picture of the Universe* (KEK) p 303
[87] Koyama K, Makishima K, Tanaka Y and Tsunemi H 1986 *Publ. Astron. Soc. Japan* **38** 121
[88] Koyama K, Ikeuchi S and Tomisaka K 1986 *Publ. Astron. Soc. Japan* **38** 503
[89] Koyama K *et al* 1989 *Nature* **339** 603
[90] Olive K A, Schramm D N, Steigman G and Walker T P 1990 *Phys. Lett.* **236B** 454
[91] Zhang W *et al* 1988 *Phys. Rev. Lett.* **61** 385
[92] Hirata K S 1991 *PhD thesis* University of Tokyo
[93] Giacconi R and Zamorani G 1987 *Astrophys. J.* **313** 20
[94] Boldt E 1987 *Phys. Rep.* **146** 215
[95] Clayton D D 1983 *Principles of Stellar Evolution and Nucleosynthesis* (Chicago, IL: University of Chicago Press)
[96] Schwarzschild M *Structure and Evolution of the Stars* (Princeton, NJ: Princeton University Press)
[97] Van den Berg D A and Bell R A 1985 *Astrophys. J. Supp.* **58** 561; 1985 *Astrophys. J. Supp.* **58** 711
[98] Bahcall J N 1966 *Phys. Rev. Lett.* **17** 398

[99] Voloshin M B, Vysotsky M and Okun L B 1986 *JETP* **64** 446

[100] Hirata K S *et al* 1991 *Phys. Rev. Lett.* **66** 9

[101] Kirsten T 1991 *Proc. 14th Int. Symp. on Neutrino Physics and Astrophysics 'NEUTRINO'90'* ed K Winter to be published

[102] Glashow S L and Krauss L M 1987 *Phys. Lett.* **190B** 199

[103] Acker A, Pakvasa S and Pantaleone J 1990 *Preprint UH-511-710-90* University of Hawaii

[104] Barger V, Phillips R J N and Whisnant K 1990 *Phys. Rev. Lett.* **65** 3084

[105] Friemann J A, Haber H E and Freese K 1988 *Phys. Lett.* **200B** 115

[106] Bahcall J N and Bethe H A 1990 *Phys. Rev. Lett.* **65** 2233

[107] Bilenky S M and Pontecorvo B M 1978 *Phys. Rep.* C **41** 225

[108] Turck-Chieze *et al* 1988 *Astrophys. J.* **335** 425

[109] Pakvasa S 1991 *Proc. XXVth Int. Conf. on High Energy Physics* (Singapore: World Scientific) to be published; *Preprint UH-511-715-90* University of Hawaii

[110] Raghavan R 1991 *Proc. XXVth Int. Conf. on High Energy Physics* (Singapore: World Scientific) to be published

[111] Sinclair D 1991 *Proc. 14th Int. Conf. on Neutrino Physics and Astrophysics 'NEUTRINO'90'* ed K Winter, to be published

[112] Aglietta M *et al* 1989 *Europhys. Lett.* **8** 611

[113] Berger Ch. *et al* 1990 *Phys. Lett.* **245B** 305

[114] Casper D *et al* 1990 *BU Preprint 90–23* Boston University, Boston

[115] Casper D 1990 *PhD Thesis* Boston University

[116] Hirata K S *et al* 1988 *Phys. Lett.* **205B** 416

[117] Takita M *PhD Thesis* University of Tokyo

[118] Quigg C, Reno M H and Walker T P 1986 *Phys. Rev. Lett.* **57** 774

[119] Reno M H and Quigg C 1988 *Phys. Rev.* D **37** 657

[120] Eichten E, Hinchliffe I, Lane K and Quigg C, *Rev. Mod. Phys.* 1984 **56** 579

[121] Giacomelli G 1991 *Proc. 14th Int. Conf. Neutrino Physics and Astrophysics 'NEUTRINO'90'* ed K Winter, to be published

[122] DeMarzo C *et al* 1984 *Proposal for a large area detector dedicated to monopole search, astrophysics, and cosmic ray physics at the Gran Sasso Laboratory Gran Sasso Laboratory*

[123] Svoboda R *et al* 1987 *Astrophys. J.* **315** 420

[124] Oyama Y *et al* 1989 *Phys. Rev.* D **39** 1481

[125] Sobel H 1991 *Proc. 14th Int. Conf. Neutrino Physics and Astrophysics 'NEUTRINO'90'* ed K Winter, to be published

[126] Bosetti P *et al* 1988 'DUMAND II, *Proposal to construct a deep ocean laboratory for the study of high-energy neutrino astrophysics and particle physics*', *University of Hawaii*

[127] Sasaki M *et al* 1988 *Proc. 2nd Workshop on Elementary-Particle Picture of the Universe* ed M Yoshimura *et al* (KEK) p 181

[128] Adams A *et al* 1990 *GRANDE 90-005* UC Irvine

[129] Baltrusaitis R M *et al* 1985 *Phys. Rev.* D **31** 2192

[130] Steigman G *et al* 1978 *Astron. J.* **83** 1050

[131] Spergel D N and Press W H 1985 *Astrophys. J.* **294** 663

[132] Press W H and Spergel D N 1985 *Astrophys. J.* **296** 679

[133] Krauss M *et al* 1985 *Astrophys. J.* **299** 1001

[134] Faulkner J and Gilliland R 1985 *Astrophys. J.* **299** 994

[135] Gaisser T K, Steigman G and Tilav S 1986 *Phys. Rev.* D **34** 2206

[136] Rubakov V A 1982 *Nucl. Phys.* B **203** 311

[137] Callan C G Jr 1982 *Phys. Rev.* D **25** 2141; 1982 **26** 2058

[138] Arafune J and Fukugita M 1983 *Phys. Lett.* **133B** 380

[139] Arafune J and Fukugita M 1983 *Phys. Rev. Lett.* **50** 1901